43 uve 650
lbf 41702

Ausgeschieden im Jahr 2025

Control in Power Electronics
Selected Problems

ACADEMIC PRESS SERIES IN ENGINEERING

Series Editor
J. David Irwin
Auburn University

This is a series that will include handbooks, textbooks, and professional reference books on cutting-edge areas of engineering. Also included in this series will be single-authored professional books on state-of-the-art techniques and methods in engineering. Its objective is to meet the needs of academic, industrial, and governmental engineers, as well as to provide instructional material for teaching at both the undergraduate and graduate level.

This series editor, J. David Irwin, is one of the best-known engineering educators in the world. Irwin has been chairman of the electrical engineering department at Auburn University for 27 years.

Published books in the series:

Supply Chain Design and Management, 2002, M. Govil and J. M. Proth
Power Electronics Handbook, 2001, M. H. Rashid, editor
Control of Induction Motors, 2001, A. Trzynadlowski
Embedded Microcontroller Interfacing for McoR Systems, 2000, G. J. Lipovski
Soft Computing & Intelligent Systems, 2000, N. K. Sinha, M. M. Gupta
Introduction to Microcontrollers, 1999, G. J. Lipovski
Industrial Controls and Manufacturing, 1999, E. Kamen
DSP Integrated Circuits, 1999, L. Wanhammar
Time Domain Electromagnetics, 1999, S. M. Rao
Single- and Multi-Chip Microcontroller Interfacing, 1999, G. J. Lipovski
Control in Robotics and Automation, 1999, B. K. Ghosh, N. Xi, and T. J. Tarn

CONTROL IN POWER ELECTRONICS
Selected Problems

Editors

MARIAN P. KAZMIERKOWSKI
Warsaw University of Technology, Warsaw, Poland

R. KRISHNAN
Virginia Tech, Blacksburg, Virginia, USA

FREDE BLAABJERG
Aalborg University, Aalborg, Denmark

ACADEMIC PRESS
An imprint of Elsevier Science

Amsterdam   Boston   London   New York   Oxford   Paris   San Diego
San Francisco   Singapore   Sydney   Tokyo

This book is printed on acid-free paper. ∞

Copyright 2002, Elsevier Science (USA).

All rights reserved.

No part of this publication may be reproduced or transmitted in any form or by any means, electronic or mechanical, including photocopy, recording, or any information storage and retrieval system, without permission in writing from the publisher.

Requests for permission to make copies of any part of the work should be mailed to: Permissions Department, Harcourt, Inc., 6277 Sea Harbor Drive, Orlando, Florida 32887-6777.

Explicit permission from Academic Press is not required to reproduce a maximum of two figures or tables from an Academic Press chapter in another scientific or research publication provided that the material has not been credited to another source and that full credit to the Academic Press chapter is given.

Academic Press
An imprint of Elsevier Science
525 B Street, Suite 1900, San Diego, California 92101-4495, USA
http://www.academicpress.com

Academic Press
84 Theobolds Road, London WC1X 8RR, UK
http://www.academicpress.com

Library of Congress Catalog Card Number: 2001098018
International Standard Book Number: 0-12-402772-5

PRINTED IN THE UNITED STATES OF AMERICA
02 03 04 05 06 07 MB 9 8 7 6 5 4 3 2 1

Contents

Preface		vii
List of Contributors		xi

Part I: PWM Converters: Topologies and Control

1. Power Electronic Converters
   Andrzej M. Trzynadlowski — 1
2. Resonant dc Link Converters
   Stig Munk-Nielsen — 45
3. Fundamentals of the Matrix Converter Technology
   C. Klumpner and F. Blaabjerg — 61
4. Pulse Width Modulation Techniques for Three-Phase Voltage Source Converters
   Marian P. Kazmierkowski, Mariusz Malinowski, and Michael Bech — 89

Part II: Motor Control

5. Control of PWM Inverter-Fed Induction Motors
   Marian P. Kazmierkowski — 161
6. Energy Optimal Control of Induction Motor Drives
   F. Abrahamsen — 209
7. Comparison of Torque Control Strategies Based on the Constant Power Loss Control System for PMSM
   Ramin Monajemy and R. Krishnan — 225
8. Modeling and Control of Synchronous Reluctance Machines
   Robert E. Betz — 251
9. Direct Torque and Flux Control (DTFC) of ac Drives
   Ion Boldea — 301
10. Neural Networks and Fuzzy Logic Control in Power Electronics
    Marian P. Kazmierkowski — 351

Part III: Utilities Interface and Wind Turbine Systems

11. Control of Three-Phase PWM Rectifiers
    Mariusz Malinowski and Marian P. Kazmierkowski   419

12. Power Quality and Adjustable Speed Drives
    Steffan Hansen and Peter Nielsen   461

13. Wind Turbine Systems
    Lars Helle and Frede Blaabjerg   483

Index   511

Preface

This book is the result of cooperation initiated in 1997 between Danfoss Drives A/S (www.danfoss.com.drives) and the Institute of Energy Technology at Aalborg University in Denmark. A four-year effort known as The International Danfoss Professor Program* was started. The main goal of the program was to attract more students to the multidisciplinary area of power electronics and drives by offering a world-class curriculum taught by renowned professors. During the four years of the program distinguished professors visited Aalborg University, giving advanced courses in their specialty areas and interacting with postgraduate students. Another goal of the program was to strengthen the research team at the university by fostering new contacts and research areas. Four Ph.D. studies have been carried out in power electronics and drives. Finally, the training and education of engineers were also offered in the program. The program attracted the following professors and researchers (listed in the order in which they visited Aalborg University):

Marian P. Kazmierkowski, *Warsaw University of Technology, Poland*
Andrzej M. Trzynadlowski, *University of Nevada, Reno, USA*
Robert E. Betz, *University of Newcastle, Australia*
Prasad Enjeti, *Texas A&M, USA*
R. Krishnan, *Virginia Tech, Blacksburg, USA*
Ion Boldea, *Politehnica University of Timisoara, Romania*
Peter O. Lauritzen, *University of Washington, USA*
Kazoo Terada, *Hiroshima City University, Japan*
Jacobus D. Van Wyk, *Virginia Tech, Blacksburg, USA*
Giorgio Spiazzi, *University of Padova, Italy*
Bimal K. Bose, *University of Tennessee, Knoxville, USA*
Jaeho Choi, *Chungbuk National University, South Korea*
Peter Vas, *University of Aberdeen, UK*

---

* F. Blaabjerg, M. P. Kazmierkowski, J. K. Pedersen, P. Thogersen, and M. Toennes, An industry-university collaboration in power electronics and drives, *IEEE Trans. on Education*, **43**, No. 1, Feb. 2000, pp. 52–57.

Among the Ph.D. students visiting the program were:

Pawel Grabowski, *Warsaw University of Technology, Poland*
Dariusz L. Sobczuk, *Warsaw University of Technology, Poland*
Christian Lascu, *Politehnica University of Timisoara, Romania*
Lucian Tutelea, *Politehnica University of Timisoara, Romania*
Christian Klumpner, *Politehnica University of Timisoara, Romania*
Mariusz Malinowski, *Warsaw University of Technology, Poland*
Niculina Patriciu, *University of Cluj-Napoca, Romania*
Florin Lungeanu, *Galati University, Romania*
Marco Matteini, *University of Bologna, Italy*
Marco Liserre, *University of Bari, Italy*

The research carried out in cooperation with the Danfoss Professor Program resulted in many publications. The high level of the research activities has been recognized worldwide and four international awards have been given to team members of the program.

Most of the research results are included in this book, which consists of the following three parts:

Part I: PWM Converters: Topologies and Control (four chapters)
Part II: Motor Control (six chapters)
Part III: Utilities Interface and Wind Turbine Systems (three chapters)

The book has strong monograph attributes, however, some chapters can also be used for undergraduate education (e.g., Chapters 4, 5, and 9–11) as they contain a number of illustrative examples and simulation case studies.

We would like to express thanks to the following people for their visionary support of this program:

Michael Toennes, *Manager of Low Power Drives, Danfoss Drives A/S*
Paul B. Thoegersen, *Manager of Control Engineering, Danfoss Drives A/S*
John K. Pedersen, *Institute Leader, Institute of Energy Technology, Aalborg University*
Kjeld Kuckelhahn, *Vice President of Product Development, Danfoss Drives A/S*
Finn R. Pedersen, *President of Fluid Division, Danfoss A/S*, former *President of Danfoss Drives A/S*
Joergen M. Clausen, *President and CEO of Danfoss A/S*

We would also like to thank the Ministry of Education in Denmark and Aalborg University for their support of the program.

We would like to express our sincere thanks to the chapter contributors for their cooperation and patience in various stages of the book preparation. Special thanks are directed to Ph.D. students Mariusz Cichowlas, Marek Jasinski, Mateusz Sikorski, and Marcin Zelechowski from the Warsaw University of Technology for their help in preparing the entire manuscript. We are grateful to our editor at Academic Press, Joel Claypool, for his patience and continuous support.

Thanks also to Peggy Flanagan, project editor, who interfaced pleasantly during copyediting and proofreading. Finally, we are very thankful to our families for their cooperation.

Marian P. Kazmierkowski, *Warsaw University of Technology, Poland*

R. Krishnan, *Virginia Tech, Blacksburg, USA*

Frede Blaabjerg, *Aalborg University, Denmark*

List of Contributors

F. Abrahamsen Aalborg, Denmark

Michael Bech Aalborg University, Aalborg, Denmark

Robert E. Betz School of Electrical Engineering and Computer Science, University of Newcastle, Callaghan, Australia

Frede Blaabjerg Institute of Energy Technology, Aalborg University, Aalborg, Denmark

Ion Boldea University Politehnica, Timisoara, Romania

Steffan Hansen Danfoss Drives A/S, Grasten, Denmark

Lars Helle Institute of Energy Technology, Aalborg University, Aalborg, Denmark

Marian P. Kazmierkowski Warsaw University of Technology, Warsaw, Poland

C. Klumpner Institute of Energy Technology, Aalborg University, Aalborg, Denmark

R. Krishnan The Bradley Department of Electrical and Computer Engineering, Virginia Tech, Blacksburg, Virginia

Mariusz Malinowski Warsaw University of Technology, Warsaw, Poland

Ramin Monajemy Samsung Information Systems America, San Jose, California

Stig Munk-Nielsen Institute of Energy Technology, Aalborg University, Aalborg, Denmark

Peter Nielsen Danfoss Drives A/S, Grasten, Denmark

Andrzej M. Trzynadlowski University of Nevada, Reno, Nevada

CHAPTER 1

Power Electronic Converters

ANDRZEJ M. TRZYNADLOWSKI
University of Nevada, Reno, Nevada

This introductory chapter provides a background to the subject of the book. Fundamental principles of electric power conditioning are explained using a hypothetical generic power converter. Ac to dc, ac to ac, dc to dc, and dc to ac power electronic converters are described, including select operating characteristics and equations of their most common representatives.

1.1 PRINCIPLES OF ELECTRIC POWER CONDITIONING

Electric power is supplied in a "raw," fixed-frequency, fixed-voltage form. For small consumers, such as homes or small stores, usually only the single-phase ac voltage is available, whereas large energy users, typically industrial facilities, draw most of their electrical energy via three-phase lines. The demand for conditioned power is growing rapidly, mostly because of the progressing sophistication and automation of industrial processes. Power conditioning involves both *power conversion*, ac to dc or dc to ac, and *control*. Power electronic converters performing the conditioning are highly efficient and reliable.

Power electronic converters can be thought of as networks of semiconductor power switches. Depending on the type, the switches can be uncontrolled, semicontrolled, or fully controlled. The state of uncontrolled switches, the *power diodes*, depends on the operating conditions only. A diode turns on (closes) when positively biased and it turns off (opens) when the conducted current changes its polarity to negative. Semicontrolled switches, the *SCRs* (silicon controlled rectifiers), can be turned on by a gate current signal, but they turn off just like the diodes. Most of the existing power switches are fully controlled, that is, they can both be turned on and off by appropriate voltage or current signals.

Principles of electric power conversion can easily be explained using a hypothetical "generic power converter" shown in Fig. 1.1. It is a simple network of five switches, S0 through S4, of which S1 opens and closes simultaneously with S2, and S3 opens and closes simultaneously with S4. These four switches can all be open (OFF), but they may not be all closed (ON) because they would short the supply source. Switch S0 is only closed when all the other switches are open. It is assumed that the switches open and close instantly, so that currents flowing through them can be redirected without interruption.

2 CHAPTER 1 / POWER ELECTRONIC CONVERTERS

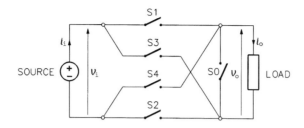

FIGURE 1.1
Generic power converter.

The generic converter can assume three states only: (1) State 0, with switches S1 through S4 open and switch S0 closed, (2) State 1, with switches S1 and S2 closed and the other three switches open, and (3) State 2, with switches S3 and S4 closed and the other three switches open. Relations between the output voltage, v_o, and the input voltage, v_i, and between the input current, i_i, and output current, i_o, are

$$v_o = \begin{cases} 0 & \text{in State } 0 \\ v_i & \text{in State } 1 \\ -v_i & \text{in State } 2 \end{cases} \tag{1.1}$$

and

$$i_i = \begin{cases} 0 & \text{in State } 0 \\ i_o & \text{in State } 1 \\ -i_o & \text{in State } 2. \end{cases} \tag{1.2}$$

Thus, depending on the state of generic converter, its switches connect, cross-connect, or disconnect the output terminals from the input terminals. In the last case (State 0), switch S0 provides a path for the output current (load current) when the load includes some inductance, L. In absence of that switch, interrupting the current would cause a dangerous impulse overvoltage, $L di_o/dt \to -\infty$.

Instead of listing the input–output relations as in Eqs. (1.1) and (1.2), the so-called *switching functions* (or *switching variables*) can be assigned to individual sets of switches. Let $a = 0$ when switch S0 is open and $a = 1$ when it is closed, $b = 0$ when switches S1 and S2 are open and $b = 1$ when they are closed, and $c = 0$ when switches S3 and S4 are open and $c = 1$ when they are closed. Then,

$$v_o = \bar{a}(b - c)v_i \tag{1.3}$$

and

$$i_i = \bar{a}(b - c)i_o. \tag{1.4}$$

The ac to dc power conversion in the generic converter is performed by setting it to State 2 whenever the input voltage is negative. Vice-versa, the dc to ac conversion is realized by periodic repetition of the State 1–State 2– . . . sequence (note that the same state sequence appears for the ac to dc conversion). These two basic types of power conversion are illustrated in Figs. 1.2 and 1.3. Thus, electric power conversion is realized by appropriate operation of switches of the converter.

Switching is also used for controlling the output voltage. Two basic types of voltage control are *phase control* and *pulse width modulation*. The phase control consists of delaying States 1

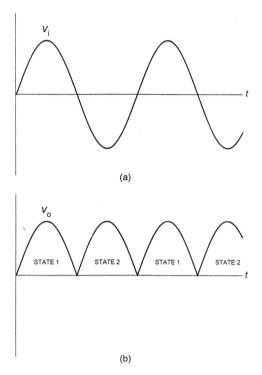

FIGURE 1.2
Ac to dc conversion in the generic power converter: (a) input voltage, (b) output voltage.

and 2 and setting the converter to State 0. Figure 1.4 shows the generic power converter operating as an ac voltage controller (ac to ac converter). For 50% of each half-cycle, State 1 is replaced with State 0, resulting in significant reduction of the rms value of output voltage (in this case, to $1/\sqrt{2}$ of rms value of the input voltage). The pulse width modulation (PWM) also makes use of State 0, but much more frequently and for much shorter time intervals. As shown in Fig. 1.5 for the same generic ac voltage controller, instead of removing whole "chunks" of the waveform, numerous "slices" of this waveform are cut out within each switching cycle of the converter. The *switching frequency*, a reciprocal of a single switching period, is at least one order of magnitude higher than the input or output frequency.

The difference between phase control and PWM is blurred in dc to dc converters, in which both the input and output frequencies are zero, and the switching cycle is the operating cycle. The dc to dc conversion performed in the generic power converter working as a *chopper* (dc to dc converter) is illustrated in Fig. 1.6. Switches S1 and S2 in this example operate with the *duty ratio* of 0.5, reducing the average output voltage by 50% in comparison with the input voltage. The duty ratio of a switch is defined as the fraction of the switching cycle during which the switch is ON.

To describe the magnitude control properties of power electronic converters, it is convenient to introduce the so-called *magnitude control ratio*, M, defined as the ratio of the actual useful output voltage to the maximum available value of this voltage. In dc-output converters, the useful output voltage is the dc component of the total output voltage of the converter, whereas in ac-output ones, it is the fundamental component of the output voltage. Generally, the magnitude control ratio can assume values in the -1 to $+1$ range.

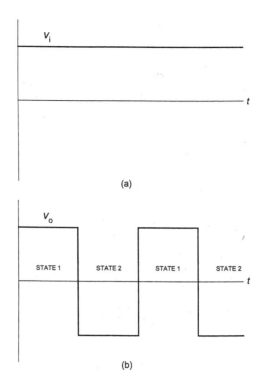

FIGURE 1.3
Dc to ac conversion in the generic power converter: (a) input voltage, (b) output voltage.

In practical power electronic converters, the electric power is supplied by voltage sources or current sources. Each of these can be of the uncontrolled or controlled type, but a parallel capacitance is a common feature of the voltage sources while a series inductance is typical for the current sources. The capacitance or inductance is sufficiently large to prevent significant changes of the input voltage or current within an operating cycle of the converter. Similarly, loads can also have the voltage-source or current-source characteristics, resulting from a parallel capacitance or series inductance. To avoid direct connection of two capacitances charged to different voltages or two inductances conducting different currents, a voltage-source load requires a current-source converter and, vice versa, a current-source load must be supplied from a voltage-source converter. These two basic source-converter-load configurations are illustrated in Fig. 1.7.

1.2 AC TO DC CONVERTERS

Ac to dc converters, the *rectifiers*, come in many types and can variously be classified as uncontrolled versus controlled, single-phase versus multiphase (usually, three-phase), half-wave versus full-wave, or phase-controlled versus pulse width modulated. Uncontrolled rectifiers are based on power diodes; in phase-controlled rectifiers SCRs are used; and pulse width modulated

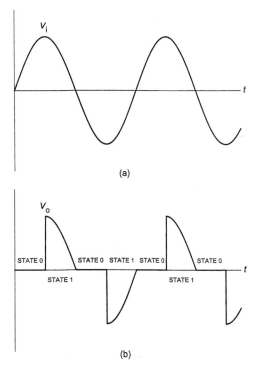

FIGURE 1.4
Phase control of output voltage in the generic power converter operating as an ac voltage controller: (a) input voltage, (b) output voltage.

rectifiers require fully controlled switches, such as *IGBTs* (insulated gate bipolar transistors) or *power MOSFETs*.

The two most common rectifier topologies are the single-phase bridge and three-phase bridge. Both are full-wave rectifiers, with no dc component in the input current. This current is the main reason why half-wave rectifiers, although feasible, are avoided in practice. The single-phase and three-phase diode rectifiers are shown in Fig. 1.8 with an RLE (resistive–inductive–EMF) load. At any time, one and only one pair of diodes conducts the output current. One of these diodes belongs to the common-anode group (upper row), the other to the common-cathode group (lower row), and they are in different legs of the rectifier. The line-to-line voltage of the supply line constitutes the input voltage of the three-phase rectifier, also known as a *six-pulse rectifier*. The single-phase bridge rectifier is usually referred to as a *two-pulse rectifier*.

In practice, the output current in full-wave diode rectifiers is continuous, that is, it never drops to zero. This mostly dc current contains an ac component (ripple), dependent on the type of rectifier and parameters of the load. Output voltage waveforms of rectifiers in Fig. 1.8 within a single period, T, of input frequency are shown in Fig. 1.9, along with example waveforms of the output current. The average output voltage (dc component), V_o, is given by

$$V_o = \frac{2}{\pi} V_{i,p} \approx 0.63 V_{i,p} \tag{1.5}$$

for the two-pulse diode rectifier and

$$V_o = \frac{3}{\pi} V_{i,p} \approx 0.95 V_{i,p} = 0.95 V_{LL,p} \tag{1.6}$$

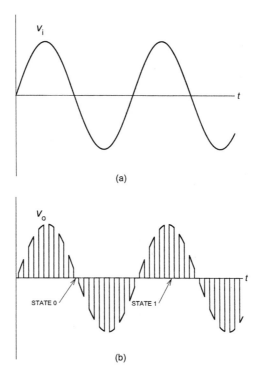

FIGURE 1.5
PWM control of output voltage in the generic power converter operating as an ac voltage controller: (a) input voltage, (b) output voltage.

for the six-pulse diode rectifier. Here, $V_{i,p}$ denotes the peak value of input voltage, which, in the case of the six-pulse rectifier, is the peak line-to-line voltage, $V_{LL,p}$.

In phase-controlled rectifiers shown in Fig. 1.10, diodes are replaced with SCRs. Each SCR must be turned on (fired) by a gate signal (firing pulse) in each cycle of the supply voltage. In the angle domain, ωt, where ω denotes the supply frequency in rad/s, the gate signal can be delayed by α_f radians with respect to the instant in which a diode replacing a given SCR would start to conduct. This delay, called a *firing angle*, can be controlled in a wide range. Firing pulses for all six SCRs are shown in Fig. 1.11. Under the continuous conductance condition, the average output voltage, $V_{o(con)}$, of a controlled rectifier is given by

$$V_{o(con)} = V_{o(unc)} \cos(\alpha_f) \tag{1.7}$$

where $V_{o(unc)}$ denotes the average output voltage of an uncontrolled rectifier (diode rectifier) of the same type. It can be seen that $\cos(\alpha_f)$ constitutes the magnitude control ratio of phase-controlled rectifiers. Example waveforms of output voltage in two- and six-pulse controlled rectifiers are shown in Fig. 1.12, for the firing angle of 45°.

As in uncontrolled rectifiers, the output current is basically of the dc quality, with certain ripple. The *ripple factor*, defined as the ratio of the rms value of the ac component to the dc component, increases with the firing angle. At a sufficiently high value of the firing angle, the continuous current waveform breaks down into separate pulses. The conduction mode depends on the load EMF, load angle, and firing angle. The graph in Fig. 1.13 illustrates that relation for a six-pulse rectifier: for a given firing angle, the continuous conduction area lies below the line representing this angle. For example, for load and firing angles both of 30°, the load EMF

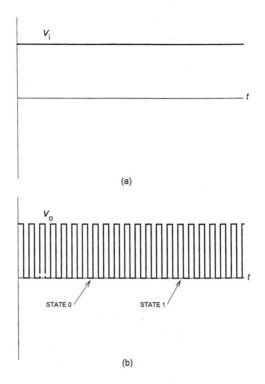

FIGURE 1.6
PWM control of output voltage in the generic power converter operating as a chopper: (a) input voltage, (b) output voltage.

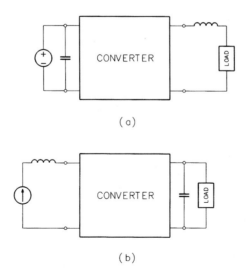

FIGURE 1.7
Two basic source-converter-load configurations: (a) voltage-source converter with a current-source load, (b) current-source converter with a voltage-source load.

8 CHAPTER 1 / POWER ELECTRONIC CONVERTERS

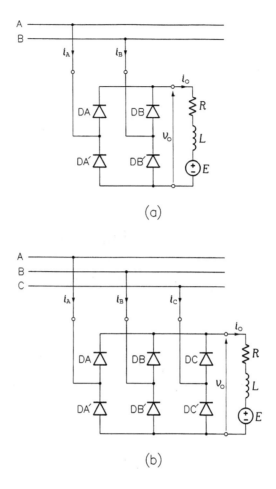

FIGURE 1.8
Diode rectifiers: (a) single-phase bridge, (b) three-phase bridge.

coefficient, defined as the ratio of the load EMF to the peak value of line-to-line input voltage, must not be greater than 0.75.

Equation (1.7) indicates that the average output voltage becomes negative when $\alpha_f > 90°$. Then, as the output current is always positive, the power flow is reversed, that is, the power is transferred from the load to the source, and the rectifier is said to operate in the inverter mode. Clearly, the load must contain a negative EMF as a source of that power.

Figure 1.14 shows four possible *operating quadrants* of a power converter. In Quadrants 1 and 3, the rectifier transfers electric power from the source to the load, while Quadrants 2 and 4 represent the inverter operation. A single controlled rectifier can only operate in Quadrants 1 and 4, that is, with a positive output current. As illustrated in Fig. 1.15a, the current can be reversed using a cross-switch between a rectifier and a load, typically a dc motor. In this way, the rectifier and load terminals can be connected directly or cross-connected. This method of extending operation of the rectifier on Quadrants 2 and 3 is only practical when the switch does not have to be used frequently as, for example, in an electric locomotive. Therefore, a much more common solution consists in connecting two controlled rectifiers in antiparallel, creating the so-called *dual converter* shown in Fig. 1.15b.

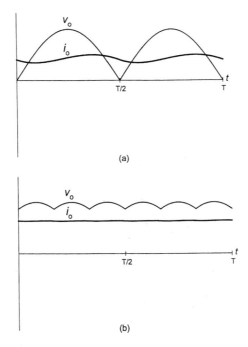

FIGURE 1.9
Output voltage and current waveforms in diode rectifiers: (a) single-phase bridge, (b) three-phase bridge.

There are two types of dual converters. Figure 1.16 shows the *circulating current-free dual converter*, that is, a rectifier in which single SCRs have been replaced with antiparallel SCR pairs. This arrangement is simple and compact, but it has two serious weaknesses. First, to prevent an interphase short circuit, only one internal rectifier can be active at a given time. For example, with TB1 and TC1′ conducting, TC2′ is forward biased and, if fired, it would short lines B and C. This can easily be prevented by appropriate control of firing signals, but when a change in polarity of the output current is required, the incoming rectifier must wait until the current in the outgoing rectifier dies out and the conducting SCRs turn off. This delay slows down the response to current control commands, which in certain applications is not acceptable. Secondly, as in all single phase-controlled rectifiers, if the firing angle is too large and/or the load inductance is too low, the output current becomes discontinuous, which is undesirable. For instance, such a current would generate a pulsating torque in a dc motor, causing strong acoustic noise and vibration.

In the *circulating current-conducting dual converter*, shown in Fig. 1.17, both constituent rectifiers are active simultaneously. Depending on the operating quadrant, one rectifier works with the firing angle, $\alpha_{f,1}$, less than 90°. The other rectifier operates in the inverter mode with the firing angle, $\alpha_{f,2}$, given by

$$\alpha_{f,2} = \beta - \alpha_{f,1} \tag{1.8}$$

where β is a controlled variable. It is maintained at a value of about 180°, so that both rectifiers produce the same *average* voltage. However, the *instantaneous* output voltages of the rectifiers are not identical, and their difference generates a current circulating between the rectifiers. If the rectifiers were directly connected as in Fig. 1.15b, the circulating current, limited by the resistance of wires and conducting SCRs only, would be excessive. Therefore, reactors are

10 CHAPTER 1 / POWER ELECTRONIC CONVERTERS

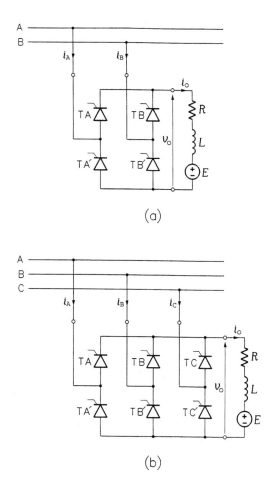

FIGURE 1.10
Phase-controlled rectifiers: (a) single-phase bridge, (b) three-phase bridge.

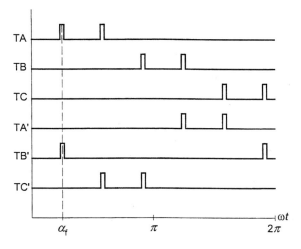

FIGURE 1.11
Firing pulses in the phase-controlled six-pulse rectifier.

1.2 AC TO DC CONVERTERS

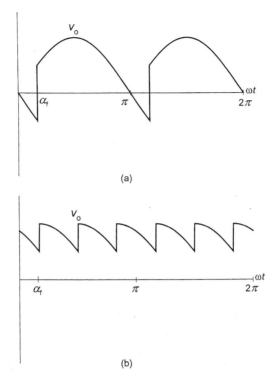

FIGURE 1.12
Output voltage waveforms in phase-controlled rectifiers: (a) single-phase bridge, (b) three-phase bridge (firing angle of 45°).

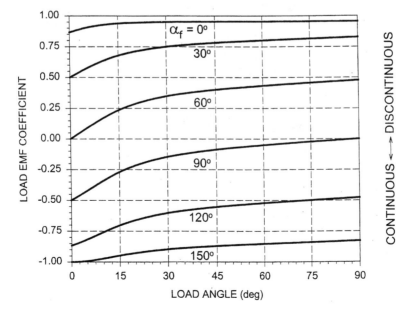

FIGURE 1.13
Diagram of conduction modes of a phase-controlled six-pulse rectifier.

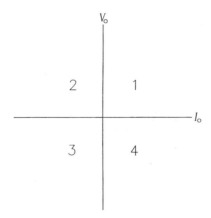

FIGURE 1.14
Operating quadrants of a controlled rectifier.

placed between the rectifiers and the load, strongly reducing the ac component of the circulating current.

The circulating current is controlled in a closed-loop control system which adjusts the angle β in Eq. (1.8). Typically, the circulating current is kept at the level of some 10% to 15% of the rated current to ensure continuous conduction of both constituent rectifiers. The converter is thus seen to employ a different scheme of operation from the circulating current-free converter. Even when the load consumes little power, a substantial amount of power enters one rectifier and the difference between this power and the load power is transferred back to the supply line by the second rectifier. Reactors L_3 and L_4 can be eliminated if the constituent rectifiers are supplied from isolated sources, such as two secondary windings of a transformer.

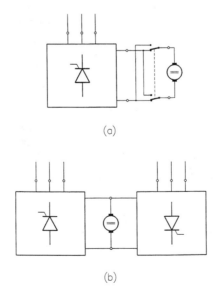

FIGURE 1.15
Rectifier arrangements for operation in all four quadrants: (a) rectifier with a cross-switch, (b) two rectifiers connected in antiparallel.

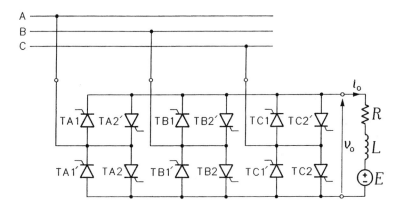

FIGURE 1.16
Circulating current-free dual converter.

Both the uncontrolled and phase-controlled rectifiers draw square-wave currents from the supply line. In addition, the input power factor is poor, especially in controlled rectifiers, where it is proportional to $\cos(\alpha_f)$. These flaws led to the development of *PWM rectifiers*, in which waveforms of the supply currents can be made sinusoidal (with certain ripple) and in phase with the supply voltages. Also, even with very low values of the magnitude control ratio, continuous output currents are maintained. Fully controlled semiconductor switches, typically IGBTs, are used in these rectifiers.

A voltage-source PWM rectifier based on IGBTs is shown in Fig. 1.18. The diodes connected in series with the IGBTs protect the transistors from reverse breakdown. Although the input current, i_a, to the rectifier is pulsed, most of its ac component come from the input capacitors, while the current, i_A, drawn from the power line is sinusoidal, with only some ripple. Appropriate control of rectifier switches allows obtaining a unity input power factor. Example waveforms of the output voltage, v_o, output current, i_o, and input currents, i_a and i_A, are shown in Figs. 1.19 and 1.20, respectively.

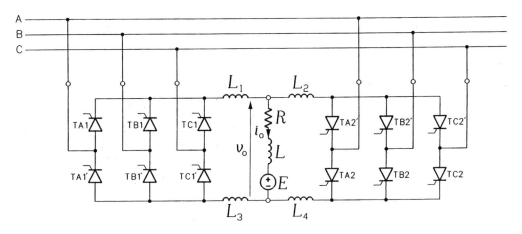

FIGURE 1.17
Circulating current-conducting dual converter.

14 CHAPTER 1 / POWER ELECTRONIC CONVERTERS

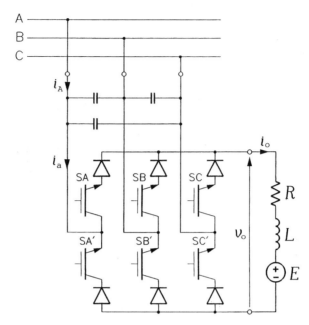

FIGURE 1.18
Voltage-source PWM rectifier.

The *voltage-source PWM rectifier* is a *buck-type converter*, that is, its maximum available output voltage (dependent on the PWM technique employed) is less than the peak input voltage. In contrast, the *current-source PWM rectifier* shown in Fig. 1.21 is a *boost-type* ac to dc converter, whose output voltage is higher than the peak input voltage. Figure 1.22 depicts example waveforms of the output voltage and current and the input current of the rectifier.

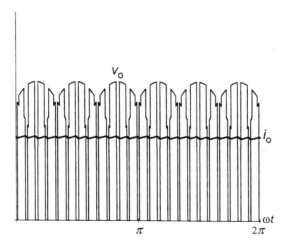

FIGURE 1.19
Output voltage and current waveforms in a voltage-source PWM rectifier.

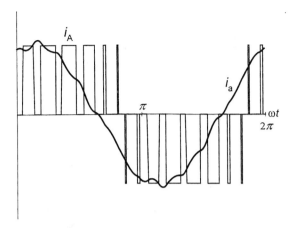

FIGURE 1.20
Input current waveforms in a voltage-source PWM rectifier.

The amount of ripple in the input and output currents of the PWM rectifiers described depends on the switching frequency and size of the inductive and capacitive components involved. In practice, PWM rectifiers are typically of low and medium power ratings.

1.3 AC TO AC CONVERTERS

There are three basic types of ac to dc converters. The simplest ones, the *ac voltage controllers*, allow controlling the output voltage only, while the output frequency is the same as the input frequency. In *cycloconverters*, the output frequency can be controlled, but it is at least one order of magnitude lower than the input frequency. In both the ac voltage controllers and cycloconverters, the maximum available output voltage approaches the input voltage. *Matrix*

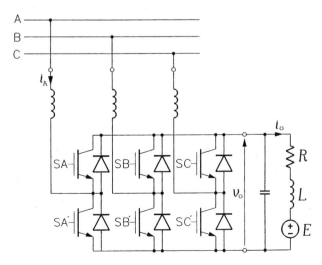

FIGURE 1.21
Current-source PWM rectifier.

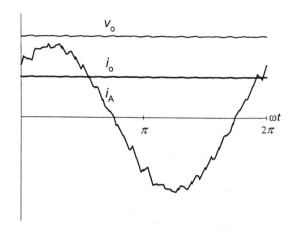

FIGURE 1.22
Waveforms of the output voltage and current and the input current in a current-source PWM rectifier.

converters are most versatile, with no inherent limits on the output frequency, but the maximum available output voltage is about 15% lower than the input voltage.

A pair of semiconductor power switches connected in antiparallel constitutes the basic building block of ac voltage controllers. Phase-controlled converters employ pairs of SCRs, SCR-diode pairs, or *triacs*. A single-phase ac voltage controller is shown in Fig. 1.23 and example waveforms of the output voltage and current in Fig. 1.24. The rms output voltage, V_o, is given by

$$V_o = V_i \sqrt{\frac{1}{\pi}\left\{\alpha_e - \alpha_f - \frac{1}{2}[\sin(2\alpha_e) - \sin(2\alpha_f)]\right\}} \qquad (1.9)$$

where α_e denotes the so-called *extinction angle*, dependent on the firing angle, α_f, and load angle, φ. The precise value of φ is usually unknown and changing. Consequently, as shown in Fig. 1.25, only an envelope of control characteristics, $M = f(\alpha_f)$, where $M = V_o/V_i$, can accurately be determined.

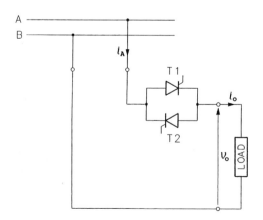

FIGURE 1.23
Phase-controlled single-phase ac voltage controller.

1.3 AC TO AC CONVERTERS

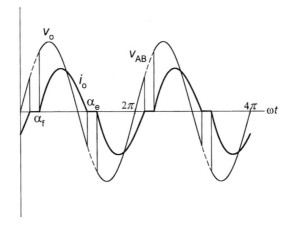

FIGURE 1.24
Output voltage and current waveforms in a phase-controlled single-phase ac voltage controller (firing angle of 60°).

The output voltage equals the input voltage, and the current is continuous and sinusoidal, when $\alpha_f = \varphi$. This can easily be done by applying a packet of narrowly spaced firing pulses to a given switch at the instant of zero-crossing of the input voltage waveform. The first pulse which manages to fire the switch appears at $\omega t \approx \varphi$, and the ac voltage controller becomes a *static ac switch*, which can be turned off by cancelling the firing pulses.

Several topologies of phase-controlled three-phase ac voltage controllers are feasible, of which the most common, *fully controlled* controller, usually based on triacs, is shown in Fig. 1.26.

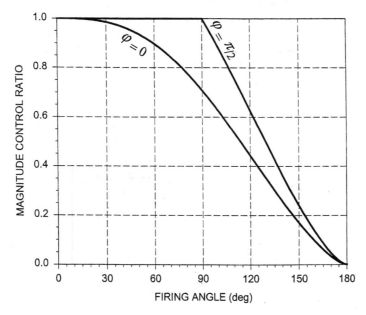

FIGURE 1.25
Envelope of control characteristics of a phase-controlled single phase ac voltage controller.

18 CHAPTER 1 / POWER ELECTRONIC CONVERTERS

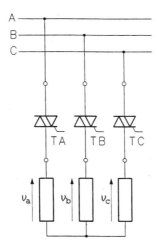

FIGURE 1.26
Phase-controlled, fully controlled three-phase ac voltage controller.

If switching functions, a, b, and c, are assigned to each triac, output voltages, v_a, v_b, and v_c, of the controller are given by

$$\begin{bmatrix} v_a \\ v_b \\ v_c \end{bmatrix} = \begin{bmatrix} a & -b & c \\ -a & b & -c \\ -a & -b & c \end{bmatrix} \begin{bmatrix} v_A \\ v_B \\ v_C \end{bmatrix} \quad (1.10)$$

where v_A, v_B, and v_C are line-to-ground voltages of the supply line. Analysis of operation of the fully controlled controller is rather difficult since, depending on the load and firing angle, the controller operates in one of three modes: (1) Mode 1, with two or three triacs conducting; (2) Mode 2, with two triacs conducting; and (3) Mode 3, with none or two triacs conducting. The output voltage waveforms are complicated, as illustrated in Fig. 1.27 for voltage v_a of a controller with resistive load and a firing angle of 30°. The waveform consists of segments of the v_A, $v_{AB}/2$, and $v_{AC}/2$ voltages. The envelope of control characteristics of the controller is shown in Fig. 1.28.

Four other topologies of the phase-controlled three-phase ac voltage controller are shown in Fig. 1.29. If ratings of available triacs are too low, actual SCRs must be used. In that case, an SCR-diode pair is employed in each phase of the controller. Such a *half-controlled* controller is shown in Fig. 1.29a. If the load is connected in delta, the three-phase ac voltage controller can have the topology shown in Fig. 1.29b. The triacs (or SCR-diode pairs) can also be connected after the load, as in Figs. 1.29c and 1.29d.

Similarly to phase-controlled rectifiers, phase-controlled ac voltage controllers draw distorted currents from the supply line, and their input power factor is poor. Again, as in the rectifiers, these characteristics can significantly be improved by employing pulse width modulation. Pulse width modulated ac voltage controllers, commonly called *ac choppers*, require fully controlled power switches capable of conducting current in both directions. Such switches can be assembled from transistors and diodes; two such arrangements are shown in Fig. 1.30.

The *PWM ac voltage controller*, also known as *ac chopper*, is shown in Fig. 1.31 in the single-phase version. For simplicity, and to stress the functional analogy to the generic converter, the bidirectional switches are depicted as mechanical contacts. When the main switch, S1, is

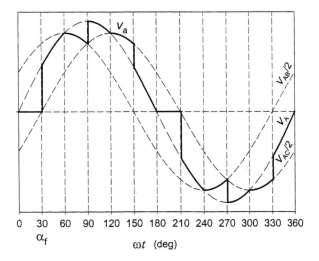

FIGURE 1.27
Waveform of the output voltage in a fully controlled three-phase ac voltage controller (resistive load, firing angle of 30°).

chopping, that is, turning on and off many times per cycle, the current drawn from the LC input filter is interrupted. Therefore, another switch, S2, is connected across the load. It plays the role of the freewheeling switch S0 in the generic power converter in Fig. 1.1. Switches S1 and S2 are operated complementarily: when S1 is turned on, S2 is turned off and vice versa. Denoting the duty ratio of switch S1 by D_1, the magnitude control ratio, M, taken as ratio of the rms output voltage, V_o, to rms input voltage, V_i, equals $\sqrt{D_1}$.

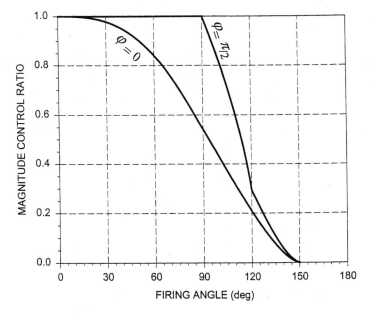

FIGURE 1.28
Envelope of control characteristics of the fully controlled three-phase ac voltage controller.

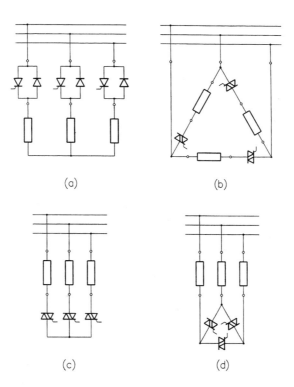

FIGURE 1.29
Various topologies of phase-controlled three-phase ac voltage controllers: (a) half-controlled, before-load, (b) delta-connected, before-load, (c) wye-connected after-load, (d) delta-connected, after-load.

Example waveforms of the output voltage, v_o, and current, i_o, of the ac chopper are shown in Fig. 1.32. The high-frequency component of the pulsed input current, i_a, is mostly supplied by the filter capacitors, so that the current, i_A, drawn from the power line is similar to that of the PWM voltage-source rectifier (see Fig. 1.20). Analogously to the single-phase ac chopper in Fig. 1.31, three-phase ac choppers can be obtained from their phase-controlled counterparts by replacing each triac with a fully controlled bidirectional switch. A similar switch must be connected in parallel to each phase load to provide an alternative path for the load current when the load is cut off from the supply source by the main switch.

The dual converter in Fig. 1.17 can be operated as a single-phase cycloconverter by varying the firing angle $\alpha_{f,1}$ in accordance with the formula

$$\alpha_{f,1}(t) = \cos^{-1}[M \sin(\omega_o t)] \qquad (1.11)$$

where the magnitude control ratio, M, represents the ratio of the peak value of the fundamental output voltage to the maximum available dc voltage of the constituent rectifiers. The output frequency, ω_o, must be significantly lower than the supply frequency, ω. Example waveforms of the output voltage of such cycloconverter are shown in Fig. 1.33 for $\omega_o/\omega = 0.2$ and two values of M: 1 and 0.5.

Two three-phase six-pulse cycloconverters are shown in Fig. 1.34. The cycloconverter with isolated phase loads in Fig. 1.34a is supplied from a single three-phase source. If the loads are interconnected, as in Fig. 1.34b, individual phases of the cycloconverter must be fed from separate sources, such as isolated secondary windings of the supply transformer. Practical

1.3 AC TO AC CONVERTERS 21

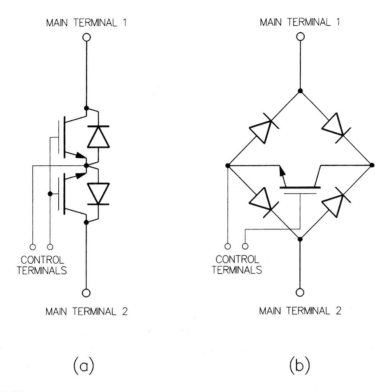

FIGURE 1.30
Fully controlled bidirectional power switch assemblies: (a) two transistors and two diodes, (b) one transistor and four diodes.

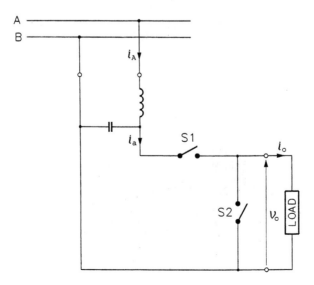

FIGURE 1.31
Single-phase ac chopper.

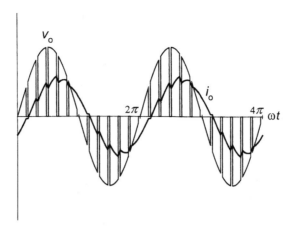

FIGURE 1.32
Output voltage and current waveforms in an ac chopper.

cycloconverters are invariably high-power converters, typically used in adjustable-speed synchronous motor drives requiring sustained low-speed operation.

The matrix converter, shown in Fig. 1.35 in the three-phase to three-phase version, constitutes a network of bidirectional power switches, such as those in Fig. 1.30, connected between each of the input terminals and each of the output terminals. In this respect, the matrix converter

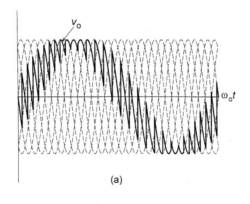

(a)

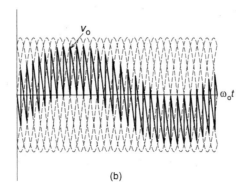

(b)

FIGURE 1.33
Waveforms of output voltage in a six-pulse cycloconverter: (a) $M = 1$, (b) $M = 0.5$ ($\omega_o/\omega = 0.2$).

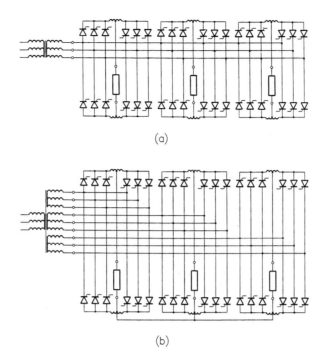

FIGURE 1.34
Three-phase six-pulse cycloconverters: (a) with isolated phase loads, (b) with interconnected phase loads.

constitutes an extension of the generic power converter in Fig. 1.1. The voltage of any input terminal can be made to appear at any output terminal (or terminals), while the current in any phase of the load can be drawn from any phase (or phases) of the supply line. An input LC filter is employed to screen the supply system from harmonic currents generated by the converter, which operates in the PWM mode. The load inductance assures continuity of the output currents. Although, with the 9 switches, the matrix converter can theoretically have 512 states, only 27 states are permitted. Specifically, at any time, one and only one switch in each row must be closed. Otherwise, the input terminals would be shorted or the output currents would be interrupted.

The voltages, v_a, v_b, and v_c, at the output terminals are given by

$$\begin{bmatrix} v_a \\ v_b \\ v_c \end{bmatrix} = \begin{bmatrix} x_{Aa} & x_{Ba} & x_{Ca} \\ x_{Ab} & x_{Bb} & x_{Cb} \\ x_{Ac} & x_{Bc} & x_{Cc} \end{bmatrix} \begin{bmatrix} v_A \\ v_B \\ v_C \end{bmatrix} \qquad (1.12)$$

where x_{Aa} through x_{Cc} denote switching functions of switches S_{Aa} through S_{Cc}, and v_A, v_B, and v_C are the voltages at the input terminals. In turn, the line-to-neutral output voltages, v_{an}, v_{bn}, and v_{cn}, can be expressed in terms of v_a, v_b, and v_c as

$$\begin{bmatrix} v_{an} \\ v_{bn} \\ v_{cn} \end{bmatrix} = \frac{1}{3} \begin{bmatrix} 2 & -1 & -1 \\ -1 & 2 & -1 \\ -1 & -1 & 2 \end{bmatrix} \begin{bmatrix} v_a \\ v_b \\ v_c \end{bmatrix}. \qquad (1.13)$$

24 CHAPTER 1 / POWER ELECTRONIC CONVERTERS

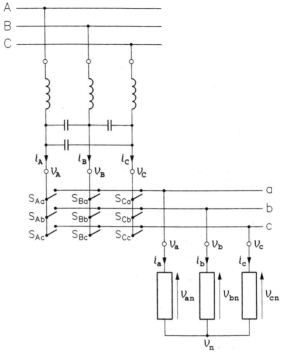

FIGURE 1.35
Three-phase to three-phase matrix converter.

The input currents, i_A, i_B, and i_C, are related to the output currents, i_a, i_b, and i_c, as

$$\begin{bmatrix} i_A \\ i_B \\ i_C \end{bmatrix} = \begin{bmatrix} x_{Aa} & x_{Ab} & x_{Ac} \\ x_{Ba} & x_{Bb} & x_{Bc} \\ x_{Ca} & x_{Cb} & x_{Cc} \end{bmatrix} \begin{bmatrix} i_a \\ i_b \\ i_c \end{bmatrix}. \tag{1.14}$$

Fundamentals of both the output voltages and input currents can successfully be controlled by employing a specific, appropriately timed sequence of the switching functions. As a result of such control, the fundamental output voltages acquire the desired frequency and amplitude, while the low-distortion input currents have the required phase shift (usually zero) with respect to the corresponding input voltages.

Example waveforms of the output voltage and current are shown in Fig. 1.36. For reference, waveforms of the line-to-line input voltages are shown, too. The output frequency, ω_o, in Fig. 1.36a is 2.8 times higher than the input frequency, ω, while the ω_o/ω ratio in Fig. 1.36b is 0.7. Respective magnitude control ratios, M, are 0.8 and 0.4.

Apart from the conceptual simplicity and elegance, matrix converters have not yet found widespread application in practice. Two major reasons are the low voltage gain, limited to $\sqrt{3}/2 \approx 0.866$, and unavailability of fully controlled bidirectional semiconductor switches.

1.4 DC TO DC CONVERTERS

Dc to dc converters, called *choppers*, are supplied from a dc voltage source, typically a diode rectifier and a *dc link*, as shown in Fig. 1.37. The dc link consists of a large capacitor connected

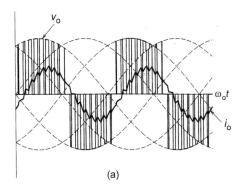

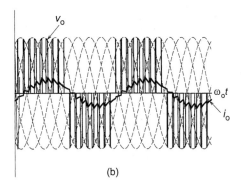

FIGURE 1.36
Output voltage and current waveforms in a matrix converter: (a) $\omega_o/\omega = 2.8, M = 0.8$; (b) $\omega_o/\omega = 0.7, M = 0.4$.

across the input terminals of the chopper and, often but not necessarily, a series inductance. The capacitor smooths the dc voltage produced by the rectifier and serves as a source of the high-frequency ripple current drawn by the chopper. The inductor provides an extra screen for the supply power system against the high-frequency currents. All choppers are pulse width modulated, the phase control being infeasible with both the input and output voltages of the dc type.

Most choppers are of the step-down (buck) type, that is, the average output voltage, V_o, is always lower than the input voltage, V_i. The *first-quadrant chopper*, based on a single fully

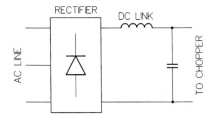

FIGURE 1.37
Dc voltage source for choppers.

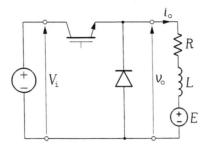

FIGURE 1.38
First-quadrant chopper.

controlled switch and a freewheeling diode, is shown in Fig. 1.38. Both the output voltage, v_o, and current, i_o, can only be positive. The average output voltage is given by

$$V_o = DV_i \tag{1.15}$$

where D denotes the duty ratio of the switch. The magnitude control ratio, M, is defined here as V_o/V_i and it equals D. Example waveforms of v_o and i_o are shown in Fig. 1.39, with M changing from 0.5 to 0.75. As in all PWM converters, the output voltage is pulsed, but the output current is continuous thanks to the load inductance. The current ripple is inverse proportional to the switching frequency, f_{sw}. Specifically, the rms value, $I_{o,ac}$, of the ac component of the output current is given by

$$I_{o,ac} = \frac{|M|(1-|M|)}{2\sqrt{3}Lf_{sw}} V_i \tag{1.16}$$

where L denotes inductance of the load.

The reason for the absolute value, $|M|$, of the magnitude control ratio appearing in Eq. (1.16) is that this ratio in choppers can assume both the positive and negative values. In particular, $M > 0$ indicates operation in the first and third quadrant (see Fig. 1.14), while $M < 0$ is specific for choppers operating in the second and fourth quadrant. The most versatile dc to dc converter, the four-quadrant chopper shown in Fig. 1.40, can, as its name indicates, operate in all four quadrants.

In the first quadrant, switch S4 is turned on all the time, to provide a path for the output current, i_o, while switch S1 is chopping with the duty ratio D_1. The remaining two switches, S2 and S3, are OFF. In the second quadrant, it is switch S2 that is chopping, with the duty ratio D_2, and all the other switches are OFF. Analogously, in the third quadrant, switch S1 is ON, switch S3 is chopping with the duty ratio D_3 and, in the fourth quadrant, switch S4 is chopping with the

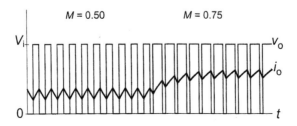

FIGURE 1.39
Example waveform of output voltage and current in a first-quadrant chopper.

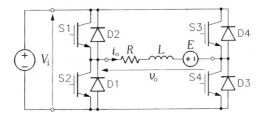

FIGURE 1.40
Four-quadrant chopper.

duty ratio D_4. When a chopping switch is OFF, conduction of the output current is taken over by a respective freewheeling diode, for instance, D1 in the first quadrant of operation. The magnitude control ratio, M, is given by

$$M = \begin{cases} D_1 & \text{in Quadrant 1} \\ 1 - D_2 & \text{in Quadrant 2} \\ -D_3 & \text{in Quadrant 3} \\ D_4 - 1 & \text{in Quadrant 4.} \end{cases} \quad (1.17)$$

If the chopper operates in Quadrants 2 and 4, the power flows from the load to the source, necessitating presence of an EMF, E, in the load. The EMF must be positive in Quadrants 1 and 2, and negative in Quadrants 3 and 4. For sustained operation of the chopper with a continuous output current, the magnitude control ratio must be limited in dependence on the ratio E/V_i as illustrated in Fig. 1.41. These limitations, as well as Eq. (1.17), apply to all choppers.

Any less-than-four-quadrant chopper can easily be obtained from the four-quadrant topology. Consider, for instance, a two-quadrant chopper, capable of producing an output voltage of both polarities, but with only a positive output current. Clearly, this converter can operate in the first and fourth quadrants. Its circuit diagram, shown in Fig. 1.42, is determined by eliminating switches S2 and S3 and their companion diodes, D2 and D3, from the four-quadrant chopper circuit in Fig. 1.40.

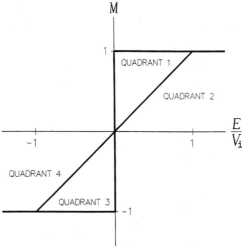

FIGURE 1.41
Allowable ranges of the magnitude control ratio in a four-quadrant chopper.

28 CHAPTER 1 / POWER ELECTRONIC CONVERTERS

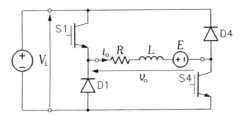

FIGURE 1.42
First-and-fourth-quadrant chopper.

A step-up (boost) chopper, shown in Fig. 1.43, produces a pulsed output voltage, whose amplitude, $V_{o,p}$, is higher than the input voltage. If a sufficiently large capacitor is connected across the output terminals, the output voltage becomes continuous, with $V_o \approx V_{o,p} > V_i$. When switch S is turned on, the input inductor, L_c, is charged with electromagnetic energy, which is then released into the load by turning the switch off. The magnitude control ratio, M, defined as $V_{o,p}/V_i$, in an ideal (lossless) step-up chopper is given by

$$M = \frac{1}{1-D} \tag{1.18}$$

where D denotes the duty ratio of the switch. In real choppers, the value of M saturates at a certain level, usually not exceeding 10 and dependent mostly on the resistance of the input inductor. Example waveforms of the output voltage and current in a step-up chopper without the output capacitor are shown in Fig. 1.44.

1.5 DC TO AC CONVERTERS

Dc to ac converters are called *inverters* and, depending on the type of the supply source and the related topology of the power circuit, they are classified as *voltage-source inverters* (VSIs) and *current-source inverters* (CSIs). The simplest, single-phase, half-bridge, VSI is shown in Fig. 1.45. The switches may not be ON simultaneously, because they would short the supply source. There is no danger in turning both switches off, but the output voltage, v_o, would then depend on the conducting diode, that is, it could not be determined without some current sensing arrangement. Therefore, only two states of the inverter are allowed. Consequently, a single switching function, a, can be assigned to the inverter. Defining it as

$$a = \begin{cases} 0 & \text{if SA} = \text{ON} \quad \text{and} \quad \text{SA}' = \text{OFF} \\ 1 & \text{if SA} = \text{OFF} \quad \text{and} \quad \text{SA}' = \text{ON}, \end{cases} \tag{1.19}$$

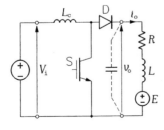

FIGURE 1.43
Step-up chopper.

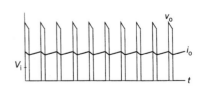

FIGURE 1.44
Output voltage and current waveforms in a step-up chopper ($D = 0.75$).

1.5 DC TO AC CONVERTERS

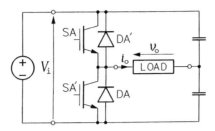

FIGURE 1.45
Single-phase, half-bridge, voltage-source inverter.

the output voltage of the inverter is given by

$$v_o = V_i \left(a - \frac{1}{2}\right) \quad (1.20)$$

where V_i denotes the dc input voltage. Only two values of v_o are possible: $V_i/2$ and $-V_i/2$. To prevent the so-called *shot-through*, that is, a short circuit when one switch is turned on and the other has not yet turned off completely, the turn-on is delayed by a few microseconds, called a *dead*, or *blanking*, *time*. The same precaution is taken in all VSIs, with respect to switches in the same leg of the power circuit.

The more common, single-phase full-bridge VSI, shown in Fig. 1.46, has two active legs, so that two switching functions, a and b, must be used to describe its operation. Notice that the topology of the inverter is identical to that of the four-quadrant chopper in Fig. 1.40. The output voltage can be expressed in terms of a and b as

$$v_o = V_i(a - b) \quad (1.21)$$

which implies that it can assume three values: V_i, 0, and $-V_i$. Thus, the maximum voltage gain of this inverter is twice as high as that of the half-bridge inverter.

Two modes of operation can be distinguished: the square-wave mode, loosely related to the phase-control mode in rectifiers, and the PWM mode. In the square-wave mode, so named because of the resultant shape of the output voltage waveform, each switch of the inverter is turned on and off only once per cycle of the output voltage. A specific sequence of inverter states is imposed, the state being designated by the decimal equivalent of ab_2. For example, if $a = 1$ and $b = 1$, the full-bridge inverter is said to be in State 3 because $11_2 = 3_{10}$. The output voltage waveform for the full-bridge inverter in the so-called optimal square-wave mode, which results in the minimum total harmonic distortion of this voltage, is shown in Fig. 1.47.

The output current, i_o, depends on the load, but generally, because of the high content of low-order harmonics (3rd, 5th, 7th, etc.) in the output voltage, it strays substantially from a sinewave.

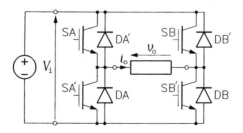

FIGURE 1.46
Single-phase, full-bridge, voltage-source inverter.

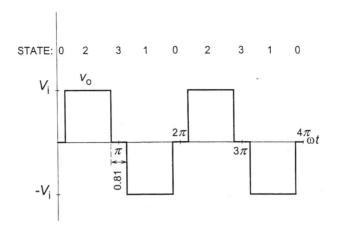

FIGURE 1.47
Output voltage waveform in a single-phase, full-bridge, voltage-source inverter in the optimal square-wave mode.

Not so in an inverter operating in the PWM mode, which results in a sinusoidal current with high-frequency ripple. Example waveforms of v_o and i_o in a PWM inverter are shown in Fig. 1.48.

Three-phase counterparts of the single-phase half-bridge and full-bridge VSIs in Figs. 1.45 and 1.46 are shown in Figs. 1.49a and 1.49b, respectively. The three-phase full-bridge inverter is one of the most common power electronic converters nowadays, predominantly used in ac adjustable speed drives and three-phase ac uninterruptable power supplies (UPSs).

The capacitive voltage-divider leg of the incomplete-bridge inverter also serves as a dc link. Two switching functions, a and b, can be assigned to the inverter, because of the two active legs of its power circuit. The line-to-line output voltages are given by

$$\begin{bmatrix} v_{AB} \\ v_{BC} \\ v_{CA} \end{bmatrix} = V_i \begin{bmatrix} 1 & -1 & 0 \\ 0 & 1 & -\frac{1}{2} \\ -1 & 0 & \frac{1}{2} \end{bmatrix} \begin{bmatrix} a \\ b \\ 1 \end{bmatrix} \quad (1.22)$$

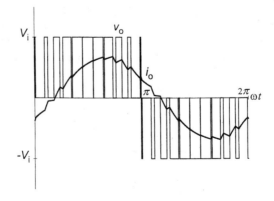

FIGURE 1.48
Output voltage and current waveforms in a single-phase, full-bridge, voltage-source inverter in the PWM mode.

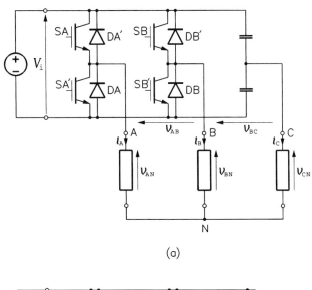

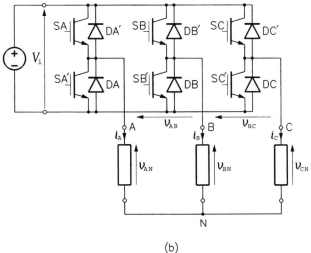

FIGURE 1.49
Three-phase voltage-source inverters: (a) incomplete-bridge, (b) full-bridge.

and the line-to-neutral voltages by

$$\begin{bmatrix} v_{AN} \\ v_{BN} \\ v_{CN} \end{bmatrix} = \frac{V_i}{3} \begin{bmatrix} 2 & -1 & -\frac{1}{2} \\ -1 & 2 & -\frac{1}{2} \\ -1 & -1 & 1 \end{bmatrix} \begin{bmatrix} a \\ b \\ 1 \end{bmatrix}. \quad (1.23)$$

Voltage v_{AB} can assume three values, $-V_i$, 0, and V_i, but v_{BC} and v_{CA} can assume only two values, $-V_i/2$ and $V_i/2$. The line-to-neutral voltages v_{AN} and v_{BN} can assume four values, $-V_i/2$, $-V_i/6$, $V_i/6$, and $V_i/2$, and v_{CN} can assume three values, $-V_i/3$, 0, and $V_i/3$. This voltage asymmetry makes the square-wave operation impractical, and the incomplete-bridge inverter can only operate in the PWM mode. The maximum voltage gain, taken as the ratio of the maximum available peak value of the fundamental line-to-line output voltage to the input voltage, is only 0.5. Therefore, in spite of the cost savings resulting from the reduced device count, the incomplete-bridge inverter is rarely used in practice.

Because of the three active legs, three switching functions, a, b, and c, are associated with the full-bridge three-phase inverter. The line-to-line and line-to-neutral output voltages are given by

$$\begin{bmatrix} v_{AB} \\ v_{BC} \\ v_{CA} \end{bmatrix} = V_i \begin{bmatrix} 1 & -1 & 0 \\ 0 & 1 & -1 \\ -1 & 0 & 1 \end{bmatrix} \begin{bmatrix} a \\ b \\ c \end{bmatrix} \quad (1.24)$$

and

$$\begin{bmatrix} v_{AN} \\ v_{BN} \\ v_{CN} \end{bmatrix} = \frac{V_i}{3} \begin{bmatrix} 2 & -1 & -1 \\ -1 & 2 & -1 \\ -1 & -1 & 2 \end{bmatrix} \begin{bmatrix} a \\ b \\ c \end{bmatrix}. \quad (1.25)$$

Each line-to-line voltage can assume three values, $-V_i$, 0, and V_i, and each line-to-neutral voltage five values, $-2V_i/3$, $-V_i/3$, 0, $V_i/3$, and $2V_i/3$. In the PWM mode, the maximum voltage gain is 1, that is, the maximum attainable peak value of the fundamental line-to-line voltage equals the dc supply voltage.

As in single-phase inverters, the state of the full-bridge inverter can be defined as the decimal equivalent of abc_2. The 5-4-6-2-3-1... state sequence, each state lasting one-sixth of the desired period of the output voltage, results in the square-wave mode of operation, illustrated in Fig. 1.50. In this mode, the fundamental line-to-line output voltage has the highest possible peak value, equal to $1.1V_i$, yielding a maximum voltage gain 10% higher than that in the PWM mode. At the same time, in the inverter, there is no possibility of magnitude control of the output voltage, and the voltage waveforms are rich in low-order harmonics, spoiling the quality of output currents. Therefore, in practical energy conversion systems involving inverters, the square-wave mode is used sparingly, only when a high value of the output voltage is necessary.

Example waveforms of switching functions and output voltages of the three-phase full-bridge inverter in the PWM mode are shown in Fig. 1.51. The cycle of output voltage is divided here into 12 equal *switching intervals*, pulses of switching functions located at centers of the intervals. Thus, the switching frequency, f_{sw}, is 12 times higher than the output frequency, f. In practical inverters, the switching frequency is usually maintained constant and independent of the output frequency, at a level representing the best trade-off between the switching losses and quality of the output currents.

The so-called *voltage space vectors*, an idea originally conceived for analysis of three-phase electrical machines, are a useful tool for analysis and control of three-phase power converters as well. Denoting individual phase voltages of an inverter by v_a, v_b, and v_c (they can be line-to-line, line-to-ground, or line-to-neutral voltages), the voltage space vector, **v**, is defined as

$$\mathbf{v} = v_d + jv_q \quad (1.26)$$

where

$$\begin{bmatrix} v_d \\ v_q \end{bmatrix} = \begin{bmatrix} 1 & -\frac{1}{2} & -\frac{1}{2} \\ 0 & \frac{\sqrt{3}}{2} & -\frac{\sqrt{3}}{2} \end{bmatrix} \begin{bmatrix} v_a \\ v_b \\ v_c \end{bmatrix}. \quad (1.27)$$

Reduction of three phase voltages, v_a, v_b, and v_c, to two components, v_d and v_q, of the voltage vector is only valid when $v_a + v_b + v_c = 0$. Then, only two of these voltages are independent

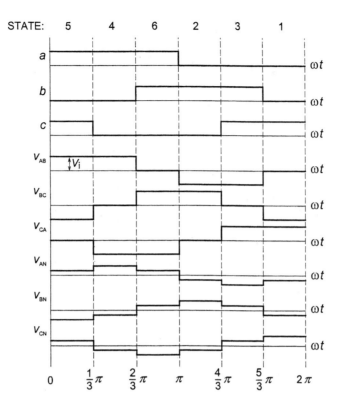

FIGURE 1.50
Waveforms of switching functions and output voltages in a three-phase, full-bridge inverter in the square-wave operation mode.

variables, so that the amount of information carried by v_a, v_b, and v_c is the same as that carried by v_d and v_q.

If

$$\begin{bmatrix} v_a \\ v_b \\ v_c \end{bmatrix} = V_p \begin{bmatrix} \cos(\omega t + \varphi) \\ \cos(\omega t + \varphi - \frac{2}{3}\pi) \\ \cos(\omega t + \varphi - \frac{4}{3}\pi) \end{bmatrix}, \qquad (1.28)$$

then

$$\begin{bmatrix} v_d \\ v_q \end{bmatrix} = \frac{3}{2} V_p \begin{bmatrix} \cos(\omega t + \varphi) \\ \sin(\omega t + \varphi) \end{bmatrix}, \qquad (1.29)$$

that is,

$$\mathbf{v} = \frac{3}{2} V_p e^{j(\omega t + \varphi)}. \qquad (1.30)$$

Thus, as the time, t, progresses, the voltage space vector, $\mathbf{v}$, revolves with the angular velocity ω in a plane defined by a set of orthogonal coordinates d and q.

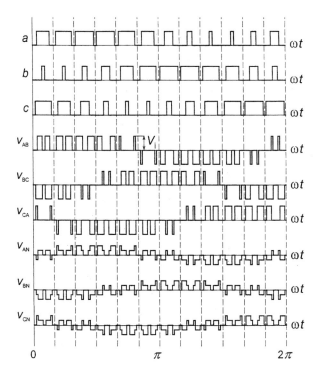

FIGURE 1.51
Waveforms of switching functions and output voltages in a three-phase, full-bridge inverter in the PWM mode.

In application to VSIs, the revolving voltage vector describes the fundamental output voltages. Each state of the inverter produces a specific stationary voltage space vector, and the revolving vector, **v**, which is to follow a reference vector, **v***, must be synthesized from the stationary vectors in a time-averaging process. The maximum possible value of **v** determines the maximum voltage gain of the inverter.

The four stationary voltage vectors, $\mathbf{V}_0$ through $\mathbf{V}_3$, of the three-phase incomplete-bridge inverter, corresponding to its four allowable states, are shown in Fig. 1.52 in the per-unit format. The input voltage, V_i, is taken as the base voltage. In the process of pulse width modulation, the vector, **v**, of fundamental output voltage is synthesized as

$$\mathbf{v} = \sum_{S=0}^{3} D_S \mathbf{V}_S \qquad (1.31)$$

where D_S denotes the duty ratio of State S ($S = 0 \ldots 3$). Each switching interval, a small fraction of the period of output voltage, is divided into several nonequal subintervals, constituting durations of individual states of the inverter. Not all four states must be used; three are enough. The vectors employed depend on the angular position of the synthesized vector **v**. Since the sum of all the duty ratios involved equals 1, the maximum available voltage vector to be generated is limited. It can be shown that the circle shown in Fig. 1.52 represents the locus of that vector. In other words, the maximum magnitude of **v** is $\sqrt{3}/4\, V_i$, which corresponds to the peak value of the line-to-line fundamental output voltage being equal to one-half of the dc supply voltage of the inverter.

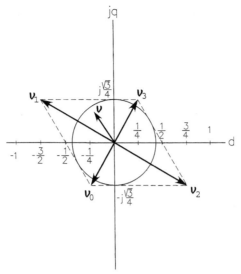

FIGURE 1.52
Space vectors of line-to-neutral output voltage in a three-phase incomplete-bridge voltage-source inverter.

Space vectors of line-to-line output voltage of the full-bridge inverter are shown in Fig. 1.53. There are six nonzero vectors, $\mathbf{V}_1$ through $\mathbf{V}_6$, whose magnitude equals the dc input voltage, V_i, and two zero vectors, $\mathbf{V}_0$ and $\mathbf{V}_7$. In general,

$$\mathbf{v} = \sum_{S=0}^{7} D_S \mathbf{V}_S, \qquad (1.32)$$

but in practice, only a zero vector and two nonzero vectors framing the output voltage vector are used. For instance, the vector $\mathbf{v}$ in Fig. 1.53 is synthesized from vectors $\mathbf{V}_2$ and $\mathbf{V}_3$ and a zero vector, $\mathbf{V}_0$ or $\mathbf{V}_7$. The radius of circular locus of the maximum output voltage vector indicates a voltage gain twice as high as that of the incomplete-bridge inverter. Specifically, the maximum

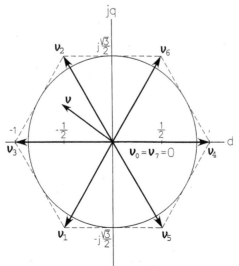

FIGURE 1.53
Space vectors of line-to-neutral output voltage in a three-phase full-bridge voltage-source inverter.

available peak value of the fundamental line-to-line output voltage equals V_i. Notice that going clockwise around the vector diagram yields the state sequence 4-6-2-3-1-5..., characteristic for square-wave operation with positive phase sequence, A-B-C. The counterclockwise state sequence 4-5-1-3-2-6..., would result in a negative phase sequence, A-C-B.

The voltage-source inverters described can be termed "two-level," because each output terminal, temporarily connected to either of the two dc buses, can only assume two voltage levels. Recently, *multilevel inverters* have been receiving increased attention. Using the same power switches, they have higher voltage ratings than their two-level equivalents. Also, their output voltage waveforms are distinctly superior to these in two-level inverters, especially in the square-wave mode.

The most common, *three-level neutral-clamped* inverter is shown in Fig. 1.54. Each leg of the inverter is comprised of four semiconductor power switches, S_1 through S_4, with freewheeling diodes, D_1 through D_4, and two clamping diodes, D_5 and D_6, that prevent the dc-link capacitors from shorting. The total of 12 semiconductor power switches implies a high number of possible inverter states. In practice, 27 states are employed only, as each leg of the inverter is allowed to assume only the three following states: (1) S1 and S2 are ON, S3 and S4 are OFF, (2) S2 and S3 are ON, S1 and S4 are OFF, and (3) S1 and S2 are OFF, S3 and S4 are ON. It can be seen that the dc input voltage, V_i, is always applied to a pair of series-connected switches, which explains the

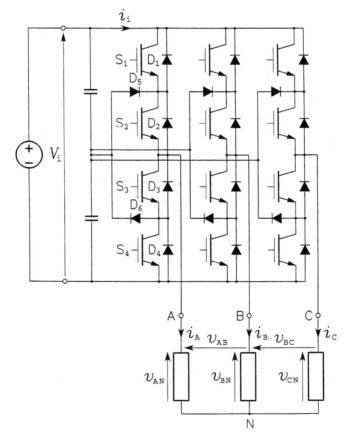

FIGURE 1.54
Three-level neutral-clamped inverter.

already-mentioned advantage of multilevel inverters with respect to the voltage rating. It can be up to twice as high as the rated voltage of the switches.

Stationary space vectors of output voltage of the three-level inverter are shown in Fig. 1.55. The maximum voltage gain of the inverter is the same as that of the two-level full-bridge inverter, but the availability of 27 stationary voltage vectors allows for higher quality of the output voltage and current. For instance, the square-wave mode of operation results in five-level line-to-line output voltages and seven-level line-to-neutral voltages. Waveforms of output voltages in the three-level inverter operating in the square-wave mode are shown in Fig. 1.56. PWM techniques produce output currents with very low ripple even when medium switching frequencies are employed.

All the inverters described so far are *hard-switching* converters, switches of which turn on and off under nonzero voltage and nonzero current conditions. This results in switching losses, which at high switching frequencies can be excessive. High rates of change of voltages (dv/dt) and currents (di/dt) cause a host of undesirable side effects, such as accelerated deterioration of stator insulation and rotor bearings, conducted and radiated electromagnetic interference (EMI), and overvoltages in cables connecting the inverter with the motor. Therefore, for more than a decade, significant research effort has been directed toward development of practical *soft-switching* power inverters.

In soft-switching inverters, switches are turned on and off under zero-voltage or zero-current conditions, taking advantage of the phenomenon of electric resonance. As is well known, transient currents and voltages having *ac-type* waveforms can easily be generated in low-resistance inductive-capacitive (LC) *dc-supplied* circuits. Thus, a resonant LC circuit (or circuits) constitutes an indispensable part of the inverter. Two classes of soft-switching inverters have emerged. In the first class, no switches are added to initiate the resonance. A representative of this class, the classic *resonant dc link* (RDCL) inverter with voltage pulse clipping, is shown in Fig. 1.57.

The resonant circuit consists of an inductor, L_r, and capacitor, C_r, placed between the dc link (not shown) and the inverter. Low values of the inductance and capacitance make the resonance

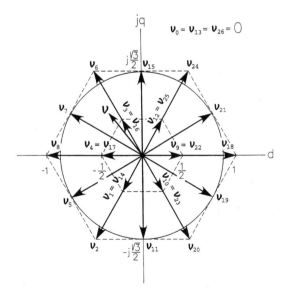

FIGURE 1.55
Space vectors of line-to-neutral output voltage in a three-level neutral-clamped inverter.

38 CHAPTER 1 / POWER ELECTRONIC CONVERTERS

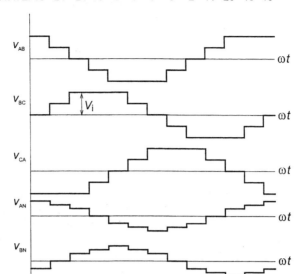

FIGURE 1.56
Output voltage waveforms in a three-level neutral-clamped inverter in the square-wave mode.

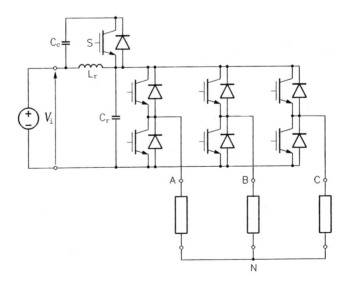

FIGURE 1.57
Resonant dc link inverter.

frequency several orders of magnitude higher than the output frequency of the inverter. The resonance is initiated by turning on, for a short period of time, both switches in a leg of the inverter bridge. This loads L_r with electromagnetic energy, which is then released into C_r when one of the switches in question is turned off. Because of the low resistance of the resonant circuit, the voltage across the capacitor acquires a sinusoidal waveform. Should the clamping circuit, based on switch S and capacitor C_c, be inactive, the peak value of the output voltage would approach $2V_i$. When the voltage drops to zero, the freewheeling diodes of the inverter become forward biased and they short the dc buses of the power circuit. This creates zero-voltage conditions for inverter switches, allowing lossless switching of these switches.

Output voltage waveforms in the RDCL inverter consist of packets of narrowly spaced resonant pulses. The clamping circuit clips these pulses in order to increase voltage density. A packet of clipped voltage pulses is shown in Fig. 1.58. It represents the equivalent of a single rectangular pulse of output voltage in a hard-switching inverter. It can be seen that the voltage pulses have low dv/dt, which alleviates certain undesirable side effects of inverter switching.

In the other class of soft-switching inverters, auxiliary switches trigger the resonance, so that the main switches are switched under zero-voltage conditions. One phase (phase A) of such a converter, the *auxiliary resonant commutated pole* (ARCP) inverter, is shown in Fig. 1.59. To minimize high dynamic stresses on main switches SA and SA', a resonant snubber, based on the inductor L_A and capacitors C_{A1} and C_{A2}, is employed. The resonance is initiated by turning on the bidirectional switch, composed of auxiliary switches S_{A1} and S_{A2} and their antiparallel diodes. The auxiliary switches are turned on and off under zero-current conditions.

ARCP inverters, typically designed for high-power applications, provide highly efficient power conversion. In contrast to the RDCL inverter, whose output voltage waveforms consist of packets of resonant pulses, the ARCP inverter is capable of true pulse width modulation, but with the voltage pulses characterized by low dv/dt. Typically, IGBTs or GTOs are used as main switches, while MCTs or IGBTs serve as auxiliary switches.

The voltage source supplying a VSI provides a fixed dc input voltage. A battery pack or, more often, a rectifier (usually uncontrolled) with a dc link, such as the dc source for choppers shown in Fig. 1.37, is used. Consequently, polarity of the input current depends on the direction of power transfer between the source and the load of the inverter, which explains the necessity of the freewheeling diodes. Voltage sources are more "natural" than current sources. However, a fair representation of a current source can be obtained combining a current-controlled rectifier

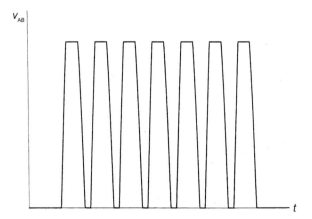

FIGURE 1.58
Packet of output voltage pulses in a resonant dc link inverter.

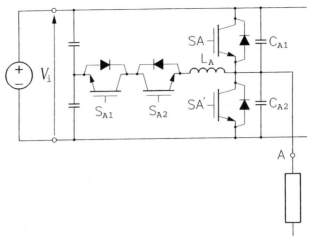

FIGURE 1.59
One phase of the auxiliary resonant commutated pole inverter.

and a large series inductor, as illustrated in Fig. 1.60. Such a source can be used for supplying dc power to a current-source inverter. In that case, it is the input current, I_i, that is maintained constant, and reversal of the power transfer is accompanied by a change in polarity of the input voltage.

The three-phase current-source inverter, shown in Fig. 1.61, differs from its full-bridge voltage-source counterpart by the absence of freewheeling diodes, which, because of the unidirectional input current, would be superfluous. Analogously, the single-phase CSI can be obtained from the single-phase full-bridge VSI in Fig. 1.46 by removing the freewheeling diodes.

In contrast to VSIs, both switches in the same leg of a CSI can be ON simultaneously. Because of the current-source characteristics of the supply system, no overcurrent will result. It is interruption of the current that is dangerous, because of the large dc-link inductance. Therefore, when changing the state of the inverter, both switches in one phase are kept closed for a short period of time, an analogy to the dead time in VSIs.

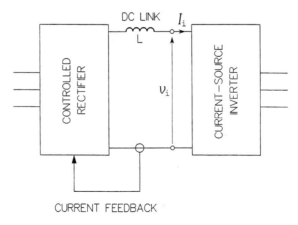

FIGURE 1.60
Supply arrangement for the current-source inverter.

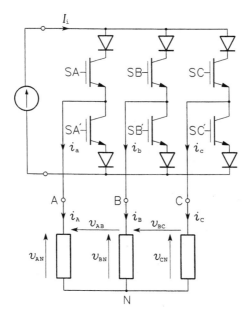

FIGURE 1.61
Three-phase current-source inverter.

Because both switches in the same phase can be ON or OFF simultaneously and one switch can be ON while the other is OFF, switching functions for individual phases would have to be of the quaternary type (having values of 0, 1, 2, or 3). This would be inconvenient; therefore, binary switching functions, a, a', b, b', c, and c', are assigned to individual switches, SA through SC', instead. This makes for 64 possible states of the inverter, of which, however, only 9 are employed, to avoid the uncertainty resulting from supplying a multipath network from a current source. Only one path of the output current is permitted, for example, from terminal A to terminal C, that is, with $i_A = -i_C$.

The output line currents in the three-phase CSI can be expressed as

$$\begin{bmatrix} i_A \\ i_B \\ i_C \end{bmatrix} = I_i \left(\begin{bmatrix} a \\ b \\ c \end{bmatrix} - \begin{bmatrix} a' \\ b' \\ c' \end{bmatrix} \right). \tag{1.33}$$

Defining the state of a CSI as $aa'bb'cc'_2$, only States 3, 6, 9, 12, 18, 24, 33, 36, and 48 are used, States 3, 12, and 48 producing zero output currents. Space vectors of the CSI are shown in Fig. 1.62. To facilitate interpretation of individual vectors, terminals of the inverter passing the output current are also marked. For example, current vector $\mathbf{i}_{33}$, produced in State 33, represents the situation when the load current flows between terminals A and C, that is, $i_A = I_i$ and $i_C = -I_i$.

Analogously to the VSI, when the state sequence corresponding to sequential current vectors $\mathbf{i}_{36}, \mathbf{i}_{33}, \mathbf{i}_9, \mathbf{i}_{24}, \mathbf{i}_{18}, \mathbf{i}_6, \ldots$, is imposed, with each state lasting one-sixth of the desired period of the output voltage, the CSI operates in the square-wave mode illustrated in Fig. 1.63. In practice, the rate of change of the current at the leading and trailing edge of each pulse is limited. Still, the load inductance generates spikes of the output voltage at these edges which, in addition to the nonsinusoidal shape of current waveforms, constitutes a disadvantage of CSIs. Sinusoidal currents (with a ripple) are produced in the PWM CSI, which is obtained by adding capacitors

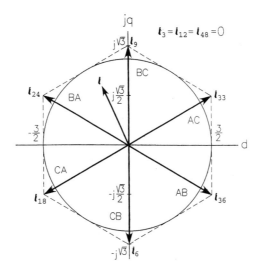

FIGURE 1.62
Space vectors of line current in a three-phase current-source inverter.

between the output terminals. These capacitors shunt a part of the harmonic content of square-wave currents, so that the load currents resemble those of the VSI.

1.6 CONCLUSION

The described variety of power electronic converters allows efficient conversion and control of electrical power. Pulse width modulated converters offer better operating characteristics than the phase-controlled ones, but the very process of high-frequency switching creates undesirable side effects of its own. It seems that phase-controlled rectifiers and ac voltage controllers will

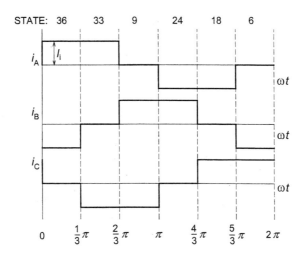

FIGURE 1.63
Output current waveforms in a current-source inverter in the square-wave mode.

maintain their presence in power electronics for years to come. The same observation applies to hard-switching converters, although the soft-switching ones will certainly increase their market share.

Switching functions, which stress the discrete character of power electronic converters, and space vectors of voltage and current, are convenient tools for analysis and control of these converters. It is also worth noting that the progress in speed and efficacy of information and power processing in modern PWM converters makes them nearly ideal power amplifiers.

CHAPTER 2

Resonant dc Link Converters

STIG MUNK-NIELSEN
Institute of Energy Technology, Aalborg University, Aalborg, Denmark

Applying soft switching reduces device-switching losses compared to hard-switched voltage source inverter (VSI), making it an interesting alternative. In contrast to the PWM-VSI, where the snubberless main circuit design is dominant, there is no resonant circuit configuration that has a dominant position. There are a variety of different circuits that are able to realize device soft switching; every circuit has its own merits and demerits. This chapter will present a few different converter configurations which are considered to be basic. The basic configuration has inspired a wide variety of converters; a few of those are also presented. Finally, discrete modulators are presented.

2.1 OVERVIEW OF RESONANT DC LINK CONVERTERS

2.1.1 Parallel Resonant dc Link

Reduction of the switching losses in the PWM-VSI may be done by a snubber circuit, which is robust and simple to realize. However, snubber circuits are designed to dissipate switching power in a resistor and the total losses are increased. The resonant circuit is ideally a nondissipative circuit and therefore an interesting alternative to snubber circuits.

The parallel resonant dc link (RDCL) converters have an oscillating link voltage that oscillates between zero voltage and a peak voltage. Figure 2.1 shows a parallel resonant converter. The switching of the transistors in the converter must be synchronized with the zero voltage periods of v_{do} [1] to obtain zero voltage switching (ZVS). This strategy eliminates the possibility of high-resolution PWM, and instead discrete pulse modulation (DPM) must be used. In [2] DPM is described, and it is concluded that the DPM has a performance comparable to that of PWM-VSI if the resonant link frequency is more than 6 times higher than the PWM switching frequency. When a comparison of the output waveforms is done in [3, 2], it is shown that DPM converter's spectrum performance relative to PWM is lower at modulation indexes below 0.3–

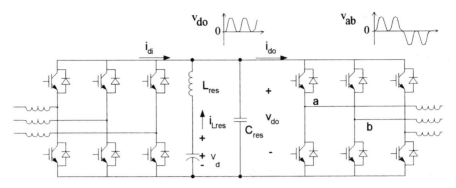

FIGURE 2.1
Parallel resonant dc link converter.

0.5. Compared to hard switching converters with a stiff dc voltage link, the voltage peak-to-peak amplitude seen by the transistors can be more than twice the dc voltage, V_d. The peak voltage of v_{do} is often limited by an auxiliary clamp circuit [4]. A high peak voltage across the terminals has several disadvantages: high voltage rating of converter devices, and stress on load-machine insulation, which may cause insulation breakdown.

2.1.2 Series Resonant dc Link

The series resonant dc link converter (Fig. 2.2) uses the principle of zero current switching (ZCS), where lossless switching is obtained. The converter is closely related to the thyristor converter and the link voltage is bipolar, which demands switches with symmetrical voltage blocking capability. The dc link current is oscillating between zero and, at a minimum, twice the dc link current, which is supplied by a dc inductor, L_d. There must always be a current path for the inductive dc link current and a capacitor filter is therefore necessary.

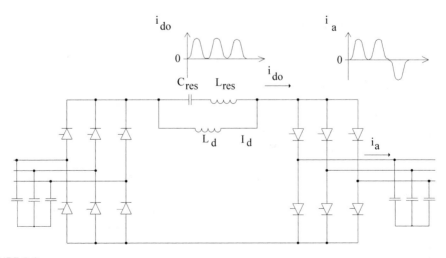

FIGURE 2.2
Series resonant dc link converter.

The converter is born with the possibility of rectification with unity power factor and bidirectional power flow. When using a parallel resonant topology, the ac voltage rectification is often done by diodes that eliminate the controlled switches, the bidirectional power flow, and the unity power factor correction options. With series resonant converters only 12 thyristors are needed for a full-bridge three-phase ac to ac conversion, whereas in parallel resonant converters 12 transistors and 12 diodes are used.

The firing of the thyristors must be synchronized with the zero current periods in the link, and again DPM is used. Spectral performance is dependent on the switching frequency. Normally the switching frequency is limited to 30 kHz because of the relative slow switching times of thyristors [5].

The link current stress is minimum twice the dc inductor current, and the conduction losses are thus relatively high compared to parallel resonant converters.

One general drawback of the serial converter is the necessary filter capacitance on the ac sides [6]. The interaction of the filter capacitance and the motor load inductance causes high-frequency oscillations on the load current. Further on the ac capacitor is bulky. A passive first-order filter can be used to reduce the high-frequency oscillation to an acceptable level at the expense of extra components and ohmic power dissipation [6]. This solution makes the size of the converter dependent on the load.

2.1.3 Pole Commutated dc Link

The pole commutated converter (Fig. 2.3) has a stiff dc link voltage, but the converter switches are switched under zero voltage conditions, and therefore low switching losses are obtained. To obtain ZVS an auxiliary resonant circuit is used. Each converter branch uses one circuit and the auxiliary circuit has four terminals. Three are connected to the dc link terminals and the fourth terminal is connected to the branch terminal [7, 8].

Unlike the parallel and series resonant dc link converters, the pole commutated converter is able to perform PWM. Another advantage is that the main load current is not flowing through the resonant elements, and in this way the current stress on the resonant inductor is relatively small.

Compared to hard switching converters the voltage stress on the converter switches is almost the same and with less output voltage dv/dt. There is a trade-off between a low dv/dt and a small

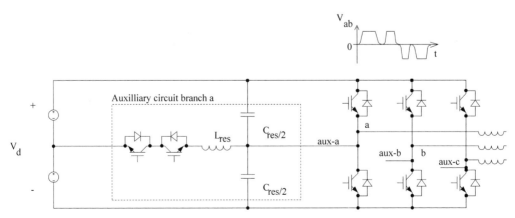

FIGURE 2.3
Auxiliary pole commutated converter.

minimum pulse-width duration. An increased resonant frequency, given by L_{res} and C_{res}, increases the dv/dt but lowers the limits on the pulse width. This will give a better spectral performance. The pole commutated converters obtain a spectral performance close to that of the PWM-VSI for a given switching frequency [7].

2.2 PARALLEL RESONANT CONVERTERS

To limit the peak voltage of the parallel resonant dc link converters a clamping method must be used. Often an additional clamp circuit is used, but it is also possible to limit the peak voltage simply by controlling the inverter switches. Three different methods are described in the following.

2.2.1 Passive Clamp

The passively clamped parallel resonant dc link converter is described in [9]. The link clamp circuit is a transformer with a diode as shown in Fig. 2.4. Ideally the transformer clamp the link voltage v_{do} to a clamp level of 2 times the dc link voltage V_d. In practice stray inductance causes a clamping level higher than twice the V_d. A link voltage clamp factor of 2.02 is obtained in [9]. The link voltage amplitude is 1252 V with a link voltage $V_d = 620$ V.

2.2.2 Active Clamp

In the active clamped resonant dc link (ACRDCL) converter the link voltage amplitude, v_{do}, is limited below 2 times V_d by a clamp circuit. The ACRDCL [10] is shown in Fig. 2.5.

The link voltage amplitude v_{do} is clamped by the voltages, $V_k + V_d$ when diode D_1 conducts. During the period diode D_1 conducts, the inductor L_{res} discharges. In order to enable the next resonant cycle to reach zero voltage, the inductor L_{res} must be recharged. Switch S_1 is turned on during the recharge of the inductor.

A control strategy for the active clamp is found by energy considerations. Energy flowing into the clamp source V_k must be equal to the energy flowing out. The active clamp circuit uses an ideal voltage source, in a realization of the clamp circuit voltage sources are often avoided, due to

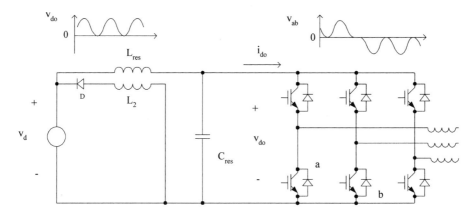

FIGURE 2.4
Passive clamped resonant converter.

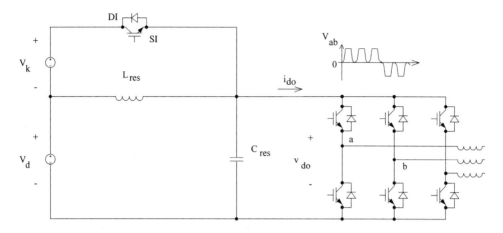

FIGURE 2.5
Active clamped resonant converter.

the circuit complexity and costs. In [10, 11] several suggestions from the literature describe how to realize the clamp circuit.

2.2.3 Voltage Peak Control

A voltage peak higher than $2V_d$ is generated if the resonant inductor L_{res}, shown in Fig. 2.1, is suddenly discharged and the inductor discharge through the resonant capacitor. This situation is shown in Fig. 2.6a. An IGBT in the inverter is turned off and the link current i_{do} abruptly changes amplitude. The following charge of the resonant capacitor is responsible for the high voltage peak. The high voltage peak can be prevented. By turning the IGBT off a short time before the zero dc link voltage condition the charge of the capacitor can be controlled. Turning the IGBT off at a link voltage level of ΔV_{do} causes the next resonant voltage peak to be twice the dc link voltage. The principle is illustrated in Fig. 2.6b.

The voltage peak control (VPC) strategy used to control the resonant dc link voltage is derived in [12]. The strategy can be formulated using

$$\Delta V_{do} = V_d \left(1 - \cos\left(a \sin\left(\frac{\Delta i_{do} Z_{res}}{2 V_d} \right) \right) \right)$$

where V_d is the dc link voltage, Z_{res} is the resonant impedance $= \sqrt{L_{res}/C_{res}}$, and Δi_{do} is the link current change.

The ΔV_{do} is the resonant voltage level where the dc link current must change, to make the next resonant voltage peak twice the dc link voltage. Experimental results are shown in Fig. 2.7.

2.3 PARALLEL RESONANT PWM CONVERTERS

This section deals with the type of parallel resonant converters which use PWM and have zero voltage switching of the main switches. Three converters are presented: the notch commutated three-phase PWM converter [13, 14], the zero switching loss PWM converter with resonant circuit [15], and the modified ACRDCL converter for PWM operation [16].

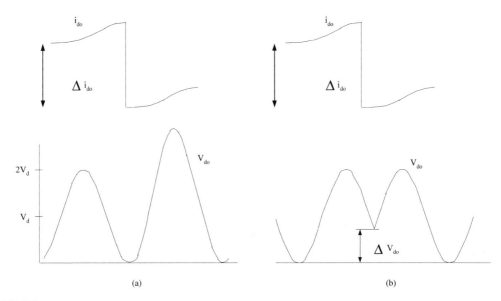

FIGURE 2.6
Principle of the voltage peak control strategy. (a) dc link current change i_{do} causes an increased resonant voltage peak of v_{do}. (b) Timing the link current change the increased voltage peak of v_{do} is avoided.

2.3.1 Notch Commutated Three-Phase PWM Converter

In the notch commutated three-phase PWM converter the zero voltage periods do not happen at discrete instants but are synchronized with a pulse from the pulse width modulator. Figure 2.8 shows the converter.

The converter link circuit makes synchronization with the converter switches possible. Until a commutation is wanted, the switch S_1 is off. Then S_1 turns on and decreases the resonant inductor current to an initial value, which ensures a resonant period with ZVS. The clamp voltage, V_k, and the size of the inductor L_{res} determine the time it takes to reach the initial current.

The energy in C_{res} is dissipated in S_1 at turn-on, and the clamp level is therefore low. In [14] the clamp level is about 1.2.

The converter has the desirable features of PWM, but there are turn-on losses when S_1 is turned on due to a discharge of the C_{res} capacitors.

2.3.2 Zero Switching Loss PWM Converter with Resonant Circuits

This converter offers PWM and zero voltage switching like the notch commutated converter described earlier, but the voltage stress on the converter components is lower. The converter is presented in [15]. Figure 2.9 shows the converter.

The voltage stress on the converter switches is similar to PWM-VSI, but the topology requires 2 switches and 3 diodes compared to the notch converters with 1 transistor and 1 diode. The resonant circuit is different from the other described resonant circuits. It uses two resonant states, determined by changing the resonant capacitor value. Changing a resonant state makes a link oscillation possible without any dc-link voltage overshoot ($V_k = 0$). However, there is a higher current stress on the converter components.

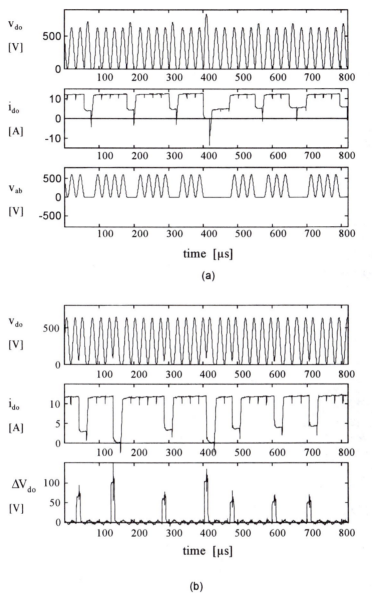

FIGURE 2.7
(a) Experimental results of the RDCL converter. dc link voltage $V_d = 300$ V. (b) Experimental results of the RDCLVPC converter. dc link voltage $V_d = 300$ V.

2.3.3 Modified ACRDCL for PWM Operation

This converter offers zero voltage switching of the converter switches and PWM operation of the converter [16]. It is an extension of the ACRDCL with an extra switch and diode. Figure 2.10 shows the converter.

The voltage stress on the converter switches is limited to V_d, until converter switching is needed. Before the converter switching takes place, the energy of the resonant inductor L_{res} must

52   CHAPTER 2 / RESONANT DC LINK CONVERTERS

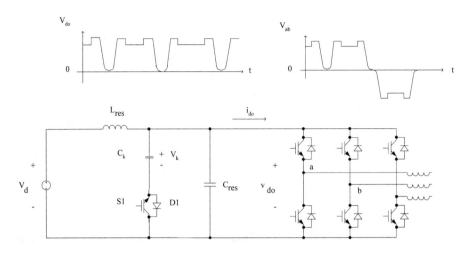

FIGURE 2.8
The notch commutated converter.

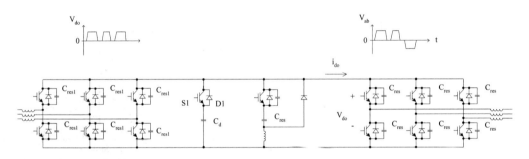

FIGURE 2.9
The zero switching loss PWM converter with resonant circuits.

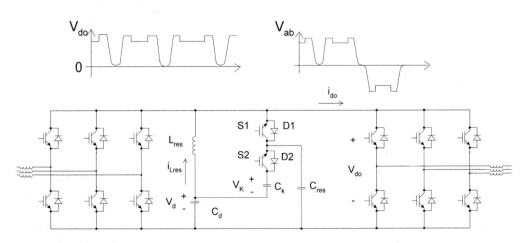

FIGURE 2.10
The modified ACRDCL for PWM operation.

increase. The resonant inductor energy is supplied from the clamp source V_k. During the charging interval of the resonant inductor the link voltage is $v_{do} = V_d + V_k$.

Finishing the resonant cycles it is necessary to clamp the voltage at $V_d + V_k$. During this clamp interval the energy applied during the charging interval of the inductor is transferred back to the clamp circuit.

2.4 MAINTAINING RESONANCE

This section describes how the resonance of the resonant circuit is kept on. At first sight the problem appears simple; after some time one realizes this is wrong. The maintaining of the resonance is a key problem that must be solved. Proper operation of the resonance ensures zero or low voltage switching of the inverter switches. During a design procedure there are several things to consider:

1. How to initiate the resonance? This can be done in a rough or a gentle way as described later.
2. The resonance must be error tolerant, meaning that if an error occurs and the resonance is prevented from completing, the resonant circuit control must be able to restore the resonance in the next resonant cycle. It is not acceptable if the whole converter stops because of electrical noise.
3. The resonant circuit stores reactive power, and in order to keep resonant circuit losses low, the level of reactive power must be kept low.
4. Finally, the circuit that controls the resonance should be capable of operating the converter at dc link voltages as high as 500 V.

Two methods are presented. The first is known as the short circuit method. The second is proposed based on the experience of working with the first described method. It is a bit more complex to describe theoretically than the short circuit method and it apparently requires a few extra components.

2.4.1 Short Circuit Method

The resonance is maintained by a short circuit of the inverter bridge legs, during the zero voltage interval. The short circuit interval ensures sufficient energy storage in the resonant inductor to overcome the circuit losses and therefore ensures that the link voltage resonates to zero voltage in the next time interval.

Several major loss elements are present in a resonant circuit, where the most significant are the serial resistance of the resonant inductor. The total resistive losses are determined from an experiment where a measurement of succeeding resonant voltage amplitudes is used to determine the quality factor, Q. With the knowledge of Z_{res} the equivalent serial resistor is $R = Z_{res}/Q$. An equation may be found which describes the level of current in L_{res} needed to overcome the resistor losses:

$$\Delta I_{Lres} = \sqrt{\frac{R\left(\frac{V_d}{Z_{res}}\right)^2}{f_{res} L_{res}}}.$$

2.4.2 Realization of the Short Circuit Method

The short circuit of the resonant inverter is done by turning all the inverter switches on when the resonant link voltage v_{do} reaches zero voltage. During the first part of the short circuit period L_{res} is properly discharged and the antiparallel diode conducts. After the resonant inductor is discharged, the current flow turns and the charging of the resonant inductor begins. The short circuit lasts until the inductor current has increased to $\Delta I_{L_{res}} + i_{phase}$. Since the beginning of the zero period is easy to detect, turning the short circuit on is straightforward, but turning it off is more difficult.

Several methods can be used to decide when to turn the short circuit off. A measurement of the resonant inductor current is the most direct way, but a high-speed, low-inductance current shunt is needed. Or, measure the on-state IGBT voltage drop and, from this, determine how long the IGBTs should be conducting. Implementing the method is relatively simple, but its accuracy is not good because of the on-state voltage dependency on conducted current.

One may decide to use a direct current measurement method. The advantage of this method is better accuracy, but the implementation is quite complex.

In Fig. 2.11 link current and resonant voltage are shown for a dc link voltage of 150 V. The initial inductor current is kept close to 1.5 A; a lower initial current was tried, but caused instability. An increase in the dc link voltage increases the initial resonant inductor current and the reactive losses.

2.4.3 A Non-Short Circuit Method

Transferring energy to the resonant circuit is necessary in order to compensate for loss elements. In the short circuit method this is done at the beginning of the resonant period. Another method is to transfer the energy inductively, which can be done by a secondary winding on the resonant coil, making it a transformer. The secondary side energy source is a current generator that produces pulses with the frequency of the resonant circuit as shown in Fig. 2.12. If galvanic isolation is not required the transformer may be omitted.

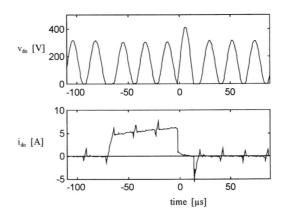

FIGURE 2.11
Measured resonant link voltage and current. The resonance is maintained by the short circuit method.

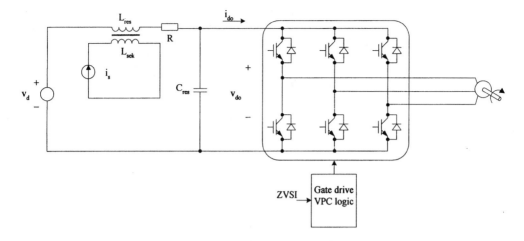

FIGURE 2.12
Non-short circuit method to maintain the resonance, using a transformer and current generator.

The current source is typically a square wave, in phase with the resonant inductor voltage. The advantage of zero-phase displacement is that the transition of current happens at zero voltage, and therefore the current source switches have low switching losses.

If there is excessive energy in the resonant circuit, it is transferred to the voltage source V_d during the conducting interval of the antiparallel diodes. If the voltage v_{do} fails to reach zero, the current generator will continue to supply the resonant circuit with energy and eventually the voltage reaches zero. This method of energizing the resonant circuit makes the resonance robust.

With an approximation assuming $R \ll Z_{res}$, the current amplitude of the current generator may be calculated:

$$\hat{i}_s = R \frac{V_d}{Z^2}.$$

In Table 2.1 two current equations are shown in the non-short circuit method box. The amplitude of i_s covers the case of a sinusoidal current source and i_{ds} is used for square wave currents.

2.4.4 A Laboratory Test of the Non-Short Circuit Method

Since there were stability problems with the short circuit method operating at dc link voltages above 300 V, the converter was first tested at the 300-V level, then a 500-V level was used. Stable converter operation was a fact, using the non-short circuit method.

Looking at Fig. 2.13 it can be seen that energy is stored in the resonant inductor at the end of a resonant period. The energy is transferred to the dc link voltage capacitor during the antiparallel diode conducting interval.

Based on simulation and laboratory experience the following is concluded about the non-short circuit method:

1. The resonance is started without current stress of the inverter switches, or any excessive stress at all.
2. If an error occurs and the resonant voltage v_{do} does not resonate below, e.g., 15 V, the resonance is not terminated, the energy transfer to the resonant circuit is continued, and the converter operation is not affected. This makes the converter operation robust.

Table 2.1 Calculation of Initial Current Using the Short Circuit Method and Current Amplitude Using the Non-Short Circuit Method

$V_d = 500\,\text{V}$, $L_{res} = 150\,\mu\text{H}$, $C_{res} = 100\,\text{nC}$, $Z_{res} = 38.7\,\Omega$, $f_{res} = 41.09\,\text{kHz}$, $R = 0.35\,\Omega$.

Short circuit method	Non-short circuit method
$\Delta i_{Lres} = \sqrt{\dfrac{R\left(\dfrac{V_d}{Z_{res}}\right)^2}{f_{res} L_{res}}}$	$\hat{i}_s = R\dfrac{V_d}{Z_{res}^2}$
	$i_{ds} = \hat{i}_s \dfrac{\pi}{4}$
$\Delta i_{Lres} = 3.1\,\text{A}$	$i_{ds} = 92\,\text{mA}$

3. The resonant converter is operated at a dc link voltage of 500 V, and there is no reason why this should not be increased.

Another advantage of the method is that no phase current or link current measurement is required.

2.5 CONVERTER MODULATION STRATEGIES

In this section discrete pulse modulation strategies are presented. Three different strategies are described briefly. The switching of the converter devices happens at discrete instants in synchronization with the link voltage zero intervals. It is not realistic to use a standard PWM strategy, since the resonant frequency is much lower than the normal clock frequency of the

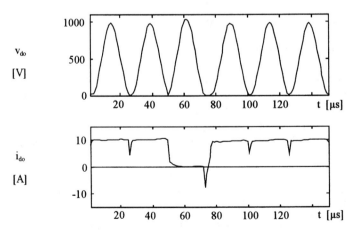

FIGURE 2.13
Measured link voltage and link current using the non-short circuit method.

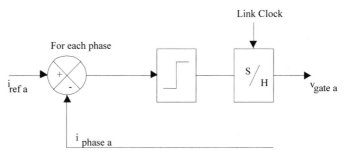

FIGURE 2.14
Delta current modulator.

PWM timer, which is in the megahertz region. Fortunately, discrete pulse modulation is available; in contrast to PWM, the switching frequency is not constant, causing an output voltage spectrum with the harmonics spread out between the fundamental and the resonant frequency. The most common modulators are described in [2, 11, 17, 18].

2.5.1 Delta Current Modulator

The delta current modulator (DCM), shown in Fig. 2.14, is a single-phase modulator with zero hysteresis comparator; the phase–phase voltage changes polarity relatively often compared to PWM. Usually PWM only allows one branch switchover between two succeeding active vectors. The resonant link current changes polarity often; therefore, the link stress is relatively high.

2.5.2 Adjacent State Current Modulator

The adjacent state current modulator (ASCM), shown Fig. 2.15, is a modification of the delta current modulator. The adjacent state modulator allows only the converter to generate succeeding active vectors that are adjacent vectors.

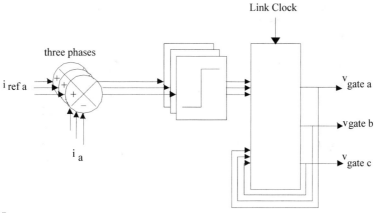

FIGURE 2.15
Adjacent state current modulator.

58 CHAPTER 2 / RESONANT DC LINK CONVERTERS

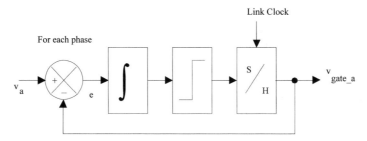

FIGURE 2.16
Single phase sigma delta modulator (SDM).

The vector sequence $100 \to 110 \to 010$ is allowed, but the vector sequence $100 \to 010$ is not allowed. If the succeeding vector to an active results in more than one BSO, a zero voltage vector is selected. The zero voltage vector that is closest to the preceding vector is selected.

The use of zero voltage vectors limits the link stresses considerably because the link current reversals are eliminated.

2.5.3 Sigma Delta Modulator

Sigma delta modulators, shown in Figs. 2.16 and 2.17, are simpler to realize than the DCM and ASCM because they require only a voltage reference and no feedback from the load. The sigma delta modulator has lower dynamic performance compared to the DCM, but it has superior THD performance.

2.6 CONCLUSIONS

Resonant converters offer the advantage of soft switching, thus decreasing the switching losses relative to hard-switched converters, but this is no guarantee that the efficiency of the resonant converter is higher, because the resonant circuit is not lossless. The output voltage quality of the

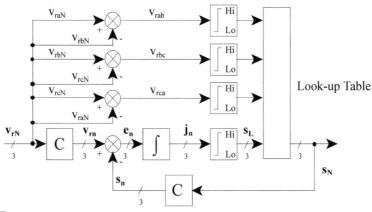

FIGURE 2.17
Space vector sigma delta modulator for a three-phase inverter.

resonant converters is comparable with that of the standard PWM-VSI; however, the harmonic content is smeared over a wide frequency range between the fundamental and the resonant frequency. Resonant converters are more complex than the PWM-VSI in terms of circuit complexity.

REFERENCES

[1] D. M. Divan, The resonant dc-link inverter—a new concept in static power conversion. Proc. of IAS '86, 1986, pp. 648–655.
[2] A. Mertens and H.-C. Skudelny, Calculations on the spectral performance of discrete pulse modulation strategies. Proc. of PESC 1991, pp. 357–365.
[3] M. Dehmlow, K. Heumann, and R. Sommer, Comparison of resonant converter topologies. Proc. of ISIE 1993, Budapest, pp. 765–770.
[4] A. Mertens and D. M. Divan, A high frequency resonant dc link inverter using IGBTs. Proc. of IPEC 1990, pp. 152–160.
[5] Y. Murai, S. G. Abeyrante, T. A. Lipo, and P. Caldeira, Dual-flow pulse trimming concept for a series resonant dc link power conversion. Proc. of PESC 1991, pp. 254–260.
[6] Y. Murai and T. A. Lipo, High-frequency series-resonant dc link power conversion. *IEEE Trans. Indust. Appl.* **28**, pp. 1277–1285 (1992).
[7] R. W. De Doncker and J. P. Lyons, The auxiliary resonant commutated pole converter. Proc. of IAS 1990, pp. 1228–1235.
[8] R. W. De Doncker and J. P. Lyons, The auxiliary quasi-resonant dc link inverter. Proc. of PESC 1991, pp. 248–253.
[9] G. L. Skibinski, The design and implementation of a passive clamp resonant dc link inverter for high power applications. Ph.D. thesis, University of Wisconsin-Madison, 1992.
[10] D. M. Divan and G. L. Skibinski, Zero switching loss inverters for high power applications. Proc. of IAS 1987, pp. 627–634.
[11] A. Petterteig, Development and control of a resonant dc-link converter for multiple motor drives. Ph.D. thesis, 1992, University of Trondheim, ISBN 82-7119-359-7.
[12] S. Munk-Nielsen, F. Blaabjerg, and J. K. Pedersen, A new robust and simple three phase resonant converter. Proc. of IAS 1997, pp. 1667–1672.
[13] V. G. Agelidis, P. D. Ziogas, and Deza Joos, Optimum use of dc side commutation in PWM inverters. Proc. of PESC 1991, pp. 277–282.
[14] V. G. Agelidis, P. D. Ziogas, and D. Joos, An optimum modulation strategy for a novel notch commutated 3-Φ PWM inverter. Proc. of IAS 1991, pp. 809–818.
[15] J. W. Choi and S. K. Sul, Resonant link bi-directional power converter without electrolytic capacitor. Proc. of PESC 1993, pp. 293–299.
[16] S. Salama and Y. Tadros, Quasi resonant 3-phase IGBT inverter. Proc. of PESC 1995, pp. 28–33.
[17] V. G. Vekataramanan, Topology, analysis and control of a resonant dc link power converter. Ph.D. thesis, University of Wisconsin-Madison, 1992.
[18] T. G. Habetler and D. M. Divan, Performance characterization of a new discrete pulse modulated current regulator. Proc. of IAS '88, 1988, pp. 395–405.

CHAPTER 3

Fundamentals of the Matrix Converter Technology

C. KLUMPNER and F. BLAABJERG
Institute of Energy Technology, Aalborg University, Aalborg, Denmark

This chapter presents the state of the art in matrix converter technology. The introduction presents the basic diagrams and the permitted switching states of the converter and the transfer functions of the output voltage and input current. Because this converter employs bidirectional switches, the implementation and the specific bidirectional switch commutation aspects are presented. Different modulation strategies and advanced control strategies applied to matrix converters proposed in the literature are briefly presented. Implementation issues regarding the design of the input filter, of the clamp circuit, as well as a proposed strategy for ride-through operation and the realization of an integrated motor drive based on a matrix converter, are also included. The summary will point out the outstanding contributions of matrix converter technology.

3.1 OVERVIEW

A matrix converter consists of nine bidirectional switches, arranged in three groups of three, each group being associated with an output line. This arrangement of bidirectional switches connects any of the input line a, b, or c to any of the output line A, B, or C, as it is shown in Fig. 3.1a. A bidirectional switch is able to control the current and to block the voltage in both directions. If the input and the output three-phase systems are orthogonally disposed, the converter diagram becomes similar to a matrix, with the rows consisting of the three input lines (a, b, c), the columns consisting of the three output lines (A, B, C), and bidirectional switches connecting each row to each column, which in Fig. 3.1b are symbolized with circles. There are 512 possible combinations of switches in a three-phase to three-phase matrix converter. In order to provide safe operation of the converter, when operating with bidirectional switches, two basic rules have to be followed:

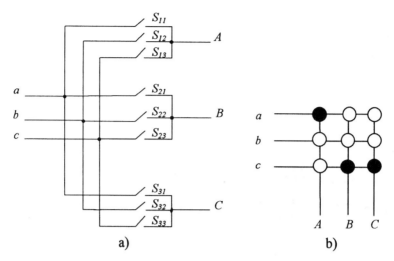

FIGURE 3.1
Basic topology of a matrix converter: (a) electric scheme; (b) symbol.

- **Do not** connect two different input lines to the same output line (short-circuit of the mains, which causes overcurrents)
- **Do not** disconnect the output line circuits (interrupt inductive currents, which causes overvoltages)

Therefore, an output line must be connected all the time to a single input line. This allows us to symbolize the state of the matrix converter by using a group of three letters, which give the input lines connected to the output lines in the following order: *A-B-C*. For example, "*acc*" means that the output lines *A*, *B*, and *C* are respectively connected to input lines *a*, *c*, and *c*.

If the basic rules mentioned before applies, the maximum number of permitted switching states of the matrix converter is reduced to 27, and these are shown in Fig. 3.2. Six switching states provide a direct connection of each output line to a different input line, producing a rotating voltage vector with amplitude and frequency similar to the input voltage system and direction dependent on the sequence: synchronous or inverse. Another 18 switching states produce active vectors of variable amplitude, depending on the selected line-to-line voltage, but at a stationary position. The last three switching states produce a zero vector, by connecting all the output lines to the same input line.

The transfer matrix *T* usually represents the state of the converter switches:

$$T = \begin{bmatrix} T_{11} & T_{12} & T_{13} \\ T_{21} & T_{22} & T_{23} \\ T_{31} & T_{32} & T_{33} \end{bmatrix} \qquad (3.1)$$

where $T_{ij} = \{-1, 0, 1\}$ are the possible conduction states of the bidirectional switches.

Each row shows the state of the switches connected on the same input line and each column shows the state of the switches connected on the same output line. Because of the instantaneous power transfer of the matrix converter, the electrical parameters (voltage, current) in one side may be reconstructed from the corresponding parameters in the other side, at any instant. The input phase voltages are given, because the matrix converter is connected to the grid. Therefore,

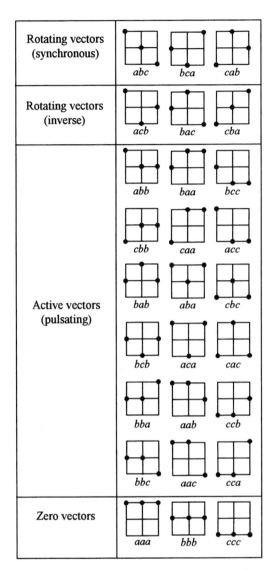

FIGURE 3.2
Permitted switching states (27) in a three-phase to three-phase matrix converter.

by applying the direct transformation for the input phase voltages $U_{a,b,c}$, the output line-to-line voltages $U_{AB, BC, CA}$ are found:

$$\begin{bmatrix} U_{AB} \\ U_{BC} \\ U_{CA} \end{bmatrix} = \begin{bmatrix} T_{11} & T_{12} & T_{13} \\ T_{21} & T_{22} & T_{23} \\ T_{31} & T_{32} & T_{33} \end{bmatrix} \cdot \begin{bmatrix} U_a \\ U_b \\ U_c \end{bmatrix} \quad \text{or} \quad U_{out} = T \times U_{in}. \tag{3.2}$$

The output currents are a result of applying the previously determined output voltages to a given load. By applying the inverse transformation for the output currents $I_{A, B, C}$, the input currents $I_{a, b, c}$ are found:

$$\begin{bmatrix} I_a \\ I_b \\ I_c \end{bmatrix} = \begin{bmatrix} T_{11} & T_{21} & T_{31} \\ T_{12} & T_{22} & T_{32} \\ T_{13} & T_{23} & T_{33} \end{bmatrix} \cdot \begin{bmatrix} I_A \\ I_B \\ I_C \end{bmatrix} \quad \text{or} \quad I_{in} = T^T \times I_{out}. \tag{3.3}$$

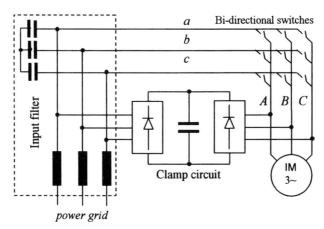

FIGURE 3.3
The practical scheme of a matrix converter drive.

Therefore, by knowing the load parameters, the input voltages, and the switching states, given by a proper modulation strategy, the output voltages and the input currents are found. This is a simple method of solving a matrix converter drive system in a simulation program.

The matrix converter acts as a current source inverter at the mains side, and as a voltage source inverter in the load side. Therefore, an input LC filter is necessary to filter the high-frequency ripple from the input currents. For protection purposes, it was shown in [1, 2] that a clamp circuit is needed to provide safe shutdown of the converter during faulty situations as overcurrent on the output side or voltage disturbances on the input side. In Fig. 3.3, a practical topology of a matrix converter drive is shown, including the input filter and the clamp circuit.

3.2 ANALYSIS OF BIDIRECTIONAL SWITCH TOPOLOGIES

In a synthesis paper written in 1988 [3], one of the first presenting the expectations for future development in ac drives, the matrix converter was credited with great potential. This was before important advances in this technology were achieved, such as reaching the highest limit of the voltage transfer ratio (0.86), proposing space vector modulation for matrix converters, and proposing a semisoft commutation strategy of bidirectional switches. At that time, it was considered that the MOS controlled thyristor (MCT) will evolve toward a reverse blocking capability and will provide the "ideal" switch for matrix converters. Later, because of the soft-commutation advantage, it has been shown that higher efficiency compared to a standard voltage source inverter (VSI) may be reached at higher switching frequencies, of more than 10 kHz. As the MCT is a slow device, the solution to implement a bidirectional switch remains the IGBT device. Other synthesis papers published later [4–8] give the same credit to the matrix converter technology, as the trend now is toward improving the interaction with the power grid, providing bidirectional power flow, and increasing the efficiency of the drive while operating at higher switching frequency, decreasing the drive size, and integrating more complex silicon structures in power modules.

Attempts were reported in the literature [9] and a patent was issued [10], proposing a new power device for matrix converter applications: the reverse blocking IGBT (RBIGBT), which decreases to one the number of semiconductor devices per load phase path. This will create

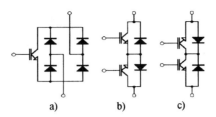

FIGURE 3.4
Bidirectional switch topologies using unidirectional switches: (a) diode-embedded switch; (b) common-emitter switch (CE); (c) common-collector switch (CC).

conditions to increase the efficiency of the matrix converters above the diode-bridge VSI, because the conduction losses will be produced only by a single RBIGBT per phase. In year 2000 [11], the first commercial RBIGBT device was reported to be available on the market.

The industrial development of the matrix converter has been obstructed by the lack of a true force-commutated bidirectional switch. Using unidirectional devices available on the market, there are three ways to obtain a bidirectional switch: the diode embedded unidirectional switch (Fig. 3.4a), the two common-emitter (CE) unidirectional switches (Fig. 3.4b), or the two common-collector unidirectional switches (CC) (Fig. 3.4c). The diode-embedded switch requires only one gate driver and one active switch, which is more convenient for implementation than the other two configurations. A comprehensive analysis [1, 9, 12, 13] shows that the embedded switch topology causes higher conduction losses because the current path consists of two fast recovery diodes (FRDs) and one IGBT and higher switching losses because all commutations are hard switched. The two topologies based on antiparallel connection of two unidirectional switches (CE and CC) allow for lower conduction losses because the current path consists only of one FRD and one IGBT and for semisoft switching (half soft switching and half hard switching), as will be presented in the following section.

It is expected that when the RBIGBT device becomes available for mass production, it will minimize the device count and the conduction losses in a matrix converter, while the control circuits will remain the same.

3.3 BIDIRECTIONAL SWITCH COMMUTATION TECHNIQUES

Similar to VSI, where dead-time commutation is necessary to eliminate the risk of shoot-through (short-circuit of the dc-link capacitors through an inverter leg) caused by nonideal commutation characteristics, specific commutation techniques must be implemented when bidirectional switches are operated. Figure 3.5a shows the basic circuit and Fig. 3.5b shows the ideal command signals when the output line *out* is switched from one input line x to the other input line y, by operating two bidirectional switches S_x and S_y. The commutation technique depends on the type of bidirectional switches employed in the matrix converter hardware.

In the case of ideal true four-quadrant switches, there are two possibilities for performing the commutation:

- The dead-time current commutation, referred to as "break before make," is shown in Fig. 3.5c. This method consists of turning off the off-going switch while the on-coming switch is still disconnected, to avoid short-circuit of the inputs and thereby eliminate the

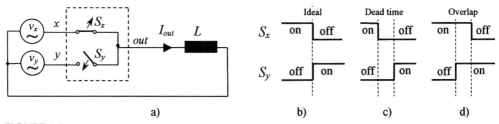

FIGURE 3.5
Commutation of the output phase *out* from input phase *x* to input phase *y*: (a) principle diagram; (b) ideal commutation; (c) dead-time commutation; (d) overlap commutation.

overcurrent risk. This will cause overvoltage on the output side; therefore a clamp circuit connected to the output to provide a continuity of the load current is necessary, but it will cause high switching losses. The use of a regenerative clamp circuit has been reported in [14].

- The overlap current commutation, referred to as "make before break," is shown in Fig. 3.5d. This method consists of turning on the on-coming switch while the off-going switch is still conducting, to provide continuity for the output line circuit in order to eliminate the risk of overvoltage. This will cause high circulating currents between the input phases, which have to be limited during the commutation, by adding extra chokes in the inputs to decrease di/dt.

Both methods require extra reactive elements and produce high losses. Because of practical implementation reasons, a bidirectional switch uses antiparalleled unidirectional switches, which provide independent control for each direction of the current. Therefore, other commutation strategies have been proposed [15–22] to provide safe commutation. Basically, the operation principle of such a method consists in a two- or four-step commutation technique, which avoids the disadvantages of the methods discussed before because it does not interrupt the load circuit or short-circuit the power grid lines, and therefore it does not require any additional reactive elements to provide safe commutation.

There are two ways to perform the so-called safe commutation: one is based on the sign of the output current [15–21] and the other one is based on the sign of the line-to-line voltage between the switches involved in the commutation process [22]. In the case of an output current sign detection based control, the first action is to disable the current path for circulating currents and then to apply overlapping commutation for the on-coming switch with the off-going switch. Therefore, the risk of short circuit on the input side is eliminated, and semisoft commutation, which means that half of the switch commutations are performed naturally, is achieved. In the case of a line-to-line voltage sign detection based control, the first action is to close a unidirectional path in the on-coming switch, which will not cause circulating currents, to ensure that it is a path for the load current before opening a unidirectional path in the off-going switch; this is identical for the other direction of current. The difference between the two methods is that for the output current sign control, a maximum of two unidirectional switches will be closed at any instant, whereas for the line-to-line voltage sign control, three unidirectional switches are closed at some instants. Because the load is usually inductive, while on the input side an LC circuit is usually used, it is considered that current sign detection offers safer and more stable operation even though it requires current transducers on the output side. For the other method, voltage transducers are necessary anyway for the control, but this signal may be noisy. Here, only the current sign controlled commutation will be presented.

3.3.1 Current Sign Four-Step Commutation Strategy

This strategy [15–18] operates the switches in such a way that after the commutation is completed, the switch acts as a four-quadrant bidirectional switch, so that the load current can freely reverse direction. The commutation takes place in four steps:

Step 1: Turn off the off-going non-conducting switch. This way, current direction is not able to change sign.

Step 2: Turn on the on-coming conducting switch. Now, there is unidirectional connection between input lines, but no circulating current may occur. In cases where there is a condition for natural commutation between the off-going switch and on-coming switch, that starts at this moment.

Step 3: Turn off the off-going conducting switch. At this time, in the case of hard commutation, the current is forced to switch from the off-going switch to the on-coming switch.

Step 4: Turn on the on-coming nonconducting switch. This is a passive step, with the purpose of reestablishing the four-quadrant characteristic of the ac switch, so the currents can change sign naturally.

This is shown in Fig. 3.6 where the commutation scheme takes into account the current sign and also the possibility to switch from one steady switching state to the other steady switching state.

In Fig. 3.7 the possible path for the output current allowed by the four-step commutation strategy is illustrated for both current signs. The duration of the "passive" commutation steps (1 and 4) is not critical, because it is supposed that the devices that are switched will or will not conduct; therefore it may change state faster. The duration of the "active" commutation (2 or 3) is critical and should be chosen in agreement with the switching characteristics of the devices employed in the antiparallel topology.

A complicated solution to adapt the duration of this "active" commutation has been proposed in [18], which modifies the period of the commutation clock with respect to the output current magnitude, but implies more complicated hardware. However, by adjusting the commutation clock period according to the maximum current magnitude, a safe and low-loss commutation is achieved, and for smaller output currents the commutation will take place faster than the clock period, which does not cause any problem. The variation of the duty-cycle duration caused by the

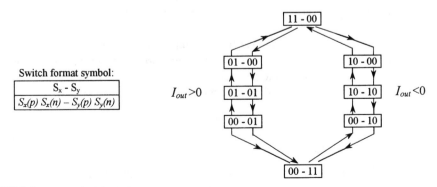

FIGURE 3.6
Four-step commutation scheme depending on the current sign.

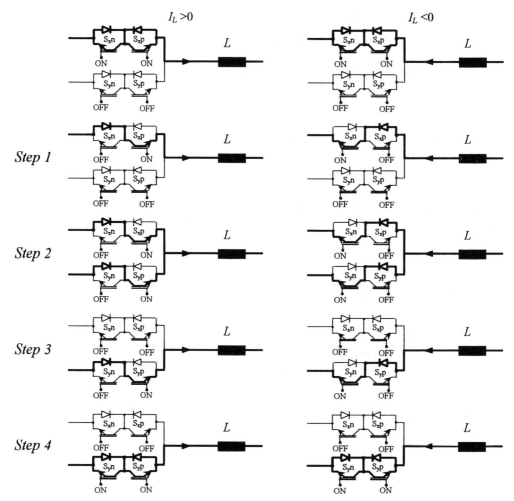

FIGURE 3.7
Illustration of current path allowed during the four-step commutation.

load current, according to the real commutation situation, may be satisfactorily compensated for in the processor control program as with the dead time in a voltage source inverter (VSI). This will be dependent on the load current, which is measured, and the device parameters, which may be given.

Another problem is the commutation near to zero output current, where errors in the current sign may cause dead-time commutation. However, the energy in the inductance of the load, at the very low current levels that are susceptible to being affected by offset, does not cause dangerous overvoltage. This may be handled by the clamp circuit, without causing an additional increase in the commutation losses.

3.3.2 Two-Step Commutation Strategy

This strategy [19–21] has been developed in order to reduce the number of steps and the complexity of the commutation control unit. To perform the commutation in only two steps, the bidirectional current path of the switch in the turn-on steady state is normally disabled, but the

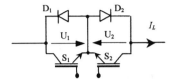

If $I_L > 0$ then S_1 and D_2 are conducting, D_1 and S_2 are reverse biased
$U_1 = +1.5$ V, $U_2 = -0.7$ V
If $I_L < 0$ then D_1 and S_2 are conducting, S_1 and D_2 are reverse biased
$U_1 = -0.7$ V, $U_2 = +1.5$ V

FIGURE 3.8
Logic algorithm based on the voltage drop to detect the current direction [19].

strategy is performed similarly to the four-step commutation, while the steady states of the bidirectional switches correspond to step 1, respectively step 3, presented in Fig. 3.7. At low current levels where the current transducer may be affected by offset, it has been proposed to enable bidirectional current path of the on-state switch. In order to avoid the risk of a short circuit of the input lines during the commutations, a true dead-time current commutation has to be used.

3.3.3 Current Sign Detection Methods

In order to control the bidirectional switch commutation properly, current transducers such as Hall transducers or shunts have been used. Also, a few other methods, which claim not to be susceptible to offset error and exploit the characteristic of the bidirectional switch topology, have been reported in the literature:

- Comparing the magnitude of the voltage across the unidirectional switches in a CE bidirectional topology [19], as shown in Fig. 3.8. This method has the advantage that it does not require any transducer and provides good precision, but it requires components to handle the peak line-to-line voltage which appears across nonconducting unidirectional switches (IGBT+FRD);
- Measuring the voltage drop across two antiparallel diodes connected on the output line [21]. This method is very simple, but causes voltage drop on the output side, in an application where the voltage transfer ratio is a sensitive issue. However, for a normal (slow) diode, which causes a 0.7–1 V voltage drop, the decrease of the voltage transfer ratio is only 0.2–0.3%.

3.4 MODULATION TECHNIQUES FOR MATRIX CONVERTERS

The first modulator proposed for matrix converters, known as the Venturini modulation, used a complicated scalar model that gave a maximum voltage transfer ratio of 0.5 [23, 24]. An injection of a third harmonic of the input and output voltage (3.4) was proposed in order to fit the reference output voltage in the input voltage system envelope, and the voltage transfer ratio reached the maximum value of 0.86 [25–29]. This is shown in Fig. 3.9.

$$v_{\text{ref}} = v_{\text{out}} \cdot \sin(\omega_{\text{out}} \cdot t) - \frac{v_{\text{out}}}{6} \cdot \sin(3 \cdot \omega_{\text{out}} \cdot t) + \frac{v_{\text{in}}}{4} \cdot \sin(3 \cdot \omega_{\text{in}} \cdot t). \qquad (3.4)$$

Next, indirect modulation was proposed [14, 30, 31]. This approach simplified the modulator model by making it possible to implement classical PWM modulation strategies in matrix converters [14]. Modulation models using space vectors (SVM) [32–36] or the direct torque control (DTC) [37] simplified the modulator model, making it easier to control the converter under unbalanced and distorted power supply conditions or to implement high-performance

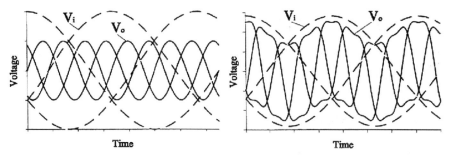

FIGURE 3.9
Output voltage reference waveforms fitting in the input voltage system: (left) sinusoidal (V_{out}/V_{in} = max 0.5.); (right) injection of third harmonics (V_{out}/V_{in} = max 0.86). (Reprinted with permission from [1], Fig. 3.3, p. 36).

control of the induction motors. A few modulation techniques are briefly presented, while more details are given for space vector modulation.

3.4.1 Scalar Modulation

In order to provide balance for the output voltages and also for the input currents, it is necessary that the modulation strategy use the input voltages equally when producing the output voltages. Scalar modulation, also referred to as Venturini modulation, establishes independent relations for each output, by sampling and distributing slides of input voltages in such a way that the average result follows the reference output phase voltage and the average input currents are sinusoidal. Therefore [12], (3.2) and (3.3), which define relations between output and input at any instant, become (3.5) and (3.6), in order to provide balance during the switching period:

$$\begin{bmatrix} V_A(t) \\ V_B(t) \\ V_C(t) \end{bmatrix} = \begin{bmatrix} m_{11}(k) & m_{12}(k) & m_{13}((k) \\ m_{21}(k) & m_{22}(k) & m_{23}(k) \\ m_{31}(k) & m_{32}(k) & m_{33}(k) \end{bmatrix} \cdot \begin{bmatrix} V_a(t) \\ V_b(t) \\ V_c(t) \end{bmatrix} \quad (3.5)$$

$$\begin{bmatrix} I_a(t) \\ I_b(t) \\ I_c(t) \end{bmatrix} = \begin{bmatrix} m_{11}(k) & m_{21}(k) & m_{31}(k) \\ m_{12}(k) & m_{22}(k) & m_{32}(k) \\ m_{13}(k) & m_{23}(k) & m_{33}(k) \end{bmatrix} \cdot \begin{bmatrix} I_A(t) \\ I_B(t) \\ I_C(t) \end{bmatrix}, \quad (3.6)$$

where $m_{ij}(k)$ represents the duty cycles of a switch connecting output line i to input line j in the k switching period.

The input phase voltages and the output currents are considered constant during the switching period. At any time, $0 \leq m_{ij}(k) \leq 1$, and also, because the output phase circuit should not remain disconnected from an input phase (rule 2):

$$\sum_{j=1}^{3} m_{ij}(k) = 1. \quad (3.7)$$

In order to provide maximum voltage transfer ratio, injection of third harmonics is needed (3.4) and the reference output voltage becomes

$$\begin{bmatrix} V_A(t) \\ V_B(t) \\ V_C(t) \end{bmatrix} = \sqrt{2} \cdot V_O \cdot \begin{bmatrix} \cos(\omega_O \cdot t) \\ \cos(\omega_O \cdot t - 2\pi/3) \\ \cos(\omega_O \cdot t - 4\pi/3) \end{bmatrix} - \sqrt{2} \cdot \frac{V_O}{6} \cdot \begin{bmatrix} \cos(3\omega_O \cdot t) \\ \cos(3\omega_O \cdot t) \\ \cos(3\omega_O \cdot t) \end{bmatrix}$$
$$+ \sqrt{2} \cdot \frac{V_I}{4} \cdot \begin{bmatrix} \cos(3\omega_I \cdot t) \\ \cos(3\omega_I \cdot t) \\ \cos(3\omega_I \cdot t) \end{bmatrix}. \quad (3.8)$$

where V_O and V_I are the RMS values of the output and input voltage systems, and ω_O and ω_I are the angular frequencies of the output and input voltage systems.

If the reference voltage vector, given in (3.8), is replaced in the transfer function of the output voltages depending on the duty cycles and the input voltages (3.5), a complicated model appears. However, if a simplification is introduced, as zero angle displacement between the input current and voltage, the duty cycles [12] are given by

$$m_{ij} = \frac{1}{3} \cdot \left\{ 1 + 2 \cdot \frac{V_O}{V_I} \cdot \cos\left(\omega_I t - 2 \cdot (j-1)\frac{\pi}{3}\right) \right.$$
$$\times \left[\cos\left(\omega_O t - 2 \cdot (i-1)\frac{\pi}{3}\right) - \frac{1}{6} \cdot \cos(3\omega_O t) + \frac{1}{2\sqrt{3}} \cdot \cos(3\omega_I t) \right] \quad (3.9)$$
$$\left. - \frac{2}{3\sqrt{3}} \cdot \frac{V_O}{V_I} \cdot \left[\cos\left(4\omega_I t - 2 \cdot (j-1)\frac{\pi}{3}\right) - \cos\left(2\omega_I t - 2 \cdot (1-j)\frac{\pi}{3}\right) \right] \right\}.$$

Because of the complexity of the duty-cycle computation, this algorithm is time consuming and requires nine commutations in the switching period. It is possible to reduce the number of sequences inside the switching period, if the three zero vectors (*aaa, bbb, ccc*) theoretically generated separately by the Venturini method are compressed into a single sequence, and this is placed at the beginning of the switching pattern [12]. Because the modulation is scalar, the switching state for each output phase is established independently, and both types of vectors, rotating and active, are inherently generated. Other modulator models have been derived by employing only one type of switching-state vector, which simplifies the mathematical models, such as space vector modulation by using only active vectors [32–36], or by using only rotating vectors [38, 39].

3.4.2 Modulation with Rotating Vectors

In this situation only rotating vectors of both direct and inverse sequence are used, in conjunction with the zero vector, in order to vary smoothly the amplitude and the instantaneous frequency of the output voltage [38, 39]. Investigations have tested a combination of direct rotating vectors with zero vectors and combination of inverse rotating vectors with zero vectors, both situations presenting similar performance on the output side, but uncontrolled displacement of the input current vector has been obtained. It has been reported that the use of both directions of rotating vectors provides proper control of the displacement of the input current vector [39].

3.4.3 Indirect Modulation

The indirect modulation model uses only active (pulsating) vectors. The main idea of the indirect modulation technique is to consider the matrix converter as a two-stage transformation converter: a rectification stage to provide a constant virtual dc-link voltage U_{pn} during the switching period by mixing the line-to-line voltages in order to produce sinusoidal distribution of the input currents, and an inverter stage to produce the three output voltages [14]. Figure 3.10 shows the converter model when the indirect modulation technique is used.

By multiplying the rectification stage matrix R by the inversion stage matrix I, the converter transfer matrix T is obtained:

$$T = I \cdot R \tag{3.10}$$

$$\begin{bmatrix} T_{11} & T_{12} & T_{13} \\ T_{21} & T_{22} & T_{23} \\ T_{31} & T_{32} & T_{33} \end{bmatrix} = \begin{bmatrix} I_1 \\ I_2 \\ I_3 \end{bmatrix} \cdot [R_1 \ R_2 \ R_3], \tag{3.11}$$

where $R_i = \{-1, 0, 1\}$ are the switch states of the rectification stage and $I_j = \{0, 1\}$ are the switch states of the inversion stage.

In this way, it is possible to implement known PWM strategies in both the rectifier and the inverter stage. The first implementation of indirect modulation, reported in [14], used scalar control. A modulation function to combine the two line-to-line input voltages of the highest magnitude provides virtual constant dc-link voltage and sinusoidal sharing of the virtual dc-link current in the unfiltered input currents. The command switching signals, corresponding to this virtual rectifying stage, are sent to the corresponding row of matrix converter switches, while the signal for the columns is generated by a scalar PWM, according to the motor control requirements. By combining the signals for the rows and columns with AND-logic gates, the gate signals for each bidirectional switch are generated.

3.4.4 Indirect Space Vector Modulation

A method to generate the desired PWM pattern is to use the space vector modulation (SVM) technique [32–36]. The detailed space vector modulation theory will be presented in Chapter 4 and therefore it will not be explained here.

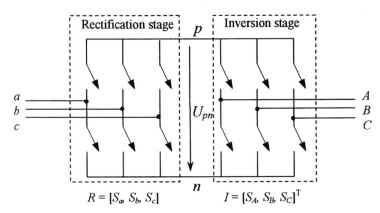

FIGURE 3.10
The matrix converter model when an indirect modulation technique is used.

3.4 MODULATION TECHNIQUES FOR MATRIX CONVERTERS

This technique uses a combination of the two adjacent vectors and a zero vector to produce the reference vector. The proportion between the two adjacent vectors gives the direction and the zero-vector duty cycle determines the magnitude of the reference vector. The input current vector I_{in} that corresponds to the rectification stage (Fig. 3.11a) and the output voltage vector U_{out} that corresponds to the inversion stage (Fig. 3.11b) are the reference vectors.

In order to implement the SVM, it is necessary to determine the position of the two reference vectors. The input reference current vector I_{in} is given by the input voltage vector if an instantaneous unitary power factor is desired, or it is given by a custom strategy to compensate for unbalanced and distorted input voltage system. The output reference voltage vector U_{out} may be produced with a given V/Hz dependence or may be a result of a vector control scheme.

When the absolute positions of the two reference vectors θ_{in} and θ_{out} are known, the relative positions inside the corresponding sector θ^*_{in} and θ^*_{out}, as well as the sectors of the reference vectors, are determined:

$$in_{sec} = \text{trunc}\left(\frac{\theta_{in}}{\pi/3}\right) \qquad \theta^*_{in} = \theta_{in} - \pi/3 \cdot in_{sec} \qquad (3.12)$$

$$out_{sec} = \text{trunc}\left(\frac{\theta_{out}}{\pi/3}\right) \qquad \theta^*_{out} = \theta_{out} - \pi/3 \cdot out_{sec}. \qquad (3.13)$$

The duty cycles of the active switching vectors are calculated for the rectification stage by using (3.14) and (3.15):

$$d_\gamma = m_I \cdot \sin\left(\frac{\pi}{3} - \theta^*_{in}\right) \qquad (3.14)$$

$$d_\delta = m_I \cdot \sin(\theta^*_{in}) \qquad (3.15)$$

and for the inversion stage by

$$d_\alpha = m_U \cdot \sin\left(\frac{\pi}{3} - \theta^*_{out}\right) \qquad (3.16)$$

$$d_\beta = m_U \cdot \sin(\theta^*_{out}) \qquad (3.17)$$

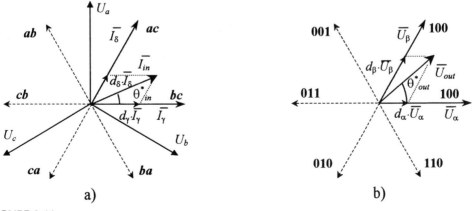

FIGURE 3.11
Generation of the reference vectors using SVM: (a) rectification stage; (b) inversion stage.

where m_I and m_U are the rectification and inversion stage modulation indexes, and θ^*_{in} and θ^*_{out} are the angles within their respective switching hexagon of the input current and output voltage reference vectors. Usually,

$$m_I = 1 \quad \text{and} \quad m_U = \sqrt{3} \cdot U_{out}/U_{pn}, \tag{3.18}$$

where U_{out} is the magnitude of the output reference voltage vector and U_{pn} is the virtual dc-link voltage.

In the ideal sinusoidal and balanced input voltages, the virtual dc-link voltage remains constant:

$$U_{pn} = d_\gamma \cdot U_{line-\gamma} + d_\delta \cdot U_{line-\delta} = 0.86 \cdot \sqrt{2} \cdot U_{line} \tag{3.19}$$

where $U_{line-\gamma}$ and $U_{line-\delta}$ are the instantaneous values of the two line-to-line voltages to be combined in the switching period to produce the virtual dc-link and U_{line} is the magnitude of the line-to-line voltage system.

To obtain a correct balance of the input currents and the output voltages, the modulation pattern should be a combination of all the rectification and inversion duty-cycles $(\alpha\gamma - \alpha\delta - \beta\delta - \beta\gamma - 0)$. The duty cycle of each sequence is determined as a product of the corresponding duty cycles:

$$d_{\alpha\gamma} = d_\alpha \cdot d_\gamma; \quad d_{\alpha\delta} = d_\alpha \cdot d_\delta; \quad d_{\beta\delta} = d_\beta \cdot d_\delta; \quad d_{\beta\gamma} = d_\beta \cdot d_\gamma, \tag{3.20}$$

The duration of the zero vector is calculated by

$$d_0 = 1 - (d_{\alpha\gamma} + d_{\alpha\delta} + d_{\beta\delta} + d_{\beta\gamma}). \tag{3.21}$$

Finally, the duration of each sequence is calculated by multiplying the corresponding duty cycle by the switching period. It is possible to optimize the switching pattern by changing the position of sequences inside the pattern. Therefore, the number of switchings could be reduced in order to provide a single switch commutation per sequence, as proposed in [35].

Figs. 3.12–3.15 show typical experimental waveforms for a matrix converter drive, which uses indirect space vector modulation (ISVM) and is running with a 4 kHz switching frequency. The converter feeds a 4 kW/2880 rpm induction motor, loaded with 50% of nominal torque, at 30 Hz output frequency (modulation index was 0.6).

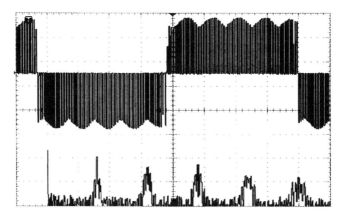

FIGURE 3.12
The output line voltage (250 V/div, 4 ms/div), and FFT (20 dB/div, 2.5 kHz/div) at 30 Hz output frequency.

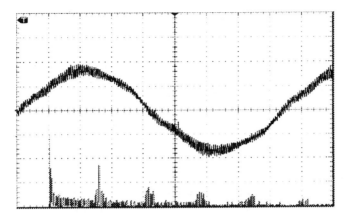

FIGURE 3.13
The output current (5 A/div, 4 ms/div), and FFT (20 dB/div, 2.5 kHz/div) at 30 Hz output frequency.

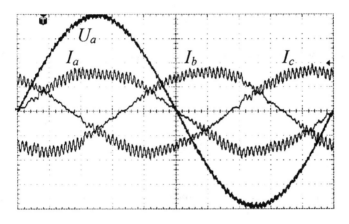

FIGURE 3.14
Input phase voltage U_a (80 V/div) and the three input currents I_a, I_b, I_c (2.5 A/div) vs time (4 ms/div).

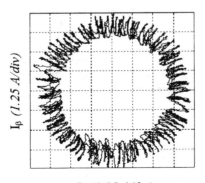

FIGURE 3.15
Input phase current locus of a matrix converter.

3.5 DTC APPLIED TO A MATRIX CONVERTER FED INDUCTION MOTOR DRIVE

In [37] an extended application of the DTC principle was proposed to a matrix converter fed induction motor drive. Additional to the classical DTC model, based on a commutation table with two entries to select the inverter stage vector, depending on the flux and torque error signals, a new entry was added to select the rectifying stage vector: the error of the angle of the input current vector. The diagram of the estimator for the motor flux and torque and for the converter input current angle sine is presented in Fig. 3.16a. The control scheme of the matrix converter is presented in Fig. 3.16b.

The control uses the input voltages and the output currents information, as well as the matrix converter switching state, in order to reconstitute the output voltages and the input currents, because of the instantaneous power transfer characteristic. The output voltages and currents are used to estimate the induction motor flux (using the voltage flux model) and the torque. The input current is software filtered to eliminate the ripple. The input current vector is compared with the input voltage vector to establish the displacement sine angle. It may be concluded that this scheme uses the indirect modulation scheme, as the displacement sine angle selects which line-to-line voltage will be chosen as a "dc-link voltage," giving the rectification stage vector. After that, the classical DTC applies: the flux and torque error selects the appropriate inversion stage vector. Combining the two vectors, the matrix converter switching state is determined.

The authors claim that this DTC scheme gives good performance in the high-speed range, proven with simulations, but no investigations have been made to determine how the limit of the voltage transfer ratio is affected.

3.6 STRATEGIES TO COMPENSATE FOR UNBALANCED AND DISTORTED INPUT VOLTAGES

Because direct power conversion implies instantaneous power transfer, the matrix converter performance is affected by unbalance and distortion of the input voltage system. In the case of input voltage unbalance, the virtual dc-link voltage given in (3.19) is no longer constant and the output voltages are no longer symmetrical. Therefore, distortion of the output voltages is caused, which produces distorted output currents. Because of the backward transformation of the output currents to the input currents, distorted input currents are produced. The overall effect is that distorted input voltage causes distorted and unbalanced input current. Therefore, research work has been directed to investigate different modulation strategies to compensate for these effects [35, 36]. As a general remark, all these methods are effective only if the locus of the output voltage vector can fit inside the input voltage locus. Considering the indirect modulation model, there are two possibilities to compensate the overall influence of unbalance and low-order harmonics supply for a matrix converter drive:

- By correcting the reference angle of the input current to reduce the harmonics content in the input currents
- By correcting the modulation index in the inversion stage with respect to the virtual dc-link voltage, to provide constant magnitude of the output voltage vector

In order to improve the motor side performance, it is possible to compensate the influence of unbalanced and distorted input voltages, which causes the virtual dc-link voltage to fluctuate.

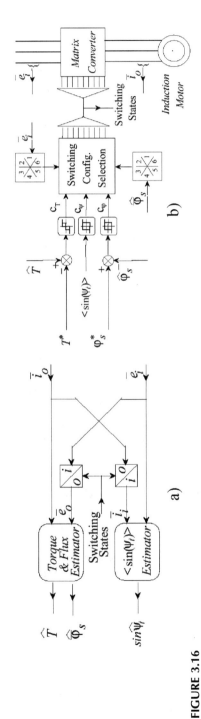

FIGURE 3.16
DTC of induction motor using a matrix converter: (a) estimator for the motor torque and flux and for the input current displacement angle; (b) the control scheme of the matrix converter. (Reprinted with permission from [37], Fig. 8.)

78 CHAPTER 3 / FUNDAMENTALS OF THE MATRIX CONVERTER TECHNOLOGY

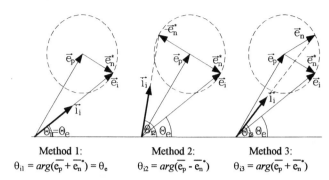

FIGURE 3.17
Detection of the input current reference angle in unbalanced supply condition. (Reprinted from [36], Fig. 1.)

This method is similar to the dc-link ripple compensation in standard converters and it has been presented in [35]. However, distortion of the input current is inherently caused.

Three strategies to modulate the input current vector reference for unbalance conditions have been reported in [36] in the case of unbalanced supply conditions, and they are presented in Fig. 3.17. In the case of ideal sinusoidal and balanced supply conditions, all the three modulation methods give identical performance. The detection of the positive e_p and negative e_n sequence is therefore necessary. Because real supply condition also implies low-order harmonics, the compensation process becomes more complicated.

The first compensation method (Method 1) is equivalent to the instantaneous power factor and is easy to implement, but causes the highest THD, while harmonics of positive sequence are produced. The second compensation method (Method 2), with no physical meaning, completely removes the harmonics of the input current, but causes positive and negative sequences of the fundamental (unbalance). The third compensation method (Method 3) requires rotating the input reference current vector with constant velocity, which is easy to implement by incrementing a counter, while the reset of the counter is synchronized with the zero crossing of input phase-to-neutral voltage on input line a. This causes current harmonics of both positive and negative sequence, but with half the amplitude of the first compensation method.

3.7 IMPLEMENTATION ASPECTS OF MATRIX CONVERTERS

It was presented in Section 3.1 that the silicon structure of nine bidirectional switches needs a few reactive elements in order to work properly, and these are found in the input filter and in the clamp circuit (see Fig. 3.3). Below, a few aspects regarding the hardware implementation, such as auxiliary circuits needed for safe matrix converter operation and design criteria for these circuits, will be presented.

3.7.1 The Input Filter

This has to reduce the input current ripple with minimum installed energy on the reactive elements. The most used topology is an LC series circuit. The use of more complex topologies has been recommended in the literature in order to achieve higher attenuation at the switching frequency, but they are not practical.

The design of the input filter has to accomplish the following:

- Produce an input filter with a cutoff frequency lower than the switching frequency,

$$L_{in} \cdot C_{in} = \frac{1}{\omega_0^2}, \qquad (3.22)$$

where L_{in}, C_{in} are the value of the inductance and the capacitor of the input filter and $\omega_0 = 2\pi \cdot f_0$ is the resonance pulsation of the input filter.

- Maximize the displacement angle φ_{min-in} for a given minimum output power [1],

$$\frac{C_{in}}{P_n} = k_{min} \cdot \tan \varphi_{min-in} \frac{1}{3 \cdot \omega_n \cdot U_n^2}, \qquad (3.23)$$

where $P_n \cong 3 \cdot U_n \cdot I_n$ is the input active power (considering a close to unity power factor at full load); U_n, I_n are the rated input phase voltage and current of the converter; $P_{min} = k_{min} \cdot P_n$ is the minimum power level where the displacement angle L_{min-in} reaches its limit and $\omega_n = 2 \cdot \pi \cdot f_n$ is the pulsation of the power grid.

- Minimize the input filter volume or weight for a given reactive power, by taking into account the energy densities which are different for film capacitors than for iron chokes:

$$\frac{S_L}{S_C} = \frac{1}{[3 \cdot \omega_0 \cdot U_n^2]^2} \cdot \left(\frac{P_n}{C_{in}}\right)^2 \qquad (3.24)$$

where $S_L = \omega_n \cdot L_{in} \cdot I_n^2$ (choke), $S_C = \omega_n \cdot C_{in} \cdot U_n^2$ (capacitor) are the installed VA in reactive components and $\omega_0 = 2 \cdot \pi \cdot f_0 = (L_{in} \cdot C_{in})^{-1/2}$.

- Minimize the voltage drop on the filter inductance at the rated current in order to provide the highest voltage transfer ratio:

$$\frac{\Delta U}{U_n} = 1 - \sqrt{1 - (\omega_n \cdot L_{in})^2 \cdot \left(\frac{I_n}{U_n}\right)^2} = 1 - \sqrt{1 - l_{in}^2}, \qquad (3.25)$$

where ΔU_n is the drop in voltage magnitude due to the influence of the input filter and l_{in} is the filter inductance in p.u.

Usually, the cutoff frequency of the LC input filter ω_0 is chosen to provide a given attenuation at the switching frequency. In addition, the value of the capacitor or the inductance is chosen based on one of the previous criteria.

3.7.2 The Clamp Circuit

Similar to standard diode bridge VSI, the matrix converter topology needs to be protected against overvoltage and overcurrent. Furthermore, this topology is more sensitive to disturbances and therefore more susceptible to failures due to the lack of an energy storage element in the dc link. Disturbances, which may cause hardware failures, are:

- Faulty interswitch commutations, such as internal short circuit of the mains or disconnecting the circuit of the motor currents
- Shutdown of the matrix converter during an overcurrent situation on the motor side
- Possible overvoltage on the input side caused by the converter power-up or by voltage sags

The protection issues for matrix converters have received increased attention, in order to build a reliable prototype. A solution to solve some of the problems consists of connecting a clamp circuit on the output side [1, 40], one patent being issued for the necessity of clamping the output lines [30]. The clamp circuit consists of a B6 fast recovery diode rectifier and a capacitor to store the energy accumulated in the inductance of the load, as shown in Fig. 3.3, caused by the output currents. The worst case regarding the energy level stored in the leakage inductance occurs when the output current reaches the overcurrent protection level, causing a converter shutdown. Transferring the energy safely from the leakage inductance in the clamp capacitor gives the design criteria for choosing the value of the capacitance:

$$\frac{3}{4} \cdot i_{max}^2 \cdot (L_{\delta S} + L_{\delta R}) = \frac{1}{2} \cdot C_{clamp} \cdot (U_{max}^2 - 2 \cdot U_{line}^2) \qquad (3.26)$$

where i_{max} is the current level which triggers the overcurrent protection, $L_{\delta S} + L_{\delta R}$ is the overall leakage inductance of the induction motor, C_{clamp} is the value of the clamp capacitor, and U_{max} is the maximum allowable overvoltage. Solutions to clamp the inductive currents on the input filter capacitors have been proposed in [28] and tested in [41], showing a potential for reducing the component count. Another solution [42] proposed was to dissipate the energy of inductive currents in varistors and in the semiconductors by employing active gate drivers.

3.7.3 The Power-Up Circuit

The purpose of this circuit is to provide a fast and safe power-up of the matrix converter, by eliminating the transients with higher overvoltage levels, characteristic of LC-series circuits. Damping of the inductance is the most convenient for implementation. Series damping eliminates the transients, but after power-up, the damping resistors should be bypassed in order to avoid higher power loss and therefore, the bypass relay should be able to handle the nominal current of the converter continuously. The principle diagram is presented in Fig. 3.18a.

The voltage across the clamp capacitor controls the bypass relay, which has normal open (NO) armatures: if the voltage is low, the relay is deenergized, and therefore the damping resistors are introduced in the main circuit, in series with the filter inductance. The value of the series damping resistors R_s should satisfy (3.27), in order to eliminate the transients:

$$R_s \geq 2\sqrt{\frac{L_{in}}{C_{in}}}. \qquad (3.27)$$

In order to reduce the level of oscillating energy that accumulates in the filter inductance during the transients, by-passing the inductors with parallel damping resistors is another solution. The principle diagram is presented in Fig. 3.18b. The value of the parallel damping resistor R_p should be smaller than the reactance of the choke, calculated at the cut-off frequency ω_0 of the input filter.

$$R_p \leq \omega_0 \cdot L_{in} \qquad (3.28)$$

The voltage across the clamp capacitor controls the relay. If the voltage is low, the relay is deenergized and the normal closed (NC) armatures introduce the damping resistors in the circuit, by-passing the input filter inductance. If the voltage increases above a certain level, the relay is energized and the damping resistors are disconnected. The relay contacts are used only during the power-up. The in-rush current into the clamp circuit is higher, but it is similar to the situation when no overvoltage reduction scheme is used. In this case, the disadvantage is that the diodes from the clamp circuit should be able to handle large currents during power-up. If no neutral

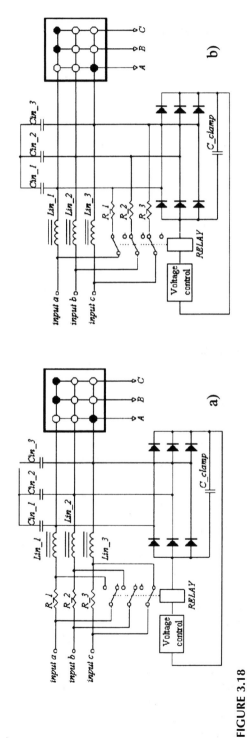

FIGURE 3.18
Power-up circuits for matrix converters: (a) using series damping resistors; (b) using parallel damping resistors.

current flows in the input filter circuit, it is possible to simplify these schemes, by damping the input filter only in two phases, which will require less components.

3.8 POWER MODULE WITH BIDIRECTIONAL SWITCHES FOR MATRIX CONVERTER APPLICATIONS

The necessity to employ special designed power modules in matrix converter applications has been presented in a few publications [1, 2, 18, 21, 40]. The main reason is that integrating the bidirectional switches, which consist of IGBTs and FRDs, in a module decreases notably the stray inductance in the power stage, and thereby the switching losses and stress. Supplementary production costs may be decreased by having a low number of components necessary to build the power stage, as well as having a reduced number of connections. In addition, the physical size of the power module is given according to the thermal stress per switch group, because in a matrix converter, compared to a standard VSI, the thermal stress is shared by more switches.

The configuration of the bidirectional switch should be chosen in order to minimize the overall cost of the assembly. By using CC bidirectional switches to build the power module, the number of insulated power supplies is reduced to six. Packing the switches in a 3-phase/1-phase ($3\phi/1\phi$) topology, as shown in Fig. 3.19a, is more convenient because it requires few changes from a standard six-pack VSI topology to produce it, and symmetry is achieved. Also, keeping all switches that belong to an output phase inside the same enclosure will minimize the stray inductance. Only three identical power modules are necessary to build a matrix converter.

A power module with CC bidirectional switches has been developed employing 25 A/1200 V devices and is housed in a 22-pin power module, with physical dimensions of $63 \times 48 \times 12.5$ mm. The placement of the IGBTs and the FRDs, the connections inside the power module, and the pin designation are presented in Fig. 3.19b. The nonconnected terminals are placed around the power pins (U1, U2, U3, and I1) to ensure the clearance required by the standards. In Fig. 3.19c two generations of power modules built to develop matrix converter prototypes are presented for size comparison purposes.

3.9 LIMITED RIDE-THROUGH CAPABILITY FOR MATRIX CONVERTERS

Compared to diode-bridge VSI technology, which is robust to grid disturbance because of the energy storage capability provided by the dc-link capacitors, the matrix converter seems to be inferior in terms of ride-through capability. Also, in the research field, only one attempt has been made to investigate the possibility to provide the matrix converter with the capability to operate during voltage sags or momentary power interruptions [43], and it will be briefly described next.

Because the matrix converter is based on galvanic connections between inputs and outputs, the ride-through control strategy has to provide separation of the motor circuit from the power grid. Disconnecting all the matrix converter switches, but providing continuity for the motor currents by using the clamp circuit or connecting all the output phases to a single input phase (zero-vector), does the required separation. By applying a zero vector, the motor current increases and the value of the energy stored in the leakage inductance increases. The stator flux stops moving, but the rotor flux is still moving because of rotor motion. When the rotor flux starts to lead, the electromagnetic torque changes sign and the increase of energy in the leakage inductance is based on mechanical energy conversion. Disconnecting all the active switches causes the conduction of the clamp circuit diodes. The stator current decreases and the energy

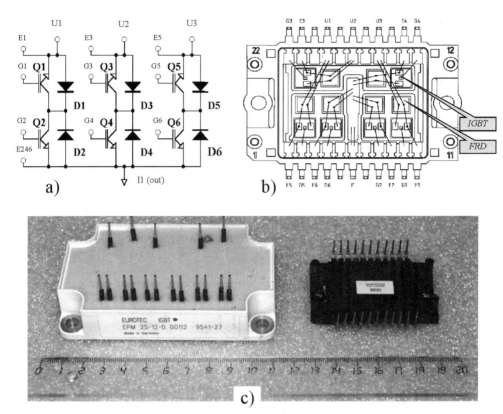

FIGURE 3.19
Three-phase to one-phase power module with bidirectional switches using common collector (CC) connected IGBTs. (a) Internal scheme; (b) placement of the devices and internal structure; (c) old (1995) and new (1999) generation of power modules.

stored in the leakage inductance is transferred to the clamp circuit capacitor. By alternating the two permitted matrix converter states during ride-through operation, it is possible to control the motor currents and to transfer energy from the rotor inertia to the clamp circuit capacitor, as long as the residual flux and the shaft speed are not zero. It is not possible to control the magnitude of the rotor flux ψ_r during ride-through operation, because it is not possible to apply an active vector to the motor, which may increase the flux. The ride-through capability depends critically on the motor parameters and on the flux level before starting the ride-through operation.

The equivalent scheme of a matrix converter during the proposed ride-through strategy is presented in Fig. 3.20. The induction motor acts as a synchronous machine with the EMF produced by the decaying rotor flux. The motor leakage inductance, the matrix converter, and the clamp circuit form a boost converter, which works in ac and uses bidirectional switches. The energy is boosted up into the clamp circuit, which acts as a resistive load because the 180° displacement angle between the phase voltage U_{mot} and current I_{mot} fundamentals, as shown in Fig. 3.21.

Figure 3.22 shows the test of the ride-through strategy on a 3 kW induction motor drive. Initially, the matrix converter was running at 30 Hz. A 40% of the motor rated torque is applied on the motor shaft, while the load profile is linearly dependent with the speed. A 200 ms ride-through operation is imposed in the control to emulate a power failure. The motor speed is decreasing from 866 rpm to 465 rpm because of the load torque and rotor inertia. A successful

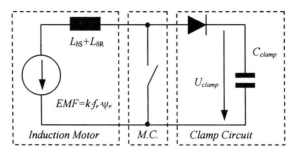

FIGURE 3.20
The equivalent scheme of the MCM during ride-through operation.

restart from nonzero motor flux and speed condition is performed and the steady state before the simulated power failure is reestablished within 300 ms from restart.

3.10 THE MATRIX CONVERTER MOTOR: THE NEXT GENERATION OF INTEGRATED MOTOR DRIVES?

The low volume, sinusoidal input current, bidirectional power flow, and lack of bulky and limited-lifetime electrolytic capacitors recommend this topology for this application. Furthermore, because the motor and the drive are a single unit, by matching the nominal voltage of the motor to the maximum voltage transfer ratio of the matrix converter, the main drawback of the matrix converter topology has been overcome.

In order to evaluate this solution, a 4-kW prototype has been built [44], using the enclosure of a standard integrated motor drive. Even though the technology to manufacture this prototype was not industrial, the task has been successfully completed, as shown in Fig. 3.23.

The implementation was based on power modules with bidirectional switches described in Section 3.8, on the implementation of the 4-step commutation method in a single chip

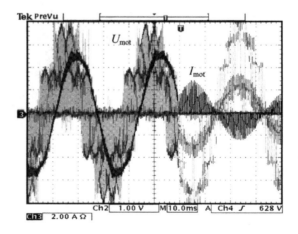

FIGURE 3.21
Measured motor phase voltage U_{mot} (100 V/div) and current I_{mot} (2 A/div) during transition from normal operation to ride-through operation. Time: 10 ms/div.

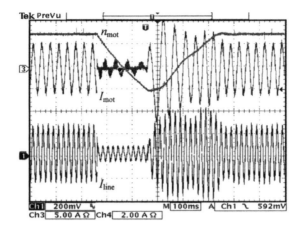

FIGURE 3.22
Illustration of 200 ms ride-through operation with successful restart and reestablishing the steady state shaft speed: shaft speed n_{mot} (170 rpm/div), motor current I_{mot} (5 A/div), and line phase current I_{line} (2 A/div). Time: 100 ms/div.

programmable logic device, on a 16-bit microcontroller mounted on a mini-module. Other implementation details are given in Tables 3.1 and 3.2. Here, the volume used with reactive components is given in two situations: an industrial drive based on standard diode-bridge VSI and for the prototyped MCM. It is seen that in this implementation, the MCM requires more space for reactive components, but this situation will change with increased switching frequency. In addition, it should be noted that the MCM is a regenerative drive with sinusoidal input current capability, and a similar performance implemented with a back-to-back VSI topology will involve higher line inductance and dc-link capacitance.

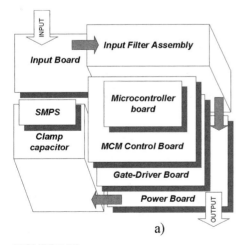
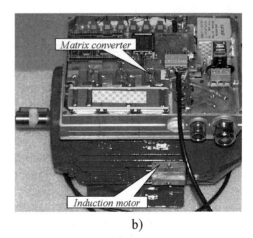

FIGURE 3.23
Implementation of the 4 kW matrix converter motor: (a) block diagram and power flow on the converter; (b) the assembled prototype.

Table 3.1 The Volume of the Reactive Components in a Standard FCM Enclosure (Danfoss FCM304, 4 kW)

dc chokes	$60 \times 50 \times 40$ mm	2 pcs.	240 cm^3
dc-link capacitors	$\phi 35 \times 47$ mm	4 pcs.	230 cm^3
Total:	470 cm^3 (20.7% of the enclosure volume)		

Table 3.2 The Volume of Reactive Components in an MCM Enclosure Based on Matrix Converter (4 kW)

ac chokes	$60 \times 50 \times 40$ mm	3 pcs. (1.4 mH)	360 cm^3
Input filter capacitors	$21 \times 38 \times 41$ mm	3 pcs. (4.7 µF)	98 cm^3
Decoupling capacitors	$11 \times 20.5 \times 26.5$ mm	9 pcs. (0.68 µF)	54 cm^3
Clamp capacitor	$42.5 \times 46.5 \times 55$ mm	1 pcs. (25 µF)	109 cm^3
Total:	621 cm^3 (27.3% of the enclosure volume)		

3.11 SUMMARY

In this chapter, a brief description of the matrix converter technology is given. The most important components of this technology presented are topologies of bidirectional switches and their specific commutation techniques, modulation methods for matrix converters to provide a sine-wave-in sine-wave-out operation, strategies to compensate for the influence of unbalanced power grid, implementation aspects, and a strategy for ride-through operation. Finally, a specific application, which will take advantage of the high power density and the high efficiency of this topology, is presented: the matrix converter motor, a bidirectional power flow integrated motor drive.

REFERENCES

[1] P. Nielsen, The matrix converter for an induction motor drive. Industrial Ph.D. Fellowship EF 493, ISBN 87-89179-14-5, Aalborg University, Denmark, August 1996.
[2] C. Klumpner, New contributions to the matrix converter technology. Ph.D. Thesis, "Politehnica" University of Timisoara, Romania, December 2000.
[3] T. A. Lipo, Recent progress in the development in solid-state AC motor drives. *IEEE Trans. Power Electron.* **32**, 105–117 (1988).
[4] B. K. Bose, Power Electronics—A technology review. *Proc. IEEE* **80**, 1303–1334 (1992).
[5] R. Kerkman, G. L. Skibinski, and D. W. Schengel, AC drives: Year 2000 (Y2K) and beyond. Proc. of APEC '99, vol. 1, pp. 28–39, 1999.
[6] P. Thoegersen and F. Blaabjerg, Adjustable speed drives in the next decade. The next step in industry and academia. Proc. of PCIM '00, Intelligent Motion, pp. 95–104, 2000.
[7] R. D. Lorentz, The future of electric drives: where are we headed? Proc. of PEVD '00, pp. 1–6, 2000.
[8] K. Phillips, Power electronics: Will our current technical vision take us to the next level of AC drive product performance? Proc. of IAS Annual Meeting, Plenary session 1, CD-ROM version, 2000.
[9] S. Bernet, T. Matsuo, and T. A. Lipo, A matrix converter using reverse blocking NPN-IGBTs and optimized pulse patterns. Proc. of PESC '96, vol. 1, pp. 107–113, 1996.
[10] H.-H. P. Li, Bidirectional lateral insulated gate bipolar transistor having increased voltage blocking capability. U.S. Patent No. 5,977,569 (1999).
[11] IXYS, IXRH 50N60, IXRH 50N120: High voltage RBIGBT. Forward and reverse blocking IGBT. Advanced technical information, http://www.ixys.net/l400.pdf, 2000.

[12] L. Zhang, C. Watthanasarn, and W. Shepherd, Analysis and comparison of control techniques for AC-AC matrix converters. *IEE Proc. Electr. Power App.*, **145**, 284–294 (1998).

[13] S. Sunter and H. Altun, A method for calculating semiconductor losses in the matrix converter. Proc. of MELECON '98, Vol. 2, pp. 1260–1264, 1998.

[14] C. L. Neft and C. D. Shauder, Theory and design of a 30-hp matrix converter. *IEEE Trans. Indust. Appl.* **28**, 546–551 (1992).

[15] N. Burany, Safe control of 4-quadrant switches. Proc. of IAS '89, Vol. 2, pp. 1190–1194, 1989.

[16] J. H. Youm and B.-H. Kwon, Switching technique for current-controlled AC-to-AC converters. *IEEE Trans. Indust. Electron.* **46**, 309–318 (1999).

[17] A. Christensson, Switch-effective modulation strategy for matrix converters. Proc. of EPE '97, pp. 4.193–4.198, 1997.

[18] J. Chang, Adaptive overlapping commutation control of modular AC-AC converter and integration with device module of multiple AC-AC switches. U.S. Patent no. 5,892,677 (1999).

[19] L. Empringham, P. Wheeler, and J. C. Clare, Intelligent commutation of matrix converter bi-directional switch cells using novel gate drive techniques. Proc. of PESC '98, pp. 707–713, 1998.

[20] M. Ziegler and W. Hofmann, Seminatural two step commutation strategy for matrix converters. Proc. of PESC '98, pp. 727–731, 1998.

[21] K. G. Kerris, P. W. Wheeler, F. Clare, and L. Empringham, Implementation of a matrix converter using P-channel MOS-controlled thyristors. Proc. of PEVD '00, pp. 35–39, 2000.

[22] J. Oyama, T. Higuchi, E. Yamada, T. Koga, and T. Lipo, New control strategy for matrix converter. Proc. of PESC '89, Vol. 1, pp. 360–367, 1989.

[23] M. Venturini, A new sine wave in, sine wave out conversion technique eliminates reactive elements. Proc. of Powercon, E3-1–E3-15, 1980.

[24] M. Venturini and A. Alesina, The generalised transformer: a new bidirectional sinusoidal waveform frequency converter with continuously adjustable input power factor. Proc. of PESC '80, pp. 242–252, 1980.

[25] M. Venturini and A. Alesina, Analysis and design of optimum-amplitude nine-switch direct AC-AC converters. *IEEE Trans. Power Electron.* **4**, 101–112 (1989).

[26] D. G. Holmes and T. A. Lipo, Implementation of a controlled rectifier using AC-AC matrix converter theory. Proc. of PESC '89, **1**, pp. 353–359, 1989.

[27] G. Roy and G. E. April, Cycloconverter operation under a new scalar control algorithm. Proc. of PESC '89, Vol. 1, pp. 368–375, 1989.

[28] R. R. Beasant, W. C. Beattie, and A. Refsum, An approach to realisation of a high-power Venturini converter. Proc. of PESC '90, pp. 291–297, 1990.

[29] P. Wheeler and D. A. Grant, A low loss matrix converter for AC variable-speed drives. Proc. of EPE '93, pp. 27–32, 1993.

[30] C. L. Neft, AC power supplied static switching apparatus having energy recovery capability. U.S. Patent No. 4,697,230 (1987).

[31] C. D. Schauder, Matrix converter circuit and commutation method. U.S. Patent No. 5,594,636 (1997).

[32] L. Huber and D. Borojevic, Space vector modulator forced commutated cyclo-converters. Proc. of IAS '89, Vol. 1, pp. 871–876, 1989.

[33] L. Huber and D. Borojevic, Space vector modulated three-phase to three-phase matrix converter with input power factor correction. *IEEE Trans. Indust. Appl.* **31**, 1234–1245 (1995).

[34] D. Casadei, G. Grandi, G. Serra, and A. Tani, Space vector control of matrix converters with unity input power factor and sinusoidal input/output waveforms. Proc. of EPE '93, **7**, pp. 170–175, 1993.

[35] P. Nielsen, F. Blaabjerg, and J. K. Pedersen, Space vector modulated matrix converter with minimized number of switchings and feedforward compensation of input voltage unbalance. Proc. of PEDES '96, Vol. 2, pp. 833–839, 1996.

[36] P. Nielsen, D. Casadei, G. Serra, and A. Tani, Evaluation of the input current quality by three different modulation strategies for SVM controlled matrix converters with input voltage unbalance. Proc. of PEDES '96, Vol. 2, pp. 794–800, 1996.

[37] D. Casadei, G. Serra, and A. Tani, The use of matrix converter in direct torque control of induction machines. Proc. of IECON '98, pp. 744–749, 1998.

[38] S. Halasz, I. Schmidt, and T. Molnar, Matrix converter for induction motor drive. Proc. of EPE '95, Vol. 2, pp. 2.664–2.669, 1995.

[39] M. Milanovic and B. Dobaj, A novel unity power factor correction principle in direct AC to AC matrix converters. Proc. of PESC '98, Vol. 1, pp. 746–752, 1998.

[40] P. Nielsen, F. Blaabjerg, and J. K. Pedersen, New protection issues of a matrix converter—design considerations for adjustable speed drives. *IEEE Trans. Indust. Appl.* **35**, 1150–1161 (1999).
[41] A. Schuster, A matrix converter without reactive clamp elements for an induction motor drive system. Proc. of PESC '98, Vol. 1, pp. 714–720, 1998.
[42] J. Mahlein and M. Braun, A matrix converter without diode clamped over-voltage protection. Proc. of PIEMC '00, Vol. 2, pp. 817–822 (2000).
[43] C. Klumpner, I. Boldea, and F. Blaabjerg, Short term ride-through capabilities for direct frequency converters. Proc. of PESC '00, Vol. 1, pp. 235–241, 2000.
[44] C. Klumpner, P. Nielsen, I. Boldea, and F. Blaabjerg, A new matrix converter-motor (MCM) for industry applications. Proc. of IAS '00, Vol. 3, pp. 1394–1402 (2000).

CHAPTER 4

Pulse Width Modulation Techniques for Three-Phase Voltage Source Converters

MARIAN P. KAZMIERKOWSKI and MARIUSZ MALINOWSKI
Warsaw University of Technology, Warsaw, Poland

MICHAEL BECH
Aalborg University, Aalborg, Denmark

4.1 OVERVIEW

Application areas of power converters still expand thanks to improvements in semiconductor technology, which offer higher voltage and current ratings as well as better switching characteristics. On the other hand, the main advantages of modern power electronic converters, such as high efficiency, low weight, small dimensions, fast operation, and high power densities, are being achieved through the use of the so-called *switch mode operation*, in which power semiconductor devices are controlled in *ON/OFF* fashion (no operation in the active region). This leads to different types of *pulse width modulation* (PWM), which is a basic energy processing technique applied in power converter systems. In modern converters, PWM is a high-speed process ranging—depending on the rated power—from a few kilohertz (motor control) up to several megahertz (resonant converters for power supply).

Historically, the best-known triangular carrier-based (CB) sinusoidal PWM (also called suboscillation method) for three-phase static converter control was proposed by Schönung and Stemmler in 1964 [18]. However, with microprocessor developments, the space vector modulation (SVM) proposed by Pfaff, Weschta, and Wick in 1982 [60] and further developed by van der Broeck, Skudelny, and Stanke [4] becomes a basic power processing technique in three-phase PWM converters.

In the three-phase isolated neutral load topology (Fig. 4.1), the phase currents depend only on the voltage difference between phases. Therefore, a common term can be added to the phase voltages in the CB-SPWM, thus shifting their mean value, without affecting AC side currents. Based on this observation Hava, Kerkman, and Lipo [117] as well as Blasko [110] have proposed in 1997 so called *generalized* or *hybrid* PWM, which use a zero sequence system of triple

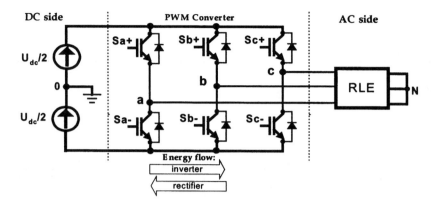

FIGURE 4.1
Three-phase voltage source PWM converter.

harmonic frequency as the common term. This extends the liner region of the modulator and improves its performance in terms of average switching frequency (flat-top or discontinuous PWM) and current ripple reduction. Parallel to the deterministic modulators, whose spectrum consists of discrete frequency components, random PWM has also been suggested by Trzynadlowski, Kirlin, and Legowski [141]. Because of the redistribution of the spectral power over a wider frequency range, the random PWM for a three-phase inverter can affect secondary issues of PWM converter systems, such as acoustic and electromagnetic (EMI) noise.

A detailed review of the state of the art of PWM in power converters is beyond the scope of this chapter. For reviews, [8] and [11] can be suggested. Therefore, this chapter reviews the principal PWM techniques that have been developed during recent years.

In industry, the most often used power converter topology of Fig. 4.1 can work in two modes:

- *Inverter*: When energy is converted from the dc side to the ac side. This mode is used in variable-speed drives and ac power supplies, including uninterruptible power supplies (UPS).
- *Rectifier*: When the energy of the mains (50 or 60 Hz) is converted from the ac side to the dc side. This mode has application in power supplies with unity power factor (UPF).

The most commonly used PWM methods for three-phase converters impress either the voltages or the currents into the ac side (Fig. 4.2). The first part of this chapter focuses on open-loop PWM voltage control techniques (Fig. 4.2a), including space vector and carrier based with zero-sequence signal PWM. Subsequently, PWM current controllers operating in a closed-loop fashion (Fig. 4.2b) are presented.

4.2 OPEN-LOOP PWM

4.2.1 Basic Requirements and Definitions

The PWM converter should meet some general demands such as the following:

- Wide range of linear operation
- Minimal number of switchings to maintain low switching losses in power components

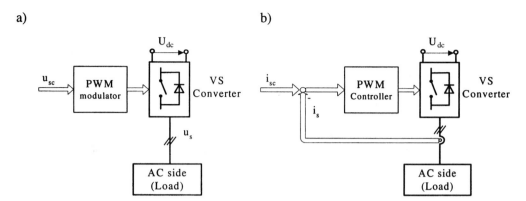

FIGURE 4.2
Basic PWM schemes: (a) open loop–voltage control, (b) closed loop–current control.

- Minimal content of harmonics in voltage and current, because they produce additional losses and noise in load
- Elimination of low-frequency harmonics (in the case of motors it generates torque pulsation)
- Operation in overmodulation region including square wave

Additionally, investigations are led with the purpose of:

- Simplification because the modulator is one of the most time-consuming parts of the control algorithm
- Reduction of acoustic noise (random modulation)
- Reduction of common-mode voltage

Table 4.1 presents basic definitions and parameters which characterize PWM methods.

Table 4.1 Basic Parameters of PWM

	Name of parameter	Symbol	Definition	Remarks
1	Modulation index	M	$M = U_{1m}/U_{1(six-step)}$ $= U_{1m}/(2/\pi)U_{dc}$	Two definitions of modulation index are used. For sinusoidal modulation $0 \leq M \leq 0,785$ or $0 \leq m \leq 1$
		m	$m = U_m/U_{m(t)}$	
2	Max. linear range	M_{max}	$0 \ldots 0.907$	Depends on shape of modulation signal
		m_{max}	$0 \ldots 1.154$	
3	Overmodulation		$M > M_{max}$ $m > m_{max}$	Nonlinear range used for increase of U_{out}
4	Frequency modulation ratio (pulse number)	m_f	$m_f = f_s/f_1$	For $m_f > 21$ asynchronous modulation is used
5	Switching frequency (number)	$f_s\ (l_s)$	$f_s = f_T = 1/T_s$ T_s—sampling time	Constant
6	Total harmonic distortion	THD	$THD\% = 100*(I_h/I_{s1})$	Used for voltage and current
7	Current distortion factor	d	$\mathbf{I}_{h(rms)}/\mathbf{I}_{h(six-step)(rms)}$	Independent of load parameters
8	Polarity consistency rule	PCR		Avoids ±1 dc voltage transition

4.2.2 Carrier-Based PWM

4.2.2.1 Sinusoidal PWM. Sinusoidal modulation is based on a triangular carrier signal. By comparison of the common carrier signal with three reference sinusoidal signals U_a^*, U_b^*, U_c^*, the logical signals, which define the switching instants of the power transistor (Fig. 4.3), are generated. Operation with constant carrier signal concentrates voltage harmonics around the switching frequency and multiples of the switching frequency. The narrow range of linearity is a limitation for classical CB-SPWM because the modulation index reaches $M_{max} = \pi/4 = 0.785$ ($m = 1$) only; e.g., the amplitude of reference signal and carrier are equal. The overmodulation region occurs above M_{max}, and a PWM converter, which is treated like a power amplifier, operates in the nonlinear part of the characteristic (see Fig. 4.15).

4.2.2.2 CB-PWM with Zero Sequence Signal (ZSS). If the neutral point on the ac side of the power converter N is not connected with the dc side midpoint 0 (in Fig. 4.1), the phase current depends only on the voltage difference between phases. Therefore, it is possible to insert an additional zero sequence signal (ZSS) of third harmonic frequency, which does not produce phase voltage distortion U_{aN}, U_{bN}, U_{cN} and without affecting load average currents (Fig. 4.4). However, the current ripple and other modulator parameters (e.g., extending of the linear region to $M_{max} = \pi/2\sqrt{3} = 0.907$, reduction of the average switching frequency, current harmonics) are changed by the ZSS. Added ZSS occurs between N and 0 points and is visible like a U_{N0} voltage and can be observed in U_{a0}, U_{b0}, U_{c0} voltages (Fig. 4.5).

Figure 4.5 presents different waveforms of additional ZSS, corresponding to different PWM methods. It can be divided in two groups: *continuous* and *discontinuous* modulation (DPWM). The most known method of continuous modulation is method with sinusoidal ZSS. With 1/4 amplitude it corresponds to the minimum of output current harmonics and with 1/6 amplitude it corresponds to the maximal linear range [112]. The triangular shape of ZSS with 1/4 amplitude corresponds to conventional (analog) space vector modulation with symmetrical placement of zero vectors in sampling time (see Section 4.2.3). Discontinuous modulation is formed by unmodulated 60° segments (converter power switches do not switch) shifted from 0 to $\pi/3$

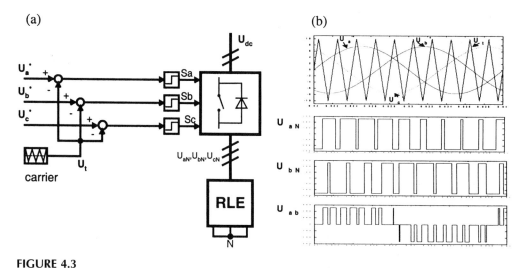

FIGURE 4.3
(a) Block scheme of carrier-based sinusoidal modulation (CBS-PWM). (b) Basic waveforms.

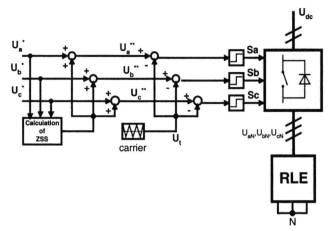

FIGURE 4.4
Block scheme of modulator based on additional zero sequence signal (ZSS).

[different shift Ψ gives a different type of modulation (Fig. 4.6)]. It finally gives lower (average 33%) switching losses. Detailed descriptions of different kinds of modulation based on ZSS can be found in [117].

4.2.3 Space Vector Modulation (SVM)

4.2.3.1 Basics of SVM. The SVM strategy is based on space vector representation of the converter ac side voltage (Fig. 4.7a) and has become very popular because of its simplicity [4, 60]. A three-phase, two-level converter provides eight possible switching states, made up of six active and two zero switching states. Active vectors divide the plane for six sectors, where a reference vector U^* is obtained by switching on (for the proper time) two adjacent vectors. It can be seen that vector U^* (Fig. 4.7a) is possible to implement by different switch on/off sequences of U_1 and U_2, and that zero vectors decrease the modulation index. The allowable length of the U^* vector, for each α angle, is $U^*_{max} = U_{dc}/\sqrt{3}$. Higher values of output voltage (reaching the six-step mode), up to the maximal modulation index ($M = 1$), can be obtained by an additional nonlinear overmodulation algorithm (see Section 4.2.5).

Contrary to CB-PWM, in the SVM there are no separate modulators for each phase. Reference vector U^* is sampled with fixed clock frequency $2f_s = 1/T_s$, and next $U^*(T_s)$ is used to solve equations that describe times t_1, t_2, t_0, and t_7 (Fig. 4.7b). Microprocessor implementation is described with the help of a simple trigonometrical relationship for the first sector (4.1a and 4.1b), and recalculated for the next sectors (n):

$$t_1 = \frac{2\sqrt{3}}{\pi} MT_s \sin(\pi/3 - \alpha) \tag{4.1a}$$

$$t_2 = \frac{2\sqrt{3}}{\pi} MT_s \sin\alpha. \tag{4.1b}$$

After t_1 and t_2 calculation, the residual sampling time is reserved for zero vectors U_0 and U_7 with the condition that $t_1 + t_2 \leq T_s$. The equations (4.1a) and (4.1b) are identical for all variants of

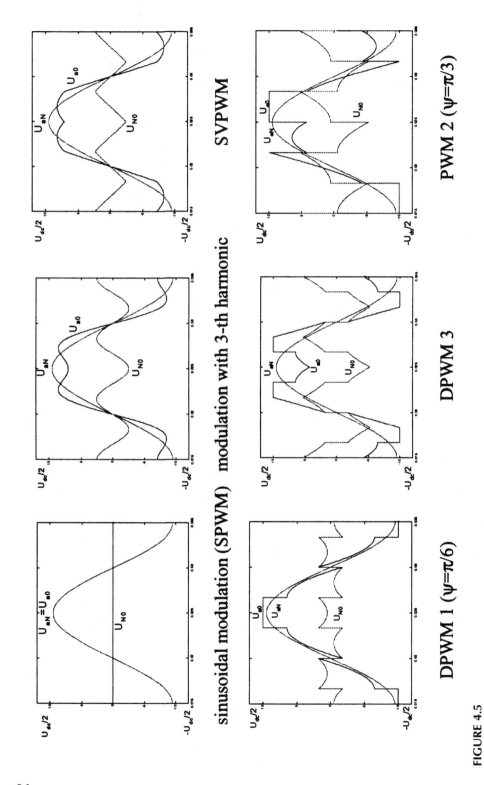

FIGURE 4.5
Variants of PWM modulation methods in dependence on shape of ZSS.

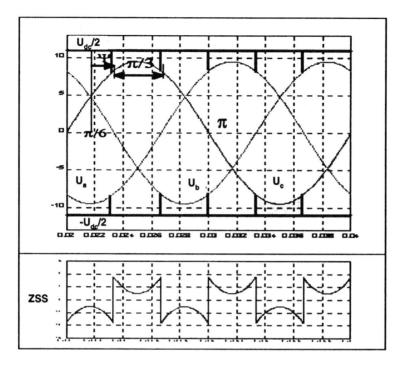

FIGURE 4.6
Generation of ZSS for the DPWM method.

SVM. The only difference is in different placement of zero vectors $U_0(000)$ and $U_7(111)$. It gives different equations defining t_0 and t_7 for each method, but the total duration time of zero vectors must fulfill the conditions

$$t_{0,7} = T_s - t_1 - t_2 = t_0 + t_7. \tag{4.2}$$

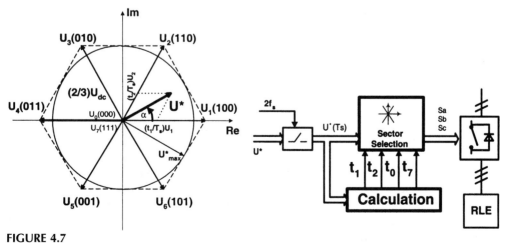

FIGURE 4.7
(a) Space vector representation of three-phase converter. (b) Block scheme of SVM.

Table 4.2 Voltages between a, b, c and N, 0 for Eight Converter Switching States

	U_{a0}	U_{b0}	U_{c0}	U_{aN}	U_{bN}	U_{cN}	U_{N0}
U_0	$-U_{dc}/2$	$-U_{dc}/2$	$-U_{dc}/2$	0	0	0	$-U_{dc}/2$
U_1	$U_{dc}/2$	$-U_{dc}/2$	$-U_{dc}/2$	$2U_{dc}/3$	$-U_{dc}/3$	$-U_{dc}/3$	$-U_{dc}/6$
U_2	$U_{dc}/2$	$U_{dc}/2$	$-U_{dc}/2$	$U_{dc}/3$	$U_{dc}/3$	$-2U_{dc}/3$	$U_{dc}/6$
U_3	$-U_{dc}/2$	$U_{dc}/2$	$-U_{dc}/2$	$-U_{dc}/3$	$2U_{dc}/3$	$-U_{dc}/3$	$-U_{dc}/6$
U_4	$-U_{dc}/2$	$U_{dc}/2$	$U_{dc}/2$	$-2U_{dc}/3$	$U_{dc}/3$	$U_{dc}/3$	$U_{dc}/6$
U_5	$-U_{dc}/2$	$-U_{dc}/2$	$U_{dc}/2$	$-U_{dc}/3$	$-U_{dc}/3$	$2U_{dc}/3$	$-U_{dc}/6$
U_6	$U_{dc}/2$	$-U_{dc}/2$	$U_{dc}/2$	$U_{dc}/3$	$-2U_{dc}/3$	$U_{dc}/3$	$U_{dc}/6$
U_7	$U_{dc}/2$	$U_{dc}/2$	$U_{dc}/2$	0	0	0	$U_{dc}/2$

The neutral voltage between N and 0 points is equal (see Table 4.2) to

$$U_{N0} = \frac{1}{T_s}\left(-\frac{U_{dc}}{2}t_0 - \frac{U_{dc}}{6}t_1 + \frac{U_{dc}}{6}t_2 + \frac{U_{dc}}{2}t_7\right) = \frac{U_{dc}}{2}\frac{1}{T_s}\left(-t_0 - \frac{t_1}{3} + \frac{t_2}{3} + t_7\right). \quad (4.3)$$

4.2.3.2 Three-phase SVM with Symmetrical Placement of Zero Vectors (SVPWM).
The most popular SVM method is modulation with symmetrical zero states (SVPWM):

$$t_0 = t_7 = (T_s - t_1 - t_2)/2. \quad (4.4)$$

Figure 4.8a shows the formation of gate pulses for (SVPWM) and the correlation between duty time T_{on}, T_{off} and the duration of vectors t_1, t_2, t_0, t_7. For the first sector commutation delay can be computed as:

$$\begin{aligned}
T_{aon} &= t_0/2 & T_{aoff} &= t_0/2 + t_1 + t_2 \\
T_{bon} &= t_0/2 + t_1 & T_{boff} &= t_0/2 + t_2 \\
T_{con} &= t_0/2 + t_1 + t_2 & T_{coff} &= t_0/2.
\end{aligned} \quad (4.5)$$

For conventional SVPWM times t_1, t_2, t_0 are computed for one sector only. Commutation delay for other sectors can be calculated with the help of a matrix:

$$\begin{bmatrix} T_{aoff} \\ T_{boff} \\ T_{coff} \end{bmatrix} = \begin{bmatrix} \overset{sector1}{1\ 1\ 1} & \overset{sector2}{1\ 1\ 1} & \overset{sector3}{1\ 1\ 1} & \overset{sector4}{1\ 1\ 1} & \overset{sector5}{1\ 1\ 1} & \overset{sector6}{1\ 1\ 1} \\ 1\ 0\ 0 & 1\ 1\ 0 & 0\ 1\ 0 & 0\ 1\ 1 & 0\ 0\ 1 & 1\ 0\ 1 \\ 1\ 1\ 0 & 0\ 1\ 0 & 0\ 1\ 1 & 0\ 0\ 1 & 1\ 0\ 1 & 1\ 0\ 0 \end{bmatrix}^T \begin{bmatrix} 0.5T_0 \\ T_1 \\ T_2 \end{bmatrix}.$$

$$(4.6)$$

4.2.3.3 Two-phase SVM.
This type of modulation proposed in [127] was developed in [110,113] and is called *discontinuous* pulse width modulation (DPWM) for CB technique with an additional zero sequence signal (ZSS) in [117]. The idea is based on the assumption that only two phases are switched (one phase is clamped by 60^0 to the lower or upper dc bus). It gives only one zero state per sampling time (Fig. 4.8b). Two-phase SVM provides a 33% reduction of effective switching frequency. However, switching losses also strongly depend on the load power factor angle (Fig. 4.16). It is a very important criterion, which allows further reduction of switching losses up to 50% [117].

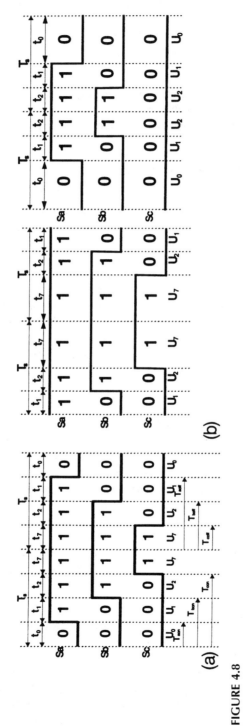

FIGURE 4.8
Vector placement in sampling time: (a) conventional SVPWM ($t_0 = t_7$). (b) DPWM ($t_0 = 0$ and $t_7 = 0$).

Figure 4.9a shows several kinds of two-phase SVM. It can be seen that sectors are adequately moved on 0°, 30°, 60°, 90° and denoted as PWM(0), PWM(1), PWM(2), and PWM(3), respectively. Figure 4.9b presents phase voltage U_{aN}, pole voltage U_{a0}, and voltage between neutral points U_{N0} for these modulations. Zero states description for PWM(1) can be written as

$$t_0 = 0 \Rightarrow t_7 = T_s - t_1 - t_2 \text{ when } 0 \leq \alpha < \pi/6$$
$$t_7 = 0 \Rightarrow t_0 = T_s - t_1 - t_2 \text{ when } \pi/6 \leq \alpha < \pi/3. \tag{4.7}$$

4.2.3.4 Variants of Space Vector Modulation. From Eqs. (4.1)–(4.3) and knowledge of the U_{N0} voltage shape (Fig. 4.9b), it is possible to calculate the duration of zero vectors t_0, t_7. An evaluation and properties of different modulation methods is shown in Table 4.3.

4.2.4 Carrier-Based PWM versus Space Vector PWM

Comparison of CB-PWM methods with additional ZSS to SVM is shown in Fig. 4.10. The upper part shows pulse generation through comparison of reference signals U_a^{**}, U_b^{**}, U_c^{**} with a triangular carrier signal. The lower part of the figure shows gate pulse generation in SVM (obtained by calculation of duration time of active vectors U_1, U_2 and zero vectors U_0, U_7). It is visible that both methods generate identical gate pulses. It can also be observed from Figs. 4.9 and 4.10 that the degree of freedom represented in the selection of the ZSS waveform in CB-PWM corresponds to different placement of the zero vectors $U_0(000)$ and $U_7(111)$ in sampling time $T_s = 1/2f_s$ of the SVM. Therefore, there is no difference between CB-PWM and SVM (CB-DPWM1=PWM (1)-SVM with one zero state in sampling time). The difference is only in the treatment of three-phase quantities: CB-PWM operates in terms of three-phase natural components, whereas SVM uses an artificial (mathematically transformed) vector representation.

4.2.5 Overmodulation

In CB-PWM, by increasing the reference voltage beyond the amplitude of the triangular carrier signal, some switching cycles are skipped and the voltage of each phase remains clamped to one of the dc buses. This range shows a high nonlinearity between reference and output voltage amplitude and requires infinite amplitude of reference in order to reach a six-step output voltage.

In SVM-PWM the allowable length of reference vector U^* which provides linear modulation is equal to $U^*_{max} = U_{dc}/\sqrt{3}$ (circle inscribed in hexagon $M = 0.906$) (Fig. 4.11). To obtain higher values of output voltage (to reach the six-step mode) up to the maximal modulation index $M = 1$, an additional nonlinear overmodulation algorithm has to be applied. This is because the minimal pulse width become shorter than critical (mainly dependent on power switch characteristics—usually in the range of a few microseconds) or even negative. Zero vectors are never used in this type of modulation.

4.2.5.1 Algorithm Based on Two Modes of Operation. Two overmodulation regions are considered (Fig. 4.12). In region I the magnitude of the reference voltage is modified in order to keep the space vector within the hexagon. It defines the maximum amplitude that can be reached for each angle. This mode extends the range of the modulation index up to 0.95. Mode II starts from $M = 0.95$ and reaches the six-step mode $M = 1$. Mode II defines both the magnitude and the angle of the reference voltage. To implement both modes an approach based on lookup tables or neural networks (see Section 10.5.1 in Chapter 10) can be applied.

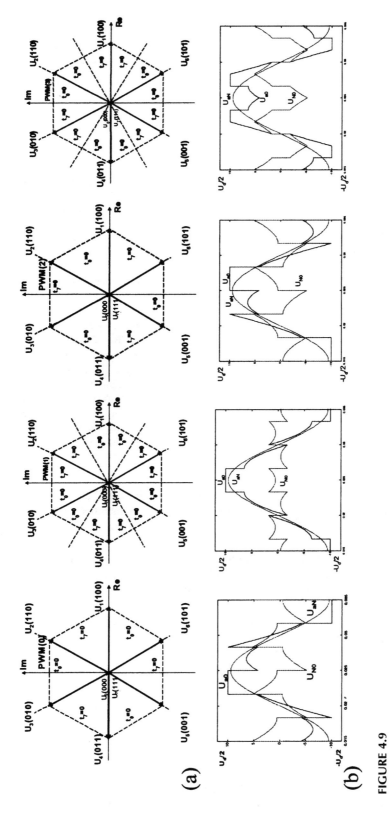

FIGURE 4.9
(a) Placement of zero vectors in two-phase SVM. Succession: PWM(0)= 0^0, PWM(1)= 30^0, PWM(2)= 60^0, and PWM(3)= 90^0. (b) Phase voltage U_{aN}, pole voltage U_{a0}, and voltage between neutral points U_{N0} for each modulation.

Table 4.3 Variants of Space Vector Modulation

Vector modulation methods	Calculation of t_0 and t_7	Remarks
Vector modulation with $U_{N0} = 0$	$t_0 = \dfrac{T_s}{2}(1 - \dfrac{4}{\pi}M\cos\alpha)$ $t_7 = T_s - t_0 - t_1 - t_2$	• Equivalent of classical CB-SPWM (no difference between U_{aN} and U_{b0} voltages) • Linear region $M_{\max} = 0.785$
Vector modulation with third harmonic	$t_0 = \dfrac{T_s}{2}(1 - \dfrac{4}{\pi}M(\cos\alpha - \dfrac{1}{6}\cos 3\alpha)$ $t_7 = T_s - t_0 - t_1 - t_2$	• Low current distortions • More complicated calculation of zero vectors • Extended linear region: $M = 0.907$
Three-phase SVM with symmetrical zero states (SVPWM)	$t_0 = t_7 = (T_s - t_1 - t_2)/2$	• Most often used in microprocessor technique (zero state vectors symmetrical in sampling time $2T_s$) • Current harmonic content almost identical to that in previous method
Two-phase SVM	$t_0 = 0 \Rightarrow t_7 = T_s - t_1 - t_2$ when $0 \leq \alpha < \pi/6$ $t_7 = 0 \Rightarrow t_0 = T_s - t_1 - t_2$ when $\pi/6 \leq \alpha < \pi/3$ (for DPWM1)	• Equivalent of DPWM methods in CB-PWM technique • 33% switching frequency and switching losses reduction • Higher current harmonic content at low modulation index • Only one zero state per sampling time, simple calculation (Fig. 4.8)

Overmodulation Mode I: Distorted Continuous Reference Signal. In this range, the magnitude of the reference vector is changed while the angle is transmitted without any changes ($\alpha_p = \alpha$). However, when the original reference trajectory passes outside the hexagon, the time average equation gives an unrealistic duration for the zero vectors. Therefore, to compensate for reduced fundamental voltage, i.e., to track with the reference voltage U^*, a modified reference voltage trajectory U is selected (Fig. 4.13a). The reduced fundamental components in the region where the reference trajectory surpasses the hexagon is compensated for by a higher value in the corner [120].

The on the time durations, for the region where modified reference trajectory is moved along the hexagon, are calculated as

$$t_1 = T_S \frac{\sqrt{3}\cos\alpha - \sin\alpha}{\sqrt{3}\cos\alpha + \sin\alpha} \tag{4.8a}$$

$$t_2 = T_S - t_1 \tag{4.8b}$$

$$t_0 = 0. \tag{4.8c}$$

Overmodulation Mode II: Distorted Discontinuous Reference Signal. Operation in this region is illustrated in Fig. 4.13b. The trajectory changes gradually from a continuous hexagon to

4.2 OPEN-LOOP PWM

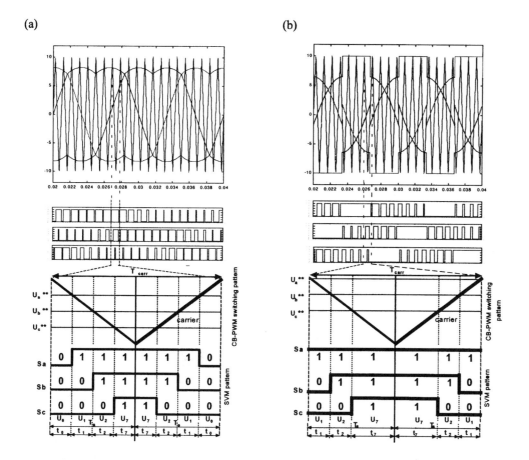

FIGURE 4.10
Comparison of CB-PWM with SVM. (a) SVPWM; (b) DPWM. From top: CB-PWM with pulses; short segment of reference signal at high carrier frequency (when reference signals are straight lines); formation of pulses in SVM.

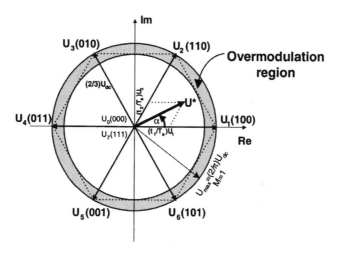

FIGURE 4.11
Overmodulation region in space vector representation.

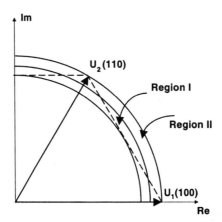

FIGURE 4.12
Subdivision of the overmodulation region.

the six-step operation. To achieve control in overmodulation mode II, both the reference magnitude and reference angle (from α to α_p) are changed:

$$\alpha_p = \begin{cases} 0 & 0 \leq \alpha \leq \alpha_h \\ \dfrac{\alpha - \alpha_h}{\pi/6 - \alpha_h}\dfrac{\pi}{6} & \alpha_h \leq \alpha \leq \pi/3 - \alpha_h \\ \pi/3 & \pi/3 - \alpha_h \leq \alpha \leq \pi/3. \end{cases} \quad (4.9)$$

The modified vector is held at a vertex of the hexagon for holding angle α_h over a particular time and then partly tracks the hexagon sides in every sector for the rest of the switching period. The holding angle α_h controls the time interval when the active switching state remains at the

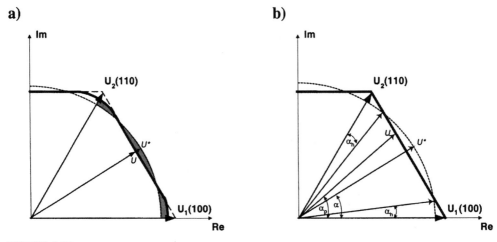

FIGURE 4.13
Overmodulation. (a) Mode I ($0.907 < M < 0.952$); (b) mode II ($0.952 < M < 1$). U^*, reference trajectory (dashed line); U, modified reference trajectory (solid line).

vertices. It is a nonlinear function of the modulation index, which can be piecewise linearized as [125]

$$\alpha_h = 6.4 \cdot M - 6.09 \quad (0.95 \leq M \leq 0.98)$$
$$\alpha_h = 11.75 \cdot M - 11.34 \quad (0.98 \leq M \leq 0.9975) \quad (4.10)$$
$$\alpha_h = 48.96 \cdot M - 48.43 \quad (0.9975 \leq M \leq 1.0).$$

The six-step mode is characterized by selection of the switching vector, which is closest to the reference vector for one-sixth of the fundamental period. In this way the modulator generates the maximum possible converter voltage. For a given switching frequency, the current distortion increases with the modulation index. The distortion factor strongly increases when the reference waveform becomes discontinuous in mode II.

4.2.5.2 *Algorithm Based on Single Mode of Operation.* In a simple technique proposed in [114], the desired voltage angle is held constant when the reference voltage vector is located outside of hexagon. The value at which the command angle is held is determined by the intersection of the circle with the hexagon (Fig. 4.14). The angle at which the command is held (hold angle) depends on the desired modulation index (M) and can be found from Eq. (4.11) (max circuit trajectory is related to the maximum possible fundamental output voltage $2/\pi U_{dc}$, not to $2/3 U_{dc}$—see Fig. 4.11):

$$\alpha_1 = \arcsin\left(\frac{\sqrt{3}}{2M'}\right) \quad (4.11\text{a})$$

$$M' = \left(\frac{2\sqrt{3}-3}{2\sqrt{3}-\pi}\right)M + \left(\frac{3-\pi}{2\sqrt{3}-\pi}\right) \quad (4.11\text{b})$$

$$\alpha_2 = \frac{\pi}{3} - \alpha_1. \quad (4.11\text{c})$$

For a desired angle between 0 and α_1, the commanded angle tracks its value. When the desired angle increases over α_1, the commanded angle stays at α_1 until the desired angle becomes $\pi/6$. After that, the commanded angle jumps to the value $\alpha_2 = \pi/3$ of α_1. The commanded value of α is kept constant at α_2 for any desired angle between $\pi/6$ and α_2. For a desired angle between α_2 and $\pi/3$, the commanded angle tracks the value of desired angle, as in Fig. 4.14. The advantage

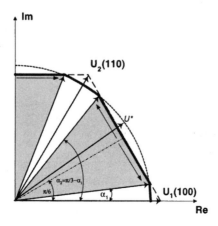

FIGURE 4.14
Overmodulation: single mode of operation.

of linearity and easy implementation is obtained at the cost of somewhat higher harmonic distortion.

4.2.6 Performance Criteria

Several performance criteria are considered for selection of suitable modulation method [8, 9, 11]. Some of them are defined in Table 4.1. In this subsection, further important criteria such as range of linear operation, distortion factor, and switching losses are discussed.

4.2.6.1 Range of Linear Operation.
The range of the linear part of the control characteristic for sinusoidal CB-PWM ends at $M = \pi/4 = 0.785$ ($m = 1$) of the modulation index (Fig. 4.15), i.e., when reference and carrier amplitudes become equal. The SVM or CB-PWM with ZSS injection provides extension of the linear range up to $M_{max} = \pi/2\sqrt{3} = 0.907$ ($m_{max} = 1.15$). The region above $M = 0.907$ is the nonlinear overmodulation range.

4.2.6.2 Switching Losses.
Power losses of the PWM converter can be generally divided into *conduction* and *switching losses*. Conduction losses are practically the same for different PWM techniques, and they are much lower than switching losses. For the switching loss calculation, linear dependency of the switching energy loss on the switched current is assumed. This also was proved by the measurement results [123]. Therefore, for high switching frequency, the total average value of the transistor switching power losses, for continuous PWM, can be expressed as

$$P_{sl(c)} = \frac{1}{2\pi} \int_{-(\pi/2)+\varphi}^{(\pi/2)+\varphi} k_{TD} \cdot i \cdot f_s d\alpha = \frac{k_{TD} I f_s}{\pi} \tag{4.12}$$

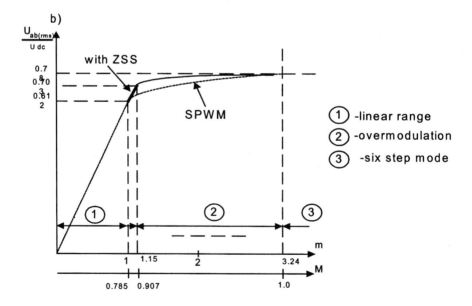

FIGURE 4.15
Control characteristic of PWM converter.

where $k_{TD} = k_T + k_D$, the proportional relation of the switching energy loss per pulse period to the switched current for the transistor and the diode.

In the case of discontinuous *PWM* the following properties hold from the symmetry of the pole voltage:

$$P_{sl}(-\varphi) = P_{sl}(\varphi)$$
$$P_{sl}(\varphi) = P_{sl}(\pi - \varphi) \quad \text{where} \quad 0 < \varphi < \pi. \tag{4.13}$$

Therefore, it is sufficient to consider the range from 0 to $\pi/2$ for the *DPWM* as follows [123]:

$$PWM(1) \Rightarrow P_{sl}(\varphi) = \begin{cases} P_{sl(c)} \cdot (1 - \frac{1}{2}\cos\varphi) & \text{for } 0 < \varphi < \pi/3 \\ P_{sl(c)} \cdot \left(\frac{\sqrt{3}\sin\varphi}{2}\right) & \text{for } \pi/3 < \varphi < \pi/2 \end{cases} \tag{4.14a}$$

$$PWM(0) \Rightarrow P_{sl}(\varphi) = P_{sl(PWM(1))} \cdot \left(\varphi - \frac{\pi}{6}\right) \tag{4.14b}$$

$$PWM(2) \Rightarrow P_{sl}(\varphi) = P_{sl(PWM(1))} \cdot \left(\varphi + \frac{\pi}{6}\right) \tag{4.14c}$$

$$PWM(3) \Rightarrow P_{sl}(\varphi) = \begin{cases} P_{sl(c)} \cdot (1 - \frac{\sqrt{3}-1}{2}\cos\varphi) & \text{for } 0 < \varphi < \pi/6 \\ P_{sl(c)} \cdot \frac{\sin\varphi + \cos\varphi}{2} & \text{for } \pi/6 < \varphi < \pi/3 \\ P_{sl(c)} \cdot \left(1 - \frac{\sqrt{3}-1}{2}\sin\varphi\right) & \text{for } \pi/3 < \varphi < \pi/2 \end{cases} \tag{4.14d}$$

Switching losses depend on the type of discontinuous modulation and power factor angle that is shown in Fig. 4.16 (comparison to continuous modulation). Since the switching losses increase with the magnitude of the phase current (approximately linearly), selecting a suitable modulation can significantly improve performance of the converter. Switching losses are on average reduced

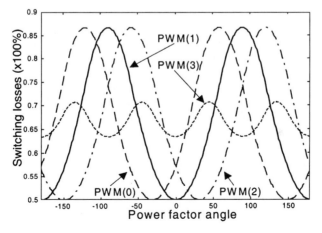

FIGURE 4.16
Switching losses ($P_{sl}\varphi/P_{sl(c)}$) versus power factor angle.

about 33%. In favorable conditions, when modulation is clamped in phase conducting maximum current, switching losses decrease up to 50%.

4.2.6.3 Distortion and Harmonic Copper Loss Factor.
The rms harmonic current, defined as

$$I_{h(rms)} = \sqrt{\frac{1}{T}\int_0^T [i_L(t) - i_{L1}(t)]^2 dt}, \qquad (4.15)$$

depends on the type of PWM and also on the value of the ac side impedance. To eliminate the influence of ac side impedance parameters, the *distortion factor* is commonly used (see Table 4.1):

$$d = I_{h(rms)}/I_{h(six\text{-}step)(rms)}. \qquad (4.16)$$

For the six-step operation the distortion factor is $d = 1$. It should be noted that harmonic copper losses in the ac-side load are proportional to d^2. Therefore, d^2 can be considered as a *loss factor*. Values of distortion factor can be computed for different modulation methods [8, 123]. It depends on the switching frequency, the modulation index M, and the shape of the ZSS (Fig. 4.17):

- For continuous modulation:

$$\text{SPWM} \quad d = \frac{4M}{\sqrt{6}\pi k_{fSB}}\sqrt{1 - \frac{32M}{\sqrt{3}\pi^2} + \frac{3M^2}{\pi}}, \; M \in \left[0, \frac{\pi}{4}\right] \qquad (4.17a)$$

$$\text{SVPWM} \quad d = \frac{4M}{\sqrt{6}\pi k_{fSB}}\sqrt{1 - \frac{32M}{\sqrt{3}\pi^2} + \frac{9M^2}{2\pi}\left(1 - \frac{3\sqrt{3}}{4\pi}\right)}, \; M \in \left[0, \frac{\pi}{2\sqrt{3}}\right]. \qquad (4.17b)$$

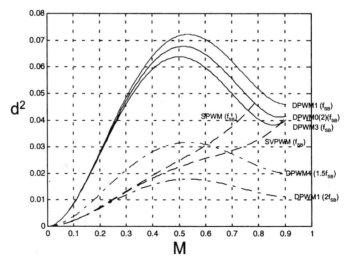

FIGURE 4.17
Square of the current distortion factor as a function of the modulation index.

- For discontinuous modulation (DPWM):

DPWM1 $$d = \frac{4M}{\sqrt{6}\pi k_{f_{SB}}}\sqrt{4 - \frac{4M}{\sqrt{3}\pi^2}(8 + 15\sqrt{3}) + \frac{9M^2}{2\pi}\left(2 + \frac{\sqrt{3}}{2\pi}\right)}, \ M \in \left[0, \frac{\pi}{2\sqrt{3}}\right]$$ (4.18a)

DPWMO(2) $$d = \frac{4M}{\sqrt{6}\pi k_{f_{SB}}}\sqrt{4 - \frac{140M}{\sqrt{3}\pi^2} + \frac{9M^2}{2\pi}\left(2 + \frac{3\sqrt{3}}{4\pi}\right)}, \ M \in \left[0, \frac{\pi}{2\sqrt{3}}\right]$$ (4.18b)

DPWM3 $$d = \frac{4M}{\sqrt{6}\pi k_{f_{SB}}}\sqrt{4 - \frac{4M}{\sqrt{3}\pi^2}(62 - 15\sqrt{3}) + \frac{9M^2}{2\pi}\left(2 + \frac{\sqrt{3}}{\pi}\right)}, \ M \in \left[0, \frac{\pi}{2\sqrt{3}}\right],$$ (4.18c)

where $k_{f_{SB}}$ is defined as the ratio of carrier frequency to base of carrier frequency.

Figure 4.17 shows that, for the same carrier frequency, discontinuous PWM possesses higher current harmonic content than continuous methods. The harmonic content is similar for both methods at high PWM index only. However, we should remember that discontinuous modulation possesses lower switching losses. Therefore, the carrier frequency can be increased by a factor of 3/2 for a 33% switching loss reduction, or doubled for a 50% switching loss reduction. This gives lower current distortion at the same switching losses as for continuous PWM.

4.2.7 Adaptive Space Vector Modulation (ASVM)

The concept of adaptive space vector modulation (ASVM) provides the following [126]:

- Full control range including overmodulation and six-step operation
- Maximal (up to 50%) reduction of switching losses at 33% reduction of average switching frequency

The above features are achieved by the use of three different modes of SVM with instantaneous tracking of the ac current peak. PWM operation modes are distributed in the range of modulation index (M) as follows (Fig. 4.18a):

A: $0 < M \leq 0.5$ Conventional SVM with symmetrical zero switching states
B: $0.5 < M \leq 0.908$ Discontinuous SVM with one zero state per sampling time (two-phase or flat top PWM)
C: $0.908 < M \leq 0.95$ Overmodulation mode I (see Section 4.2.5.1)
D: $0.95 < M \leq 1$ Overmodulation mode II

The combination of regions A with B without current tracing, suggested in [110, 117], is known as hybrid PWM. In the region B of discontinuous PWM, for maximal reduction of switching losses, the peak of the load current should be located in the center of the "flat" parts. Therefore, it is necessary to observe the position of the load peak current. Stator oriented components i_α, i_β of the measured load current are transformed into polar coordinates and compared with voltage reference angle (Eq. 4.19). This allows identification of load power factor

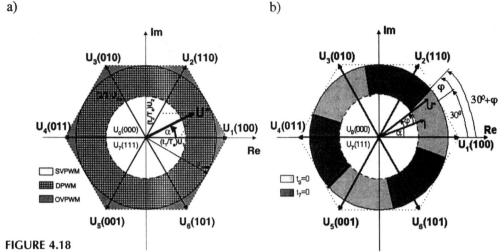

FIGURE 4.18
Adaptive modulator. (a) Effect of modulation index; (b) effect of load angle.

angle φ, which decides the placement of the clamped region. Thus the ring from Fig. 4.18b will be adequately moved (φ). For each sector:

$$\begin{aligned} \text{If } \alpha < \varphi + \kappa &\Rightarrow t_0 = 0 \\ \text{If } \alpha > \varphi + \kappa &\Rightarrow t_7 = 0 \end{aligned} \quad (4.19)$$

where α is the reference voltage angle, φ is the power sector angle, and κ is adequate for successive sectors $\pi/6$, $\pi/2$, $5\pi/6$, $7\pi/6$, $3\pi/2$, $11\pi/6$. This provides tracking of the power sector angle in the full range from $-\pi$ to π and provides maximal reduction of switching losses (Fig. 4.19).

The full algorithm of the adaptive modulator is presented in Fig. 4.20. Typical waveforms illustrating operation of the adaptive modulator in transition regions are presented in Fig. 4.21.

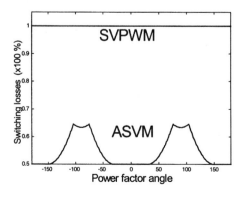

FIGURE 4.19
Switching losses versus load power factor angle for conventional SVPWM and ASVM.

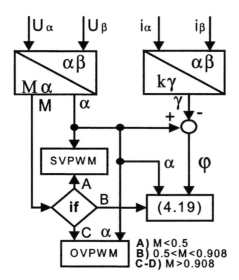

FIGURE 4.20
Algorithm of adaptive SVM.

4.2.8 Random PWM

To encircle the motivation for studying random PWM, it may initially be noted that PWM techniques generate a time-periodic switching function, which maps into the frequency domain as a spectrum consisting of discrete frequency components [9]. We denote PWM methods having these characteristics as deterministic modulators. Apart from the frequency components coming from the reference signal, all other components (harmonics) are generally unwanted as they may cause current and voltage distortion, extra power losses and thermal stress, electromagnetic interference (EMI), torque ripple in rotating machines, mechanical vibrations, and radiation of acoustic noise; see [122] for a discussion of especially EMI-related secondary issues.

The traditional method to alleviate such problems is to insert filters that trap, for example, the current harmonics to places where they are less harmful. For example, some kind of high-frequency filtering is almost obligatory to ensure that the level of conducted EMI emission from a hard-switched converter does not exceed limits set up by legislative bodies. In a similar manner, output filters may be used to reduce the impact of harmonics in adjustable-speed drives, although this option is often waived because of the costs and space requirements associated with such filters.

Now, random PWM has been suggested as an alternative to deterministic PWM for two distinct reasons, which both aim at reducing the impact of harmonics in systems based on pulse-width modulated hard-switched converters:

- *Reduction of subjective acoustic noise*: It has been demonstrated that random PWM may reduce the subjective noise emitted from whistling magnetics in the audible frequency range. Investigations have focused on noise from converter-fed ac machines [128–130], but noise from dc machines [131] and ac reactors in line-side converters [132] has also been studied.
- *Alleviation of electromagnetic noise*: Compliance with standards defining limits for emission of conducted and radiated EMI may be obtainable with less filtering and shielding efforts, if deterministic PWM is replaced by random PWM [133–136].

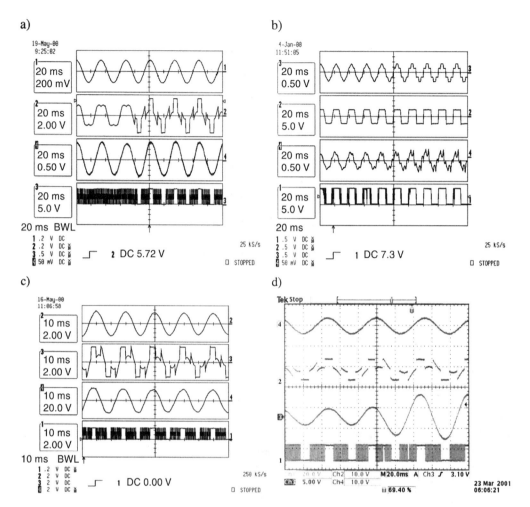

FIGURE 4.21
Experimental waveforms of *VS* converter with adaptive modulation: (a) transition from three- to two-phase SVM, (b) transition from overmodulation to six-step operation, (c) two-phase SVM-peak current tracing during reactive power changing, (d) two-phase SVM-peak current tracing during load changing. From the top: phase voltage u_{SaN} and pole voltage u_{Sa0} (estimated from U_{dc} and switching state), current i_a, pulses S_a.

Hence, the reasons for the current interest in random PWM among power electronic engineers are the prospects of alleviating such diverse problems as perception of tonal acoustic noise from electromagnetics and, on the other hand, EMI. To this end, an important feature of random PWM is that those issues are addressed in a way that does not involve analog power-level filters, which tend to be bulky and costly; only modifications to the control circuitry are needed in order to randomize an existing deterministic modulator.

The randomization can be accomplished in many different ways, but to facilitate the presentation, the general characteristics of random PWM may be summarized as follows:

- *Time-domain properties*: In contrast to deterministic modulators, a randomized pulse-width modulator generates switching functions which are nonrepetitive in the time domain, even during steady-state operation. Despite the randomization, precise synthesis of the reference signal may still be maintained.

- *Spectral properties*: The power carried by the discrete frequency components associated with deterministic PWM during steady-state operation is (partially) transferred into the continuous density spectrum. This means that a spectrum originally consisting of narrow-band harmonics is mapped into a spectrum whose power is more evenly distributed over the frequency range. Because of the redistribution of the spectral power over a wider frequency range, it may be understood that random PWM affects secondary issues in PWM-based converter systems including acoustic noise and EMI. The precise impact depends heavily on the modulator and on the system in question.

4.2.8.1 Fundamentals of Random Modulation. The starting point is the definition of the duty ratio d. To streamline the notation, it is assumed that observation interval at hand begins at time zero, i.e.,

$$d = \frac{1}{T}\int_0^T q(t)\, dt. \qquad (4.20)$$

It should be recalled that the duty ratio relates to the macroscopic time scale (a per carrier-period average), but the state of the switching function q must be defined for all time instants, i.e. on the microscopic time scale.

Now, the overall objective of a modulator is to map a reference value for d into a switching function $q(t)$ for $2 = [0; T]$ in such a way that Eq. (4.20) is fulfilled. Clearly, this mapping is not one-to-one: for a certain value of d, no unique q can be determined. To circumvent this ambiguity, constraints are added in classic modulators in order to get a unique correlation between d and q: typically, the value of the T is fixed and, furthermore, a certain pulse position is specified as well (leading-edge, lagging-edge, or center-aligned). Imposing such constraints, a unique course for q in Eq. (4.20) may be determined for a specific value of d. Figure 4.22a shows an example using center-aligned pulses.

The fundamental idea behind random PWM is to relax the standard constraints of fixed carrier period and, say, center-aligned pulse positions in order to leave room for the randomization. It follows directly from Eq. (4.20) that within a carrier interval T the switching function q may be altered in the following ways without distorting the value of d:

- *Random pulse position (RPP) modulation [137, 138]*: The position of the on-state pulse of width dT is randomized within each interval of constant duration T. The displacement should be selected so that the edges of the pulse do not extend into the adjacent PWM intervals to prevent overlap and to ensure that the switching function is implementable in real-time systems.

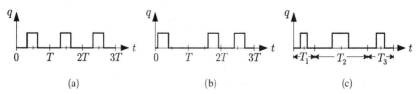

FIGURE 4.22
Fundamental principles of random pulse-width modulation. Illustration of (a) deterministic, (b) random pulse position, and (c) random carrier frequency modulation. The duty ratio is the same in all examples.

- *Random carrier frequency (RCF) modulation [128, 131, 133, 139]*: The duration of the carrier frequency, or, equivalently, the carrier period T is randomized while the duty ratio d is kept constant, i.e., the width of the pulse is changed in proportion to the instantaneous value of T. Also, the pulse position is fixed, e.g., the leading pulse edge is aligned with the beginning of each T.

The two complementary ways to randomize a modulator are illustrated in Figs. 4.22b and 4.22c, respectively. It is important to note that both methods guarantee that the average of the q waveform equals d in each carrier period, i.e., the volt-second balance is maintained on the macroscopic scale.

4.2.8.2 Aspects of Implementation of Random Modulation.
To facilitate the discussion, the principles for implementation of random PWM techniques are briefly described below. It is impossible to give detailed guidelines, because the implementation depends strongly on the hardware used for the pulse-width modulator. However, irrespective of the hardware, it is evident that apart from the normal input signals to the modulator, a randomized modulator must rely on at least one additional input in order to quantify the randomization. This extra input is a variable x, and, in addition, x should be a random variable in the sense known from the theory of stochastic processes.

To fix the ideas, the block diagram shown in Fig. 4.23 for the RCF technique introduces some of the key elements. The modulator should convert the duty ratio d and the carrier period T into a switching function q, which governs the state of the converter. Clearly, the details of how the modulator interfaces with the converter are hardware specific; the hardware used may very well contain specialized peripheral circuits such as dedicated PWM timers, which can be programmed through a set of registers, but many other solutions do also exist.

Once the duty ratio d in Fig. 4.23 has been calculated by a modulator, the next step for the RCF technique is to determine the value T for the carrier period. To fulfill this task, two functions are needed:

1. **A random number generator (RNG).** In each new carrier period, the RNG calculates a value for the random variable x, which is used below. Normally, an RNG generates uniform deviates in the $0 < x < 1$ range, i.e., all values between zero and one occur with the same probability.
2. **A probability density function (pdf).** The value x is now used to get the value for the carrier period in the particular PWM period at hand. The precise mapping from x to T is determined by a pdf as indicated in Fig. 4.23.

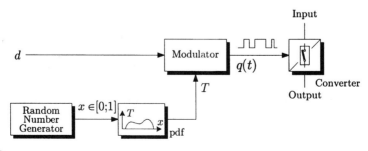

FIGURE 4.23
Example of an overall implementation of an RCF-based pulse-width modulator.

The block diagram in Fig. 4.23 only serves as an example of the implementation of an RCF modulator. Over the years, many different implementations have been suggested (both software- and hardware-based), but it is important to recall that they share the property that a noise source is needed to generate a random variable, which then is processed in some way to get T.

4.2.8.3 Examples of Spectra.

To exemplify the spectral characteristics for random PWM, Fig. 4.24 shows a set of measurement results recorded on a fully operating 1.1 kVA three-phase converter. As may be seen, the nonfundamental voltage spectrum becomes much more evenly distributed over the whole frequency axis when random carrier frequency (RCF) PWM is used compared to fixed carrier frequency (FCF) operation. To a large extent, this is also true for the acoustic noise emitted from the induction motor fed from the converter.

Much more information on random PWM may be found in references such as [140, 142].

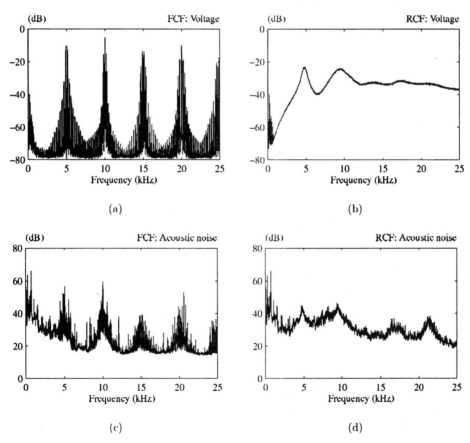

FIGURE 4.24
Examples of measured spectra on a three-phase voltage-source converter: Voltage spectra for (a) fixed-frequency PWM and (b) random carrier frequency PWM. (c) and (d): The corresponding acoustic noise spectra.

4.2.9 Summary

This chapter has reviewed basic PWM techniques developed during recent years. Key conclusions include the following:

- Parameters of *PWM* converter (linear range of operation, current harmonic, and switching losses in power components) depends on the zero vectors placement in *SVM*, and on the shape of zero sequence signal (*ZSS*) in *CB-PWM*.
- There is no one method of *PWM* that provides minimal current distortion in the whole range of control.
- Three-phase *SVM* with symmetrical zero states (*SVPWM*) should be used in the low range of the modulation index and two-phase *SVM* with one zero state in sampling time (*DPWM*) should be used in the high range of the modulation index.
- Maximal reduction of switching losses in *DPWM* is achieved when the peak of the line current is located in the center of the clamped (not switching) region.
- *SVPWM* and *DPWM* should be applied for industrial applications because both methods have low time-consuming algorithms and wide linear region.
- Adaptive space vector modulation (*ASVM*) is a universal solution for three-phase PWM converter. Among its main features are: full control range, including overmodulation and six-step operation; tracking of the peak current for instantaneous selection of two-phase PWM (this guarantees maximal reduction of switching losses up to 50%); and higher efficiency of the converter.

4.3 CLOSED-LOOP PWM CURRENT CONTROL

4.3.1 Basic Requirements and Definitions

Most applications of three-phase voltage-source PWM converters—ac motor drives, active filters, high power factor ac/dc converters, uninterruptible power supply (UPS) systems, and ac power supplies—have a control structure comprising an internal *current feedback loop*. Consequently, the performance of the converter system largely depends on the quality of the applied current control strategy. In comparison to conventional open-loop voltage PWM converters (see Section 4.2), the *current controlled PWM (CC-PWM) converters* have the following advantages:

- Control of instantaneous current waveform, high accuracy
- Peak current protection
- Overload rejection
- Extremely good dynamics
- Compensation of effects due to load parameter changes (resistance and reactance)
- Compensation of the semiconductor voltage drop and dead times of the converter
- Compensation of the dc link and ac side voltage changes

The main task of the control scheme in a CC-PWM converter (Fig. 4.25) is to force the currents in a three-phase ac load to follow the reference signals. By comparing the command i_{Ac} (i_{Bc}, i_{Cc}) and measured i_A (i_B, i_C) instantaneous values of the phase currents, the CC generates the switching states S_A (S_B, S_C) for the converter power devices which decrease the

4.3 CLOSED-LOOP PWM CURRENT CONTROL

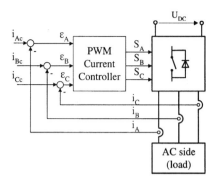

FIGURE 4.25
Basic block diagram of current controlled PWM converter.

current errors ε_A (ε_B, ε_C). Hence, in general the CC implements two tasks: *error compensation* (decreasing ε_A, ε_B, ε_C) and *modulation* (determination of switching states S_A, S_B, S_C).

4.3.1.1 Basic Requirements and Performance Criteria.
The accuracy of the CC can be evaluated with reference to basic requirements, valid in general, and to specific requirements, typical of some applications. Basic requirements are:

- No phase and amplitude errors (ideal tracking) over a wide output frequency range
- High dynamic response of the system
- Limited or constant switching frequency to guarantee safe operation of converter semiconductor power devices
- Low harmonic content
- Good dc-link voltage utilization

Note that some of the requirements, e.g., fast response and low harmonic content, contradict each other.

The evaluation of CC may be done according to performance criteria, which include static and dynamic performance. Table 4.4 presents the static criteria in two groups:

Table 4.4 Performance Criteria

Criteria definition[a]	Comments
RMS $= [1/T \int (\varepsilon_\alpha^2 + \varepsilon_\beta^2) dt]^{1/2}$	• The rms vector error
$J = 1/T \int [(\varepsilon_\alpha^2 + \varepsilon_\beta^2)]^{1/2} dt$	• The vector error integral
$N = \sum \mathrm{imp}|_{t \in (0,T)}$	• Number of switchings (also for nonperiodical)
$I_{hrms} = [1/T \int (i_{(t)} - i_{1(t)})^2 dt]^{1/2}$	The rms harmonic current
$d = I_{hrms}/I_{hrms\ six\text{-}step}$	The distortion factor
$d = [\sum h_i^2(k \cdot f_1)]^{1/2}$, $k \neq 1$	Synchronized PWM case
$d = [\int h_d^2(f) df]^{1/2}$, $f \neq f_1$	Nonsynchronized PWM case

[a] $h_i(k \cdot f_1)$, discrete current spectra; $h_d(f)$, density current spectra.

- Those valid also for open loop voltage PWM (see, e.g., [1, 8, 9, 16])
- Those specific for CC-PWM converters based on current error definition (denoted •).

The following parameters of the CC system dynamic response can be considered: dead time, settling time, rise time, time of the first maximum, and overshoot factor. The foregoing features result both from the PWM process and from the response of the control loop. For example, for dead time the major contributions arise from signal processing (conversion and calculation times) and may be appreciable especially if the control is of the digital type. On the other hand, rise time is mainly affected by the ac side inductances of the converter. The optimization of the dynamic response usually requires a compromise, which depends on the specific needs. This may also influence the choice of the CC technique according to the application considered.

In general, the compromise is easier as the switching frequency increases. Thus, with the speed improvement of today's switching components (e.g., IGBTs), the peculiar advantages of different methods lose importance and even the simplest one may be adequate. Nevertheless, for some applications with specific needs, such as active filters, which require very fast response, or high-power converters, where the number of commutations must be minimized, the most suitable CC technique must be selected.

4.3.1.2 Presentation of CC Techniques. Existing CC techniques can be classified in different ways [3, 8, 9, 11–13, 15, 27]. In this section, the CC techniques are divided into two main groups (Fig. 4.26):

- Controllers with open-loop PWM block (Fig. 4.27a)
- On–off controllers (Fig. 4.27b)

In contrast to the on–off controllers (Fig. 4.27b), schemes with open-loop PWM block (Fig. 4.27a) have clearly separated *current error compensation* and *voltage modulation* parts. This concept allows us to exploit the advantages of open-loop modulators (sinusoidal PWM, space vector modulator, optimal PWM): constant switching frequency, well-defined harmonic

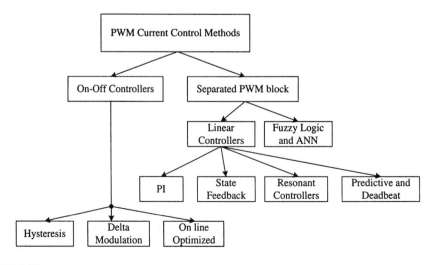

FIGURE 4.26
Current control techniques.

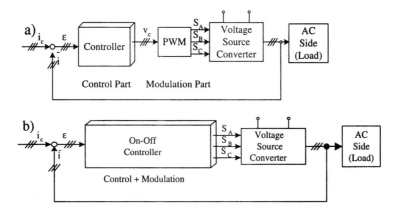

FIGURE 4.27
(a) Controller with open-loop PWM block. (b) On–off controller.

spectrum, optimum switch pattern, and good dc link utilization. Also, full independent design of the overall control structure as well as open-loop testing of the converter and load can be easily performed.

4.3.2 Introduction to Linear Controllers

4.3.2.1 Basic Structures of Linear Controllers. Two main tasks influence the control structure, when designing a current control scheme: reference tracking and disturbance rejection abilities.

A. Conventional PI Controller. The input–output relation of the control scheme presented in Fig. 4.28a can be described by

$$y(s) = \frac{C(s)G(s)}{1+C(s)G(s)}r(s) + \frac{G(s)}{1+C(s)G(s)}d(s), \qquad (4.21)$$

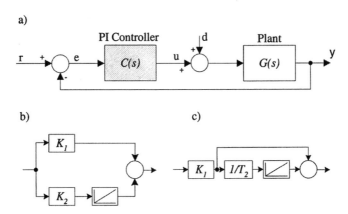

FIGURE 4.28
(a) Feedback controller; (b) and (c) two forms of PI controller structure.

or in the form

$$y(s) = T(s)r(s) + S(s)d(s),$$

where $C(s)$ is the controller transfer function (Figs. 4.28b and 4.28c), here

$$C(s) = K_1 + \frac{K_2}{s} = K_1 \frac{1 + sT_2}{sT_2}, \qquad (4.22)$$

with $T_2 = K_1/K_2$, where $G(s)$ is the plant transfer function, $T(s)$ is the reference transfer function, $S(s)$ is the disturbance transfer function, K_1 is the proportional gain, K_2 is the integral gain, T_2 is the integrating time, r is the reference signal, d is the disturbance signal, y is the output signal, and s is the Laplace variable. For good reference tracking it should be

$$T(s) = \frac{C(s)G(s)}{1 + C(s)G(s)} \approx 1, \qquad (4.23)$$

and for effective disturbance rejection,

$$S(s) = \frac{1}{1 + C(s)G(s)} \approx 0. \qquad (4.24)$$

The preceding conditions can be fulfilled for the low-frequency range. However, in the higher frequency range performance deteriorates. Moreover, PI controller parameters influence both reference tracking and disturbance rejection performance, and it is not possible to influence the characteristics separately.

B. *Internal Model Controller.* The internal model controller (IMC) structure belongs to the *robust control* methods [26]. This structure (Fig. 4.29) uses an internal model $\hat{G}(s)$ in parallel with the controlled plant $G(s)$. If the internal model is ideal, i.e., $\hat{G}(s) = G(s)$, there is no feedback in Fig. 4.29, and the transfer function of the closed-loop is expressed as

$$G_C(s) = C(s)G(s) \qquad (4.25)$$

Hence, the closed-loop system is stable if and only if $G(s)$ and $C(s)$ each are stable. In this ideal case, for $C(s) = G^{-1}(s)$, one obtains $G_C(s) = 1$, i.e., the plant dynamics will be cancelled and $y(s) = r(s)$. However, the disturbance will not be rejected, $y(s) = G(s)d(s)$.

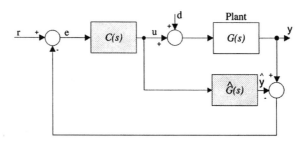

FIGURE 4.29
Internal model controller.

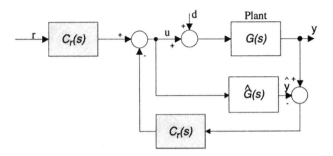

FIGURE 4.30
Two degrees of freedom controller.

C. Two Degrees of Freedom Controller Based on IMC Structure. For the control scheme of Fig. 4.30 the input–output relation (in the scalar case) is described as

$$y(s) = \frac{C_r(s)G(s)}{1 + [G(s) - G^{-1}(s)]C_y(s)} r(s) + \frac{1 - \hat{G}(s)C_y(s)}{1 + [G(s) - G^{-1}(s)]C_y(s)} G(s)d(s). \quad (4.26)$$

For the ideal plant model $\hat{G}(s) = G(s)$, one obtains

$$y(s) = C_r(s)G(s)r(s) + [1 - G(s)C_y(s)]G(s)d(s) = T(s)r(s) + S(s)G(s)d(s), \quad (4.27)$$

where $T(s) = G(s)C_r(s)$, $S(s) = 1 - G(s)C_y(s)$.

The controller of Fig. 4.30 is called a *two degrees of freedom* (TDF) controller, because one can design reference $T(s)$ and disturbance $S(s)$ transfer functions separately by selecting controllers $C_r(s)$ and $C_y(s)$, respectively.

D. State Feedback Controller. Although the mathematical description of control processes in the form of input–output relations (transfer functions) has a number of advantages, it does not make it possible to observe and control all the internal phenomena involved in the control process. Modern control theory is therefore based on the *state space method*, which provides a uniform and powerful representation in the time domain of multivariable systems of arbitrary order, linear, nonlinear, or time varying coefficients. Also, the initial conditions are easy to take into account. For linear constant-coefficient multivariable continuous-time dynamic systems, the "state space equations" can be written in vector form as follows:

$$\begin{aligned} \dot{x}(t) &= Ax(t) + Bu(t) + Ed(t), \\ y(t) &= Cx(t), \end{aligned} \quad (4.28)$$

where u is the input vector ($1 \times p$), y is the output vector ($1 \times q$), x is the state variable vector ($1 \times n$), d is the disturbance vector ($1 \times g$), A is the system (process) matrix ($n \times m$), B is the input matrix ($n \times p$), C is the output matrix ($q \times n$), E is the disturbance matrix ($n \times g$).

In the state feedback controller of Fig. 4.31 the control variable $u(t)$ can be expressed:

$$u(t) = -K^T\hat{x} + K_f u_f(t) + K_d d(t) \quad (4.29)$$

where K is the vector of the state feedback factors, K_f is the vector of the feedback controller, K_d is the vector of the disturbance controller. The feedback gain matrix K is derived by utilizing the pole assignment technique to guarantee sufficient damping. The reference tracking and disturbance rejection performance can be designed separately by selecting K_f and K_d matrixes, respectively.

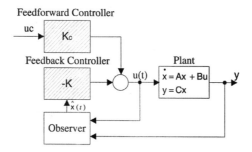

FIGURE 4.31
State feedback controller.

4.3.2.2 Standard Rules of Controller Design. The design of industrial controllers in the simplest cases of so-called parametric synthesis of linear controllers is limited to the selection of a regulator type (P, PI, PID) and the definition of optimal setting of its parameters according to the criterion adopted. This design process is normally done with "complete knowledge" of the plant. Furthermore, the plant is usually described by a linear time-invariant continuous or discrete time model. In the field of control theory, techniques have evolved to find an "optimal set of controller parameters" [12, 14].

Table 4.5 Controller Parameters According to Standard Rules (for Fast Sampling $T_S \to 0$)

Method	Plant	Proportional gain K_1	Integrating time T_2 / Integrating gain K_2	Remarks
Optimal modulus criterion (for $T_a \gg \tau_0$)	$\dfrac{K_O e^{-s\tau_0}}{1+sT_a}$	$K_1 = \dfrac{T_a}{2K_O \tau_0}$	$T_2 = T_a$ $K_2 = \dfrac{1}{2K_O \tau_0}$	• 4% overshoot in response to step change of reference • Very slow disturbance rejection
Optimal symmetry criterion (for $T_a \gg (T_b + \tau_0)$)	$\dfrac{K_O e^{-s\tau_0}}{sT_a(1+sT_b)}$	$K_1 = \dfrac{T_a}{2K_O(T_b + \tau_0)}$	$T_2 = 4(T_b + \tau_0)$ $K_2 = \dfrac{T_a}{8K_O(T_b + \tau_0)^2}$	• Fast disturbance rejection • 43% overshoot in response to step change of reference. An input filter is required ($T_F = T_2$)
Damping factor selection $\xi = 1$ (for $\tau_0 = 0$)	$\dfrac{K_O e^{-s\tau_0}}{1+sT_a}$	$K_1 = 1$	$T_2 = \dfrac{4\xi^2 T_a K_O}{(1+K_O)^2}$ $K_2 = \dfrac{(1+K_O)^2}{4\xi^2 T_a K_O}$	• Well damped
Rule of thumb		$K_1 = 1$	$T_2 = T_s$ $K_2 = \dfrac{1}{T_s}$	• Only for very rough design • T_s is sampling time

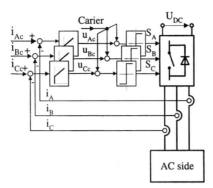

FIGURE 4.32
Ramp comparison controller.

Some of such "standard rules" commonly used in power electronics and drive control practice are given in Table 4.5. For $T_a < 4\tau_0$ the *modulus criterion* is more useful, whereas for $T_a \gg \tau_0$ it is better to apply the *symmetry criterion*. The rules of Table 4.5 are valid for continuous or fast sampled ($T_s \rightarrow 0$) discrete systems. For slow ($T_s \approx T_a$) or practical ($T_s < T_a$) sampling, the sampling time T_s has to be included in controller parameters [12]. It should be noted, however, that controller parameters calculated often on the basis of roughly estimated plant data can only be used as broad indicators of the values to be employed.

4.3.3 PI Current Controllers

4.3.3.1 Ramp Comparison Controller. The *ramp comparison current controller* uses three PI error to produce the voltage commands u_{Ac}, u_{Bc}, u_{Cc} for a three-phase sinusoidal PWM (Fig. 4.32).

In keeping with the principle of sinusoidal PWM, comparison with the triangular carrier signal generates control signals S_A, S_B, S_C for the inverter switches. Although this controller is directly derived from the original suboscillation PWM [19], the behavior is quite different, because the output current ripple is fed back and influences the switching times. The integral part of the PI compensator minimizes errors at low frequency, while proportional gain and zero placement are related to the amount of ripple. The maximum slope of the command voltage u_{Ac} (u_{Bc}, u_{Cc}) should never exceed the triangle slope. Additional problems may arise from multiple crossing of triangular boundaries. As a consequence, the controller performance is satisfactory only if significant harmonics of current commands and the load EMF are limited at a frequency well below the carrier (less than 1/9 [4]).

Example 4.1: Ramp Comparison PI Current Controller for PWM Rectifier—Simplified Design

The current control scheme under consideration is shown in Fig. 4.33a, and its block diagram in Fig. 4.33b, respectively. The design of the current controller includes:

- Selection of signal parameters: amplitude U_t and frequency f_t
- Design of PI controller

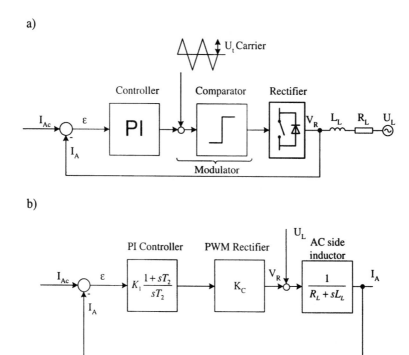

FIGURE 4.33
(a) Block scheme of current control in three-phase PWM rectifier (only phase A is shown). (b) Simplified block diagram of current control loop—small time constants are neglected.

A. Parameters of Triangular Carrier Signal. As can be seen from Fig. 4.34, a fundamental constraint of carrier-based PWM is that the maximum rate of change of the reference (output of the PI) voltage U_A should not equal or exceed that of the carrier signal. The slope condition can be found from

$$\frac{dU_A}{dt} < \frac{dU_t}{dt} \tag{4.30}$$

$$2\pi f_{Acm} + \frac{U_{Lm} + 0.5 U_{DC}}{L_L} < 4 U_{tm} f_t, \tag{4.31}$$

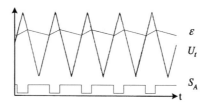

FIGURE 4.34
Typical signals in PWM regulator. ε, current error in phase A; U_t, triangular carrier signal; S_A, PWM switching signal.

4.3 CLOSED-LOOP PWM CURRENT CONTROL

For correct PWM modulator behavior, the following condition has to be fulfilled:

Slope of current error < Slope of triangle

The condition (4.30) can be transformed into

$$(I_A^* - I_A) < 4U_t f_t \qquad (4.32)$$

where I_A^* is the reference current, and I_A is the actual current.

The maximum slope of the measured current is

$$\left.\frac{dI_A}{dt}\right|_{max} = \frac{V_L + 0.5U_{DC}}{L_L} \qquad (4.33)$$

where U_L is the phase voltage, and L_L is the ac side inductance.

The maximum inclination of the reference current is

$$\left.\frac{dI_A}{dt}\right|_{max} = I_{Am}^* \omega \cos(\omega t)|_{\omega t=0} = I_{Am}^* \omega, \qquad (4.34)$$

where $\omega = 2\pi f$.

Including (4.32), (4.33), and (4.34), one obtains

$$\frac{U_L + 0.5U_{DC}}{L_L} + I_{Am}^* 2\pi f < 4U_t f_t. \qquad (4.35)$$

This equation allows one to determine the amplitude of the triangular signal U_t when the triangle frequency f_t and the ac side inductance L_L are known.

B. Parameters of PI Controller: Simplified Design According to Damping Factor Selection.
In simplified design the small time constants such as power converter dead time, feedback filter, and digital signal processing delay are neglected. So, only the dynamic of the line inductor is taken into account.

The open-loop transfer function (Fig. 4.33b) is given as

$$KG_O(s) = K_1 \frac{1 + sT_2}{sT_2} \frac{K_O}{1 + sT_L}, \qquad (4.36)$$

where

$$K_O = K_L K_C, \quad K_L = \frac{1}{R_L}, \quad T_L = \frac{L_L}{R_L}, \quad K_C = \frac{(MU_{DC})}{2U_t}. \qquad (4.37)$$

The transfer function of a PI controller is given in the form

$$G_R(s) = K_1 \frac{1 + sT_2}{sT_2}. \qquad (4.38)$$

The closed-loop transfer function calculated from Eq. (4.36) takes the form

$$KG_C(s) = \frac{I_A(s)}{I_{Ac}(s)} = \frac{s\dfrac{K_1 K_O}{T_L} + \dfrac{K_1 K_O}{T_2 T_L}}{s^2 + s\dfrac{1 + K_O K_1}{T_L} + \dfrac{K_1 K_O}{T_2 T_L}}. \qquad (4.39)$$

124 CHAPTER 4 / PULSE WIDTH MODULATION TECHNIQUES

The controller time constant is derived as a function of the damping factor ξ,

$$T_2 = \frac{4\xi^2 T_L K_1 K_O}{(1 + K_O K_1)^2}, \tag{4.40}$$

or the approximate expression ($K_O K_1 \gg 1$)

$$T_2 = \frac{4\xi^2 T_L}{K_O K_1}, \tag{4.41}$$

and for $K_1 = 1$ one obtains

$$T_2 = \frac{4\xi^2 T_L}{K_O}. \tag{4.42}$$

C. Calculations. Data:

$$U_L = 220 \text{ V} \quad f_t = 10 \text{ kHz} \quad R_L = 0.1 \quad T_S = 0.0001$$
$$I_L = 40 \text{ A} \quad U_{DC} = 700 \text{ V} \quad L_L = 10 \text{ mH}$$
$$f_L = 50 \text{ Hz}$$

Slope conditions:

$$U_T = \frac{\sqrt{2}I_L 2\pi f_L + \dfrac{\sqrt{2}U_L + 0.5 U_{DC}}{L_L}}{4 f_t} \geqslant 2V$$

Converter gain:

$$M = 0.9$$
$$K_C = \frac{M U_{DC}}{2 U_t} = 31.5$$

Load:

$$T_L = \frac{L_L}{R_L} = 0.1$$
$$K_L = \frac{1}{R_L} = 10$$

Open-loop gain:

$$K_O = K_L K_C = 315$$

Controller parameters design using damping factor selection:

$$\xi = 0.707 \Rightarrow K_1 = 1, \ T_2 = \frac{4\xi^2 T_L}{K_O} = 0.0006347, \ K_2 = \frac{4 T_L \xi^2}{K_O} = 1575$$

$$\xi = 0.1 \Rightarrow K_1 = 1, \ T_2 = \frac{4\xi^2 T_L}{K_O} = 0.00127, \ K_2 = \frac{4 T_L \xi^2}{K_O} = 787$$

D. Simulation Results. The main disadvantage of the ramp comparison controller is an inherent steady-state tracking (amplitude and phase) error. This can be observed in Fig. 4.35 where i_x and i_y components in a synchronous rotated coordinates x, y are shown. To achieve compensation, use of additional PLL circuits [24] or feedforward correction [29, 38] is also made.

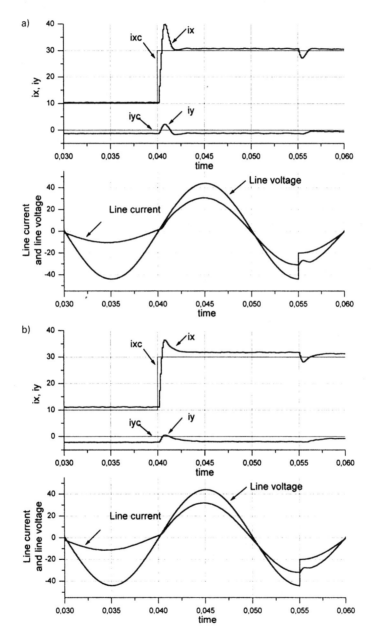

FIGURE 4.35
Simulated transient to the step change of reference current (at 0.04 s): 10 A→30 A, and the line voltage drop (at 0.055 s). (a) Damping factor selection $\xi = 0.707$, (b) $\xi = 1$. i_x and i_y current components in synchronously rotated coordinates x, y are shown only to illustrate the steady-state tracking error.

4.3.3.2 Stationary Vector Controller. In three-phase isolated neutral load topology (Fig. 4.1), the three phase currents must add to zero. Therefore, only two PI controllers are necessary and the three-phase inverter reference voltage signals can be established algebraically using two-to-three phase conversion blocks $\alpha\beta/ABC$. Figure 4.36 shows the block diagram of a PI current controller based on stationary coordinate α, β variables. The main disadvantage of the PI controller acting on ac components, namely the nonzero steady-state current error, still remains.

4.3.3.3 Synchronous Vector Controller (PI). In many industrial applications an ideally impressed current is required, because even small phase or amplitude errors cause incorrect system operation (e.g., vector-controlled ac motors, active power filters). In such cases the control schemes based on space vector approach are applied. Figure 4.37a illustrates the *synchronous controller*, which uses two PI compensators of current vector components defined in *rotating synchronous coordinates x–y* [5, 12, 14, 31, 32, 35]. Thanks to the coordinate transformations, i_{sx} and i_{sy} are dc components, and PI compensators reduce the errors of the fundamental component to zero.

However, the synchronous controller of Fig. 4.37a is more complex than the stationary controller (Fig. 4.36). It requires two coordinate transformations with explicit knowledge of the synchronous frequency ω_s. Based on [34], where Schauder and Caddy demonstrated that it is possible to perform current vector control in an arbitrary coordinates, an equivalent of a *synchronous controller* working in the *stationary coordinates* α, β with ac components has been proposed by Rowan and Kerkman [33]. As shown in Fig. 4.37b by the dashed line, the inner loop of the controller (consisting of two integrators and multipliers) is a variable frequency generator which always produces reference voltage $V_{\alpha c}$, $V_{\beta c}$ for the modulator (PWM), even when in the steady states the current error signals $\varepsilon_\alpha, \varepsilon_\beta$ are zero. Hence, this controller solves the problem of nonzero steady-state error under ac components. However, the dynamic is generally worse than that of the stationary controller because of the cross coupling between α, β components.

4.3.3.4 Stationary Resonant Controller. The transfer function of a standard PI compensator used in a synchronous controller working in rotating coordinates with dc components can be expressed as

$$G(s) = K_1 + \frac{K_2}{s} = K_1 \frac{1 + sT_2}{sT_2}, \text{ where } T_2 = \frac{K_1}{K_2}. \tag{4.43}$$

As shown in [36], an equivalent single-phase stationary ac current controller which achieves the same dc control response centered around the ac control frequency can be calculated as follows:

$$G(s) = \frac{1}{2}[g(s+j\omega) + g(s-j\omega)] = K_1 + \frac{sK_2}{s^2 + \omega_s^2}. \tag{4.44}$$

The last equation can be seen to be a *resonant controller* (Fig. 4.38) with infinite gain at the resonant frequency ω_s. To compare a stationary controller, a synchronous controller, and the resonant controller in the frequency domain, the transfer functions in both the stationary and

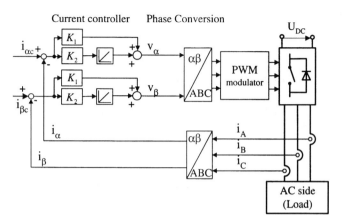

FIGURE 4.36
Stationary PI controller.

synchronous coordinates are calculated. For example, the stationary synchronous controller of Fig. 4.37b can be calculated as follows:

$$V_\alpha = K_1 \varepsilon_\alpha + u_\alpha \tag{4.45}$$

$$V_\beta = K_1 \varepsilon_\beta + u_\beta \tag{4.46}$$

where

$$u_\alpha = \frac{1}{s}(K_2 \varepsilon_\alpha - \omega_s u_\beta), \tag{4.47}$$

$$u_\beta = \frac{1}{s}(K_2 \varepsilon_\beta - \omega_s u_\alpha), \tag{4.48}$$

Inserting Eq. (4.47) and Eq. (4.48) into Eq. (4.45) and Eq. (4.46), one obtains

$$V_\alpha = \left[K_1 + \frac{sK_2}{s^2 + \omega_s^2}\right]\varepsilon_\alpha + \left[\frac{\omega_s K_2}{s^2 + \omega_s^2}\right]\varepsilon_\beta \tag{4.49a}$$

$$V_\beta = -\left[\frac{\omega_s K_2}{s^2 + \omega_s^2}\right]\varepsilon_\alpha + \left[K_1 + \frac{sK_2}{s^2 + \omega_s^2}\right]\varepsilon_\beta, \tag{4.49b}$$

with

$$G_1 = K_1 + \frac{sK_2}{s^2 + \omega_s^2}, \quad G_2 = \frac{\omega_s K_2}{s^2 + \omega_s^2}.$$

Equations (4.49a) and (4.49b) can be represented in matrix form as

$$\begin{bmatrix} V_\alpha \\ V_\beta \end{bmatrix} = \begin{bmatrix} G_1 & G_2 \\ -G_2 & G_1 \end{bmatrix} \begin{bmatrix} \varepsilon_\alpha \\ \varepsilon_\beta \end{bmatrix} \tag{4.50}$$

Similarly, transfer functions for other controllers can be calculated, and results are shown in Table 4.6. Summing up the conclusions from these expressions, we may say:

- Any current controller that is required to achieve zero steady-state error must have infinite dc gain in the synchronous rotating coordinates.

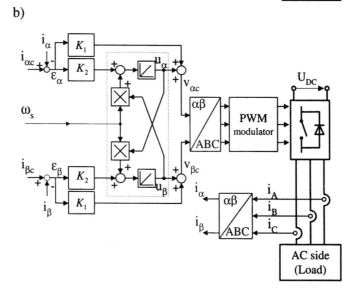

FIGURE 4.37
Synchronous PI controller (a) working in rotating coordinates x, y with dc components, (b) working in stationary coordinates α, β with ac components.

- Hence, a stationary PI controller can never achieve zero error, since its transfer function G_1 in the synchronous coordinates does not have an integral term.
- In contrast, the resonant controller includes an integral term K_2/s in the synchronous coordinates, and therefore can achieve zero error in the stationary coordinates.
- The cross-coupling terms G_2 add to the complexity of controller implementation. It suggests that the resonant controller can be simple implemented in stationary coordinates.

Example 4.2: Synchronous Current Controller for PWM Rectifier—Design Based on Standard Rules

The block diagram of a synchronous current controller working in x–y coordinates for PWM rectifier is shown in Fig. 4.39.

4.3 CLOSED-LOOP PWM CURRENT CONTROL

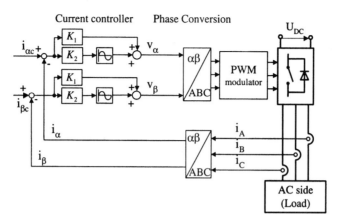

FIGURE 4.38
Stationary resonant controller.

Table 4.6 Transfer Functions of PI Current Controllers in Stationary and Synchronous Coordinates

Current controller	Stationary coordinates $\alpha-\beta$ (ac components)[a]	Synchronous (rotating) coordinates $x-y$ (dc components)
Stationary PI (Fig. 4.36)	$\begin{bmatrix} g & 0 \\ 0 & g \end{bmatrix}$	$\begin{bmatrix} G_1 & -G_2 \\ G_2 & G_1 \end{bmatrix}$
Stationary synchronous PI (Fig. 4.37b)	$\begin{bmatrix} G_1 & G_2 \\ -G_2 & G_1 \end{bmatrix}$	$\begin{bmatrix} g & 0 \\ 0 & g \end{bmatrix}$
Stationary resonant PI (Fig. 4.38)	$\begin{bmatrix} G_1 & 0 \\ 0 & G_1 \end{bmatrix}$	$\begin{bmatrix} G_1 + \dfrac{K_2}{s} & -G_2 \\ G_2 & G_1 + \dfrac{K_2}{s} \end{bmatrix}$

[a] Where $g = K_1 + \dfrac{K_2}{s}$, $G_1 = K_1 + \dfrac{sK_2}{s^2 + \omega_s^2}$, $G_2 = \dfrac{\omega_s K_2}{s^2 + \omega_s^2}$.

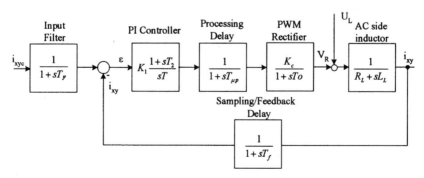

FIGURE 4.39
Block diagram of synchronous current controller.

A. Open-Loop Transfer Function. The following simplifying assumption are made:

- The cross-coupling effect between the x and y axes due to inductance L is neglected
- The dead time of the power converter (also processing and sampling) is approximated by a first-order inertia element:

$$e^{-sT_o} \approx \frac{1}{1+sT_o} \qquad (4.51)$$

- The sum of small time constants is defined as

$$\tau_\Sigma = T_{\mu p} + T_o + T_f \qquad (4.52)$$

where $T_{\mu p}$ is the processing/execution time of the algorithm, T_o is the power converter dead time, and T_f is the time delay of the feedback filter and sampling.

Note that with switching frequency f_s the statistical delay of the PWM inverter is $(0.5)/2f_s$, delay of the time discrete signal processing $1/2f_s$ and feedback delay (average) $(0.5)/2f_s$. So, the sum of the small time constant τ_Σ is the range $(1.5)/2f_s$ to $1/f_s$.

The open-loop transfer function is given by the equation

$$KG_o(s) = K_1 \frac{1+sT_2}{sT_2} \frac{1}{1+s\tau_\Sigma} \frac{K_O}{1+sT_L} \qquad (4.53)$$

where

K_1, T_2 is the proportional gain and integral time of PI controllers

$$K_O = K_C K_L, \quad K_L = \frac{1}{R_L}, \quad T_L = \frac{L_L}{R_L}, \quad \text{the gain and time constant of the line reactor} \qquad (4.54)$$

K_C is the power converter (PWM) gain,

The choice of optimal current controller parameters depends on the line reactor time constant T_L relative to the sum τ_Σ of all other small time constants (see Section 4.3.2.2).

B. For $T_L \gg \tau_\Sigma$ Controller Parameter Selection According to Symmetry Criterion (see Table 4.5).

$$T_2 = 4\tau_\Sigma, \quad K_1 = \frac{T_L}{2K_O \tau_\Sigma}, \qquad (4.55)$$

which substituted in Eq. (4.53) yields open-loop transfer functions of the form:

$$KG_o(s) = \frac{T_L}{2K_O \tau_\Sigma} \frac{1+s4\tau_\Sigma}{s4\tau_\Sigma(1+s\tau_\Sigma)} \frac{K_O}{1+sT_L} \approx \frac{T_L}{2\tau_\Sigma} \frac{1+s4\tau_\Sigma}{s4\tau_\Sigma(1+s\tau_\Sigma)sT_L} \approx \frac{1+s4\tau_\Sigma}{s^2 8\tau_\Sigma^2 + s^3 8\tau_\Sigma^3} \qquad (4.56)$$

For the closed-loop transfer function we obtain

$$KG(s) = \frac{1+s4\tau_\Sigma}{1+s4\tau_\Sigma + s^2 8\tau_\Sigma^2 + s^3 8\tau_\Sigma^3} \qquad (4.57)$$

To compensate for the forcing element in the numerator, use is made of the input inertia filter

$$G_F(s) = \frac{1}{1+s4\tau_\Sigma}, \qquad (4.58)$$

4.3 CLOSED-LOOP PWM CURRENT CONTROL

so that expression (4.57) becomes

$$KG_{CF}(s) = KG_C(s)G_F(s) = \frac{1}{1 + s4\tau_\Sigma + s^2 8\tau_\Sigma^2 + s^3 8\tau_\Sigma^3}, \qquad (4.59)$$

or approximately

$$KG_{CF}(s) \approx \frac{1}{1 + s4\tau_\Sigma} = \frac{1}{1 + sT_{eq}}, \qquad (4.60)$$

where $T_{eq} = 4\tau_\Sigma$ is the equivalent time constant of the closed current control loop optimized according to the symmetry criterion.

C. Calculations. Data:

$$U_L = 220 \text{ V} \qquad f_t = 10 \text{ kHz} \qquad R_L = 0.1$$
$$I_L = 40 \text{ A} \qquad U_{DC} = 700 \text{ V} \qquad L_L = 10 \text{ mH} \qquad T_s = \frac{1}{2f_s} = 0.0001$$
$$f_L = 50 \text{ Hz}$$

Slope conditions:

$$U_T = \frac{\sqrt{2}I_L 2\pi f_L + \frac{\sqrt{2}U_L + 0.5U_{DC}}{L_L}}{4f_t} = 2 \text{ V}$$

Converter gain:

$$M = 0.9$$
$$k_C = \frac{MU_{DC}}{2U_t} = 31.5$$

Load:

$$T_L = \frac{L_L}{R_L} = 0.1$$
$$K_L = \frac{1}{R_L} = 10$$

Sum of the small time constants:

$$\tau_\Sigma = 2T_S = 0.0002$$

Open-loop gain:

$$K_O = K_L K_C = 315$$

Controller parameters design using symmetry criterion:

$$K_1 = \frac{T_L}{2K_O\tau_\Sigma} = 0.8$$
$$T_2 = 4\tau_\Sigma = 0.0008$$
$$K_2 = \frac{T_L}{8K_o(\tau_\Sigma)^2} = 992$$

D. Simulation Results. See Fig. 4.40.

E. Decoupling Control. So far the cross-coupling effect due to line inductance L has been neglected. However, it can be easily compensated for using a decoupling network inside or outside the controller (Fig. 4.42). Figure 4.41 illustrates, in expanded time scale, improvements due to the decoupling network.

Remark: Note that the controller of Fig. 4.39 designed according to the symmetry criterion is a simplest form of the two degrees of freedom (TDF) controller (Fig. 4.30). The disturbance rejection performance is defined by PI parameters, whereas reference tracking performance can be separately adjusted by selecting input filter time constant T_F. More sophisticated TDF controller structure allows compensation for plant uncertainties, noise, and converter dead time [27].

Example 4.3: Synchronous Current Controller for Field Oriented Induction Motor— Decoupled Design Based on Internal Model Control Approach

A. Induction Motor Model. Using the induction motor model given by Eqs. (5.1)–(5.4), the following complex-valued differential equation can be derived:

$$x_\sigma T_N \frac{d\mathbf{i}_s}{dt} + r_{im}\mathbf{i}_s + j\omega_s x_\sigma \mathbf{i}_s = \frac{x_M}{x_r}\left(\frac{r_r}{x_r} - j\omega_m\right)\boldsymbol{\psi}_r + \mathbf{u}_s \tag{4.61}$$

where total resistance and total leakage reactance are expressed as

$$r_{im} = r_s + \left(\frac{x_M}{x_r}\right)^2 r_r \tag{4.62}$$

$$x_\sigma = \sigma x_s$$

Equation (4.61) can be rewritten in the matrix form

$$\begin{bmatrix} u_{sx} \\ u_{sy} \end{bmatrix} = \begin{bmatrix} sx_\sigma + r_{im} & -\omega_s x_\sigma \\ \omega_s x_\sigma & sx_\sigma + r_{im} \end{bmatrix}\begin{bmatrix} i_{sx} \\ i_{sy} \end{bmatrix} + \begin{bmatrix} -\frac{x_M r_r}{x_r^2}\psi_r \\ -\frac{x_M}{x_r}\omega_m\psi_r \end{bmatrix}, \tag{4.63}$$

where indices x,y denote components in field oriented coordinates (Fig. 5.2a).

Note that this system is *coupled* because the matrix is not diagonal. It means that any changes of the voltage component in x (y) axes results in changes in both current components (x and y). This implies that the classical design methods of linear and decoupled systems are not valid.

Remark: The rotor flux component of Eq. (4.63) can be treated as a slowly varying disturbance and will be neglected in further considerations.

Decoupling network outside of controller: Neglecting the last part of Eq. (4.63) with the rotor flux and denoting the controller output signals as v_x and v_y, one obtains from Eq. (4.63)

$$\begin{bmatrix} u_{sx} \\ u_{sy} \end{bmatrix} = \begin{bmatrix} v_x \\ v_y \end{bmatrix} + \begin{bmatrix} -\omega_s x_\sigma i_{sy} \\ \omega_s x_\sigma i_{sx} \end{bmatrix} \tag{4.64}$$

and

$$\begin{bmatrix} v_x \\ v_y \end{bmatrix} = \begin{bmatrix} sx_\sigma + r_{im} & 0 \\ 0 & sx_\sigma + r_{im} \end{bmatrix}\begin{bmatrix} i_{sx} \\ i_{sy} \end{bmatrix}. \tag{4.65}$$

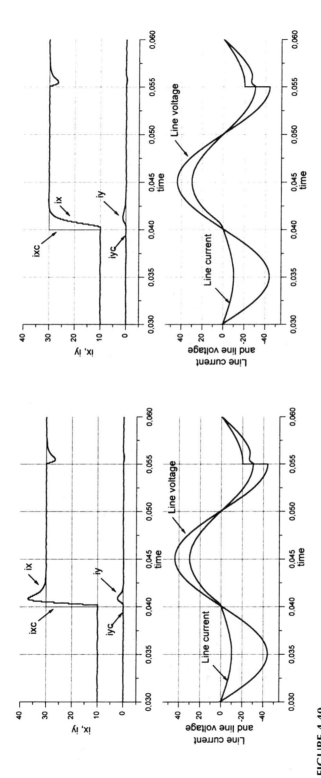

FIGURE 4.40

Simulated transient to the step change of reference current (at 0.04 s): 10 A→30 A, and the line voltage drop (at 0.055 s) (a) without input filter, (b) with input filter.

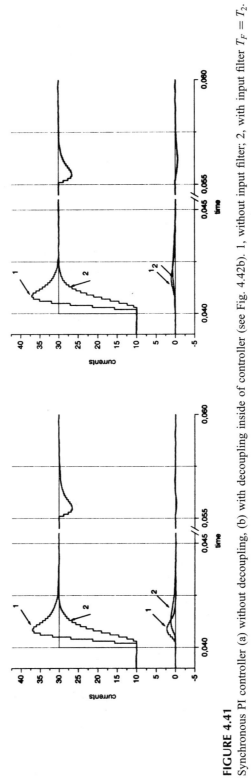

FIGURE 4.41
Synchronous PI controller (a) without decoupling, (b) with decoupling inside of controller (see Fig. 4.42b). 1, without input filter; 2, with input filter $T_F = T_2$.

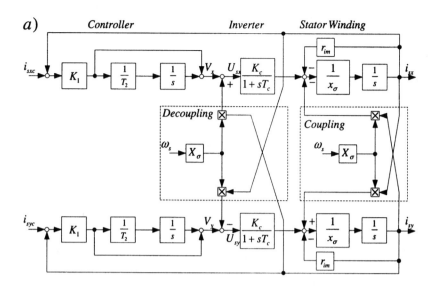

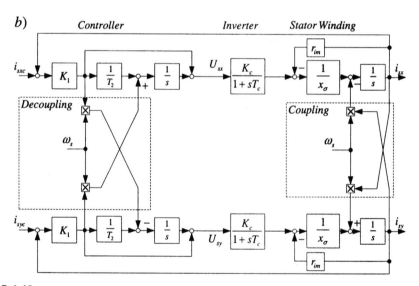

FIGURE 4.42
Two methods of decoupling in synchronous PI current controller: (a) with decoupling network as an inner feedback loop, (b) with decoupling network in the controller.

The whole current controller with decoupling is presented in Fig. 4.42a. Because of the limited dc voltage of the voltage source inverter, the PI controllers have to include *antiwindup* protections. This kind of protection is achieved by updating only the integral terms as long as saturation in the voltage is not detected.

The matrix transfer function of the decoupled current controlled induction motor is

$$G_o^{-1}(s) = \begin{bmatrix} sx_\sigma + r_{im} & 0 \\ 0 & sx_\sigma + r_{im} \end{bmatrix}. \quad (4.66a)$$

Decoupling network inside of controller: The more direct way to controller design is to use the transfer function of the plant in the form

$$G_o^{-1}(s) = \begin{bmatrix} sx_\sigma + r_{im} & -\omega_s x_\sigma \\ \omega_s x_\sigma & sx_\sigma + r_{im} \end{bmatrix} \quad (4.66b)$$

In this case the decoupling terms will be placed inside the controller as shown in Fig. 4.42b.

C. Design of Current Loop Based on Internal Model Approach. The internal model control (IMC) scheme of Fig. 4.29 presented in Section 4.3.2.1 can be viewed as a special case of the classical control structure (see Fig. 4.43), where the series controller has the following transfer function:

$$G_r(s) = [I - C(s) G_i(s)]^{-1} C(s). \quad (4.67)$$

Note that $G_O(s)$, $G_i(s)$, and $C(s)$ are transfer function matrices.

With the ideal model $G_i(s) = G_O(s)$ the closed-loop system has the transfer function

$$G_c(s) = G_o(s) C(s). \quad (4.68)$$

Note that if we assume the closed-loop transfer function as a multivariable first-order system, i.e.,

$$G_c(s) = \frac{1}{1 + \tau s} I, \quad (4.69)$$

then the behavior of the system depends only on one time constant τ, and the tuning procedure is very simple.

From Eqs. (4.67) and (4.68), assuming the closed-loop system defined by Eq. (4.68), we can obtain

$$G_r(s) = \left[I - \frac{1}{1 + \tau s} I \right]^{-1} G_o^{-1}(s) \frac{1}{1 + \tau s}. \quad (4.70)$$

Controller of Fig. 4.42a: Including the plant equation (4.66a) into the controller equation (4.70) one obtains

$$G_r(s) = \frac{x_\sigma}{\tau} \left[1 + \frac{1}{\frac{x_\sigma}{r_{im}} s} \right] I. \quad (4.71)$$

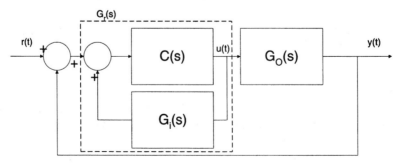

FIGURE 4.43
Equivalent scheme of the IMC in classic control structure.

Finally, the parameters of two PI controllers are

$$K_1 = \frac{x_\sigma}{\tau}, \qquad T_2 = \frac{x_\sigma}{r_{im}} T_N, \qquad (4.72)$$

where the K_1 is the gain factor, and T_2 is the integration time of the PI controller.

Controller of Fig. 4.42b: Including the plant equation (4.66b) into the controller equation (4.70) one obtains

$$G_r(s) = \begin{bmatrix} \dfrac{sx_\sigma + r_{im}}{s\tau} & -\dfrac{\omega_s x_\sigma}{s\tau} \\ \dfrac{\omega_s x_\sigma}{s\tau} & \dfrac{sx_\sigma + r_{im}}{s\tau} \end{bmatrix}. \qquad (4.73)$$

In this case the controller parameters are also given by Eq. (4.72). However, the controller structure is coupled and nonlinear (multiplication by synchronous angular speed ω_s).

D. Selection of the Closed-Loop Rise Time. For the assumed first-order closed-loop transfer function of Eq. (4.69) the 10% to 90% rise time t_r and bandwidth $b = 1/\tau$ are related as follows [26]:

$$\tau = \frac{1}{b} = \frac{t_r}{\ln 9} = \frac{t_r}{2.2} \approx 0.45 t_r. \qquad (4.74)$$

So, ideally for controller design, only one parameter should be specified: the desired closed-loop rise time t_r. In practical systems, however, two limitations have to be taken into account: sampling frequency and inverter saturation.

Selection of the sampling and switching frequencies: For good system performance the angular sampling frequency should be selected to be at least 10 times the closed-loop bandwidth

$$\omega_s = 2\pi f_s \geq 10b. \qquad (4.75)$$

Since in the system with carrier-based sinusoidal and space vector modulation, current is sampled in synchrony with inverter switching time, the switching frequency f_{sw} should not be lower than half the sampling frequency f_s. Hence:

$$2\pi f_{sw} \geq 5b. \qquad (4.76)$$

From Eqs. (4.74) and (4.76) one obtains the required switching frequency for a given rise time:

$$f_{sw} \geq 5 \frac{\ln 9}{2\pi t_r} \approx \frac{1.75}{t_r}. \qquad (4.77)$$

So, for a 1.75 ms rise time, the required switching frequency should be at least 1 kHz and sampling frequency 2 kHz.

Inverter saturation: If the rise time is selected to be too short, high controller gains will result, and the inverter will saturate during transients. This occurs, especially, at high motor speed when the stator voltage is also high. Therefore, the actual rise time will be longer as in the ideal case when inverter operates in linear region. To take into account limitation due to inverter saturation, let us calculate the controller response Δu to a reference change Δi_c as

$$\Delta u(0) = \lim_{s \to \infty} s \; G_r(s) \frac{\Delta i_c}{s} = G_r(\infty) \Delta i_c = \frac{1}{\tau} \begin{bmatrix} x_\sigma & 0 \\ 0 & x_\sigma \end{bmatrix} \Delta i_c. \qquad (4.78)$$

For the controllers given by Eqs. (4.71) and (4.73) we obtain

$$|\Delta u(0)| = \frac{x_\sigma}{\tau} |\Delta i_c|. \qquad (4.79)$$

138 CHAPTER 4 / PULSE WIDTH MODULATION TECHNIQUES

If we assume a reference change $\Delta i_c = 10\%$ of rated current I_{SN} and the voltage reserve as half of the maximum $0.5U_{SN}$, we obtain the condition

$$\tau \geq x_\sigma \frac{0.1 I_{SN}}{0.5 U_{SN}} = x_\sigma \frac{I_{SN}}{5 U_{SN}}, \qquad (4.80)$$

and with Eq. (4.74),

$$t_r[pu] \geq \frac{x_\sigma}{2.25}. \qquad (4.81)$$

For $x_\sigma = 0.2$ pu, $t_r = 0.08888$ pu, with the base frequency 50 Hz, the minimum rise time is $t_r = 0.0888/314 = 0.28$ ms. So, the limitation due to inverter saturation is not critical for the IM but can be for PMSM. Finally, the rise time selection rule can be expressed:

$$t_r = \max\left[\frac{\tau}{0.45}, \frac{10}{\omega_s}, \frac{5}{\omega_{sw}}, \frac{x_\sigma}{2.25}\right]. \qquad (4.82)$$

E. Calculation. Data:

$$U_{S(RMS)} = 400 \text{ V}; \qquad f_t = 5 \text{ kHz}; \qquad r_s = 0.0787; \qquad r_r = 0.0467;$$

$$I_{S(RMS)} = 4.9 \text{ A}; \qquad U_{DC} = 560 \text{ V}; \qquad x_s = 2.273; \qquad x_r = 2.293;$$

$$x_M = 2.1864; \qquad T_s = \frac{1}{2f_s} = 0.0001$$

Stator winding, rotor winding, total leakage reactance of induction motor:

$$x_{s\sigma} = x_S - x_M = 2.273 - 2.1864 = 0.0866$$
$$x_{r\sigma} = x_r - x_M = 2.293 - 2.1864 = 0.1066$$
$$x_\sigma = x_{s\sigma} + x_{r\sigma} = 0.0866 + 0.1066 = 0.1932$$

Total resistance:

$$r_{im} = r_s + \left(\frac{x_M}{x_r}\right)^2 r_r = 0.0787 + \left(\frac{2.1864}{2.293}\right)^2 0.0467 = 0.12116$$

For:

$$t_r = 0.001 \text{ ms}$$
$$\tau = 0.45 t_r = 0.00045$$
$$T_N = \frac{1}{2\pi f_N}$$

Controller parameters design:

$$K_1 = \frac{x_\sigma}{\tau} = \frac{0.1932}{0.00045 \cdot 314} = \frac{0.1932}{0.1413} = 1.367$$
$$T_2 = \frac{x_\sigma}{r_{im}} T_N = \frac{0.1932}{0.12116} 0.00318 = 0.00509$$

F. Simulation. See Fig. 4.44.

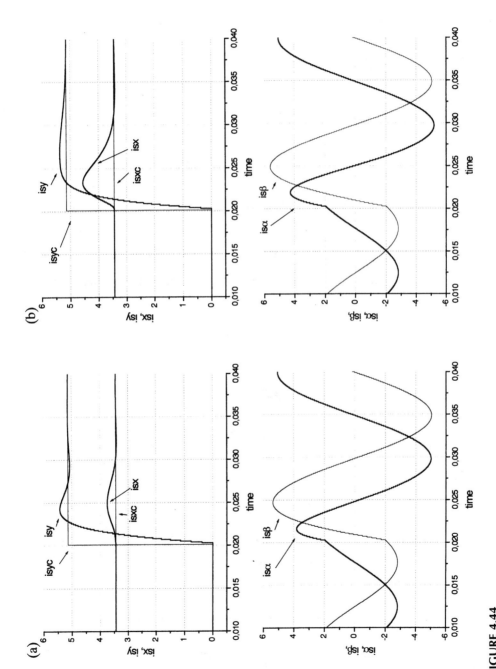

FIGURE 4.44
Simulations for decoupling inside of controller (Fig. 4.42b).

Example 4.4: Design of Stationary Resonant Current Controller for PWM Rectifier

The block diagram of a stationary resonant current controller for PWM rectifier (only phase A) is shown in Fig. 4.45.

A. Open-Loop Transfer Function. Transfer function of the resonant controller given by Eq. (4.44) can be expressed as follows:

$$G_C(s) = K_1 + \frac{sK_2}{s^2 + \omega_s^2} = \frac{c_0 + c_1 s + c_2 s^2}{s^2 + \omega_s^2} \tag{4.83}$$

where $c_0 = K_1 \omega_s^2$, $c_1 = K_2$, $c_2 = K_1$.

The open-loop transfer function (Fig. 4.45) is given by

$$KG_O(s) = \frac{c_0 + c_1 s + c_2 s^2}{s^2 + \omega_s^2} \cdot \frac{1}{R_L + sL_L}. \tag{4.84}$$

B. Controller Design Based on Naslin Polynomial. The characteristic polynomial of the closed-loop transfer function can be calculated as

$$D(s) = c_0 + c_1 s + c_2 s^2 + (R_L + sL_L)(\omega_s^2 + s^2). \tag{4.85}$$

The parameters of the controller can be computed based on the third-order Naslin polynomial:

$$P_N(s) = a_0 (1 + \frac{s}{\omega_0} + \frac{s^2}{\alpha \omega_0^2} + \frac{s^3}{\alpha^3 \omega_0^2}). \tag{4.86}$$

From the two last equations one obtains:

$$\begin{aligned} c_0 &= L_L \omega_0^3 \alpha^3 - R_L \omega_s^2 \\ c_1 &= L_L \omega_0^2 \alpha^3 - R_L \omega_s^2 \\ c_2 &= L_L \omega_0 \alpha^2 - R_L \end{aligned} \tag{4.87}$$

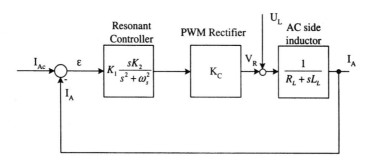

FIGURE 4.45
Simplified block diagram of current control loop with resonant controller (small time constants are neglected and $K_C = 1$).

or in the form

$$K_1 = L_L\omega_0\alpha^2 - R_L$$
$$K_2 = L_L\omega_0^2\alpha^3 - L_L\omega_s^2 \qquad (4.88)$$
$$\omega_s^2 = \alpha\omega_0^2$$

C. *Calculations.* Data:

$$R_L = 0.1$$
$$L_L = 10 \text{ mH}$$
$$\omega_s = 314 \text{ rad/s}$$

Selecting the Naslin polynomial parameter $\alpha = 2$:

$$\omega_0 = \frac{1}{\sqrt{\alpha}}\omega_s = \frac{1}{\sqrt{2}}314 \text{ rad/s} \approx 222 \text{ rad/s}$$

Controller parameters:

$$c_0 = 865424.2$$
$$c_1 = 2956.7$$
$$c_2 = 8.78$$

or in the form

$$K_1 = 8.87$$
$$K_2 = 2956.7$$

D. *Simulation Results.* See Fig. 4.46.

E. *Multiresonant Controller.* The presented design procedure can be extended for two or more different resonant angular frequencies $\omega_{s1}, \omega_{s2}, \ldots \omega_{sn}$ resulting in a *multiresonant controller.* For example, with two frequencies ω_{s1}, ω_{s2} the transfer function of the controller can be represented as

$$G_C(s) = K_1 + \frac{sK_2}{s^2 + \omega_{s1}^2} + \frac{sK_3}{s^2 + \omega_{s2}^2} = \frac{c_0 + c_1 s + c_2 s^2 + c_3 s^3 + c_4 s^4}{(s^2 + \omega_{s1}^2)(s^2 + \omega_{s2}^2)} \qquad (4.89)$$

and a fifth-order Naslin polynomial can be used for parameter calculation.

In a PWM line rectifier ω_{s2} can be tuned on third harmonic $\omega_{s2} = 3\omega_{s1}$ to compensate for the effect of line voltage harmonics [35].

4.3.3.5 State Feedback Current Controller. The conventional PI compensators in the current error compensation part can be replaced by a state feedback controller working in stationary [29] or synchronous rotating coordinates [13, 25, 28, 30].

The controller of Fig. 4.47 works in *synchronous rotating coordinates d–q* and is synthesized on the basis of linear multivariable state feedback theory. A feedback gain matrix $\mathbf{K} = [\mathbf{K}_1, \mathbf{K}_2]$ is derived by utilizing the pole assignment technique to guarantee sufficient damping. While with integral part ($\mathbf{K}_2$) the static error can be reduced to zero, the transient error may be unacceptably large. Therefore, feedforward signals for the reference ($\mathbf{K}_f$) and disturbance ($\mathbf{K}_d$) inputs are added to the feedback control law. Because the control algorithm guarantees the dynamically

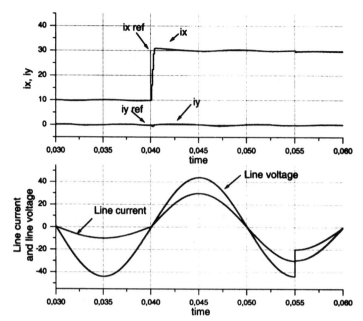

FIGURE 4.46
Simulations for resonant controller (Fig 4.45).

correct compensation for the EMF voltage, therefore, the performance of the state feedback controller is superior to those of conventional PI controllers [28]. However, the design procedure is more complex (see, for example, [30]).

4.3.3.6 Constant Switching Frequency Predictive Controller. In previous Sections 4.3.3.1–5 we discussed current control techniques which could be synthesized on the basis of the continuous-time approach and then, under the assumption of fast sampling ($T_S \to 0$),

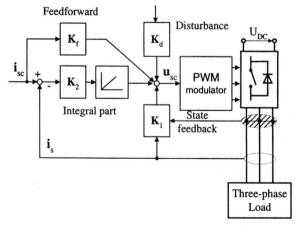

FIGURE 4.47
State feedback current controller.

4.3 CLOSED-LOOP PWM CURRENT CONTROL

implemented as discrete-time systems. In the case of constant switching frequency predictive controller, the algorithm calculates once every sample period T_S the voltage vector commands $\mathbf{u}_c(kT_S)$ which will force the actual current vector $\mathbf{i}(kT_S)$ according to its command, $\mathbf{i}_c[(k+1)T_S]$ (Fig. 4.48). The calculated voltage vector $\mathbf{u}_c(kT_S)$ is then implemented in the sinusoidal or space vector PWM algorithm (see Section 4.2). The predictive controller can be implemented in both stationary or synchronous coordinates [88, 94].

Discrete-Time Model of RLE Load. The three-phase ac side *RLE* load can be described by a space vector based voltage equation as

$$\mathbf{u} = R\mathbf{i} + L\frac{d\mathbf{i}}{dt} + \mathbf{e} \tag{4.90}$$

where

$\mathbf{u} = \frac{2}{3}[U_A(t) + \mathbf{a}U_B(t) + \mathbf{a}^2 U_C(t)]$, the voltage space vector

$\mathbf{i} = \frac{2}{3}[I_A(t) + \mathbf{a}I_B(t) + \mathbf{a}^2 I_C(t)]$, the current space vector

$\mathbf{e} = \frac{2}{3}[E_A(t) + \mathbf{a}E_B(t) + \mathbf{a}^2 E_C(t)]$, the EMF voltage space vector

$\mathbf{a} = e^{j2\pi/3}$, the complex unit vector

ABC are phase voltage, currents, and EMF, respectively.

Assuming that $\mathbf{u}$ and $\mathbf{e}$ are constant between sampling instants kT_S and $(k+1)T_S$, Eq. (4.90) can be discretized as follows:

$$\mathbf{i}[(k+1)T_S] = e^{-T_S/T_L}\mathbf{i}(kT_S) + \int_{kT_S}^{(k+1)T_S} e^{-[(k+1)T_S - t]/T_L} dt \cdot \frac{1}{L}(\mathbf{u} - \mathbf{e}) \tag{4.91}$$

where $T_L = L/R$.

Equation (4.91) can be written in the following discrete form

$$\frac{1}{\delta}\mathbf{i}(k+1) - \frac{\chi}{\delta}\mathbf{i}(k) = \mathbf{u}(k) - \mathbf{e}(k) \tag{4.92}$$

where $\delta = e^{-T_S/T_L}$, $\chi = \frac{1}{R}(1 - e^{-T_S/T_L})$.

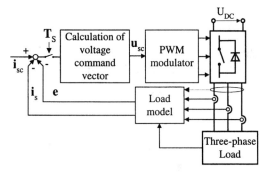

FIGURE 4.48
Constant switching frequency predictive current controller.

Discrete-Time Current Controller. With definition of the current error vector as

$$\Delta \mathbf{i} = \mathbf{i}_c(k) - \mathbf{i}(k) \tag{4.93}$$

where $\mathbf{i}_c(k)$ is the reference current vector, Eq. (4.92) can be expressed as

$$[\mathbf{i}_c(k+1) - \chi \mathbf{i}_c(k)] - \Delta \mathbf{i}(k+1) + \chi \Delta \mathbf{i}(k) = \delta[\mathbf{u}(k) - \mathbf{e}(k)]. \tag{4.94}$$

Introducing the commanded voltage vector $\mathbf{u}_c(k)$,

$$\mathbf{u}(k) = \mathbf{u}_c(k) + \Delta \mathbf{u}(k), \tag{4.95}$$

which can be expressed as

$$\mathbf{u}_c(k) = \frac{1}{\delta}[\mathbf{i}_c(k+1) - \chi \mathbf{i}_c(k)] + \mathbf{e}(k), \tag{4.96}$$

Eq. (4.94) simplifies to

$$\Delta \mathbf{i}(k+1) = \chi \Delta \mathbf{i}(k) - \delta \Delta \mathbf{u}(k) \tag{4.97}$$

Using the state feedback

$$\Delta \mathbf{u}(k) = K_1 \Delta \mathbf{i}(k) \tag{4.98}$$

one obtains from Eq. (4.97)

$$\Delta \mathbf{i}(k+1) = (\chi - \delta K_1) \Delta \mathbf{i}(k). \tag{4.99}$$

With the choice of the gain $K_1 = \chi/\delta$, the current error at the step $(k+1)$ will be zero (assuming that converter voltage $|\mathbf{u}(k)|$ is not saturated) and the predictive controller is called a *dead-beat controller*.

Current and EMF Prediction. It can be seen from Eq. (4.96) that calculation of the voltage command vector in step k requires the knowledge of the EMF vector $\mathbf{e}(k)$ and current vector commands $\mathbf{i}_c$ in steps k and $(k+1)$. Therefore, these quantities must be estimated or measured.

Based on the Lagrange interpolation formula,

$$\mathbf{i}_c(k+1) = \sum_{l=0}^{n} (-1)^{n-l} \binom{n+1}{l} \mathbf{i}_c(k+l-n), \tag{4.100}$$

one obtains one-step-ahead current vectors prediction as

$$\mathbf{i}_c(k+1) = 2\mathbf{i}_c(k) - \mathbf{i}_c(k-1) \tag{4.101a}$$

for $n = 1$, and

$$\mathbf{i}_c(k+1) = 3\mathbf{i}_c(k) - 3\mathbf{i}_c(k-1) + \mathbf{i}_c(k-2) \tag{4.101b}$$

for $n = 2$ (quadratic prediction).

Similarly, the predicted EMF value can be calculated ($n = 1$) as

$$\mathbf{e}(k+1) = 3\hat{\mathbf{e}}(k) - 3\hat{\mathbf{e}}(k-1) + \hat{\mathbf{e}}(k-2) \tag{4.102}$$

where $\mathbf{e}$ is the EMF vector estimated from Eq. (4.92),

$$\hat{\mathbf{e}}(k) = \mathbf{u}(k) - \frac{1}{\delta}[\mathbf{i}(k+1) - \chi \mathbf{i}(k)]. \tag{4.103}$$

It should be noted that prediction error increases with operating frequency of the controller. Therefore, for a wide range of frequency, at least quadratic prediction must be used.

Typical response of phase current with the predictive dead-beat controller is shown in Fig. 4.49. The following data were used in simulation: $R_A = 0.5$, $L_A = 10$ mH, $E_A = 220$ V, $U_{DC} = 560$ V, $T_S = 0.1$ ms, $f_S = 5$ kHz.

The predictive current controller is generally more difficult to implement and usually must be matched to a specific load parameters. Also, errors caused by computational delays and prediction create future problems in practical implementation, because there is no integration part in the basic algorithm. Therefore, many improved versions, which use observers or other compensation blocks, have been proposed [11, 83, 89, 95, 98].

4.3.4 Nonlinear On–Off Controllers

The on–off nonlinear current control group includes hysteresis, delta modulation, and *on-line* optimized controllers. To avoid confusion, current controllers for resonant dc link (RDCL) topology are presented in Chapter 2. Neural networks and fuzzy logic controllers which also belong to the nonlinear group, are presented in Chapter 10, Section 10.5.2.

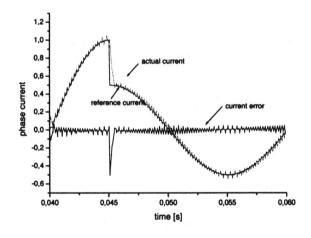

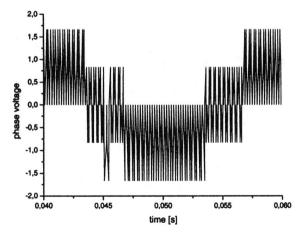

FIGURE 4.49
Simulated transients to the step change of reference current: $1 \rightarrow 0.5$ pu.

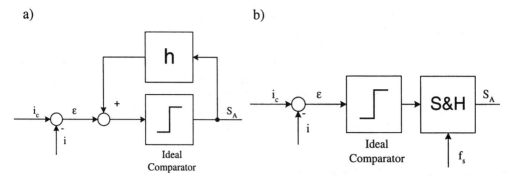

FIGURE 4.50
Two methods to limit the switching frequency of a current control system with an ideal comparator. (a) Hysteresis controller, (b) delta modulator.

4.3.4.1 Introduction.
Ideally impressed current in an inductive load could be implemented using an ideal comparator operated as an on–off controller. In such a system, however, the converter switching frequency will be infinity, and as a consequence of high switching losses, the semiconductor power devices will be damaged. Therefore, in practical schemes switching frequency is limited by introducing a hysteresis with width h or a sample and hold (S&H) block with sampling frequency f_S (Fig. 4.50). This creates two class of controllers which will be discussed in the next sections.

4.3.4.2 Hysteresis Current Controllers.
Hysteresis control schemes are based on a nonlinear feedback loop with two-level hysteresis comparators (Fig. 4.51a) [61]. The switching signals S_A, S_B, S_C are generated directly when the error exceeds an assigned tolerance band h (Fig. 4.51b).

Variable Switching Frequency Controllers. Among the main advantages of hysteresis CC are simplicity, outstanding robustness, lack of tracking errors, independence of load parameter changes, and extremely good dynamics limited only by switching speed and load time constant.

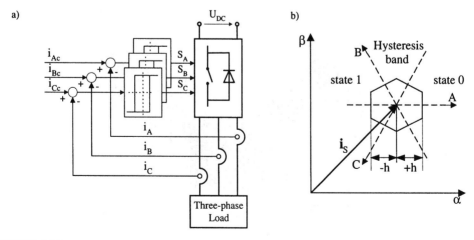

FIGURE 4.51
Two-level hysteresis controller: block scheme (a), switching trajectory (b).

However, this class of schemes, also known as free-running hysteresis controllers [16], has the following disadvantages:

- The converter switching frequency depends largely on the load parameters and varies with the ac voltage
- The operation is somewhat rough because of the inherent randomness caused by the limit cycle; therefore, protection of the converter is difficult [56]

It is characteristic of the hysteresis CC that the instantaneous current is kept exact in a tolerance band except for systems with isolated neutral where the instantaneous error can reach double the value of the hysteresis band [3, 54] (Fig. 4.52b). This is due to the interaction in the system with three independent controllers. The comparator state change in one phase influences the voltage applied to the load in two other phases (coupling). However, if all three current errors are considered as space vectors [60], the interaction effect can be compensated, and many variants of controllers known as *space vector based* can be created [41, 48, 50, 58, 63, 68]. Moreover, if three-level comparators with a lookup table are used, a considerable decrease in the inverter switching frequency can be achieved [37, 48, 50, 58, 63]. This is possible thanks to appropriate selection of zero voltage vectors [48] (Fig. 4.53).

In the synchronous rotating d–q coordinates, the error field is rectangular and the controller offers the opportunity of independent harmonic selection by choosing different hysteresis values for the d and q components [49, 62]. This can be used for torque ripple minimization in vector-controlled ac motor drives (the hysteresis band for the torque current component is set narrower than that for the flux current component) [49, 96].

Recent methods enable limit cycle suppression by introducing a suitable offset signal to either current references or the hysteresis band [45, 65, 67].

Constant Average Switching Frequency Controllers. A number of proposals have been put forward to overcome variable switching frequency. The tolerance band amplitude can be varied,

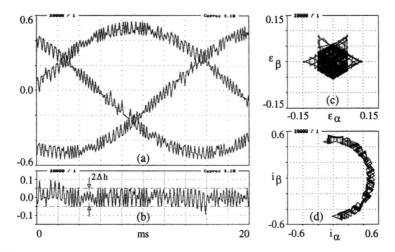

FIGURE 4.52
Hysteresis controller ($\Delta h = 0.05$): output currents (a), phase current error (b), vector current area (c), output vector current loci (d).

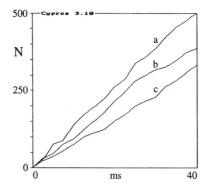

FIGURE 4.53
Number of inverter switchings N for (a) three two-level hysteresis comparators, (b) three-level comparators and lookup table working in the stationary, and (c) rotating coordinates.

according to the ac side voltage [39, 43, 47, 53–55, 57, 59, 69, 103], or by means of a PLL control (Fig. 4.54).

An approach which eliminates the interference, and its consequences, is that of decoupling error signals by subtracting an interference signal δ'' derived from the mean inverter voltage (Fig. 4.54) [54]. Similar results are obtained in the case of "discontinuous switching" operation, where decoupling is more easily obtained without estimating load impedance [55]. Once decoupled, regular operation is obtained and phase commutations may (but need not) be easily synchronized to a clock.

Although the constant switching frequency scheme is more complex and the main advantage of the basic hysteresis control—namely the simplicity—is lost, these solutions guarantee very fast response together with limited tracking error. Thus, constant frequency hysteresis controls are well suited for high-performance, high-speed applications.

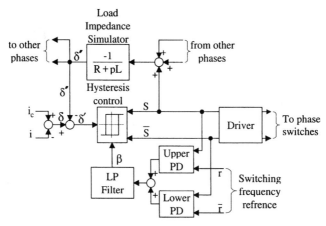

FIGURE 4.54
Decoupled, constant average switching frequency hysteresis controller [54].

4.3.4.3 Controllers with On-Line Optimization. This class of controllers performs a real-time optimization algorithm and requires complex on-line calculations, which can be implemented only on microprocessors.

Minimum Switching Frequency Predictive Algorithm. The concept of this algorithm [92] is based on space vector analysis of hysteresis controllers. The boundary delimiting the current error area in the case of independent controllers with equal tolerance band $+h$ in each of three phases makes a regular symmetrical hexagon (Fig. 4.51b). Suppose only one hysteresis controller is used—the one acting on the current error vector. In such a case, the boundary of the error area (also called the switching or error curve) might have any form (Fig. 4.55a). The location of the error curve is determined by the current command vector $\mathbf{i}_{SC}$. When the current vector $\mathbf{i}_S$ reaches a point on the error curve, seven different trajectories of the current are predicted, one for each of seven possible (six active and zero) inverter output voltage vectors. Finally, based on the optimization procedure, the voltage vector which minimizes the mean inverter switching frequency is selected. For fast transient states the strategy which minimizes the response time is applied.

Control with Field Orientation. The minimum frequency predictive CC can be implemented in any rotating or stationary coordinates. As with the three-level hysteresis controller working in d–q field oriented coordinates [49], a further switching frequency reduction can be achieved by the selection of a rectangular error curve with greater length along the rotor flux direction [96].

In practice, the time needed for the prediction and optimization procedures limits the achieved switching frequency. Therefore, in more recently developed algorithms, a reduced set of voltage vectors consisting of the two active vectors adjacent to the EMF vector and the zero voltage vector are considered for optimization without loss of quality [8].

Trajectory Tracking Control. This approach, proposed in [89, 90], combines an *off-line* optimized PWM pattern for steady-state operation with an *on-line* optimization to compensate for the dynamic tracking errors of converter currents. Such a strategy achieves very good stationary and dynamic behavior even for low switching frequencies.

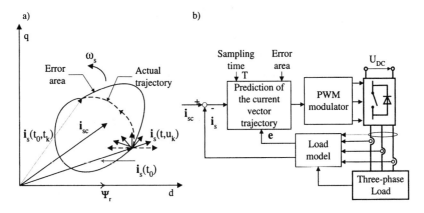

FIGURE 4.55
Minimum switching frequency predictive current controller. (a) Example of error area; (b) block scheme.

4.3.4.4 Delta Modulation (DM). The basic scheme, the *delta modulation current controller (DM-CC)* [74, 82], is shown in Fig. 4.56. It looks quite similar to that of a hysteresis CC (Fig. 4.51a), but the operating principle is quite different. In fact, only the error sign is detected by the comparators, whose outputs are sampled at a fixed rate so that the inverter status is kept constant during each sampling interval. Thus, no PWM is performed; only basic voltage vectors can be generated by the converter for a fixed time. This mode of operation gives a discretization of the inverter output voltage, unlike the continuous variation of output voltages which is a particular feature of PWM.

One effect of the discretization is that, when synthesizing periodic waveforms, a nonnegligible amount of subharmonics is generated [74, 76, 77]. Thus, to obtain comparable results, a DM should switch at a frequency about seven times higher than a PWM modulator [76]. However, DM is very simple and insensitive to the load parameters. When applied to three-phase inverters with an isolated neutral load, the mutual phase interference and the increased degree of freedom in the choice of voltage vector must be taken into account. Therefore, instead of performing independent DM in each phase control, output vectors are chosen depending not only on the error vector, but also on the previous status, so that the zero vector states become possible [73] (see Chapter 2).

Because of the S&H block applied after the ideal comparator, the switching frequency is limited to the sampling frequency f_s. The amplitude of the current harmonics is not constant but is determined by the load parameters, dc-link voltage, ac side voltage, and sampling frequency. The main advantages of DM-CC are extremely simple and tuning-free hardware implementation and good dynamics.

It is noted that the DM-CC can also be applied in the space vector based controllers working in either stationary or rotating coordinates [75, 79, 81].

Optimal Discrete Modulation Algorithm. See Chapter 10, Section 10.5.2.2.

4.3.4.5 Analog and Discrete Hysteresis. When the hysteresis controller is implemented in a digital signal processor (DSP), its operation is quite different from that in the analog scheme. Figure 4.57 illustrates typical switching sequences in analog (a) and discrete (b) implementations

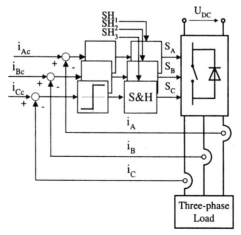

FIGURE 4.56
Delta modulation current controller: basic block scheme.

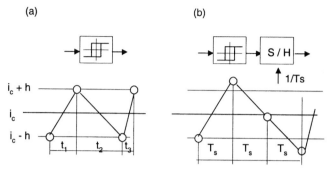

FIGURE 4.57
Operation of the analog (a) and discrete (b) hysteresis controller.

(also called sampled hysteresis). In the analog controller the current ripples are kept exactly within the hysteresis band and switching instances are not equal. In contrast, the discrete system operates at fixed sampling time T_s; however, the hysteresis controller is effective only if

$$h > \frac{di_{max}}{dt} \cdot T_s. \qquad (4.104)$$

Otherwise, the current ripple will be higher than specified by hysteresis band (as shown in Fig. 4.57b) and the controller operates rather like a delta modulator.

Figure 4.58 illustrates operation of different current controllers for the same number of switchings ($N = 26$). It is clearly seen that for hysteresis band $h = 2A$, the discrete controller (Fig. 4.58c) requires 3.3 μs sampling time (300 kHz) to exactly copy the continuous hysteresis behavior (Fig. 4.58b). With longer sampling time 330 μs (3 kHz), operation of a discrete hysteresis controller (Fig. 4.58d) is far from that of a continuous one (Fig. 4.58b).

4.4 CONCLUSIONS

Pulse width modulated (PWM) three-phase converters can operate under voltage (open loop) or current (closed loop) control. Generally, better performance and faster response is achieved in current-controlled (CC) rather than voltage-controlled systems. In ac motors, CC reduces the dependence on stator parameters and allows an immediate action on the flux and torque developed by the machine. In PWM rectifiers and active filters current must be regulated to obtain the desired active and reactive power and to minimize and/or compensate for line power factor and current harmonics.

Regarding the open-loop PWM, significant progress has been made in understanding two most commonly used methods: triangular carrier-based (CB) and space vector modulation (SVM). The degree of freedom represented in the selection of the zero sequence signal (ZSS) waveform in CB-PWM corresponds to different placement of zero vectors U_0 (000) and U_7 (111) in SVM. This important observation is used to optimize various performance factors of PWM converters resulting in different PWM algorithms. However, a single PWM method that satisfies all requirements in the full operation region of the converter does not exist; therefore, the concept of the adaptive SVM has been proposed. This concept combines the advantages of several PWM methods resulting in further reduction of switching losses.

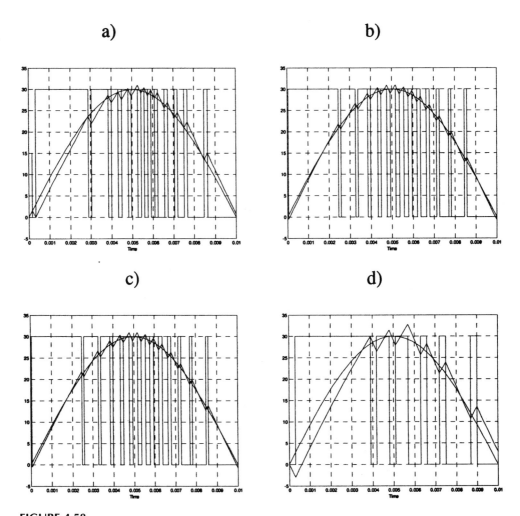

FIGURE 4.58
Current control with (a) delta modulator, $T_s = 167\,\mu s$; (b) continuous hysteresis, $h = 2A$; (c) discrete hysteresis, $h = 2A$, $T_s = 3.3\,\mu s$; (d) discrete hysteresis, $h = 2A$, $T_s = 330\,\mu s$.

Regarding closed-loop PWM current control, the main focus is put on schemes that include voltage modulators. They have clearly separated current error compensation and voltage modulation parts. This concept allows one to exploit the advantages of open-loop modulators: constant switching frequency, well-defined harmonic spectrum, optimal switching pattern, and good dc link utilization. Several linear control techniques with step-by-step examples have been presented. Also, a newly developed resonant controller that does not require coordinate transformations has been discussed. The nonlinear on–off controllers guarantee very fast response together with low tracking error; however, because of variable switching frequency, their application in commercial equipment is rather limited.

4.5 APPENDIX

The SIMULINK simulation panel was used in design examples shown in Figs. A.1–A.6.

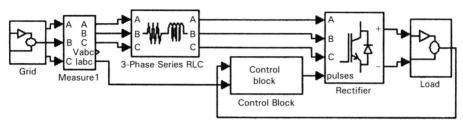

FIGURE A.1
General block diagram of three-phase PWM rectifier.

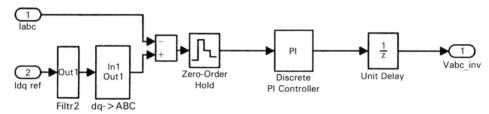

FIGURE A.2
Ramp comparison controller.

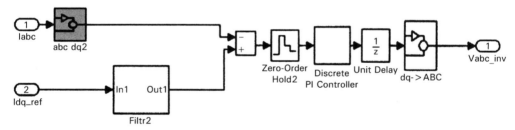

FIGURE A.3
PI synchronous controller.

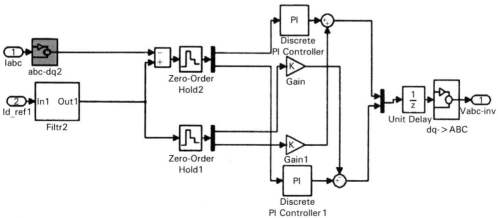

FIGURE A.4
Synchronous PI with decoupling outside of controller.

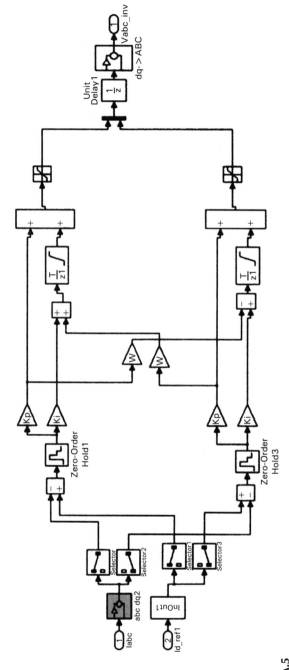

FIGURE A.5
Synchronous PI with new decoupling controller.

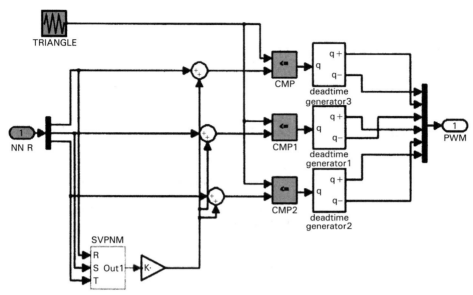

FIGURE A.6
PWM scheme.

REFERENCES

Books and Overview Papers

[1] B. K. Bose, *Power Electronics and Variable Frequency Drives*. IEEE Press, 1996.
[2] B. K. Bose, *Power Electronics and Electrical AC Drives*. Prentice Hall, Englewood Cliffs, NJ, 1986.
[3] D. M. Brod and D. W. Novotny, Current control of VSI-PWM inverters. *IEEE Trans. Indust. Appl.* **IA-21**, 562–570 (1985).
[4] H. W. van der Broeck, H. Ch. Skudelny, and G. Stanke, Analysis and realization of a pulse width modulator based on voltage space vectors. *IEEE Trans. Indust. Appl.* **24**, 142–150 (1988).
[5] H. Bühler, *Einführung in die Theorie geregelter Drehstrom-Antriebe*, Vols. 1, 2. Birkhauser, Basel, 1977.
[6] H. Ertl, J. W. Kolar, and F. C. Zach, Analysis of different current control concepts for forced commutated rectifier (FCR). In *PCI Conf. Proc.*, pp. 195–217, 1986.
[7] J. Holtz, W. Lotzkat, and A. M. Khambadadkone, On continuous control of PWM inverters in the overmodulation range including the six-step mode, in *IEEE Trans. Power Electron.*, **8**, 540–553 (1993).
[8] J. Holtz, Pulsewidth modulation for electronic power conversion. *Proc. IEEE* **82**, 1194–1214 (1994).
[9] F. Jenni and D. Wüst, *Steuerverfahren für selbstgeführte Stromrichter*. B.G. Teubner, Stuttgart, 1995.
[10] D. Jouve, J. P. Rognon, and D. Roye, Effective current and speed controllers for permanent magnet machines: A survey. In *IEEE-APEC Conf.*, pp. 384–393, 1990.
[11] M. P. Kazmierkowski and L. Malesani, Special section on PWM current regulation. *IEEE Trans. Indust. Electron.* **45**, 689–802 (1998).
[12] M. P. Kazmierkowski and H. Tunia, *Automatic Control of Converter-Fed Drives*. Elsevier, Amsterdam, 1994.
[13] D. C. Lee, S. K. Sul, and M. H. Park, Comparison of AC current regulators for IGBT inverter. *PCC'93 Conf. Rec.*, Yokohama, pp. 206–212, 1993.
[14] W. Leonhard, *Control of Electrical Drives*, 2nd ed. Springer Verlag, Berlin, 1996.
[15] L. Malesani and P. Tomasin, PWM current control techniques of voltage source converters—a survey. *IEEE IECON'93 Conf. Rec.*, Maui, Hawaii, pp. 670–675, 1993.
[16] J. D. M. Murphy and F. G. Turnbull, *Control Power Electronics of AC Motors*. Pergamon Press, 1988.

[17] D. W. Novotny and T. A. Lipo, *Vector Control and Dynamics of AC Drives*. Clarendon Press, Oxford, 1996.
[18] A. Schönung and H. Stemmler, Static frequency changers with subharmonic control in conjunction with reversible variable speed a.c. drives. *Brown Boweri Rev.* **51**, 555–577 (1964).
[19] A. M. Trzynadlowski, An overview of modern PWM techniques for three-phase voltage controlled, voltage-source inverters. In *Proc. IEEE-ISIE'96*, Warsaw, Poland, pp. 25–39, 1996.
[20] A. M. Trzynadlowski, *Introduction to Modern Power Electronics*, John Wiley, New York, 1998.

PI, Resonant, and State Feedback Controllers

[21] F. Briz, M. W. Degner, and R. D. Lorenz, Analysis and design of current regulators using complex vectors. *IEEE Trans. Indust. Appl.* **36**, 817–825 (2000).
[22] J. H. Choi and B. J. Kim, Improved digital control scheme of three phase UPS inverter using double control strategy. In *Proc. APEC*, 1997.
[23] J. W. Choi and S. K. Sul, New current control concept—Minimum time current control in 3-phase PWM converter. *IEEE PESC*, pp. 332–338, 1995.
[24] P. Enjeti, P. D. Ziogas, J. F. Lindsay, and M. H. Rashid, A novel current controlled PWM inverter for variable speed AC drives. *IEEE-IAS Conf. Rec.*, Denver, pp. 235–243, 1986.
[25] P. Feller, Speed control of an ac motor by state variables feedback with decoupling. In *Proc. IFAC on Control in Power Electronics and Electrical Drives*, Lausanne, pp. 87–93, 1983.
[26] L. Hernefors and H. P. Nee, Model-based current control of AC machines using the internal model control. *IEEE Trans. Indust. Appl.* **34**, 133–141 (1998).
[27] N. Hur, K. Nam, and S. Won, A two-degrees-of-freedom current control scheme for deadtime compensation. *IEEE Trans. Indust. Electron.* **47**, 557–564 (2000).
[28] D. C. Lee, S. K. Sul, and M. H. Park, High performance current regulator for a field-oriented controlled induction motor drive. *IEEE Trans. Indust. Appl.* **30**, 1247–1257 (1994).
[29] R. D. Lorenz and D. B. Lawson, Performance of feedforward current regulators for field oriented induction machine controllers. *IEEE Trans. Indust. Appl.* **IA-23**, 537–662 (1987).
[30] J. Moerschel, Signal processor based field oriented vector control for an induction motor drive. In *Proc. EPE-Conference*, Firenze, Italy, pp. 2.145–2.150, 1991.
[31] L. Norum, W. Sulkowski, and L. A. Aga, Compact realisation of PWM-VSI current controller for PMSM drive application using low cost standard microcontroller. In *IEEE-PESC Conf. Rec.*, Toledo, pp. 680–685, 1992.
[32] C. T. Rim, N. S. Choi, G. C. Cho, and G. H. Cho, A complete DC and AC analysis of three-phase controlled-current PWM rectifier using circuit D-Q transformation. *IEEE Trans. Power Electron.* **9** 390–396 (1994).
[33] T. M. Rowan and R. J. Kerkman, A new synchronous current regulator and an analysis of current regulated PWM inverters. *IEEE Trans. Indust. Appl.* **IA-22**, 678–690 (1986).
[34] D. Schauder and R. Caddy, Current control of voltage-source inverters for fast four-quadrant drive performance. *IEEE Trans. Indust. Appl.* **IA-18**, 163–171 (1982).
[35] Y. Sato, T. Ishizuka, K. Nezu, and T. Kataoka, A new control strategy for voltage-type PWM rectifiers to realize zero steady-state control error in input current. *IEEE Trans. Indust. Appl.* **34**, 480–486 (1998).
[36] D. N. Zmood, D. G. Holmes, and D. H. Bode, Frequency-domain analysis of three-phase linear current regulators. *IEEE Trans. Indust. Appl.* **37**, 601–610 (2001).

Hysteresis and Sliding Mode Controllers

[37] A. Ackva, H. Reinold, and R. Olesinski, A simple and self-adapting high-performance current control scheme for three-phase voltage source inverters. *IEEE PESC'92 Conf. Rec.*, Toledo, pp. 435–442, 1992.
[38] C. Andrieux and M. Lajoie-Mazenc, Analysis of different current control systems for inverter-fed synchronous machine. *EPE Conf. Rec.*, Brussels, pp. 2.159–2.165, 1985.
[39] B. K. Bose, An adaptive hysteresis-band current control technique of a voltage-fed PWM inverted for machine drive system. *IEEE Trans. Indust. Electron.* **37**, 402–408 (1990).
[40] M. Carpita and M. Marchesoni, Experimental study of a power conditioning system using sliding mode control. *IEEE Trans. Power Electron.*, **11**, 731–742 (1996).

REFERENCES

[41] T. Y. Chang and T. C. Pan, A practical vector control algorithm for μ-based induction motor drives using a new space vector controller. *IEEE Trans. Indust. Electron.* **41**, 97–103 (1994).

[42] C. Chiarelli, L. Malesani, S. Pirondini, and P. Tomasin, Single-phase, three-level, constant frequency current hysteresis control for UPS applications. *EPE'93 Conf. Rec.*, Brighton, pp. 180–185, 1993.

[43] T. W. Chun and M. K. Choi, Development of adaptive hysteresis band current control strategy of pwm inverter with constant switching frequency. *IEEE-APEC*, San José, pp. 194–199, 1996.

[44] E. Gaio, R. Piovan, and L. Malesani, Comparative analysis of hysteresis modulation methods for VSI current control. In *Proc. IEE Machines and Drives Conference* (London), pp. 336–339, 1988.

[45] V. J. Gosbell and P. M. Dalton, Current control of induction motors at low speed. *IEEE Trans. Indust. Appl.* **28**, 482–489 (1992).

[46] S. L. Jung and Y. Y. Tzou, Sliding mode control of a closed-loop regulated PWM inverter under large load variations. In *Proc. IEEE-PESC*, 1993.

[47] A. Kawamura and R. G. Hoft, Instantaneous feedback controlled PWM inverters with adaptive hysteresis. *IEEE Trans. Indust. Appl.* **IA-20**, 769–775 (1984).

[48] M. P. Kazmierkowski, M. A. Dzieniakowski, and W. Sulkowski, Novel space vector based current controllers for PWM-inverters. *IEEE Trans. Power Electron.* **6**, 158–166 (1991).

[49] M. P. Kazmierkowski and W. Sulkowski, A novel vector control scheme for transistor PWM inverted-fed induction motor drive. *IEEE Trans. Indust. Electron.* **38**, 41–47 (1991).

[50] B.-H. Kwon, T.-W. Kim, and J.-H. Youn, A novel SVM-based hysteresis current controller. *IEEE Trans. Power Electron.* **13**, 297–307 (1998).

[51] M. Lajoie-Mazenc, C. Villanueva, and J. Hector, Study and implementation of hysteresis control inverter on a permanent magnet synchronous machine. *IEEE/IAS Ann. Mtg., Conf. Rec.*, Chicago, pp. 426–431, 1984.

[52] L. Malesani, P. Mattavelli, and P. Tomasini, High-performance hysteresis modulation technique for active filters. *IEEE-APEC'96 Conf.*, pp. 939–946, 1996.

[53] L. Malesani, L. Rossetto, L. Sonaglioni, P. Tomasini, and A. Zuccato, Digital, adaptive hysteresis current control with clocked commutations and wide operating range. *IEEE Trans. Indust. Appl.* **32**, 1115–1121 (1996).

[54] L. Malesani and P. Tenti, A novel hysteresis control method for current controlled VSI PWM inverters with constant modulation frequency. *IEEE Trans. Indust. Appl.* **26**, 88–92 (1990).

[55] L. Malesani, P. Tenti, E. Gaio, and R. Piovan, Improved current control technique of VSI PWM inverters with constant modulation frequency and extended voltage range. *IEEE Trans. Indust. Appl.* **27**, 365–369 (1991).

[56] W. McMurray, Modulation of the chopping frequency in dc choppers and inverters having current hysteresis controllers. *IEEE Trans. Indust. Appl.* **IA-20**, 763–768 (1984).

[57] I. Nagy, Novel adaptive tolerance band based PWM for field oriented control of induction machines. *IEEE Trans. Indust. Electron.* **41**, 406–417 (1994).

[58] C. T. Pan and T. Y. Chang, An improved hysteresis current controller for reducing switching frequency. *IEEE Trans. Power Electron.* **9**, 97–104 (1994).

[59] E. Perssen, N. Mohan, and B. Ben Banerjee, Adaptive tolerance-band control of standby power supply provides load-current harmonic neutralization. In *IEEE-PESC Conf. Rec.* (Toledo, Spain), pp. 320–326, 1992.

[60] G. Pfaff, A. Weschta, and A. Wick, Design and experimental results of a brushless ac servo drive. *IEEE Trans. Indust. Appl.* **IA-22**, 814–821 (1984).

[61] A. B. Plunkett, A current controlled PWM transistor inverted drive. In *IEEE-IAS, Ann. Mtg., Conf. Rec.*, pp. 785–792, 1979.

[62] J. Rodriguez and G. Kästner, Nonlinear current control of an inverted-fed induction machine. *Etz-Archiv* **9**, 245–250 (1987).

[63] C. Rossi and A. Tonielli, Robust current controller for three-phase inverter using finite-state automation. *IEEE Trans. Indust. Electron.* **42**, 169–178 (1995).

[64] N. Sabanovic-Behlilovic, T. Ninomiya, A. Sabanovic, and B. Perunicic, Control of three-phase switching converters, a sliding mode approach. In *Proc. IEEE-PESC*, pp. 630–635, 1993.

[65] S. Salama and S. Lennon, Overshoot and limit cycle free current control method for PWM inverter. In *Proc. EPE'91*, Firenze, pp. 3.247–3.251, 1991.

[66] A. Tripathi and P. C. Sen, Comparative analysis of fixed and sinusoidal band hysteresis current controllers for voltage source inverters. *IEEE Trans. Indust. Electron.* **39**, 63–73 (1992).

[67] K. Tungpimolrut, M. Matsui, and T. Fukao, A simple limit cycle suppression scheme for hysteresis current controlled PWM-VSI with consideration of switching delay time. *IEEE/IAS Ann. Mtg., Conf. Rec.*, Houston, pp. 1034–1041, 1992.

[68] D. Wüst and F. Jenni, Space vector based current control schemes for voltage source inverters. *IEEE PESC'93 Conf. Rec.*, Seattle, pp. 986–992, 1993.
[69] Q. Yao and D. G. Holmes, A simple, novel method for variable-hysteresis-band current control of a three phase inverter with constant switching frequency. *IEEE/IAS Ann. Mtg. Conf. Rec.*, pp. 1122–1129, 1993.

Delta Modulation Controllers

[70] D. M. Divan, G. Venkataramanan, L. Malesani, and V. Toigo, Control strategies for synchronized resonant link inverters. *IPEC'90, Conf. Rec.*, Tokyo, pp. 338–345, 1990.
[71] M. A. Dzieniakowski and M. P. Kazmierkowski, Microprocessor-based novel current regulator for VSI-PWM inverters. *IEEE/PESC Conf. Rec.*, Toledo, pp. 459–464, 1992.
[72] P. Freere, D. Atkinson, and P. Pillay, Delta current control for vector controlled permanent magnet synchronous motors. *IEEE-IAS'92 Conf. Rec.*, Houston, pp. 550–557, 1992.
[73] T. G. Habetler and D. M. Divan, Performance characterization of a new discrete pulse modulated current regulator. *IEEE IAS'88 Conf. Rec.*, Pittsburgh, pp. 395–405, 1988.
[74] M. Kheraluwala and D. M. Divan, Delta modulation strategies for resonant link inverters. *IEEE PESC'87 Conf. Rec.*, pp. 271–278, 1987.
[75] R. D. Lorenz and D. M. Divan, Dynamic analysis and experimental evaluation of delta modulators for field oriented ac machine current regulators. *IEEE-IAS'87 Conf. Rec.*, Atlanta, pp. 196–201, 1987.
[76] A. Mertens, Performance analysis of three phase inverters controlled by synchronous delta-modulation systems. *IEEE Trans. Indust. Appl.* **30**, 1016–1027 (1994).
[77] A. Mertens and H. Ch. Skudelny, Calculations on the spectral performance of sigma delta modulators. *IEEE-PESC'91 Conf. Rec.*, Cambridge, 357–365, 1991.
[78] M. A. Rahman, J. E. Quaice, and M. A. Chowdry, Performance analysis of delta modulation PWM inverters. *IEEE Trans. Power Electron.* **2**, 227–233 (1987).
[79] G. Venkataramanan and D. M. Divan, Improved performance voltage and current regulators using discrete pulse modulation. *IEEE-PESC Conf. Rec.*, Toledo, 1992, pp. 601–606, 1991.
[80] G. Venkataramanan, D. M. Divan, and T. M. Jahns, Discrete pulse modulation stategies for high frequency inverter systems, *IEEE-PESC Conf. Rec.*, pp. 1013–1020, 1989.
[81] X. Xu and D. W. Novotny, Bus utilisation of discrete CRPWM inverters for field oriented drives. In *IEEE-IAS Ann. Mtg., Conf. Rec.*, pp. 362–367, 1988.
[82] D. Ziogas, The delta modulation technique in static PWM inverters. *IEEE Trans. Indust. Appl.* **IA-17**, 199–204 (1982).

Predictive and On-Line Optimized Controllers

[83] L. Ben-Brahim and A. Kawamura, Digital current regulation of field-oriented controlled induction motor based on predictive flux observer. In *IEEE IAS Ann. Mtg. Conf. Rec.*, pp. 607–612, 1990.
[84] L. J. Borle and C. V. Nayar, Zero average current error controlled power flow for AC-DC power converters. *IEEE Trans. Power Electron.* **10**, 725–732 (1996).
[85] K. P. Gokhale, A. Kawamura, and R. G. Hoft, Dead beat microprocessor control of PWM inverter for sinusoidal output waveform synthesis. *IEEE Trans. Indust. Appl.*, **IA-23**, 901–909 (1987).
[86] T. G. Habetler, A space vector based rectifier regulator for AC/DC/AC converters. In *Proc. EPE Conf.*, Firenze, pp. 2.101–2.107, 1991.
[87] W. Hofmann, Practical design of the current error trajectory control for PWM AC-drives. In *Proc. IEEE-APEC*, pp. 782–787, 1996.
[88] D. G. Holmes and D. A. Martin, Implementation of a direct digital predictive current controller for single and three phase voltage source inverters. *IEEE-IAS Ann. Mtg.*, San Diego, pp. 906–913, 1996.
[89] J. Holtz and B. Bayer, Fast current trajectory tracking control based on synchronous optimal pulsewidth modulation. *IEEE Trans. Indust. Appl.* **31**, 1110–1120 (1995).
[90] J. Holtz and B. Bayer, The trajectory tracking approach—a new method for minimum distortion PWM in dynamic high power drives. *IEEE Trans. Indust. Appl.* **30**, 1048–1057 (1994).
[91] J. Holtz and E. Bube, Field oriented asynchronous pulsewidth modulation for high performance ac machine drives operating at low switching frequency. *IEEE Trans. Indust. Appl.* **IA-27**, 574–581 (1991).

[92] J. Holtz and S. Stadtfeld, A predictive controller for the stator current vector of ac machines fed from a switched voltage source. In *Proc. IPEC*, Tokyo, pp. 1665–1675, 1983.

[93] M. Kassas, M. Wells, and M. Fashoro, Design and simulation of current regulators for induction motors using the error magnitude voltage vector correction (EMVVC). In *IEEE-IAS Ann. Mtg., Conf. Rec.*, pp. 132–138, 1992.

[94] T. Kawabata, T. Miyashita, and Y. Yamamoto, Dead beat control of three phase PWM inverter. *IEEE Trans. Power Electron.* **5**, 21–28 (1990).

[95] A. Kawamura, T. Haneyoshi, and R. G. Hoft, Deadbeat controlled PWM inverter with parameter estimation using only voltage sensor. *IEEE-PESC, Conf. Rec.*, pp. 576–583, 1986.

[96] A. Khambadkone and J. Holtz, Low switching frequency high-power inverter drive based on field-oriented pulse width modulation. *EPE Conf.*, pp. 4.672–677, 1991.

[97] J. W. Kolar, H. Ertl, and F. C. Zach, Analysis of on- and off-line optimized predictive current controllers for PWM converter system. *IEEE Trans. Power Electron.* **6**, 454–462, 1991.

[98] O. Kukrer, Discrete-time current control of voltage-fed three-phase PWM inverters. *IEEE Trans. Power Electron.* **11**, 260–269 (1996).

[99] H. Le-Huy and L. Dessaint, An adaptive current control scheme for PWM synchronous motor drives: analysis and simulation. *IEEE Trans. Power Electron.* **4**, 486–495 (1989).

[100] H. Le-Huy, K. Slimani, and P. Viarouge, Analysis and implementation of a real-time predictive current controller for permanent-magnet synchronous servo drives. *IEEE Trans. Indust. Electron.* **41**, 110–117 (1994).

[101] I. Miki, O. Nakao, and S. Nishiyma, A new simplified current control method for field oriented induction motor drives. In *IEEE-IAS Ann. Mtg., Conf. Rec.*, pp. 390–395, 1989.

[102] H. R. Mayer and G. Pfaff, Direct control of induction motor currents—design and experimental results. *EPE Conf.*, Brussels, pp. 3.7–3.12, 1985.

[103] A. Nabae, S. Ogasawara, and H. Akagi, A novel control scheme of current-controlled PWM inverters. *IEEE Trans. Indust. Appl.* **IA-2**, 697–701 (1986).

[104] D. S. Oh, K. Y. Cho, and M. J. Youn, Discretized current control technique with delayed voltage feedback for a voltage-fed PWM inverter. *IEEE Trans. Power Electron.* **7**, 364–373 (1992).

[105] G. Pfaff and A. Wick, Direct current control of ac drives with pulsed frequency converters. *Process Automat.* **2**, 83–88 (1983).

[106] S. K. Sul, B. H. Kwon, J. K. Kang, K. J. Lim, and M. H. Park, Design of an optimal discrete current regulator. In *IEEE-IAS Ann. Mtg., Conf. Rec.*, pp. 348–354, 1989.

[107] R. Wu, S. B. Dewan, and G. R. Slemon, A PWM ac-dc converter with fixed switching frequency. *IEEE Trans. Indust. Appl.* **26**, 880–885 (1990).

[108] R. Wu, S. B. Dewan, and G. R. Slemon, Analysis of a PWM ac to dc voltage source converter under the predicted current control with a fixed switching frequency. *IEEE Trans. Indust. Appl.* **27**, 756–764 (1991).

[109] L. Zhang and F. Hardan, Vector controlled VSI-fed AC drive using a predictive space-vector current regulation scheme. *IEEE-IECON*, pp. 61–66, 1994.

Open-Loop PWM

[110] V. Blasko, Analysis of a hybrid PWM based on modified space-vector and triangle-comparison methods. *IEEE Trans. Indust. Appl.* **33**, 756–764 (1997).

[111] S. R. Bowes and Y. S. Lai, Relationship between space-vector modulation and regular-sampled PWM. *IEEE Trans. Indust. Electron.* **44**, 670–679 (1997).

[112] G. Buja and G. Indri, Improvement of pulse width modulation techniques. *Archiv für Elektrotechnik* **57**, 281–289 (1975).

[113] D. W. Chung, J. Kim, and S. K. Sul, Unified voltage modulation technique for real-time three-phase power conversion. *IEEE Trans. Indust. Appl.* **34**, 374–380 (1998).

[114] A. Diaz and E. G. Strangas, A novel wide range pulse width overmodulation method. In *Proc. IEEE-APEC Conf.*, pp. 556–561, 2000.

[115] S. Fukuda and K. Suzuki, Using harmonic distortion determining factor for harmonic evaluation of carrier-based PWM methods. In *Proc. IEEE-IAS Conf.*, pp. 1534–1542, New Orleans, 2000.

[116] A. Haras and D. Roye, Vector PWM modulator with continuous transition to the six-step mode. In *Proc. EPE Conf.*, Sevilla, pp. 1.729–1.734, 1995.

[117] A. M. Hava, R. J. Kerkman, and T. A. Lipo, A high performance generalized discontinuous PWM algorithm. In *Proc. IEEE-APEC Conf.*, Atlanta, pp. 886–894, 1997.

[118] A. M. Hava, S. K. Sul, R. J. Kerman, and T. A. Lipo, Dynamic overmodulation characteristic of triangle intersection PWM methods. In *Proc. IEEE-IAS Conf.*, New Orleans, pp. 1520–1527, 1997.
[119] A. M. Hava, R. J. Kerman, and T. A. Lipo, Simple analytical and graphical tools for carrier based PWM methods. In *Proc. IEEE-PESC Conf.*, pp. 1462–1471, 1997.
[120] J. Holtz, W. Lotzkat, and A. Khambadkone, On continuous control of PWM inverters in the overmodulation range including the six-step mode. *IEEE Trans. Power Electron.* **8**, 546–553 (1993).
[121] F. Jenni and D. Wuest, The optimization parameters of space vector modulation. In *Proc. EPE Conf.*, pp. 376–381, 1993.
[122] R. J. Kerkman, Twenty years of PWM AC drives: When secondary issues become primary concerns. In *Proc. IEEE International Conference on Industrial Electronics, Control, and Instrumentation*, Vol. 1, pp. LVII–LXIII, 1996.
[123] J. W. Kolar, H. Ertl, and F. C. Zach, Influence of the modulation method on the conduction and switching losses of a PWM converter system. *IEEE Trans. Indust. Appl.*, **27**, 1063–1075 (1991).
[124] Y. S. Lai and S. R. Bowes, A universal space vector modulation strategy based on regular-sampled pulse width modulation. In *Proc. IEEE-IECON Conf.*, pp. 120–126 (1996).
[125] D. C. Lee and G. M. Lee, A novel overmodulation technique for space-vector PWM inverters. *IEEE Trans. Power Electron.* **13**, 1144–1151 (1998).
[126] M. Malinowski, Adaptive modulator for three-phase PWM rectifier/inverter. In *Proc. EPE-PEMC Conf.*, Kosice, pp. 1.35–1.41, 2000.
[127] H. van der Broeck, Analysis of the harmonics in voltage fed inverter drives caused by PWM schemes with discontinuous switching operation. In *Proc. EPE Conf.*, pp. 261–266, 1991.

Random PWM

[128] T. G. Habetler and D. M. Divan, Acoustic noise reduction in sinusoidal PWM drives using a randomly modulated carrier. In *Proc. IEEE Power Electronics Specialists Conference*, Vol. 2, pp. 665–671, 1989.
[129] J. K. Pedersen, F. Blaabjerg, and P. S. Frederiksen, Reduction of acoustical noise emission in AC-machines by intelligent distributed random modulation. In *Proc. EPE*, **4**, 369–375 (1993).
[130] G. A. Covic and J. T. Boys, Noise quieting with random PWM AC drives. *IEE Proc. Electric Power Appl.* **145**, 1–10 (1998).
[131] P. G. Handley, M. Johnson, and J. T. Boys, Elimination of tonal acoustic noise in chopper-controlled DC drives. *Applied Acoust.* **32**, 107–119 (1991).
[132] J. Holtz and L. Springob, Reduced harmonics PWM controlled line-side converter for electric drives. In *Proc. IEEE IAS Annual Meeting*, Vol. 2, pp. 959–964, 1990.
[133] T. Tanaka, T. Ninomiya, and K. Harada, Random-switching control in DC-to-DC converters. In *Proc. IEEE PESC*, Vol. 1, pp. 500–507, 1989.
[134] A. M. Stankovic, Random pulse modulation with applications to power electronic converters, Ph.D. thesis, Massachusetts Institute of Technology, Feb. 1993.
[135] D. C. Hamill, J. H. B. Deane, and P. J. Aston, Some applications of chaos in power converters. *Proc. IEE Colloquium on Update on New Power Electronic Techniques*, pp. 5/1–5/5, May 1997.
[136] M. Kuisma, P. Silventoinen, T. Järveläinen and T. Vesterinen, Effects of nonperiodic and chaotic switching on the conducted EMI emissions of switch mode power supplies. In *Proc. IEEE Nordic Workshop on Power and Industrial Electronics*, pp. 185–190, 2000.
[137] S. Legowski, J. Bei, and A. M. Trzynadlowski, Analysis and implementation of a grey-noise PWM technique based on voltage space vectors. In *Proc. IEEE APEC*, pp. 586–593, 1992.
[138] R. L. Kirlin, S. Kwok, S. Legowski, and A. M. Trzynadlowski, Power spectra of a PWM inverter with randomized pulse position. *IEEE Trans. Power Electron.* **9**, 463–472 (1994).
[139] Kone Osakeyhtiö, Förfarande och anordning för minskning av bullerolägenheterna vid en med chopperprincip matad elmotor (Method and apparatus for reduction of noise from chopper-fed electrical machines). Finnish Patent Application No. 861,891, filed May 6, 1986.
[140] A. M. Trzynadlowski, F. Blaabjerg, J. K. Pedersen, R. L. Kirlin, and S. Legowski, Random pulse width modulation techniques for converter fed drive systems—A review. *IEEE Trans. Indust. Appl.* **30**, 1166–1175 (1994).
[141] A. M. Trzynadlowski, R. L. Kirlin, and S. Legowski, Space vector PWM technique with minimum switching losses and a variable pulse rate. *Proc. of the 19th IEEE International Conference on Industrial Electronics, Control, and Instrumentation*, Vol. 2, pp. 689–694, 1993.
[142] M. M. Bech, Random pulse-width modulation techniques for power electronic converters, Ph.D. thesis, Aalborg University, Denmark, Aug. 2000.

CHAPTER 5

Control of PWM Inverter-Fed Induction Motors

MARIAN P. KAZMIERKOWSKI
Warsaw University of Technology, Warsaw, Poland

5.1 OVERVIEW

The induction motor, thanks to its well-known advantages of simple construction, reliability, ruggedness, and low cost, has found very wide industrial applications. Furthermore, in contrast to the commutation dc motor, it can be used in aggressive or volatile environments since there are no problems with sparks and corrosion. These advantages, however, are offset by control problems when using induction motors in speed regulated industrial drives.

The most popular high-performance induction motor control method, known as *field oriented control* (FOC) or *vector control*, has been proposed by Hasse [28] and Blaschke [22] and has constantly been developed and improved by other researchers [2, 4, 7–9, 13, 16–21, 25–46]. In this method the motor equation are (rewritten) transformed in a coordinate system that rotates with the rotor (stator) flux vector. These new coordinates are called *field coordinates*. In field coordinates—for the constant rotor flux amplitude—there is a *linear* relationship between control variables and speed. Moreover, as in a separately excited dc motor, the reference for the flux amplitude can be reduced in the field weakening region in order to limit the stator voltage at high speed.

Transformation of the induction motor equations in field coordinates has a good physical basis because it corresponds to the decoupled torque production in a separately excited dc motor. However, from the theoretical point of view other types of coordinates can be selected to achieve decoupling and linearization of the induction motor equations. That creates a basis for methods known as modern nonlinear control [5, 14, 64]. Marino *et al.* [58–60] have proposed a nonlinear transformation of the motor state variables, so that in the new coordinates, the speed and rotor flux amplitude are decoupled by feedback. This method is called *feedback linearization control* (FLC) or *input–output decoupling* [51–53, 62, 63]. A similar approach, based on a *multiscalar* model of the induction motor, has been proposed by Krzeminski [57].

162 CHAPTER 5 / CONTROL OF PWM: INVERTER-FED INDUCTION MOTORS

An approach based on the *variation* theory and energy shaping has been investigated and is called *passivity-based control* (PBC) [15]. In this case the induction motor is described in terms of the Euler–Lagrange equations expressed in generalized coordinates. When, in the mid-1980s, it appeared that control systems would be standardized on the basis of the FOC philosophy, there appeared the innovative studies of Depenbrock and Takahashi and Nogouchi (see [3] and [4] in Chapter 9), which depart from idea of coordinate transformation and the analogy with dc motor control. These innovators propose to replace motor decoupling via nonlinear coordinate transformation with bang-bang self-control, which goes together very well with ON–OFF operation of inverter semiconductor power devices. This control strategy is commonly referred as *direct torque control* (DTC) and is presented at length in Chapter 9.

5.2 BASIC THEORY OF INDUCTION MOTOR

5.2.1 Space Vector Based Equations in per Unit System

Mathematical description of the induction motor (IM) is based on complex space vectors, which are defined in a coordinate system rotating with angular speed ω_K. In per-unit and real-time representation the following equations describe the behavior of the squirrel-cage motor [7]:

$$\mathbf{u}_{sK} = r_s \mathbf{i}_{sK} + T_N \frac{d\mathbf{\psi}_{sK}}{dt} + j\omega_K \mathbf{\psi}_{sK} \tag{5.1}$$

$$0 = r_r \mathbf{i}_{rK} + T_N \frac{d\mathbf{\psi}_{rK}}{dt} + j(\omega_K - \omega_m)\mathbf{\psi}_{rK} \tag{5.2}$$

$$\mathbf{\psi}_{sK} = x_s \mathbf{i}_{sK} + x_M \mathbf{i}_{rK} \tag{5.3}$$

$$\mathbf{\psi}_{rK} = x_r \mathbf{i}_{rK} + x_M \mathbf{i}_{sK} \tag{5.4}$$

$$\frac{d\omega_m}{dt} = \frac{1}{T_M}[m - m_L] \tag{5.5}$$

$$m = \text{Im}(\mathbf{\psi}_{sK}^* i_{sK}). \tag{5.6}$$

Remarks:

- The stator and rotor quantities appearing in Eqs. (5.1)–(5.4) are complex space vectors represented in the common reference frame rotating with angular speed ω_K (hence the indices at these quantities); the way they are related to the natural components of a three phase IM can be represented (e.g., or currents) by

$$\mathbf{i}_{sK} = \tfrac{2}{3}[1 i_A(t) + \mathbf{a} i_B(t) + \mathbf{a}^2 i_C(t)] \cdot e^{-j\omega_K t} \tag{5.7a}$$

$$\mathbf{i}_{rK} = \tfrac{2}{3}[1 i_A(t) + \mathbf{a} i_B(t) + \mathbf{a}^2 i_C(t)] \cdot e^{-j(\omega_K - \omega_m)t} \tag{5.7b}$$

where i_A, i_B, i_C are instantaneous per unit values of the stator winding currents, and i_a, i_b, i_c the instantaneous per unit values of rotor winding currents referred to the stator circuit. Similar formulae hold for voltages $\mathbf{u}_{sK}$ and $\mathbf{u}_{rK}$, and for the flux linkages $\mathbf{\psi}_{sK}, \mathbf{\psi}_{rK}$.
- The motion Equation (5.5) is a real equation.
- Application of p.u. system results in:

  In the voltage equations, the factor $T_N = 1/\Omega_{SN}$ appears next to the flux linkage derivatives, which results from the real-time representation.
  In view of the identity $l = x$, the reactances are used in the flux–current equations.

The factor 3/2 and the number of pole pairs p_b have both disappeared from the electromagnetic torque equation; also, the p.u. shaft speed ω_m is independent of p_b. In the motion equation, the mechanical time constant appears as $T_M = J\Omega_{mN}/M_N$.

- Thanks to the transformation of the equations to a common reference frame, the IM parameters can be regarded as independent of rotor position.
- The electromagnetic torque formula (5.6) is independent of the choice of coordinate system, which the space vectors are represented. This is because for any coordinate system

$$\mathbf{\psi}_{sK} = \mathbf{\psi}_s e^{-j\omega_K t}, \qquad \mathbf{i}_{sK} = \mathbf{i}_s e^{-j\omega_K t}. \tag{5.8}$$

Including Eq. (5.8) in the electromagnetic torque formula (5.6) one obtains

$$m = \mathrm{Im}(\mathbf{\psi}_{sK}^* \mathbf{i}_{sK}) = \mathrm{Im}(\mathbf{\psi}_s^* e^{j\omega_K t} \cdot \mathbf{i}_s e^{-j\omega_K t}) = \mathrm{Im}(\mathbf{\psi}_s^* \mathbf{i}_s). \tag{5.9}$$

- Owing to the use of complex space vectors, and assuming that symmetric sinewaves are involved, it is possible to employ the symbolic method going over to the steady state, and thus to obtain a convenient bridge to the classical theory of IM.

5.2.2 Block Diagrams

The relation described by Eqs. (5.1)–(5.6) can be illustrated as block diagrams in terms of space vectors in complex form [4, 30] or, following resolution into two-axis components, in real form [7, 9]. When resolving vector equations, one may, in view of the motor symmetry, adopt an arbitrary coordinate reference frame. Moreover, taking advantage of the linear dependency between flux linkages and currents, the electromagnetic torque expression can also be written in a number of ways. It follows *that there is not just one block diagram of an IM*, but instead on the basis of the set of vector equations (5.1)–(5.6), one may construct various versions of such a diagram [7]. In going over to the two-axis model, essential differences between the two models depends on:

- Speed and position of reference coordinates
- Input signals
- Output signals

By the way of illustration we should consider two examples.

Example 5.1: Voltage Controlled IM in Stator—Fixed System of Coordinates (α, β, 0)

The popular induction-cage motor representation is based on a fixed coordinate system ($\omega_K = 0$), in which the complex state-space vectors can be resolved into components α and β:

$$\mathbf{u}_s = u_{s\alpha} + ju_{s\beta} \tag{5.10a}$$
$$\mathbf{i}_s = i_{s\alpha} + ji_{s\beta} \tag{5.10b}$$
$$\mathbf{i}_r = i_{r\alpha} + ji_{r\beta} \tag{5.10c}$$
$$\mathbf{\psi}_s = \psi_{s\alpha} + j\psi_{s\beta} \tag{5.10d}$$
$$\mathbf{\psi}_r = \psi_{r\alpha} + j\psi_{r\beta}. \tag{5.10e}$$

Taking the foregoing equations into account, the set of machine equations (5.1)–(5.5) can be written as

$$u_{s\alpha} = r_s i_{s\alpha} + T_N \frac{d\psi_{s\alpha}}{dt} \tag{5.11a}$$

$$u_{s\beta} = r_s i_{s\beta} + T_N \frac{d\psi_{s\beta}}{dt} \tag{5.11b}$$

$$0 = r_r i_{r\alpha} + T_N \frac{d\psi_{r\alpha}}{dt} + \omega_m \psi_{r\beta} \tag{5.12a}$$

$$0 = r_r i_{s\beta} + T_N \frac{d\psi_{r\beta}}{dt} - \omega_m \psi_{r\alpha} \tag{5.12b}$$

$$\psi_{s\alpha} = x_s i_{s\alpha} + x_M i_{r\alpha} \tag{5.13a}$$

$$\psi_{s\beta} = x_s i_{s\beta} + x_M i_{r\beta} \tag{5.13b}$$

$$\psi_{r\alpha} = x_s i_{r\alpha} + x_M i_{s\alpha} \tag{5.14a}$$

$$\psi_{r\beta} = x_s i_{r\beta} + x_M i_{s\beta} \tag{5.14b}$$

$$\frac{d\omega_m}{dt} = \frac{1}{T_M}[\psi_{s\alpha} i_{s\beta} - \psi_{s\beta} i_{s\alpha} - m_L]. \tag{5.15}$$

These equations constitute the basis for constructing the block diagram of an induction machine, as depicted in Fig. 5.1. The mathematical model thus obtained, (5.11)–(5.15), corresponds directly to the *two-phase motor description*. It can be seen from Fig. 5.1 that the IM, as a control plant, has coupled nonlinear dynamic structure and two of the state variables (rotor currents and fluxes) are not usually measurable.

Moreover, IM resistances and inductances vary considerably with significant impact on both steady-state and dynamic performances.

Example 5.2: Current Controlled IM in Synchronous Coordinates (x, y, 0)

Let us adopt a coordinate system synchronous coordinates rotating with angular speed $\omega_K = \omega_{s\psi}$ such that

$$\mathbf{\psi}_r = \psi_r = \psi_{rx}. \tag{5.16}$$

This means that the system of coordinates $x, y, 0$ adopted rotates concurrently with the rotor flux linkage vector $\mathbf{\psi}_r$, where the component $\psi_{ry} = 0$ (Fig. 5.2a). Let us assume, moreover, that it is a cage motor, i.e.,

$$u_{rx} = u_{ry} = 0, \tag{5.17}$$

and that it is current controlled. Current control or supply occurs quite frequently in practical individual drive systems when an induction machine is fed by a CSI or CCPWM-transistor inverter (cf. Chapter 4, Section 4.3). When constructing a block diagram of the machine with such an assumption, a simplification can be made by omitting the stator circuit voltage equation (5.1).

5.2 BASIC THEORY OF INDUCTION MOTOR

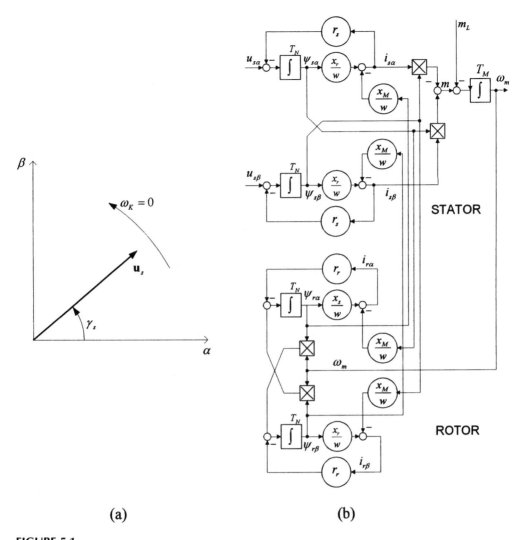

FIGURE 5.1
(a) Stator-fixed system of coordinates α, β; (b) block diagram of a voltage controlled induction motor in the system of α–β coordinates corresponding to Eqs. (5.11)–(5.15), where $w = x_s x_r - x_M^2$.

Under these assumptions, the vector equations (5.2), (5.3), and (5.4) reduce to

$$0 = r_r \mathbf{i}_r + T_N \frac{d\mathbf{\psi}_r}{dt} + j(\omega_{s\psi} - \omega_m) \cdot \mathbf{\psi}_r \tag{5.18}$$

$$\mathbf{\psi}_s = x_s \mathbf{i}_s + x_M \mathbf{i}_r \tag{5.19}$$

$$\mathbf{\psi}_r = x_r \mathbf{i}_r + x_M \mathbf{i}_s \tag{5.20}$$

$$\frac{d\omega_m}{dt} = \frac{1}{T_M} [\text{Im}(\mathbf{\psi}_s^* \mathbf{i}_s) - m_L]. \tag{5.21}$$

(a)

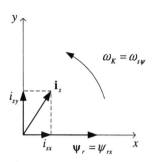

(b)

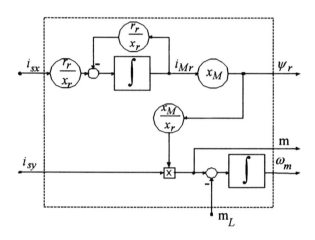

FIGURE 5.2
(a) System of synchronous coordinates x, y rotating concurrently with rotor flux linkage vector $\mathbf{\psi}_r$ (field coordinates). (b) Block diagram of a current controlled induction motor in field coordinates x, y.

Eliminating the rotor current vector $\mathbf{i}_r$ from the voltage equation (5.20) (it is inaccessible in the case of a cage motor) and substituting

$$\mathbf{i}_r = \frac{1}{x_r}\mathbf{\psi}_r - \frac{x_M}{x_r}\mathbf{i}_s, \tag{5.22}$$

we obtain from (5.18)

$$0 = \frac{r_r x_M}{x_r}\mathbf{i}_s + \frac{r_r}{x_r}\mathbf{\psi}_r + T_N \frac{d\mathbf{\psi}_r}{dt} + j(\omega_{s\psi} - \omega_m)\cdot\mathbf{\psi}_r. \tag{5.23}$$

As in the electromagnetic torque expression, the vector $\mathbf{\psi}_s$ can be eliminated:

$$m = \mathrm{Im}(\mathbf{\psi}_s^* \mathbf{i}_s) = \mathrm{Im}\left[\left(\frac{x_s - x_M^2}{x_r}\mathbf{i}_s^* + \frac{x_M}{x_r}\mathbf{\psi}_r^*\right)\mathbf{i}_s\right] = \mathrm{Im}\left(\frac{x_M}{x_r}\mathbf{\psi}_r^*\mathbf{i}_s\right). \tag{5.24}$$

Resolving the complex vectors into components x, y:

$$\mathbf{i}_s = i_{sx} + ji_{sy}, \qquad \mathbf{i}_r = i_{rx} + ji_{ry} \tag{5.25}$$

$$\mathbf{\psi}_s = \psi_{sx} + j\psi_{ry}, \qquad \mathbf{\psi}_s = \psi_{rx} + j\psi_{ry} \tag{5.26}$$

and putting

$$\omega_{s\psi} - \omega_m = \omega_r, \quad (5.27)$$

where ω_r is the angular frequency of the rotor quantities (currents, induced voltages, and flux linkages), also known as the slip frequency, one obtains the set of cage motor equations in the form

$$0 = -\frac{r_r x_M}{x_r} i_{sx} + \frac{r_r}{x_r}\psi_r + T_N \frac{d\psi_r}{dt} \quad (5.28)$$

$$0 = -\frac{r_r x_M}{x_r} i_{sy} + \omega_r \psi_r \quad (5.29)$$

$$\frac{d\omega_m}{dt} = \frac{1}{T_M}\left[\left(\frac{x_m}{x_r}\right)\psi_r i_{sy} - m_L\right]. \quad (5.30)$$

This set of equations is the basis for constructing the block diagram of Fig. 5.2b. The input quantities in this diagram are components i_{sx} and i_{sy} of the stator current vector. The output quantities are the angular shaft speed ω_m and slip frequency ω_r, while the disturbance is load torque m_L.

5.2.3 Equivalent Circuits and Phasor Diagrams

It follows from the induction motor vector equation in the synchronous coordinates, i.e., $\omega_K = \omega_s$, that under steady-state conditions all vector quantities remain constant. For that reason, the time-related derivatives in the voltage equations (5.1) and (5.2) and in the equation of motion (5.5) must be neglected. Thus one obtains a set of algebraic equations, in p.u., which describe steady-state motor operation:

$$\mathbf{u}_s = r_s \mathbf{i}_s + j\omega_s \mathbf{\psi}_s \quad (5.31a)$$

$$\mathbf{u}_r = r_r \mathbf{i}_r + j(\omega_s - \omega_M)\mathbf{\psi}_r \quad (5.31b)$$

$$\mathbf{\psi}_s = x_s \mathbf{i}_s + x_M \mathbf{i}_r \quad (5.32a)$$

$$\mathbf{\psi}_r = x_r \mathbf{i}_r + x_M \mathbf{i}_s \quad (5.32b)$$

$$0 = (\mathbf{\psi}_s^* \mathbf{i}_s) - m_L. \quad (5.33)$$

After elimination of flux linkage, the voltage equations (5.3a,b) can be written in the following general form:

$$\mathbf{u}_s = r_s \mathbf{i}_s + j\omega_s(x_s - ax_M)\mathbf{i}_s + j\omega_s a x_M \mathbf{i}_M \quad (5.34a)$$

$$a\frac{\omega_s}{\omega_r}\mathbf{u}_r = j\omega_s a x_M \mathbf{i}_M + \left[j\omega_s(a^2 x_r - ax_M) + \frac{\omega_s}{\omega_r}a^2 r_r\right]\frac{\mathbf{i}_r}{a} \quad (5.34b)$$

where $\omega_r = f_r = \omega_s - \omega_m$ is the slip frequency (5.27),

$$\mathbf{i}_M = \mathbf{i}_s + \frac{\mathbf{i}_r}{a} \quad (5.35)$$

is the magnetizing current, and a is an arbitrary constant.

Equations (5.34a) and (5.34b), first derived by Yamamura [21], demonstrate that there can be no one induction motor equivalent circuit! This is because adoption of different values of the constant a results in different equivalent circuits.

168 CHAPTER 5 / CONTROL OF PWM: INVERTER-FED INDUCTION MOTORS

Example 5.3: An Equivalent Circuit Based on Main Flux Linkage

Adopting a cage rotor motor for which $\mathbf{u}_r = 0$ and putting $a = 1$, the voltage equations (5.34) can be written as

$$\mathbf{u}_s = (r_s + j\omega_s \sigma_s x_M)\mathbf{i}_s + j\omega_s x_M \mathbf{i}_M \tag{5.36a}$$

$$0 = j\omega_s x_M \mathbf{i}_M + \left(r_r \frac{\omega_s}{\omega_r} + j\omega_s \sigma_r x_M\right)\mathbf{i}_r, \tag{5.36b}$$

while magnetizing current (5.35) can be written as

$$\mathbf{i}_M = \mathbf{i}_s + \mathbf{i}_r \tag{5.37}$$

In the foregoing,

$$\sigma_s = \frac{x_s}{x_M} - 1, \qquad \sigma_r = \frac{x_r}{x_M} - 1. \tag{5.38}$$

On the basis of the equations, one can construct the equivalent circuit depicted in Fig. 5.3a, corresponding to the familiar single-phase form of the equivalent transformer-type circuit of cage motor. It should be born in mind that:

- The diagram is valid only in steady states under sinusoidal voltage supply
- The p.u. reactance are determined for the nominal frequency $F_{sN} = 50\,\text{Hz}$ (60 Hz)
- The circuit includes constant elements, but the presence of the parameters ω_s and ω_r/ω_s emphasizes that the circuit is valid in a general case for an arbitrary stator supply angular frequency ω_s and loads ω_r
- It is only in the particular case of a motor fed with constant nominal frequency, i.e., $f_s = \omega_s = 1$, and following the introduction of slip,

$$s = \frac{\omega_s - \omega_m}{\omega_s} = \frac{\omega_r}{\omega_s} = \frac{f_r}{f_s}, \tag{5.39}$$

which is then equal to slip frequency

$$s(\omega_s = 1) = \omega_r = f_r, \tag{5.40}$$

that the circuit becomes the single-phase equivalent-transformer circuit of an induction motor as used in the classical theory of electrical machines [4, 13].

Example 5.4: An Equivalent Circuit Based on Rotor Flux Linkage

If we adopt $a = x_M/x_r$, the voltage equations (5.33a,b) for a cage motor take the form

$$\mathbf{u}_s = r_s \mathbf{i}_s + j\omega_s \sigma x_s \mathbf{i}_s + j\omega_s \left(\frac{x_M}{x_r}\right)^2 x_r \mathbf{i}_{Mr} \tag{5.41a}$$

$$0 = j\omega_s \left(\frac{x_M}{x_r}\right)^2 x_r \mathbf{i}_{Mr} + \frac{\omega_s}{\omega_r} r_r \left(\frac{x_M}{x_r}\right)^2 \mathbf{i}_r \frac{x_r}{x_M}, \tag{5.41b}$$

where $x_s - (x_M/x_r)x_M = \sigma x_s$ is the *total leakage reactance*. The magnetizing current is in this case expressed as

$$\mathbf{i}_{Mr} = \frac{\mathbf{\psi}_r}{x_M} = \mathbf{i}_s + \frac{x_r}{x_M}\mathbf{i}_r. \tag{5.42}$$

5.2 BASIC THEORY OF INDUCTION MOTOR

a)
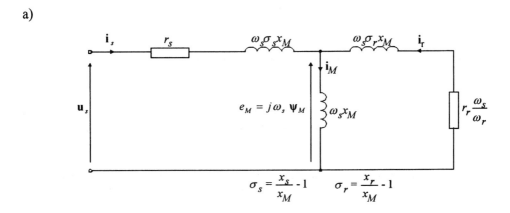

b)
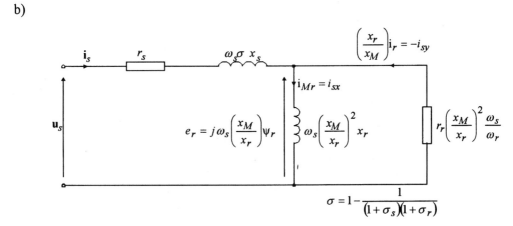

c)
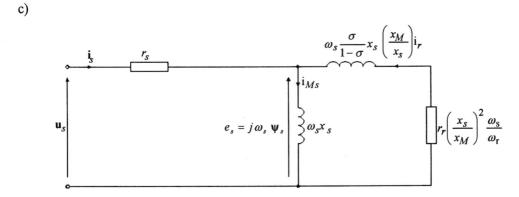

FIGURE 5.3
Steady-state equivalent circuit of induction motor based on (a) the main flux linkage $a = 1$, (b) the rotor flux linkage $a = x_M/x_r$, (c) the stator flux linkage $a = x_s/x_M$.

From Eqs. (5.41a,b) one can construct the equivalent circuit of Fig. 5.3b. The circuit has the following properties:

- There is no leakage inductance on the rotor side
- The recalculated rotor current $(x_r/x_M)\mathbf{i}_r$ is perpendicular to the magnetizing current $\mathbf{i}_{Mr}$ (cf. (5.41b))
- Because of this, the circuit illustrates decomposition of the stator current $\mathbf{i}_s$ into the rotor flux-oriented components: $i_{sx} = \mathbf{i}_{Mr}$ which forms the flux ψ_r, and $i_{sy} = -(x_r/x_M)\mathbf{i}_r$, which controls the torque developed by the motor (cf. Fig. 5.2)

The total leakage reactance σx_s appearing in the circuit of Fig. 5.3b is the sum of stator and rotor leakage reactances, which occur in the case of the equivalent circuit of Fig. 5.3a:

$$\sigma x_s = \sigma_r x_M + \sigma_s x_M. \tag{5.43}$$

The reactance σx_s, is often referred to as the *transient reactance*.

Similarly, for $a = x_s/x_M$, one obtains an equivalent circuit based on stator flux linkage without leakage inductance on the stator side (Fig. 5.3c). The phasor diagram corresponding to the equivalent circuit of Fig. 5.3 is shown in Fig. 5.4. It illustrates clearly that in the induction motor are three slightly different magnetizing currents $\mathbf{i}_M$, $\mathbf{i}_{Mr}$, and $\mathbf{i}_{Ms}$ which—when they are kept constant—correspond to stabilization of main, rotor, and stator flux linkage, respectively. All three magnetizing currents are fictitious and the real state variables are only stator voltage and current. However, which of the three magnetizing currents and hence flux linkages is kept constant influences the decomposition of the stator current $\mathbf{i}_s$ into flux- and torque-producing components, resulting in different static and dynamic torque characteristics of the induction motor.

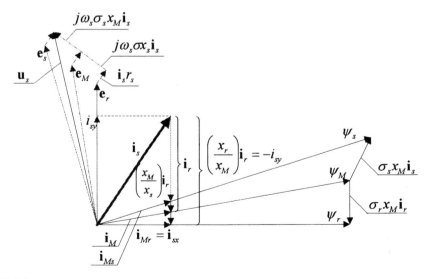

FIGURE 5.4
Phasor diagram of three equivalent circuits from Fig. 5.4.

5.2.4 Steady-State Characteristics

From the voltage and flux-current equations (5.31) and (5.32) one can find the torque developed by a machine by representing it, as has been done earlier for the stator voltage, as a function of the amplitude of only one electromagnetic variable, e.g., ψ_r, ψ_M, ψ_s, i_s, or i_r. Then we obtain

$$m = \frac{\omega_r}{r_r}\psi_r^2 \tag{5.44a}$$

$$m = \frac{r_r \omega_r}{r_r^2 + (x_r - x_M)^2 \omega_r^2}\psi_M^2 \tag{5.44b}$$

$$m = \frac{(x_M/\sigma x_s x_r)^2 r_r \omega_r}{(r_r/\sigma x_r)^2 + \omega_r^2}\psi_s^2 \tag{5.44c}$$

$$m = \frac{(1-\sigma)x_s x_r r_r \omega_r}{r_r^2 + (\omega_r x_r)^2}i_s^2 \tag{5.44d}$$

$$m = \frac{x_M^2 r_r \omega_r}{[(r_s/\omega_s)r_r - \omega_r \sigma x_s x_r]^2 + [(r_s/\omega_s)\omega_r x_r + r_r x_s]^2}\left(\frac{u_s}{\omega_s}\right)^2 \tag{5.44e}$$

$$m = \frac{r_r}{\omega_r}i_r^2. \tag{5.44f}$$

It follows from (5.44a) that the torque developed by an IM is a linear function of the slip frequency ω_r only for operation with constant rotor flux amplitude. Consequently, there is no breakdown torque and, by the same token, no stalling of the machine, From the remaining expressions (5.10) we obtain, by comparing $dm/d\omega_r$ to zero, the following breakdown slip frequencies:

$$\omega_{rk}(\psi_M = 1) = \pm\frac{r_r}{\sigma_r x_M} \tag{5.45a}$$

$$\omega_{rk}(\psi_s = 1) = \omega_{rk}(u_s = 1, r_s = 0) = \pm\frac{r_r}{\sigma x_r} \tag{5.45b}$$

$$\omega_{rk}(i_s = 1) = \pm\frac{r_r}{x_r} \tag{5.45c}$$

$$\omega_{rk}(u_s = 1) = r_r\sqrt{\frac{r_s^2 + (x_s \omega_s)^2}{(r_s x_r)^2 + (\sigma x_s x_s \omega_s)^2}}. \tag{5.45d}$$

The characteristics of torque m as a function of induction motor slip frequency ω_r which correspond to equations (5.44), are represented in Fig. 5.5. It is noteworthy that when the machine is supplied with constant amplitude current i_s, the breakdown slip frequency is $1/\sigma$ times smaller that under conditions of constant voltage amplitude ($u_s = 1$) or constant stator flux linkage ($\psi_s = 1$) operation. Moreover, under conditions of rated stator current value $i_s = 1$, the motor already operates in the high saturation range (point A in Fig. 5.5), attaining the rated flux value (e.g., $\psi_s = 1$) in the nonstable part of the characteristic (point B). Thus in contrast to the voltage-controlled IM, a current-controlled induction motor cannot operate in an open-loop system.

In many applications the IM operates not only below but also above rated speed. This is possible because most IM can be operated up to twice rated speed without mechanical problems. Typical characteristics for the 4-kW motor (data given in Appendix 2) are plotted in Fig. 5.6. Below the rated speed the flux amplitude is kept constant and, at the rated slip frequency, the motor can develop rated torque. Hence, this region is called the *constant-torque* region. Increasing the stator frequency ω_s above its rated value $\omega_s > 1$ p.u., at constant rated voltage

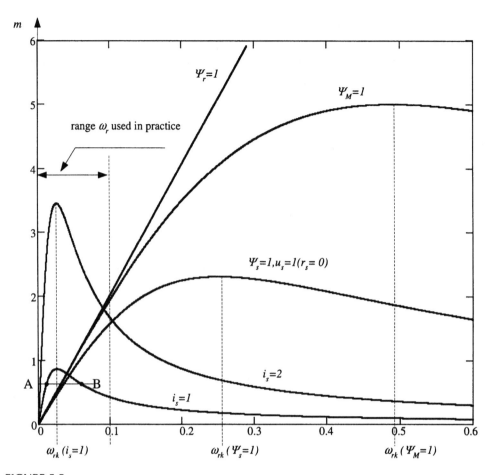

FIGURE 5.5
Torque–slip frequency characteristics for IM under different control methods (without saturation effect).

$u_s = 1$, it is possible to increase the motor speed beyond the rated speed. However, the motor flux, proportional to u_s/f_s, will be weakened. Therefore, when the slip frequency increases proportional with the stator frequency $\omega_r \sim \omega_s$, the electromagnetic power

$$p_s = \omega_m \cdot m \approx \omega_s \left(\frac{\omega_r}{r_r}\right)\left(\frac{\psi_r}{\omega_s}\right)^2$$

can be held constant, giving the name of this region (Fig. 5.6).

With constant stator voltage and increased stator frequency, the motor speed reaches the high-speed region, where the flux is reduced so much that the IM approaches its breakdown torque and slip frequency can no longer be increased. Consequently, the torque capability is reduced according to the breakdown torque characteristic $m \sim m_k \sim (1/\omega_s)^2$. This high-speed region is called the *constant-slip frequency* region (Fig. 5.6).

5.3 CLASSIFICATION OF IM CONTROL METHODS

Based on the space-vector description, the induction motor control methods can be divided into *scalar* and *vector control*. The general classification of the frequency controllers is presented in Fig. 5.7.

5.3 CLASSIFICATION OF IM CONTROL METHODS

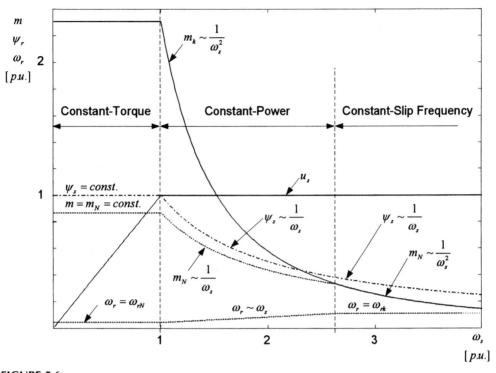

FIGURE 5.6
Control characteristics of induction motor in constant and weakened flux regions.

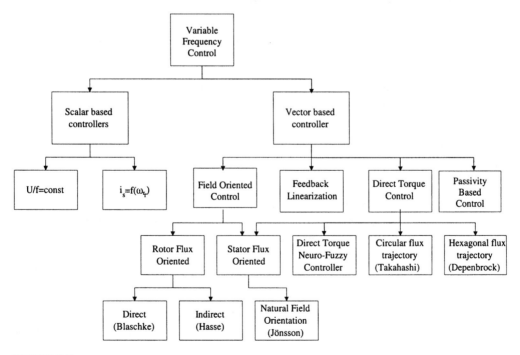

FIGURE 5.7
Classification of IM control methods.

It is characteristic for scalar control that—based on a relation valid for steady states—only the magnitude and frequency (angular speed) of voltage, currents, and flux linkage space vectors are controlled. Thus, the control system does not act on space vector position during transients. In contrast, in vector control—based on a relation valid for dynamic states—not just magnitude and frequency (angular speed), but also instantaneous positions of voltage, current, and flux space vectors are controlled. Thus, the control system acts on the positions of the space vectors and provides their correct orientation for both steady states and transients.

According to the preceding definition, vector control can be implemented in many different ways. Among the best-known strategies are field-oriented control (FOC) and direct torque control (DTC). Other, less well-known nonlinear strategies include feedback linearization control (FLC) and the recently developed passivity-based control (PBC).

It should be noted, additionally, that each of these control strategies can be implemented in different techniques (e.g., voltage and current controlled, polar or Cartesian coordinates), resulting in many variants of control schemes.

5.4 SCALAR CONTROL

As defined in Section 5.3, in scalar control schemes the phase relations between IM space vectors are not controlled during transients. The control scheme is based on steady-state characteristics, which allows stabilization of the stator flux magnitude ψ_s for different speed and torque values. In some applications variable flux operation is also considered, e.g., constant slip frequency control of a current source inverter-fed IM [7] or part-load losses minimization of voltage source inverter-fed IM drives (see Chapter 6).

5.4.1 Open-Loop Constant Volts/Hz Control

In numerous industrial applications, the requirements related to the dynamic properties of drive control are of secondary importance. This is especially the case wherever no rapid motor speed change is required and where there are no sudden load torque changes. In such cases one may just as well make use of open-loop constant voltage/Hz control systems (Fig. 5.8).

This method is based on the assumption that the flux amplitude is constant in steady-state operation. From Eq. (5.1), for $\omega_K = \omega_s$, one obtains the stator voltage vector

$$\mathbf{u}_s = r_s \mathbf{i}_s + j\omega_s \mathbf{\psi}_s \tag{5.46}$$

from which the normalized stator vector magnitude can be calculated as

$$u_s = \sqrt{(r_s i_s)^2 + (f_s \psi_s)^2}. \tag{5.47}$$

For constant stator flux linkages of $\psi_s = 1$, the applied voltage u_s versus p.u. stator frequency is shown in Fig. 5.8a. Note that for $r_s = 0$, the relationship between stator voltage magnitude and frequency is linear and Eq. (5.47) takes the form

$$\frac{u_s}{f_s} = 1, \tag{5.48}$$

giving the name of this method.

For practical implementation, however, the relation of Eq. (5.47) can be expressed as

$$u_s = u_{s0} + f_s \tag{5.49}$$

where $u_{s0} = i_s r_s$ is the offset voltage to compensate for the stator resistive drop.

5.4 SCALAR CONTROL

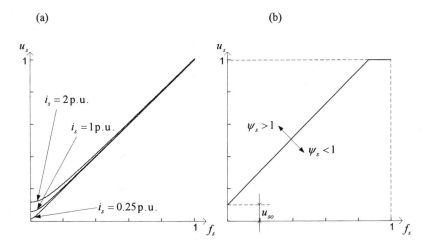

FIGURE 5.8
Stator voltage versus stator frequency for constant stator flux, $\psi_s = 1$, operation (for 4-kW induction motor with $r_s = 0.059$: (a) theoretical characteristics according to Eq. (5.47), (b) characteristic used in practice.

The block diagram of an open-loop constant V/Hz control implemented according to Eq. (5.49) for PWM-VSI fed IM drive is shown in Fig. 5.9. The control algorithm calculates the voltage amplitude, proportional to the command speed value, and the angle is obtained by the integration of this speed. The voltage vector in polar coordinates is the reference value for the

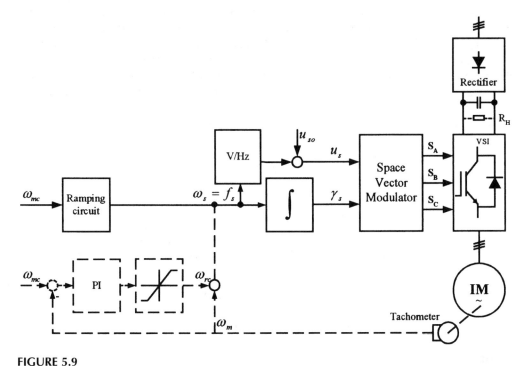

FIGURE 5.9
Constant V/Hz control scheme (dashed lines show version with limited slip frequency ω_{rc} and speed control loop).

vector modulator, which delivers switching signals to the voltage source inverter. The speed command signal ω_{mc} determines the inverter frequency $f_s = \omega_s$, which simultaneously defines the stator voltage command according to V/Hz constant.

However, the mechanical speed ω_m and hence the slip frequency $\omega_r = \omega_s - \omega_m$ are not precisely controlled. This can lead to motor operation in the unstable region of torque–slip frequency curves (Fig. 5.5), resulting in overcurrent problems. Therefore, to avoid high slip frequency values during transients a ramping circuit is added to the stator frequency control path. When speed stabilization is necessary, use may be made of speed control with slip regulation (dashed lines in Fig. 5.9). The slip frequency command ω_{rc} is generated by the speed PI controller. This signal is added to the tachometer signal and determines the inverter frequency command ω_s. Since the constant voltage/frequency control condition is satisfied, the stator flux remains constant, which guarantees the motor torque slip frequency proportionality. Thanks to limitation of slip frequency command ω_{rc} the motor will not pull out either under rapid speed command changes or under load torque changes. Rapid speed reduction results in negative slip command, and the motor goes into generator braking. The regenerated energy must then either be returned to the line by the feedback converter or dissipated in the dc-link dynamic breaking resistor R_H.

5.5 FIELD-ORIENTED CONTROL (FOC)

5.5.1 Introduction

Field-oriented control is based on decomposition of the instantaneous stator current into two components: flux current and torque-producing current. (In analogy to a separate commutator motor, the *flux current* component corresponds to the excitation current and *torque-producing current* corresponds to the armature current.) This decomposition guarantees correct orientation of the stator current vector with respect to flux linkage. However, from Fig. 5.10 it is clearly seen that selecting stator or rotor flux linkage as the basis of the reference frame will result in slightly different decomposition of the stator current vector. (Theoretically, main flux-oriented coordi-

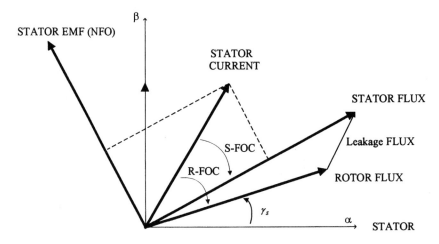

FIGURE 5.10
Vector diagram illustrating current vector position in different vector control strategies: R-FOC, orientation in respect to rotor flux; S-FOC, orientation in respect to stator flux.

nates can also be selected (compare Section 5.2.3), however, because of its low practical importance this is not considered.)

The generic idea of FOC assumes that IM is current-controlled (see Chapter 4, Section 4.3). In such cases the stator voltage equation (5.1) can be omitted (see Example 5.2). However, in the drives fed by VSI without a current control loop, both stator and rotor voltage equations have to be taken into account.

5.5.2 Rotor-Flux-Oriented Control (R-FOC)

In the case of R-FOC the angular speed of the coordinate system, in which the IM vector equations (5.1)–(5.6) are analyzed, is defined as $\omega_K = \omega_{s\psi} = \omega_s$.

5.5.2.1 Rotor Voltage Equation in Rotor-Flux Coordinates.
In the rotor-flux-oriented coordinates x–y (Fig. 5.2a), after introducing the rotor time constant

$$T_r = \frac{x_r}{r_r} T_N, \tag{5.50}$$

Eq. (5.28) and (5.29) can be rewritten as

$$\frac{d\psi_r}{dt} = -\frac{\psi_r}{T_r} + \frac{x_M}{T_r} i_{sx} \tag{5.51a}$$

$$0 = -(\omega_s - \omega_m)\frac{\psi_r}{T_N} + \frac{x_M}{T_r} i_{sy}. \tag{5.51b}$$

Equation (5.51a) describes the influence of the flux stator current components i_{sx} on the rotor flux. The motor torque, according to Eq. (5.24), can be expressed as

$$m = \frac{x_M}{x_r} \psi_r i_{sy} = \frac{x_M^2}{x_r} i_{Mr} i_{sy}, \tag{5.52}$$

where i_{Mr} is the rotor flux magnetizing current (Eq. (5.42)).

In field coordinates—for the constant rotor flux amplitude—the motor torque can be controlled by torque current i_{sy}, without any delay (Fig. 5.11a). Moreover, as in a separately excited dc motor, the flux current i_{sx} can be reduced in the field-weakening region in order to limit the stator voltage at high speed.

5.5.2.2 Stator Voltage Equation in Rotor-Flux Coordinates.
Calculating the stator flux vector from Eqs. (5.3) and (5.4), one obtains

$$\mathbf{\psi}_s = \mathbf{\psi}'_r + \sigma x_s \mathbf{i}_s, \tag{5.53}$$

where

$$\mathbf{\psi}'_r = \left(\frac{x_M}{x_r}\right) \mathbf{\psi}_r. \tag{5.54}$$

Substituting (5.53) into (5.1) yields

$$\mathbf{u}_s = r_s \mathbf{i}_s + T_N \sigma x_s \frac{d\mathbf{i}_s}{dt} + j\omega_K \sigma x_s \mathbf{i}_s + T_N \sigma x_s \frac{d\mathbf{\psi}'_r}{dt} + j\omega_K \mathbf{\psi}'_r. \tag{5.55}$$

In the rotor-flux oriented coordinates x–y (Fig. 5.2a), the foregoing equations can be resolved into two scalar equations as follows:

$$u_{sx} = r_s i_{sx} + T_N \sigma x_d \frac{di_{sx}}{dt} - \underline{\omega_s \sigma x_s i_{sy}} + T_N \frac{d\psi_r'}{dt} \tag{5.56a}$$

$$u_{sy} = r_s i_{sy} + T_N \sigma x_d \frac{di_{sy}}{dt} - \underline{\omega_s \sigma x_s i_{sx}} + \omega_s \psi_r'. \tag{5.56b}$$

These equations represent the interaction between the rotor-flux-oriented stator voltage and currents. Note that u_{sx} and u_{sy} cannot be considered as independent control variables for the rotor flux and torque, because u_{sx} includes coupling terms dependent on i_{sy} and u_{sy} includes coupling terms dependent on i_{sx} (underlined terms in Eqs. (5.56a,b)). Therefore, fast current control requires compensation of the coupling terms $\omega_s \sigma x_s i_{sy}$ and the back EMF terms $\omega_s \psi_r'$. They can be compensated in a feedforward or feedback manner (see Chapter 4, Section 4.3.3.3).

For voltage-controlled IM Eqs. (5.56a) and (5.56b) constitute a *voltage decoupler* which allows calculation of commanded voltage vector components u_{sxc}, u_{syc} based on IM parameters and the required flux current i_{sxc} and torque-producing current i_{syc}.

5.5.3 Stator-Flux-Oriented Control (S-FOC)

In the stator-flux-oriented coordinates x–y (Fig. 5.10) we have

$$\psi_{sx} = \psi_s, \quad \psi_{sy} = 0, \quad \text{and} \quad \omega_K = \omega_{s\psi} = \omega_s. \tag{5.57}$$

5.5.3.1 Rotor Voltage Equation in Stator-Flux Coordinates.
Substituting the rotor current vector from the rotor voltage Eq. (5.2) by Eq. (5.3) we obtain a differential equation for the stator flux vector:

$$\frac{d\psi_s}{dt} = -\frac{\psi_s}{T_r'} - j\omega_r \frac{\psi_s}{T_N} + \sigma x_s \frac{d\mathbf{i}_s}{dt} + \frac{x_s}{T_r'}\mathbf{i}_s + j\omega_r \frac{\sigma x_s}{T_N}\mathbf{i}_s \tag{5.58}$$

where

$$\sigma = 1 - \frac{x_M^2}{x_s x_r}$$

is the total leakage factor and the $\omega_r = \omega_s - \omega_m$ the slip frequency.

Equation (5.58) can be rewritten as

$$\frac{d\psi_s}{dt} = -\frac{\psi_s}{T_r} + \sigma x_s \frac{di_{sx}}{dt} + \frac{x_s}{T_r}i_{sx} - \omega_r \frac{\sigma x_s}{T_N}i_{sy} \tag{5.59a}$$

$$\frac{\omega_r}{T_N}(\psi_s - \sigma x_s i_{sx}) = \sigma x_s \frac{di_{sy}}{dt} + \frac{x_s}{T_r}i_{sy}. \tag{5.59b}$$

The motor torque, according to Eq. (5.6), with condition Eq. (5.57), can be expressed as

$$m = \psi_s i_{sy} = x_M i_{Ms} i_{sy} \tag{5.60}$$

with i_{Ms} the stator flux magnetizing current (Fig. 5.3c).

Equation (5.59) shows that there exists a coupling between the torque-producing stator current component i_{sy} and the stator flux ψ_s (Fig. 5.11b). Therefore, for current-controlled PWM inverters S-FOC requires a decoupling network, resulting in a more complicated control structure than R-FOC systems.

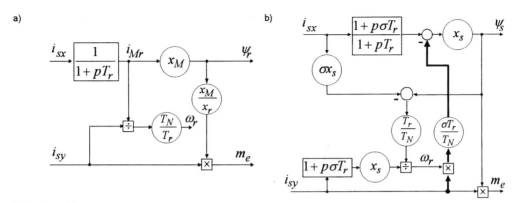

FIGURE 5.11
Torque production in current-controlled FOC: (a) R-FOC, (b) S-FOC.

5.5.3.2 Stator Voltage Equation in Stator-Flux Coordinates. Resolving Eq. (5.1) under the conditions of Eq. (5.57), one obtains two scalar equations in stator-flux coordinates x–y as follows:

$$u_{sx} = r_s i_{sx} + T_N \frac{d\psi_s}{dt} \tag{5.61a}$$

$$u_{sy} = r_s i_{sy} + \omega_s \psi_s. \tag{5.61b}$$

Note that these equations are much simpler than Eqs. (5.56a,b). For constant stator flux magnitude only the back-EMF term $\omega_s \psi_s$ has to be compensated for decoupled current control.

5.5.4 Rotor versus Stator-Flux Oriented Control

The main features and advantages of R-FOC and S-FOC are summarized in Table 5.1. The important conclusion that follows from Table 5.1 is that the R-FOC scheme can be extremely easily implemented with a current controlled PWM inverter. In contrast, the S-FOC scheme has

Table 5.1 Rotor versus Stator Flux-Oriented Control

Rotor FOC	Stator FOC
For constant flux amplitude, **linear relationship** between (i_{sx}, i_{sy}) control variables and torque (speed) (Fig. 5.11a)	For constant flux amplitude, there **exists coupling** between (i_{sx}, i_{sy}) control variables and torque (speed) (Fig. 5.11b)
Simple rotor voltage equation (5.51) in rotor-flux-oriented coordinates, thus **simple implementation with current-controlled PWM inverter** (decoupling network is not required)	Complex rotor voltage equation (5.58) in stator-flux-oriented coordinates, thus **even implementation with current-controlled PWM inverter requires decoupling network**
Complex stator voltage equation (5.55) in rotor-flux-oriented coordinates, so **decoupling network is required**	Simple stator voltage equation (5.61) in rotor-flux-oriented coordinates, so **decoupling network is extremely simple**
No critical (pull-out) slip frequency	There **exists a critical** (pull-out) slip frequency
Flux estimation: $\psi_r = (x_r/x_M)[\int(\mathbf{u}_s - r_s \mathbf{i}_s)dt - \sigma x_s \mathbf{i}_s]$ depending on two motor parameters ($r_s, \sigma x_s$)	Flux estimation: $\psi_s = \int(\mathbf{u}_s - r_s \mathbf{i}_s)dt$ depending only on one motor parameter (r_s)

the simplest implementation with a voltage controlled PWM inverter. This very important observation is confirmed by control schemes used in industrial practice.

5.5.5 Current-Controlled R-FOC Schemes

The basic problem involved in the implementation of current-controlled PWM inverters for the R-FOC scheme is the choice of a suitable current control method, which affects both the parameters obtained and the final configuration of the entire system.

In the standard version, the PWM current control loop operates in synchronous field-oriented coordinates x–y (see for details Chapter 4, Section 4.3.3.3) as shown in Fig. 5.12. The feedback stator currents i_{sx}, i_{sy} are obtained from the measured values i_A, i_B after phase conversion $ABC/\alpha\beta$:

$$i_{s\alpha} = i_{sA} \tag{5.62a}$$

$$i_{s\beta} = (1/\sqrt{3})(i_{sA} + 2i_{sB}) \tag{5.62b}$$

followed by coordinate transformation $\alpha\beta/xy$:

$$i_{sx} = i_{s\alpha} \cos \gamma_s + i_{s\beta} \sin \gamma_s \tag{5.63a}$$

$$i_{sy} = -i_{s\alpha} \sin \gamma_s + i_{s\beta} \cos \gamma_s. \tag{5.63b}$$

The PI current controllers generates voltage vector commands u_{sxc}, u_{syc} which, after coordinate transformation $xy/\alpha\beta$,

$$u_{s\alpha c} = u_{sxc} \cos \gamma_s - u_{syc} \sin \gamma_s \tag{5.64a}$$

$$u_{s\beta c} = u_{sxc} \sin \gamma_s + u_{syc\beta} \cos \gamma_s, \tag{5.64b}$$

are delivered to the space vector modulator (SVM). Finally the SVM calculates the switching signals S_A, S_B, S_C for the power transistors of PWM inverter.

The main information of the FOC scheme, namely the flux vector position γ_s necessary for coordinate transformation, can be delivered in two different ways giving generally two types of FOC schemes, called *indirect* and *direct* FOC. Most of literature adopts the following definition: indirect FOC refers to an implementation where the flux vector position γ_s is calculated from the reference values (feedforward control) and mechanical speed (position) measurement (Fig. 5.12a), whereas direct FOC refers to the case where flux vector position γ_s is measured or estimated (Fig. 5.12b) [1–4, 7–9, 12, 16].

5.5.5.1 Indirect R-FOC Scheme. The main block in this scheme is the so-called *indirect vector controller* (Fig. 5.13a), which implements the following relations derived from the IM equations in field coordinates, Eqs. (5.28)–(5.30):

$$i_{sxc} = \frac{1}{x_{Mc}}\left(\psi_{rc} + T_{rc}\frac{d\psi_{rc}}{dt}\right) \tag{5.65a}$$

$$i_{syc} = \frac{x_{rc}}{x_{Mc}}\frac{m_c}{\psi_{rc}} \tag{5.65b}$$

$$\omega_{rc} = x_{Mc}\frac{T_N}{T_{rc}}\frac{i_{syc}}{\psi_{rc}} \tag{5.66}$$

5.5 FIELD-ORIENTED CONTROL (FOC)

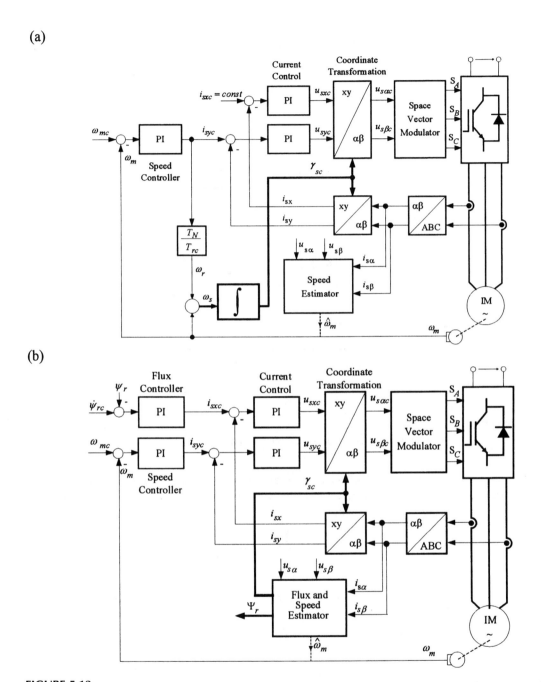

FIGURE 5.12
FOC of current-controlled PWM inverter-fed induction motor for constant flux region: (a) indirect FOC, (b) direct FOC (dashed line shows speed-sensorless operation).

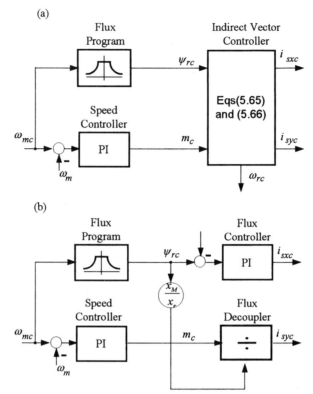

FIGURE 5.13
Extension of FOC control schemes for field-weakened operation: (a) indirect FOC, (b) direct FOC.

where indices c denote the command value of the variable and the motor parameters. They constitute basis for control in both constant and weakened field regions (Fig. 5.13a). However, in the case of constant flux $\psi_{rc} = $ const., Eqs. (5.65) and (5.66) become considerably simplified:

$$i_{sxc} = \frac{\psi_{rc}}{x_{Mc}} \tag{5.67a}$$

$$i_{syc} \sim m_c \tag{5.67b}$$

$$\omega_{rc} \sim \frac{T_N}{T_{rc}} i_{syc} \tag{5.68}$$

which corresponds to the situation shown in Fig. 5.12a. The flux vector position γ_s in respect to the stator is calculated as

$$\gamma_{sc} = \frac{1}{T_N} \int_0^t (\omega_m + \omega_{rc}) dt = \frac{1}{T_N} \int_0^t \omega_s dt \tag{5.69}$$

where the ω_m is the IM shaft speed measured by a speed sensor or estimated ($\hat{\omega}_m$) from the measured stator currents ($i_{s\alpha}, i_{s\beta}$) and voltages ($u_{s\alpha}, u_{s\beta}$).

5.5.5.2 The Effect of Parameter Changes in Indirect R-FOC.
The indirect R-FOC scheme is effective only as long as the set values of the motor parameters in the vector controller are equal to the actual motor parameter values. For the constant-rotor-flux operation region, a change

of the motor time constant T_r results in deviation in the slip frequency value ω_{rc} calculated from Eq. (5.66). The predicted rotor flux position $\gamma_{sc} = 1/T_N \int (\omega_m + \omega_{rc}) dt$ deviates from the actual position $\gamma_s = 1/T_N \int (\omega_m + \omega_r) dt$, which produces a torque angle deviation $\Delta\delta = \Delta\gamma_s = \gamma_{sc} - \gamma_s$ and, consequently, leads to incorrect subdivision of the stator current vector $\mathbf{i}_s$ into two components i_{sx}, i_{sy} (Fig. 5.14). The decoupling condition of flux and torque control cannot be achieved. This leads to:

- Incorrect rotor flux ψ_r and torque current component i_{sy} values in the steady-state operating points (for $m_c = $ const.)
- Second-order (nonlinear) system transient response to changes to torque command m_c

The effects of motor parameter detuning have been analyzed in several publications [8, 26, 39]. For a predetermined point of operation defined by the torque and flux current command values i_{syc}, i_{sxc}, it is possible to determine the effect of rotor-circuit time constant changes on real torque and rotor flux of the motor. These relations, derived from steady-state motor equations (5.51) and (5.52), can be conveniently be presented in the form

$$\frac{m}{m_c} = \frac{T_r}{T_{rc}} \frac{1 + (i_{syc}/i_{sxc})^2}{1 + [(T_r/T_{rc})(i_{syc}/i_{sxc})]^2} \tag{5.70}$$

$$\frac{\psi_r}{\psi_{rc}} = \sqrt{\frac{1 + (i_{syc}/i_{sxc})^2}{1 + [(T_r/T_{rc})(i_{syc}/i_{sxc})]^2}}. \tag{5.71}$$

It follows from these equations that the normalized torque and rotor flux values are not linear functions of the ratio of actual/predicted rotor time constant (T_r/T_{rc}) and the motor point of operation (i_{syc}/i_{sxc}).

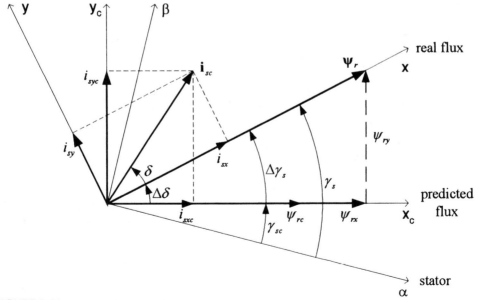

FIGURE 5.14
Incorrect orientation of predicted rotor flux ψ_{rc} ($T_{rc} > T_r$, ω_{rc} too small).

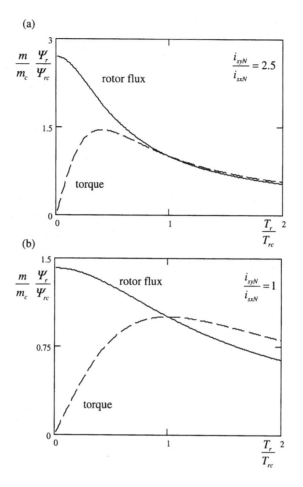

FIGURE 5.15
Detuning of steady-state effect parameters for rated flux and torque current commands: (a) high-power motors, (b) low-power motors.

For the rated values of the field-oriented current commands $i_{syc} = i_{syN}$ and $i_{sxc} = i_{sxN}$, we obtain from (5.70) and (5.71) the curves plotted in Fig. 5.15 (where the saturation effect is omitted). Note that because high-power motors have small magnetizing current i_{Mr} (at steady state $i_{sx} = i_{Mr}$) relative to the rated current i_{sN}, they are characterized by large values of $i_{syN}/i_{sxN} = 2$–2.8. For low-power motors, on the other hand, we have $i_{syN}/i_{sxN} = 1$–2.

The curves shown in Fig. 5.15 make it clear that if, for example, the actual time constant value is lower than the predicted one ($T_r/T_{rc} < 1$), the flux increases from its correct value (cf. also the vector diagram in Fig. 5.14). Both torque and flux changes depend very strongly on the i_{syN}/i_{sxN} value. Note that high-power motors are much more sensitive to the detuning of the time constant (T_r/T_{rc}) than are the low-power ones.

In a similar way, one can take into account the effect of changes in magnetizing inductance x_M introduced by magnetic circuit saturation [39, 41].

5.5.5.3 Parameter Adaptation.
As can be seen from the previous discussion, a parameter critical to the decoupling conditions of an indirect FOC system is the rotor time constant T_r. It changes primarily under the influence of temperature changes of rotor resistance (r_r) and because

of rotor reactance x_r changes brought about by the saturation effect. Whereas the temperature changes of r_r are very slow, the changes of x_r can be very fast, as for example in the course of speed reversal when the motor changes quickly between its rated speed and the field-weakening region. It is assumed that T_r changes in the $0.75T_{r0} < T_r < 1.5T_{r0}$ range, where T_{r0} is the rotor time constant at the rated load and a temperature of 75°C.

Parameter correction is effected by *on-line* adaptation. It follows from the diagrams of Fig. 5.14 that the correction signal for time constant changes $(1/\Delta T_r)$ may be found from the measured actual torque or flux values, or from such familiar quantities as torque or flux current. However, these quantities are very difficult to measure or calculate over the entire range of speed control, the difficulty being comparable to that involved in flux vector estimation (see Section 5.5.5.5) in direct FOC systems.

Figure 5.16 shows the basic idea of a T_r adaptation scheme which corresponds to the structure of the model reference adaptive system (MRAS).

The reference function F_c is calculated from command quantities (indices c) in the field coordinates x, y. The estimated function F_e is calculated from measured quantities which usually are expressed in the stator-oriented coordinates α, β. The error signal $\varepsilon = F_c - F_e$ is delivered to the PI controller, which generates the necessary correction of the rotor time constant $(1/\Delta T_r)$. This correction signal is added to an initial value $(1/T_{r0})$ giving the updated time constant $(1/T_{rc})$, which finally is used for calculation of the slip frequency ω_{rc}. In steady state, when $\varepsilon \to 0$ then $T_{rc} \to T_r$. A variety of criterion function (F) has been suggested for identification of T_r changes (see Table 5.2). Most of them work neither for no-load conditions nor for zero speed. Therefore, in the near zero speed region and no-load operation, the output signal of error calculator ε must be blocked. The last value of $\Delta(1/T_r)$ is stored in the PI T_r controller. In the present-day DSP-based implementation the stator voltage sensors are avoided and voltage vector components $u_{s\alpha}, u_{s\beta}$ are calculated from the inverter switching signals and measured dc-link voltages according to Eqs. (5.75a,b).

The *correlation* method proposed in [9] and further developed in [24] has the advantages that no additional measurements and only 30–40% additional computing time is needed. The response time of the identification algorithm also depend on IM load. The disadvantages are, first, low-updating time (because of the required averaging time to build up a meaningful correlation function), and second, the torque ripple introduced by the pseudorandom binary

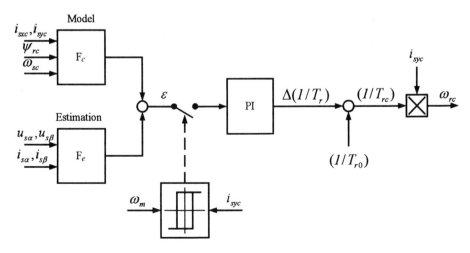

FIGURE 5.16
Principle of T_r adaptation based on model reference adaptive scheme (MRAS).

Table 5.2 T_r Adaptation Algorithms (Fig. 5.16)

	F_c	F_e	Parameter sensitivity	Author(s)	Remarks
1	$-\left(\dfrac{x_M}{x_r}\right)\psi_{rc}\omega_{sc}i_{sxc}$	$(u_{sx}i_{s\beta} - u_{s\beta}i_{sx}) +$ $-\sigma x_s(pi_{sz}i_{s\beta} - pi_{s\beta}i_{sz})$	σx_s	Garces [26]	No pure integration problem
2	$-\left(\dfrac{x_M}{x_r}\right)\psi_{rc}\omega_{sc}i_{sxc}$	$(u_{sx}i_{sy} - u_{sy}i_{sx}) +$ $-\sigma x_s\omega_s(i_{sx}^2 - i_{sy}^2)$	σx_s	Koyama et al. [38]	u_{sx} and u_{sy} are outputs of current controllers
3	$\left(\dfrac{x_M}{x_r}\right)\psi_{rc}i_{syc}$	$\psi_{sz}i_{s\beta} - \psi_{s\beta}i_{sz}$	r_s	Lorenz and Lawson [40]	
4	i_{syc}	$\dfrac{\psi_{sz}i_{s\beta} - \psi_{s\beta}i_{sz}}{\dfrac{x_r}{x_M}\sqrt{\psi_{sz}^2 + \psi_{s\beta}^2}}$	r_s	Rowan et al. [45]	Initial condition and drift problem (pure integration)
5	0	$u_{sx} - r_s i_{sx} + \omega_s \sigma x_s i_{sy}$	$r_s, \sigma x_s$	Okuyama [42], Schumacher [46]	Simple, good convergence rate

sequence (PRBS) used as a test signal, which is not allowed in practical applied drives. This method can thus track temperature-dependent variation in the rotor time constant, but not changes due to saturation effects introduced when going rapidly into the field-weakening region.

Other methods of *on-line* parameter identification based on observer technique have also been proposed [41, 43].

5.5.5.4 Direct R-FOC Scheme. The main block in this scheme is the flux vector estimator (or sensor), which generates position γ_s and magnitude ψ_r of the rotor flux vector $\boldsymbol{\psi}_r$. The flux magnitude ψ_r is controlled by a closed loop, and the flux controller generates the flux current command i_{sxc}. Above the rated speed, field weakening is implemented by making the flux command ψ_{rc} speed dependent, using a flux program generator, as shown in Fig. 5.13b. In the field-weakening region, the torque current command i_{syc} is calculated in the flux decoupler from the torque and flux commands m_c and ψ_{rc} according to Eq. (5.65b).

If the estimated torque signal m is available, the flux decoupler can be replaced by a PI torque controller which generates the torque current command i_{syc}. In both cases the influence of variable flux on torque control is compensated. However, the stator current vector magnitude has to be limited as

$$\sqrt{i_{sxc}^2 + i_{syc}^2} \leq i_{smax}. \tag{5.72}$$

5.5.5.5 Flux Vector Estimation. To avoid the use of additional sensors or measuring coils in the IM, methods of indirect flux vector generation have been developed, known as *flux models* or *flux estimators*. These are models of motor equations which are excited by appropriate easily measurable quantities, such as stator voltages and/or currents ($\mathbf{u}_s, \mathbf{i}_s$), angular shaft speed ($\omega_m$), or position angle ($\gamma_m$). There are many types of flux vector estimators, which usually are classified in terms of the input signals used [7]. Recently, only estimators based on stator currents and voltages have been used, because they avoid the need for mechanical speed and/or position sensors.

Stator flux vector estimators. Integrating the stator voltage equations represented in stationary coordinates α, β (5.11a,b), one obtains the stator flux vector components as

$$\psi_{s\alpha} = \frac{1}{T_N} \int_0^t (u_{s\alpha} - r_s i_{s\alpha})dt \tag{5.73a}$$

$$\psi_{s\beta} = \frac{1}{T_N} \int_0^t (u_{s\beta} - r_s i_{s\beta})dt. \tag{5.73b}$$

The block diagram of the stator flux estimator according to Eqs. (5.73a,b) is shown in Fig. 5.17a. Higher accuracy can be achieved if the stator flux is calculated in the scheme of Fig. 5.17b operated with polar coordinates $\boldsymbol{\psi}_s = [\psi_s, \gamma_s]$. In this scheme, use is made of coordinate transformation $\alpha\beta/xy$ (Eqs. (5.63a,b)) and voltage equations (5.61a,b) in field coordinates. To avoid the dc-offset problem of the open-loop integration, the pure integrator ($y = \frac{1}{s}x$) can be rewritten as

$$y = \frac{1}{s + \omega_c}x + \frac{\omega_c}{s + \omega_c}y \tag{5.74}$$

where x and y are the system input and output signals, and ω_c is the cutoff frequency. The first part of Eq. (5.74) represents a low-pass filter, whereas the second part implements a feedback used to compensate for the error in the output. The block diagram of the improved integrator

(a)

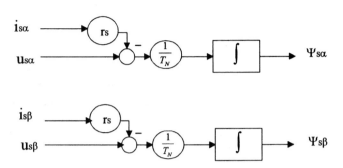

(b)

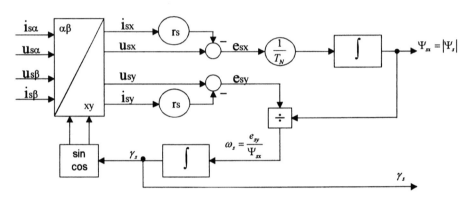

FIGURE 5.17
Stator flux vector estimators: (a) stator flux estimator in Cartesian coordinates; (b) stator flux estimator in polar coordinates.

according to Eq. (5.74) is shown in Fig. 5.18. It includes saturation block which stops the integration when the output signal exceeds the reference stator flux magnitude.

In DSP-based implementation, the voltage vector components are calculated from the inverter switching signals S_A, S_B, S_C and measured dc-link voltage U_{DC} as follows:

$$u_{s\alpha} = \frac{2}{3} U_{DC} \left(S_A - \frac{1}{2}(S_B + S_C) \right) \quad (5.75a)$$

$$u_{s\beta} = \frac{\sqrt{3}}{3} U_{DC}(S_B - S_C). \quad (5.75b)$$

However, in very low speed operation, the effect of inverter nonlinearities (dead time, dc link, and power semiconductor voltage drop) has to be compensated [31].

5.5 FIELD-ORIENTED CONTROL (FOC)

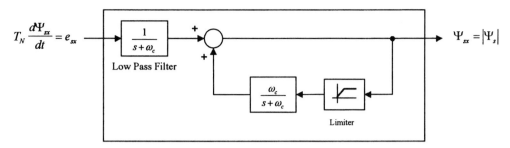

FIGURE 5.18
Improved integrator for amplitude calculation in Fig. 5.17b.

Rotor flux vector estimator. When the stator flux vector $\boldsymbol{\psi}_s$ is known, the rotor flux vector can be easily calculated from Eqs. (5.53) and (5.54) as

$$\boldsymbol{\psi}_r = \frac{x_r}{x_m}(\boldsymbol{\psi}_s - \sigma x_s \mathbf{i}_s). \tag{5.76}$$

This equation is represented in Fig. 5.19 as a block diagram in stationary α, β coordinates.

Many other methods of rotor flux estimation based on speed or position measurement have been developed. Also, observer technique is widely applied. Good review and evaluation is reported in [32] and [33].

5.5.6 Voltage Controlled S-FOC Scheme: Natural Field Orientation (NFO)

As discussed in Section 5.5.4, the implementation of S-FOC is much simpler for voltage- than for current-controlled PWM inverters. Further simplification can be achieved when instead of stator flux, the stator EMF is used as a basis for the current and/or voltage orientation (Fig. 5.10). This avoids the integration necessary for flux calculation. (Generally, EMF oriented control can be used in connection with rotor or stator flux orientation [7, 34].) Such a control scheme (Fig. 5.20), known as natural field orientation (NFO), is commercially available as an ASIC [35]. Note that the NFO scheme is developed from the stator flux model of Fig. 5.17b for $e_{sx} = 0$. The lack of current control loops and only r_s-dependent stator EMF estimation make the NFO scheme very attractive for low-cost speed, sensorless applications. However, as shown in

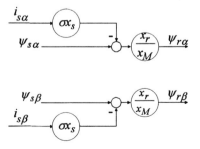

FIGURE 5.19
Rotor flux estimator based on stator flux.

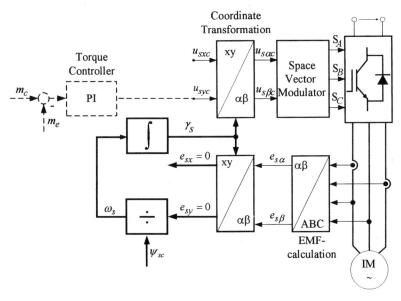

FIGURE 5.20
Voltage-controlled S-FOC scheme: NFO (dashed lines shows variant with outer torque control loop).

the oscillograms of Fig. 5.21, the torque control dynamic is limited by natural behavior of the IM (mainly by the rotor time constant, which for medium- and high-power motors can be in the range 0.2–1 s). Therefore, NFO can be attractive for low-power motors (up to 10 kW) or for low dynamic performance applications (such as V/Hz constant). An improvement can be achieved with an additional torque control loop (Fig. 5.21), which requires *on-line* torque estimation. So, the final control scheme configuration becomes like SVM-DTC [23].

5.6 MODERN NONLINEAR CONTROL OF IM

The FOC methods presented in Section 5.5 were proposed more than 30 years ago [22, 28] prior to a formal formulation of some new nonlinear control methods developed during the past 10 years [5, 14], such as the *exact linearization, passivity-based*, and *backstepping* designs. Therefore, these methods are generally called *nonlinear control schemes* [15, 64]. Some of them are briefly presented in this section.

5.6.1 Feedback Linearization Control (FLC)

The design based on exact linearization consist of two steps: In the first step, a nonlinear compensation which cancels the nonlinearities included in the IM is implemented as an inner feedback loop. In the second step, a controller which ensures stability and some predefined performance is designed based on conventional linear theory, and this linear controller is implemented as an outer feedback loop.

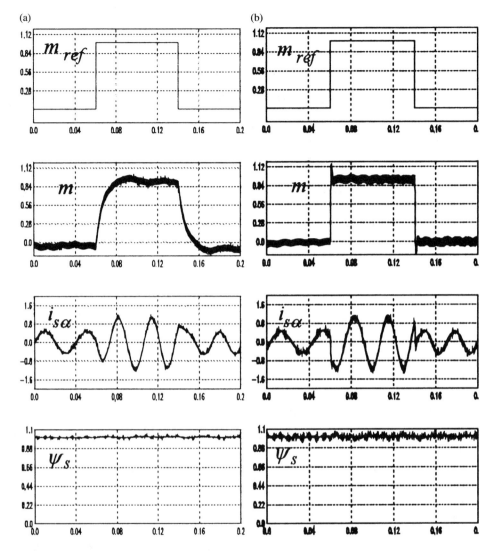

FIGURE 5.21
Torque transients in NFO control schemes for constant flux operation: (a) conventional, (b) with outer torque control loop.

5.6.1.1 Basic Principles and Block Scheme.
Selecting the vector of state variables as

$$x = [\psi_{r\alpha}, \psi_{r\beta}, i_{s\alpha}, i_{s\beta}, \omega_m]^T \tag{5.77}$$

and using p.u. time we can write the IM equations (5.11)–(5.15) in the form

$$\dot{x} = f(x) + u_{s\alpha} g_\alpha + u_{s\beta} g_\beta \tag{5.78}$$

where

$$f(x) = \begin{bmatrix} -\alpha\psi_{r\alpha} - \omega_m\psi_{r\beta} + \alpha x_M i_{s\alpha} \\ \omega_m\psi_{r\alpha} - \alpha\psi_{r\beta} + \alpha x_M i_{s\beta} \\ \alpha\beta\psi_{r\alpha} + \beta\omega_m\psi_{r\beta} - \gamma i_{s\alpha} \\ -\beta\omega_m\psi_{r\alpha} + \alpha\beta\psi_{r\beta} - \gamma i_{s\beta} \\ \mu(\psi_{r\alpha}i_{s\beta} - \psi_{r\beta}i_{s\alpha}) - \dfrac{m_L}{\tau_M} \end{bmatrix} \quad (5.79)$$

$$g_\alpha = \begin{bmatrix} 0, & 0, & \dfrac{1}{\sigma x_s}, & 0, & 0 \end{bmatrix}^T \quad (5.80)$$

$$g_\beta = \begin{bmatrix} 0, & 0, & 0, & \dfrac{1}{\sigma x_s}, & 0 \end{bmatrix}^T \quad (5.81)$$

and

$$\alpha = \frac{r_r}{x_r}; \quad \beta = \frac{x_M}{\sigma x_s x_r}; \quad \gamma = \frac{x_r^2 r_s + x_M^2 r_r}{\sigma x_s x_r^2}; \quad \mu = \frac{x_M}{\tau_M x_r}; \quad \sigma = 1 - \frac{x_M^2}{x_s x_r}.$$

Note that ω_m, $\psi_{r\alpha}$, $\psi_{r\beta}$ are independent of control signals $u_{s\alpha}$, $u_{s\beta}$. In this case it is easy to choose two variables dependent on x only. For example, we can define [5, 55, 63]

$$\phi_1(x) = \psi_{r\alpha}^2 + \psi_{r\beta}^2 = \psi_r^2 \quad (5.82)$$
$$\phi_2(x) = \omega_m \quad (5.83)$$

Let $\phi_1(x)$, $\phi_2(x)$ be the output variables. The aim of control is to obtain

- Constant flux amplitude
- Reference angular speed

Part of the new state variables we can choose according to Eqs. (5.82) and (5.83). So the full definitions of the new coordinates are given by [59, 60]

$$\begin{aligned} \dot{z}_1 &= \phi_1(x) \\ \dot{z}_2 &= L_f \phi_1(x) \\ \dot{z}_3 &= \phi_2(x) \\ \dot{z}_4 &= L_f \phi_2(x) \\ z_5 &= \arctan\left(\frac{\psi_{r\beta}}{\psi_{r\alpha}}\right) \end{aligned} \quad (5.84)$$

Note that the fifth variable cannot be linearizable and the linearization can be only partial. Denote

$$\phi_3(x) = \dot{z}_5. \quad (5.85)$$

Then the dynamic of the system is given by

$$\begin{aligned}
\dot{z}_1 &= z_2 \\
\dot{z}_2 &= L_f^2\phi_1(x) + L_{g\alpha}L_f\phi_1(x)u_{s\alpha} + L_{g\beta}L_f\phi_1(x)u_{s\beta} \\
\dot{z}_3 &= z_4 \\
\dot{z}_4 &= L_f^2\phi_2(x) + L_{g\alpha}L_f\phi_2(x)u_{s\alpha} + L_{g\beta}L_f\phi_2(x)u_{s\beta} \\
\dot{z}_5 &= L_f^2\phi_3(x).
\end{aligned} \quad (5.86)$$

We can rewrite the remaining system Eq. (5.86) in the form

$$\begin{bmatrix} \ddot{z}_1 \\ \ddot{z}_2 \end{bmatrix} = \begin{bmatrix} L_f^2\phi_1 \\ L_f^2\phi_2 \end{bmatrix} + D \begin{bmatrix} u_{s\alpha} \\ u_{s\beta} \end{bmatrix}. \quad (5.87)$$

D is given by

$$D = \begin{bmatrix} L_{g\alpha}L_f\phi_1 & L_{g\beta}L_f\phi_1 \\ L_{g\alpha}L_f\phi_2 & L_{g\beta}L_f\phi_2 \end{bmatrix}$$

It is easy to show that if $\phi_1 \neq 0$, then $\det(D) \neq 0$.

In this case we can define linearizing feedback as

$$\begin{bmatrix} u_{s\alpha} \\ u_{s\beta} \end{bmatrix} = D^{-1} \left\{ \begin{bmatrix} -L_f^2\phi_1 \\ -L_f^2\phi_2 \end{bmatrix} + \begin{bmatrix} v_1 \\ v_2 \end{bmatrix} \right\}. \quad (5.88)$$

The resulting system is described by the equations

$$\begin{aligned}
\dot{z}_1 &= z_2 \\
\dot{z}_2 &= v_1 \\
\dot{z}_3 &= z_4 \\
\dot{z}_4 &= v_2.
\end{aligned} \quad (5.89)$$

A block diagram of an induction motor with new control signals is presented in Fig. 5.22. Control signals v_1, v_2 can be calculated using linear feedback:

$$v_1 = k_{11}(z_1 - z_{1\mathrm{ref}}) - k_{12}z_2 \quad (5.90)$$
$$v_2 = k_{21}(z_3 - z_{3\mathrm{ref}}) - k_{22}z_4 \quad (5.91)$$

where coefficients k_{11}, k_{12}, k_{21}, k_{22} are chosen to determinate closed loop system dynamic. Control algorithm consist of two steps:

- calculations v_1, v_2 according to Eqs. (5.90) and (5.91),
- calculations $u_{s\alpha}$ and $u_{s\beta}$ according to Eq. (5.88).

The decoupling performance achieved in the FLC scheme of Fig. 5.23 is shown in Fig. 5.24. Note, that v_1 and v_2 influence only torque and flux amplitude changes, respectively.

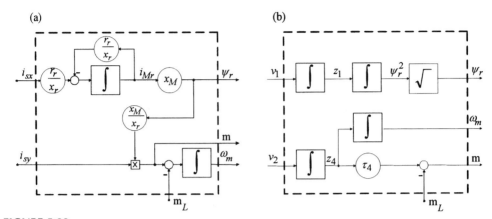

FIGURE 5.22
Block diagrams of induction motor: (a) in x–y field coordinates, (b) with new control signals v_1, v_2 (feedback linearization).

5.6.1.2 FLC versus FOC.
The simulated oscillograms obtained for FLC and FOC schemes with linear speed and rotor flux amplitude controllers are shown in Fig. 5.25. These oscillograms show the speed reversal over the constant and field-weakening regions when the motor is fed from a VSI inverter with sinusoidal PWM. As can be seen from Fig. 5.25b, the FOC does not guarantee full decoupling between speed and flux amplitude of the induction motor.

With a linear speed controller (without the flux decoupler in Fig. 5.13), the FOC scheme implements torque current limitation, whereas the FLC scheme limits the motor torque only (see Fig. 5.25a). Therefore, in the FOC scheme the torque is reduced in the field-weakening region, and the speed transient is slower than in the FLC scheme. To achieve full decoupling in the FOC scheme working in the field weakening region, similarly to in dc motor drives [7, 8], a PI speed controller with nonlinear flux decoupler (controller output signal should be divided by the rotor flux amplitude m_c/ψ_{rc} as shown in Fig. 5.13b) has to be applied. This division compensates for

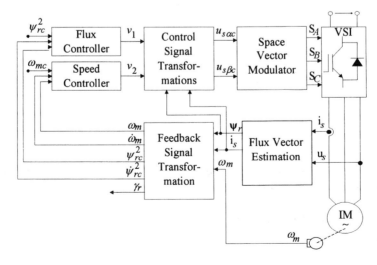

FIGURE 5.23
Feedback linearized control of PWM inverter-fed induction motor.

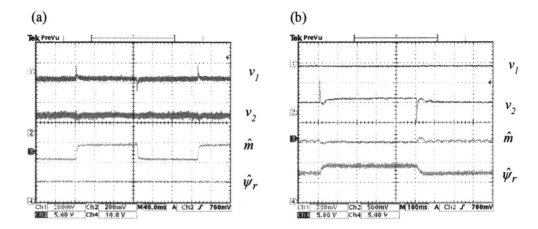

FIGURE 5.24
Experimental oscillograms illustrating decoupling performance in the FLC scheme of Fig. 5.23: (a) torque tracking, (b) flux amplitude tracking.

the internal multiplication ($m = (x_M/x_r)\psi_r i_{sy}$) needed for motor torque production in rotor-field-oriented coordinates (Fig. 5.22a). With such a nonlinear speed controller very similar dynamic performances to the FLC scheme can be achieved (see Figs. 5.25a and 5.25c).

The main features and advantages of FOC and FLC systems can be summarized as follows:

- With control variables v_1, v_2 the FLC scheme guarantees exact decoupling of the motor speed and rotor flux amplitude control in both dynamic and steady states. Therefore, a high-performance drive system working over the full speed range including field weakening can be implemented using linear speed and flux controllers.
- With control variables i_{sx}, i_{sy} the FOC scheme cannot guarantee the exact decoupling of the motor speed and rotor flux amplitude control in dynamic states.
- FLC is implemented in a state feedback fashion and needs more complex signal processing (full information about motor state variables and load torque is required). Also, the transformation and new control variables v_1, v_2 used in FLC have no such direct physical meaning as i_{sx}, i_{sy} (flux and torque current, respectively) in the case of the FOC scheme.

5.6.2 Multiscalar Control

The exact input–output linearization from stator voltages to torque, speed, and square of rotor flux amplitude has been proposed in [57].

5.6.2.1 Multiscalar Model of IM.
New variables for description of the IM are selected as follows:

$$q_{11} = \omega_r \quad \text{(mechanical speed)} \tag{5.92}$$

$$q_{12} = \text{Im}[\boldsymbol{\psi}_r^* \mathbf{i}_s] \quad \text{(electromagnetic torque)} \tag{5.93}$$

$$q_{21} = \psi_r^2 \quad \text{(square of rotor flux)} \tag{5.94}$$

$$q_{22} = \text{Re}[\boldsymbol{\psi}_r^* \mathbf{i}_s] \quad \text{(reactive torque).} \tag{5.95}$$

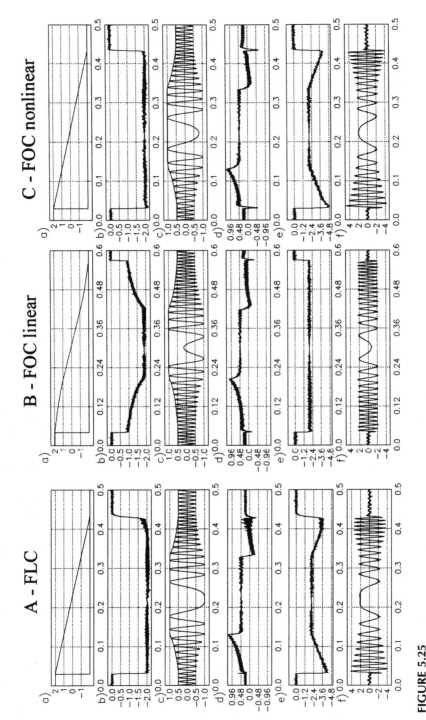

FIGURE 5.25
Feedback linearization and field-oriented control of induction motor (speed reversal including field-weakening range): (a) actual and reference speed (ω_{mref}, ω_m); (b) torque m; (c) flux component and amplitude ($\psi_{r\alpha}$, ψ_r); (d) flux current i_{sx}; (e) torque current i_{sy}; (f) current component $i_{s\beta}$.

So defined variables q_{12}, q_{21}, q_{22} are scalars and are independent of the coordinate system in which the current and flux vectors are represented. Therefore, the IM model based on variables given by Eqs. (5.93), (5.94), and (5.95) is called a *multiscalar model*.

The variables q_{12}, q_{21}, and q_{22} can be expressed in terms of field-oriented components as

$$q_{11} = \psi_{rx} i_{sy} \tag{5.96}$$

$$q_{21} = \psi_{rx}^2 \tag{5.97}$$

$$q_{22} = \psi_{rx} i_{sx}. \tag{5.98}$$

After differentiation of (5.96)–(5.98), the IM equations (5.1–5.6) can be written in the form

$$\dot{q}_{11} = \mu q_{12} - \frac{m_L}{\tau_M} \tag{5.99}$$

$$\dot{q}_{12} = -\frac{1}{\tau_{rs}} q_{12} - q_{11}(q_{22} + \beta q_{21}) + \frac{u_1}{\sigma x_2} \tag{5.100}$$

$$\dot{q}_{21} = -\frac{2}{\tau_r} q_{21} + \frac{2 x_M}{\tau_r} \tag{5.101}$$

$$\dot{q}_{22} = -\frac{1}{\tau_{rs}} q_{22} + q_{11} q_{12} + \frac{\beta}{\tau_r} q_{21} + \frac{x_M}{\tau_r} \frac{q_{12}^2 + q_{22}^2}{q_{21}} + \frac{u_1}{\sigma x_s} \tag{5.102}$$

where

$$\tau_r = \frac{x_r}{r_r}; \quad \tau_{rs} = \frac{\sigma x_s x_r}{r_r x_s + r_s x_r}; \quad \tau_M = \frac{1}{J}; \quad \mu = \frac{x_M}{\tau_M x_r}; \quad \beta = \frac{x_M}{\sigma x_s x_r} \tag{5.103}$$

$$u_1 = \text{Im}[\boldsymbol{\psi}_r^* \mathbf{u}_s] \tag{5.104}$$

$$u_2 = \text{Re}[\boldsymbol{\psi}_r^* \mathbf{u}_s]. \tag{5.105}$$

Note that

$$\frac{q_{12}^2 + q_{22}^2}{q_{21}} = i_s^2. \tag{5.106}$$

There are only four differential equations (5.99)–(5.102) describing the dynamic and static properties of the IM. The exact positioning of the rotor flux and stator current space vectors, as in the FOC schemes, is not important. This is possible because the IM torque does not depend on the position of the stator current or rotor flux in any coordinates, but on their mutual position (see Section 5.1). The mutual position and the values of the two vectors can be determinated by their scalar and vector products and the magnitude of one of them. This is a basis for interpretation of the variables q_{12}, q_{21}, q_{22}. The variables u_1 and u_2 are vector and scalar products of the rotor flux and stator voltage, and—in a similar way—they describe the mutual positions of these vectors.

5.6.2.2 Decoupling and Linearization of Multiscalar Model.
The multiscalar model of the IM developed in the previous section includes two linear equations without control variables (5.99) and (5.101), and two nonlinear equations, (5.100) and (5.102), with control variables u_1, u_2. Defining the new control variables m_1, m_2 which fulfill the nonlinear equations

$$\begin{bmatrix} u_1 \\ u_2 \end{bmatrix} = \sigma x_s \begin{bmatrix} q_{11}(q_{22} + \beta q_{21}) \\ -q_{11} q_{12} - \frac{\beta}{\tau_r} q_{21} - \frac{x_M}{\tau_r} \frac{q_{12}^2 + q_{22}^2}{q_{21}} \end{bmatrix} + \frac{1}{\tau_{rs}} \begin{bmatrix} m_1 \\ m_2 \end{bmatrix}, \tag{5.107}$$

the nonlinear IM equations (5.99)–(5.102) can be transformed into two *linear* subsystems:

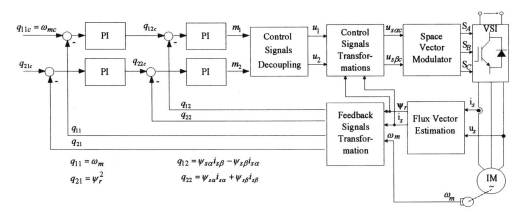

FIGURE 5.26
Multiscalar control of PWM inverter-fed induction motor.

Mechanical subsystem:

$$\dot{q}_{11} = \mu q_{12} - \frac{m_L}{\tau_M} \qquad (5.99)$$

$$\dot{q}_{12} = -\frac{1}{\tau_{rs}} q_{12} + m_1. \qquad (5.108)$$

Electromagnetic subsystem:

$$\dot{q}_{21} = -\frac{2}{\tau_r} q_{21} + \frac{2x_M}{\tau_r} \qquad (5.101)$$

$$\dot{q}_{22} = -\frac{1}{\tau_{rs}} q_{22} + m_2. \qquad (5.109)$$

Note that variables q_{12} and q_{22} have first-order dynamics described by the same time constant τ_{rs}. The block diagram of the control system designed according to Eqs. (5.107)–(5.109) is shown in Fig. 5.26. The control scheme corresponds to a cascade structure which allows simple limitation of internal variables, e.g., torque (q_{12}) can easily be limited on the output of the speed controller.

5.6.3 Passivity-Based Control (PBC)

In contrast to the feedback linerazation and multiscalar control, which results from a purely mathematical approach, the passivity-based control (also known as *energy shaping design*) has evolved from consideration of physical properties such as energy saving and passivity [15].

The main idea of passivity-based design is to reshape the energy of the IM in such a way that the required asymptotic output tracking properties will be achieved. The key point with this method is the identification of terms, called *workless forces*, which appear in the dynamic equation of the IM, but do not influence the energy balance expressed by the Euler–Lagrange description of the IM. These terms do not affect the stability properties of the IM and, therefore, there is no need to cancel or offset them with feedback control. This results in simplified design and enhanced robustness. It is characteristic for the PBC applied to the nonlinear system that the closed-loop dynamics remain nonlinear.

5.6.3.1 Model of IM in Generalized Coordinates.

The energy balance equations of an electric motor can be obtained from the force balance equations written using the generalized coordinates as follows [15]:

$$Q_e = u - R_e \dot{q}_e \Rightarrow \frac{d}{dt}\dot{q}_e \qquad (5.110)$$

$$Q_m = R_m \dot{q}_m \qquad (5.111)$$

where $q_e = \int i_k(t)dt + q_{ok}$ is the electrical coordinate: electric charge; q_m is the mechanical coordinate: shaft position; Q_e, Q_m are electric and mechanical force; and u is the stator voltage.

Decomposition of the IM into electrical and mechanical subsystems is shown in Fig. 5.27.

The dynamic equations derived by direct application of the Euler–Lagrange approach results in the following description of the IM [15]:

$$\boldsymbol{D}_e(q_m)\ddot{\boldsymbol{q}}_e + \boldsymbol{W}_1(q_m)\dot{q}_m\dot{\boldsymbol{q}}_e + \boldsymbol{R}_e\dot{\boldsymbol{q}}_e = \boldsymbol{M}_e u \qquad (5.112)$$

$$D_m \ddot{q}_m = m(\dot{\boldsymbol{q}}_e, q_m) - m_L \qquad (5.113)$$

$$m(\dot{\boldsymbol{q}}_e, q_m) = \frac{1}{2}\dot{\boldsymbol{q}}_e^T \boldsymbol{W}_1(q_m)\dot{\boldsymbol{q}}_e \qquad (5.114)$$

where

$$\boldsymbol{D}_e(q_m) = \begin{bmatrix} x_s \boldsymbol{1}_2 & x_{sr} e^{Jq_m} \\ x_{sr} e^{-Jq_m} & x_r \boldsymbol{1}_2 \end{bmatrix} \quad \text{(mutual matrix)}$$

$$\boldsymbol{R}_e = \begin{bmatrix} r_s \boldsymbol{1}_2 & 0 \\ 0 & r_r \boldsymbol{1}_2 \end{bmatrix} \quad \text{(resistance matrix)}$$

$$e^{Jq_m} = \begin{bmatrix} \cos q_m & -\sin q_m \\ \sin q_m & \cos q_m \end{bmatrix} \quad \text{(rotation matrix)}$$

$$\boldsymbol{M}_e = \begin{bmatrix} \boldsymbol{1}_2 \\ 0 \end{bmatrix}, \quad \boldsymbol{J} = \begin{bmatrix} 0 & -1 \\ 1 & 0 \end{bmatrix}, \quad \boldsymbol{1}_2 = \begin{bmatrix} 1 & 0 \\ 0 & 1 \end{bmatrix}$$

$$\boldsymbol{W}_1(q_m) = \frac{d\boldsymbol{D}_e(q_m)}{dq_m} = \begin{bmatrix} 0 & x_{sr} \boldsymbol{J} e^{Jq_m} \\ -x_{sr} \boldsymbol{J} e^{-Jq_m} & 0 \end{bmatrix};$$

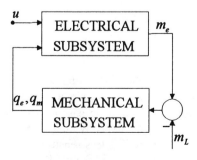

FIGURE 5.27
Decomposition of the IM into electrical and mechanical subsystems.

$\dot{q}_e = [\dot{q}_{s1}, \dot{q}_{s2}, \dot{q}_{r1}, \dot{q}_{r2}]^T$ is the current vector, $\dot{q}_m$ is the rotor angular speed, $D_m > 0$ is the rotor inertia, $\mathbf{u} = [u_1, u_2]^T$ is the stator voltage (control signals) vector, m is electromagnetic torque, m_L is load torque, and x_s, x_r, x_{sr} are stator, rotor, and mutual reactances, respectively.

5.6.3.2 Observerless PBC.
The design procedure to obtain a PBC suitable for the IM consists of three distinct steps.

- To achieve strict passifiability of the electrical subsystem, a nonlinear damping term is "injected." In [15] it has been demonstrated that the damping term is given by

$$K(\dot{q}_m) = \frac{x_{sr}^2}{4\varepsilon} \dot{q}_m^2 \tag{5.115}$$

  where $0 < \varepsilon < \min\{r_s, r_r\}$. Notice that in actual motors r_s and r_r vary with temperature. However, since this relationship gives a minimum condition for the damping term, and because of the nature of such a term, practical design can be quickly obtained imposing a high gain value to the forward subsystem.

- Currents tracking via energy shaping: this part of the design procedure consists of the choice of a set of desired currents $\dot{q}_e^*$ such that

$$\dot{q}_e^* - q_{e0} \to 0 \quad \text{as} \quad t \to 0,$$

  pursuing a relationship between inputs (electric forces) and outputs.

- The third step consist of torque tracking, i.e., the choice of $\dot{q}_e^*$ such that

$$m_c - m_0 \to 0 \quad \text{as} \quad t \to 0$$

where m_c is the reference torque.

Finally, the PBC equation is given as

$$\mathbf{u} = \underbrace{x_s \ddot{q}_{sc} + x_{sr} e^{Jq_m} \ddot{q}_{rc} + x_{sr} J e^{Jq_m} \dot{q}_m \dot{q}_{rc} + R_s \dot{q}_{sc}}_{\text{commanded dynamics}} - \underbrace{K(\dot{q}_m) \Delta \dot{q}_s}_{\text{nonlinear damping term}} - \underbrace{K_{1s} \int_0^t \Delta \dot{q}_s dt}_{\text{integral term}} \tag{5.116}$$

where

$$\dot{q}_{ec} = \begin{bmatrix} \dot{q}_{sc} \\ \dot{q}_{rc} \end{bmatrix} = \begin{bmatrix} \frac{1}{x_{sr}} \left[\left(1 + \tau_r \frac{\dot{\psi}_{rc}}{\psi_{rc}}\right) \mathbf{1}_2 + \frac{x_r}{\psi_{rc}^2} m_c J \right] \cdot e^{Jq_m} \cdot \psi_{rc} \\ -\left(\frac{m_c}{\psi_{rc}^2} J + \frac{\dot{\psi}_{rc}}{r_r \psi_{rc}} \mathbf{1}_2\right) \cdot \psi_{rc} \end{bmatrix}$$

$$\Delta \dot{q}_e = \begin{bmatrix} \Delta \dot{q}_s \\ \Delta \dot{q}_r \end{bmatrix} = \begin{bmatrix} \dot{q}_s - \dot{q}_{sc} \\ \dot{q}_r - \dot{q}_{rc} \end{bmatrix}$$

$$m_c = D_m \ddot{q}_m + m_L.$$

The block diagram of the PBC scheme for the PWM voltage-source inverter-fed IM is shown in Fig. 5.28. It is assumed that a position sensor is available; however, a flux sensor is not required.

In spite of the complex controller equations, the passivity-based design leads to a control scheme with global stability which is theoretically and experimentally proved. Also, as demonstrated in [49, 50, 54, 56, 61], the PBC is less sensitive to IM parameter changes.

When a current-controlled PWM inverter is used for the IM supply, the PBC control scheme is identical to the indirect FOC system.

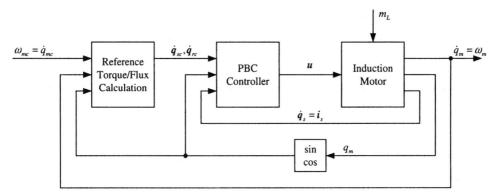

FIGURE 5.28
Block diagram of PBC scheme for PWM-inverter-fed induction motor.

5.7 CONCLUSION

This chapter has reviewed control strategies developed for PWM inverter-fed induction motor (IM) drives. Starting from the space vector description of the IM, the control strategies are generally divided into scalar and vector methods.

- Scalar control is based on linearizations of the nonlinear IM equations at steady-state operating points and typically is implemented in schemes where amplitude and frequency of stator voltage is adjusted in an open-loop fashion keeping V/Hz constant. However, such a scheme applied to a multivariable system such as IM cannot perform decoupling between inputs and outputs, resulting in problems with independent control of outputs, e.g., torque and flux. Therefore, to achieve decoupling in a high-performance IM drive, vector control, also known as field-oriented control (FOC), has been developed. FOC is now a de facto standard in high dynamic IM industrial drives.
- In FOC schemes the stator current vector is controlled with respect to flux vector position. It follows from basic theory that in the IM there are three slightly different magnetizing currents which correspond to main, rotor, and stator flux linkages, respectively. This gives three possible implementations of FOC schemes.
- The rotor-flux-oriented control (R-FOC) is easily implemented in connection with a current-controlled PWM inverter.
- The stator-flux-oriented control (S-FOC) has a very simple structure when IM is supplied from a voltage-controlled (open-loop) PWM inverter and is known as natural field orientation (NFO).
- For good low-speed operation performance, the indirect R-FOC with speed/position sensor is recommended. This scheme, however, is sensitive to changes of rotor time constant which have to be adapted on-line.
- For sensorless operation the direct R-FOC or NFO scheme can be advised.
- The group of modern nonlinear controls (Section 5.6) offers a new, interesting perspective for future research. However, from the present industrial point of view, they represent only an alternative solution to existing FOC and DTC schemes.

5.8 APPENDIX 1: PER UNIT SYSTEMS

Per unit systems are defined in terms of *base units*, which most frequently correspond to rated motor parameters. Commonly adopted base units are U_b, I_b, and Ω_{mb}, and from these the so-called derivative base quantities are determined. It is standard procedure to use capital letters to denote absolute physical quantities and to represent the relevant quantities, expressed in relative units, by small letters. Thus, for example, if

$$u = \frac{U}{U_b}, \tag{A.1}$$

then U is the physical value in V, U_b the base quantity (usually rated) in V, and u the dimensionless relative (p.u.).

The *p.u. for the induction motor* is referred to by the following basic quantities:

phase voltages

$$U_b = |\mathbf{U}_{sN}| = U_{sN\max} = \sqrt{2} U_{s(rms)N} \quad [\text{V}], \tag{A.2a}$$

phase current

$$I_b = |\mathbf{I}_{sN}| = I_{sN\max} = \sqrt{2} I_{s(rms)N} \quad [\text{A}], \text{ and} \tag{A.2b}$$

angular speed

$$\Omega_b = \Omega_{sN} \quad [\text{s}^{-1}] = 2\pi F_{sN} \quad [\text{Hz}], \tag{A.2c}$$

where $|\mathbf{U}_{sN}|$ and $|\mathbf{I}_{sN}|$ denote, respectively, stator voltage and current space vector moduli, $U_{sN\max}$ and $I_{sN\max}$ are stator phase voltage and current amplitudes, $U_{s(rms)N}$ and $I_{s(rms)N}$ the phase voltage and current rms values, N the relevant rated values, and Ω_{sN} and F_{sN} the stator rated angular speed and frequency, respectively.

From these basic quantities the following basic derivative quantities are obtained:

impedances and resistances

$$Z_b = \frac{U_b}{I_b} = \frac{U_{sN\max}}{I_{sN\max}}, \tag{A.3a}$$

flux linkages

$$\psi_b = \frac{U_b}{\Omega_b} = \frac{U_{sN\max}}{\Omega_{sN}}, \tag{A.3b}$$

inductances

$$L_b = \frac{\psi_b}{I_b} = \frac{U_{sN\max}}{I_{sN\max}\Omega_{sN}}, \tag{A.3c}$$

powers

$$S_b = \frac{2}{3} U_b I_b = \frac{2}{3} U_{sN\max} I_{sN\max} = 3 U_{s(rms)N} I_{s(rms)N}, \tag{A.3d}$$

mechanical angular speed

$$\Omega_{mb} = \frac{\Omega_b}{p_b} = \frac{\Omega_{sN}}{p_b}, \tag{A.3e}$$

torque

$$M_b = \frac{S_b}{\Omega_{mb}} = \frac{3}{2} p_b U_{sN\max} \frac{I_{sN\max}}{\Omega_{sN}}. \tag{A.3f}$$

It follows from the derivation of p.u. that:

1. The identity $l = x$ holds, since

$$l = \frac{L[\mathrm{H}]}{L_b[\mathrm{H}]} = L[\mathrm{H}]\Omega_b \frac{I_b}{U_b} = \Omega_b \frac{L[\mathrm{H}]}{Z_b[\Omega]} = \frac{X[\Omega]}{Z_b[\Omega]} = x, \tag{A.4}$$

2. The electromagnetic torque is referred to as $M_b = p_b S_N / \Omega_{sN}$, which means that torque expressed in p.u. attains the value $m = 1$ for the case when there is no reactive power input to the machine; this corresponds to the condition of parallelism between the inner voltage space vector ($\mathbf{U}_{sN} - R_s \mathbf{I}_{sN}$) and the current vector $\mathbf{I}_{sN}$ (i.e., $\varphi_{sN} = 0$) or, which amounts to the same, $\mathbf{\Psi}_{sN} \perp \mathbf{I}_{sN}$. However, if the machine operates at power factor $\cos \varphi_{sN} < 1$, i.e., $S_N > P_N$, then at the rated operation point the torque expressed in p.u. is less than unity ($m < 1$).

5.9 APPENDIX 2: INDUCTION MOTOR DATA (P.U.)

$$P_N = 4 \text{ kW} \tag{A.5}$$
$$r_s = 0.059 \tag{A.6}$$
$$r_r = 0.048 \tag{A.7}$$
$$x_s = x_r = 1.92 \tag{A.8}$$
$$x_M = 1.82 \tag{A.9}$$

5.10 NOMENCLATURE

Basic Principles

1. Complex numbers and complex space vectors are denoted by boldface, nonitalic letters, e.g., $\mathbf{U}_r$, $\mathbf{i}_s$, and their moduli by U_r, i_s, respectively.
2. Matrices and vectors are denoted by boldface italics, e.g., *A*, *u*.
3. Capital letters are used for time-dependent absolute values, especially complex space vectors and time independent physical quantities (e.g., impedances).
4. Lower case letters are used for time-dependent, dimensionless (p.u.) quantities, especially complex space vectors and their components; also per unit impedances are denoted by lower case letters.
5. Time t is an exception to rules 3 and 4, and is denoted by lower case italic. Also, angular frequency ω and frequency f are traditionally denoted by lower case italics.

Main Symbols

F_N	nominal frequency (50 or 60 Hz)
f	frequency
I	current, absolute value
i	current, p.u. value
J	moment of inertia
L	inductance, absolute value
l	inductance, p.u. value
M	torque, absolute value
m	torque, p.u. value
p_b	number of pole pairs
$p = \dfrac{d}{dt}$	derivative
R	resistance, absolute value
r	resistance, p.u. value
s	slip
T	time constant, absolute value
U	voltage, absolute value
u	voltage, p.u. value
X	reactance, absolute value calculated for nominal frequency F_N (50 Hz)
x	reactance, p.u. value calculated for nominal frequency
$\boldsymbol{x}$	state variables vector
Z	impedance, absolute value
z	impedance, p.u. value
δ	load angle
$\sigma = 1 - x_M^2/x_r x_s$	total leakage factor
Ψ	flux linkage, absolute value
ψ	flux linkage, p.u. value
Ω	angular speed, absolute value
ω	angular speed, p.u. value

Indices

c	command value, reference value
K	rotated with arbitrary speed coordinate system
M	main, magnetizing
m	mechanical
N	nominal value, rated value
r	rotor
s	stator
rms	root-mean-square value

Rectangular Coordinate Systems

α, β stator-oriented (stationary) coordintes
x, y field-oriented (rotated) coordinates

Abbreviations

DTC   direct torque control
FLC   feedback linearization control
FOC   field-oriented control
IM   induction motor
NFO   natural field orientation
PBC   passivity-based control
PWM   pulse width modulation
VSI   voltage source inverter

REFERENCES

Books

[1] I. Boldea and S. A. Nasar, *Electric Drives*. CRC Press, Boca Raton, FL, 1999.
[2] B. K. Bose, *Modern Power Electronics and AC Drives*. Prentice-Hall, Englewood Cliffs, 2001.
[3] B. K. Bose (ed.), *Power Electronics and Variable Frequency Drives*. IEEE Press, 1997.
[4] H. Buhler, *Einführung in die Theorie geregelter Drehstrom-Antriebe*. Vols. 1, 2. 2nd ed., Birkhauser, Basel, 1977.
[5] A. Isidori, *Nonlinear Control Systems, Communications and Control Engineering*. 3rd ed., Springer Verlag, Berlin, 1995.
[6] F. Jenni and D. Wüst, *Steuerverfahren für selbstgeführte Stromrichter*. B. G. Teubner, Stuttgart, 1995.
[7] M. P. Kazmierkowski and H. Tunia, *Automatic Control of Converter Fed Drives*. Elsevier, Amsterdam, 1994.
[8] R. Krishnan, *Electric Motor Drives*. Prentice-Hall, Englewood Cliffs, NJ, 2001.
[9] W. Leonhard, *Control of Electrical Drives*, 2nd ed., Springer Verlag, Berlin, 1996.
[10] N. Mohan, T. M. Undeland, and B. Robbins, *Power Electronics*. J. Wiley, New York, 1989.
[11] N. Mohan, *Advanced Electric Drives*. MNPERE, Minneapolis, 2001.
[12] J. M. D. Murphy and F. G. Turnbull, *Power Electronic Control of AC Motors*. Pergamon, Oxford, 1988.
[13] D. W. Novotny and T. A. Lipo, *Vector Control and Dynamics of AC Machines*. Clarendon Press, Oxford, 1996.
[14] H. Nijmeijer and van der Schaft, *Nonlinear Dynamical Control Systems*. Springer-Verlag, New York, 1990.
[15] R. Ortega, A. Loria, P. J. Nicklasson, and H. Sira-Ramirez, *Passivity-based Control of Euler–Lagrange Systems*. Springer Verlag, London, 1998.
[16] K. Rajashekara, A. Kawamura, and K. Matsue, *Sensorless Control of AC Motor Drives*. IEEE Press, 1996.
[17] A. M. Trzynadlowski, *The Field Orientation Principle in Control of Induction Motors*. Kluwer Academic Publisher, Boston, 1994.
[18] A. M. Trzynadlowski, *Control of Induction Motors*. Academic Press, New York, 2000.
[19] P. Vas, *Vector Control of AC Machines*. Clarendon Press, Oxford, 1990.
[20] P. Vas, *Sensorless Vector and Direct Torque Control*. Clarendon Press, Oxford, 1998.
[21] S. Yamamura, *AC Motors for High-Performance Applications*. Marcel Dekker Inc., New York, 1986.

Field-Oriented Control

[22] F. Blaschke, Das Verfahren der Feldorientirung zur Regleung der Asynchronmaschine. *Siemens Forschungs und Entwicklungsberichte*, **1**(1), 184–193 (1972).
[23] D. Casadei, G. Serra, and A. Tani, Constant frequency operation of a DTC induction motor drive for electric vehicle, in *Proc. ICEM Conf.*, 1996, Vol. 3, pp. 224–229.
[24] P. J. Costa Branco, A simple adaptive scheme for indirect field orientation of an induction motor. *ETEP* **7**, 243–249 (1997).
[25] R. De Doncker and D. W. Novotny, The universal field-oriented controller. *IEEE Trans. Indust. Appl.* **30**, 92–100 (1994).
[26] L. Garces, Parameter adaptation for the speed controlled static AC drive with a squirrel cage induction motor. *IEEE Trans. Indust. Appl.* **IA-16**, 173–178 (1980).
[27] F. Harashima, Power electronics and motion control—A future perspective. *Proc. IEEE* **82**, 1107–1111 (1994).
[28] K. Hasse, Drehzahlgelverfahren fur schnelle Umkehrantriebe mit stromrichtergespeisten Asynchron—Kurzschlusslaufermotoren. *Reglungstechnik* **20**, 60–66 (1972).
[29] J. Holtz, Speed estimation and sensorless control of AC drives. *IEEE/IECON'93 Conf. Rec.*, pp. 649–654, 1993.
[30] J. Holtz, The representation of AC machines dynamic by complex signal flow graphs. *IEEE Trans. Indust. Electron.* **42**, 263–271 (1995).
[31] J. Holtz, Sensorless vector control of induction motors at very low speed using a nonlinear inverter model and parameter identification. *IEEE-IAS Annual Meeting*, pp. 2614–2621, 2001.
[32] P. L. Jansen, R. D. Lorenz, and D. W. Novotny, Observer based direct field orientation and comparison of alternative methods. *IEEE Trans. Indust. Appl.* **IA-30**, 945–953 (1994).
[33] P. L. Jansen and R. D. Lorenz, A physically insightful approach to design and accuracy assessment of flux observers for field oriented induction machine drives. *Proc. IEEE IAS Annual Meeting*, pp. 570–577, 1992.
[34] R. Jötten and G. Mader, Control methods for good dynamic performance induction motor drives based on current and voltage as measured quantities. *Proc. IEEE/PESC'82*, pp. 397–407, 1982.
[35] R. Jönsson and W. Leonhard, Control of an induction motor without a mechanical sensor, based on the principle of "Natural Field Orientation" (NFO). *Proc. IPEC'95*, Yokohama, 1995.
[36] M. P. Kazmierkowski and W. Sulkowski, Novel vector control scheme for transistor PWM inverter-fed induction motor drive. *IEEE Trans. Indust. Electron.* **38**, 41–47 (1991).
[37] R. J. Kerkman, B. J. Seibel, and T. M. Rowan, A new flux and stator resistance identifier for AC drive systems. *IEEE Trans. Indust. Appl.* **32**, 585–593 (1996).
[38] M. Koyama, M. Yano, I. Kamiyama, and S. Yano, Microprocessor-based vector control system for induction motor drives with rotor constant identification function. *IEEE Trans. Indust. Appl.* **IA-22**, 453–459 (1986).
[39] R. Krishnan and F. C. Doran, Study of parameter sensitivity in high performance inverter-fed induction motor drive system. *Proc. IEEE/IAS'84*, pp. 510–224, 1984.
[40] R. D. Lorenz and D. B. Lawson, Simplified approach to continous on-line tuning of field-oriented induction machine drives. *Proc. IEEE-IAS Annual Meeting*, pp. 444–449, October 1988.
[41] R. Nilsen and M. P. Kazmierkowski, Reduced-order observer with parameter adaption for fast rotor flux estimation in induction machines. *Proc. IEE*, Pt. D, Vol. 136, No. 1, pp. 35–43, 1989.
[42] T. Okuyama, H. Nagase, Y. Kubota, H. Horiuchi, K. Miyazaki and S. Ibori, High performance AC speed control system using GTO converters. *Proc. IPEC-Tokyo*, pp. 720–731, 1983.
[43] T. Orłowska-Kowalska, Application of extented Luenberger observer for flux and rotor time-constant estimation in induction motor drives. *IEEE Proc.* **136**, Pt. D, 324–330 (1989).
[44] A. Rachid, On induction motors control. *IEEE Trans. Control Syst. Technol.* **5**, 380–382 (1997).
[45] T. M. Rowan, R. J. Kerkman, and D. Leggate, A simple on-line adaptation for indirect field orientation of an induction machine. *IEEE Trans. Indust. Appl.* **27**, 720–727 (1991).
[46] W. Schumacher, Mikrorechner-geregelter Asynchron-Stellatrieb. Disseration, Technische Universitat Braunschweig, 1985.

Modern Nonlinear Control

[47] M. Bodson, J. Chiasson, and R. Novotnak, High performance induction motor control via input–output linearization. *IEEE Control Syst.*, Vol. 2, 25–33 (1994).

[48] M. Bodson, A systematic approach to selecting flux references for torque maximization in induction motors. *IEEE Trans. Control Syst. Technol.* **3**, 388–397 (1995).
[49] C. Cecati, Position control of the induction motor using a passivity-based controller. *IEEE Trans. Indust. Appl.* **36**, 1277–1284 (2000).
[50] C. Cecati, Torque and speed regulation of induction motors using the passivity theory approach. *IEEE Trans. Indust. Electron.* **46**, 119–127 (1999).
[51] J. Chiasson, A. Chaudhari, and M. Bodson, Nonlinear controllers for the induction motor. *IFAC Nonlinear Control System Design Symp.*, Bordeaux, France, pp. 150–155, 1992.
[52] A. Djermoune and P. Goureau, Input–output decoupling of nonlinear control for an induction nachine. *Proc. IEEE Int. Symp. Industrial Electronics*, Warsaw, pp. 879–884, 1996.
[53] Frick, E. Von Westerholt, and B. de Fornel, Non-linear control of induction motors via input–output decoupling. *ETEP* **4**, 261–268 (1997).
[54] L. U. Gökdere and M. A. Simaan, A passivity-based method for induction motor control. *IEEE Trans. Indust. Electron.* **44**, 688–695 (1997).
[55] M. P. Kazmierkowski and D. L. Sobczuk, Sliding mode feedback linearizcd control of PWM inverter-fed induction motor. *Proc. IEEE/IECON'96*, Taipei, pp. 244–249, 1996.
[56] Ki-Chul Kim and R. Ortega, Theoretical and experimental comparison of two nonlinear controllers for current-fed induction motors. *IEEE Trans. Control Syst. Technol.* **5**, 338–348 (1997).
[57] Z. Krzeminski, Nonlinear control of induction motors. *Proc. 10th IFAC World Cong.*, Munich, pp. 349–354, 1987.
[58] R. Marino, Output feedback control of current-fed induction motors with unknown rotor resistance. *IEEE Trans. Control Syst. Technol.* **4**, 336–347 (1996).
[59] R. Marino, S. Peresada, and P. Valigi, Adaptive partial feedback linearization of induction motors. *Proc. 29th Conf. Decision and Control*, Honolulu, pp. 3313–3318, Dec. 1990.
[60] R. Marino and P. Valigi, Nonlinear control of induction motors: a simulation study. *Eur. Control Conf.*, Grenoble, France, pp. 1057–1062, 1991.
[61] E. Mendes, Experimental comparison between field oriented control and passivity based control of induction motors. *IEEE Catalog No. 97th 8280*, ISIE'97—Guimaraes, Portugal, 1997.
[62] M. Pietrzak-David and B. de Fornel, Non-linear control with adaptive observer for sensorless induction motor speed drives. *EPE J.* **11**, 7–13 (2001).
[63] D. L. Sobczuk, Nonlinear control for induction motor. *Proc. PEMC'94*, pp. 684–689, 1994.
[64] D. G. Taylor, Nonlinear control of electric machines: An overview. *IEEE Control Systems*, 41–51 (1994).

CHAPTER 6

Energy Optimal Control of Induction Motor Drives

F. ABRAHAMSEN
Aalborg, Denmark

The function of an induction motor drive is to operate with the speed and torque which at any time is required by the operator of the motor drive. But having fulfilled that, there is an extra degree of freedom in the motor control, namely the selection of flux level in the motor, and this influences the losses generated in the drive. The most common strategy is to keep a constant motor flux level. Another strategy for selection of flux level is to reduce the drive losses to a minimum, which is here called energy optimal control. After a general introduction to energy optimal control, this chapter presents different ways of realizing energy optimal control and evaluates its benefits with respect to conventional constant V/Hz-control and to drive size.

6.1 MOTOR DRIVE LOSS MINIMIZATION

There is in the literature some confusion about what to call the control principle that is here called energy optimal control. Others have called it, for example, efficiency optimized control, loss minimum control, and part load optimization. When it is chosen here mainly to use the term "energy optimal control" it is because the goal in the end is to save energy. This is equivalent to loss minimum control but not necessarily to efficiency optimized control. There may, for example, be a control strategy that can optimize the efficiency in any steady-state load situation but does not minimize the energy consumption when it comes to a real application with time-varying load. An example of this is given in [1] where search control is used in a pump system.

Although many different terms are used to describe the same control principle it seldom leads to a misunderstanding, and that is why some of them are also used here in conjunction with "energy optimal control."

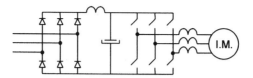

FIGURE 6.1
The drive configuration used in this chapter: three-phase IGBT converter with squirrel-cage induction motor.

6.1.1 Motor Drive Losses

Only one drive configuration is considered here: a converter consisting of a three-phase diode rectifier, a dc-link filter, a two-level three-phase IGBT inverter, and a three-phase squirrel-cage induction motor, as shown in Fig. 6.1. The control principles, however, can be extended to many variants of converters and induction motors.

Four loss components are associated with the transport of energy from the grid, through the drive and out to the motor shaft: grid loss, converter loss, motor loss, and transmission loss (see Fig. 6.2). The following list shows which loss components depend on the motor flux level and which do not.

Grid loss: The three-phase diode rectifier has a $\cos(\phi)$ which is constantly near unity, but on the other hand, the input current has an important harmonic content which generates extra losses in the grid. This loss is practically not influenced by the selection of motor flux level, but it is determined by the dc-link filter and by the conditions in the grid to which the drive is connected. In [2] it is shown how the grid harmonics can be reduced by mixing single-phase and three-phase loads.

Converter loss: The loss components that are influenced by the motor flux level are mainly switching and conduction losses of the inverter switches, and copper losses in output chokes and dc-link filters. The inverter losses are furthermore determined by the modulation strategy but this matter is not treated here. For the harmonic motor losses, see Section 6.2.1. In addition, there are the losses which do not depend on the motor control strategy, such as rectifier loss and power supply for the control electronics.

Motor loss: The only motor losses which do not depend on the control strategy are the windage and friction losses. All copper losses and core losses depend on the selection of motor flux level.

Transmission loss: The transmission loss does not depend on the motor control strategy although the loss is not negligible [3]. Worm gears should be avoided. Of the belt types, the synchronous belt with teeth has the lowest loss while the V-belts have poorer performance and rely on good maintenance. The best solution is a direct shaft coupling.

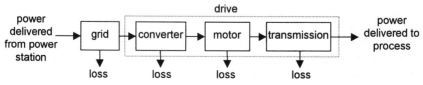

FIGURE 6.2
Overview of power flow through an electrical motor drive.

6.1.2 Loss Minimization by Motor Flux Adaptation

The overview of the drive losses reveals that only the motor and converter losses are influenced by the motor flux level selection, so energy optimal control is here constrained to consider these two loss components only. The principle of loss reduction by flux adaptation is explained by Fig. 6.3, which depicts a vector diagram of a low loaded motor at three different rotor flux levels: nominal, medium and low. The developed torque, represented by the hatched area, is proportional to $\psi_r I_r$ and is the same in all three cases. At nominal flux (Fig. 6.3a) the stator current is large and the rotor current is small, so both core losses and stator copper losses are high while the rotor copper losses are low. In Fig. 6.3b the rotor flux is reduced to 50% of its nominal value and the rotor current is doubled. This reduces the core losses and increases the rotor copper losses. The magnetization current is more than halved because the core has gone out of saturation, so the stator copper losses are also reduced considerably. In total, the motor loss in Fig. 6.3b is smaller than in Fig. 6.3a. If the rotor flux is reduced even more the core losses are still reduced, but as both the rotor and stator copper losses increase again, the total motor loss has also increased. The conclusion is that for a given load there exists a flux level that minimizes the motor loss. The optimal flux level depends primarily on the load torque. If there were no core losses in the motor the optimal flux level would be independent of the speed, but as the core losses are indeed present and they depend on the speed, the optimal flux level also depends on the speed.

The converter losses were not considered here. They primarily depend on the stator current amplitude and are also reduced in Fig. 6.3b, but to some degree they also depend on $\cos(\varphi)$ and the modulation index (see Section 6.2.2). An example of loss minimization is shown in Fig. 6.4 for a 2.2 kW motor drive. In such a low-power drive it is mainly the motor that benefits from the loss minimization and to a much lesser degree the converter. The situation is somewhat different at higher power levels where the motor loss is relatively smaller and comparable to the converter loss (see Section 6.2.4). It is clear that the loss reduction appears at low load torque.

The major disadvantage with flux reduction is illustrated in Fig. 6.5. When the flux is reduced for a given load torque and speed (operating point), the stator frequency is increased and the pull-out torque of the motor is reduced so that it becomes more sensitive to a sudden load disturbance. What happens if the load is suddenly increased depends on the type of motor control. For a vector controlled drive with speed feedback the speed can drop considerably and may possibly only be able to recover when the motor field has been restored. For an open-loop controlled drive the motor may even pull out so the control of the motor is lost and it must be stopped. An important task when designing energy optimal control is to ensure that the drive can withstand the load disturbances.

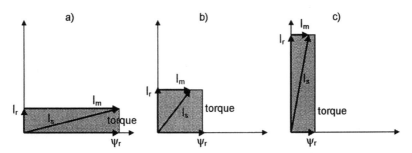

FIGURE 6.3
Illustration of the torque production at low load with different flux levels: (a) nominal flux; (b) medium; (c) low. The shaded areas denote the developed torque.

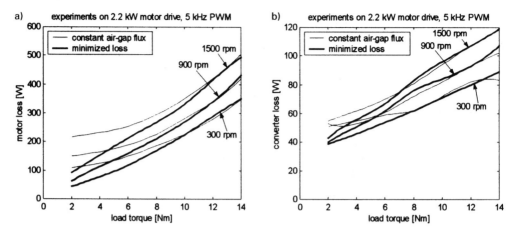

FIGURE 6.4
Measured losses of a 4-pole 2.2 kW drive operating at constant air-gap flux and with minimized drive loss: (a) motor; (b) converter. Nominal torque is 14 Nm. Switching frequency is 5 kHz.

6.2 CRITERIA FOR FLUX LEVEL SELECTION

The first problem with loss minimization is how to select the correct flux level. On the one hand one can wish to absolutely minimize the drive losses, and on the other hand one may wish not to reduce the flux so much that the drive becomes too sensitive to load disturbances and degrades the dynamic performance too much.

The second problem is how to obtain the desired flux level. If all losses in the drive were known exactly, it would be possible to calculate the desired operating point and control the drive in accordance to that, but that is not possible in practice for the following reasons:

- A number of losses are difficult to predict, including stray load losses, core losses in the case of saturation changes and harmonic content, and copper losses because of temperature changes.
- The information about the drive is incomplete. In most industrial drives only the stator currents and the dc-link voltage are measured. Additional sensors for measurement of speed

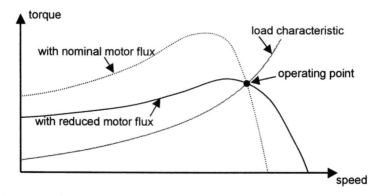

FIGURE 6.5
Torque–speed curves with a motor working in the marked operating point. The pull-out torque is reduced with reduced flux.

and input voltage/current are generally avoided as they increase cost and decrease reliability unnecessarily.

This section describes which drive losses it is necessary to consider for energy optimal control. Then follows a presentation of which criteria it is possible to use as indicators for energy optimization and how stability issues can be incorporated into the optimization strategies.

6.2.1 Motor Loss Model

An investigation is done in [4] of how the losses of a motor and converter can be modeled with a compromise between accuracy and simplicity that is adequate for energy optimal control. The models were tested with 2.2 kW, 22 kW, and 90 kW drives within the area of 0.2–1 p.u. speed and 0–1 p.u. load torque.

6.2.1.1 Fundamental Frequency Motor Losses The very well-known single phase motor model in Fig. 6.6 has appeared to be sufficient to model the fundamental frequency motor losses. It includes stator and rotor copper losses, core losses (eddy current and hysteresis), and mechanical losses (friction and windage). The stray load losses are not separately represented but included in the stator copper loss.

The stator resistance is compensated for temperature rise but skin effect is not taken into account. The rotor resistance is compensated for temperature rise and can possibly be made slip frequency dependent.

The core losses are modeled with the classical Steinmetz formula, including the dependency of both frequency and magnetization. The expression is applied for both the rotor core and the stator core:

$$P_{core} = k_h \cdot \psi_m^v \cdot f + k_e \cdot \psi_m^2 \cdot f^2 \tag{6.1}$$

where k_h is the hysteresis coefficient given by the material and design of the motor, k_n is a coefficient that depends on the magnetic material, k_e is the eddy current coefficient given by the material and design of the motor, ψ_m is the air-gap flux linkage, and f is the fundamental frequency. The mechanical losses are modeled as a function of speed.

6.2.1.2 Harmonic Motor Losses The pulse width modulated converter creates current ripple which generates harmonic losses in the motor. The harmonic current runs through the stator and rotor conductors and gives rise to harmonic copper losses, and the harmonic flux confines itself to the surface of the cores and creates harmonic core losses in the stator and rotor teeth [5]. Theoretical studies have only succeeded in characterizing the frequency dependence of

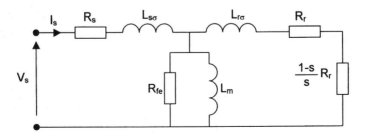

FIGURE 6.6
The motor diagram used to model the induction motor for energy optimal control.

the harmonic losses qualitatively. Measurements of harmonic motor losses as function of switching frequency are presented in Fig. 6.7. It shows that while the converter loss increases linearly with frequency, the harmonic motor losses initially decrease and then stay constant in the area from 4–10 kHz of switching frequency. In this specific case the optimal switching frequency around 3–4 kHz would be the best from a loss point of view.

Measurements in [4] showed that the dependency of harmonic losses with motor flux level is so small that it does not justify to include harmonic losses in the flux level selection, and they are therefore not treated further.

6.2.2 Converter Loss Model

The main converter loss components are power supply, rectifier conduction loss, dc-choke copper loss, inverter conduction and switching losses, and output choke conduction loss. It is primarily the losses in the inverter and output chokes that depend on the motor control strategy, which is why focus is put on them here. Whereas it is sufficient to model the loss in the output chokes with ideal resistances, the inverter loss is more complex. In [6] it is shown how the inverter loss can be calculated with high precision, based on measurements of on-state voltage drop and turn-on/turn-off losses for the diodes and transistors. The loss model which is presented here is based on the same approach, but by using some approximations the expressions are simplified, thereby enabling the loss model to be implemented in real-time calculations [7].

6.2.2.1 Inverter Conduction Loss The on-state diode and transistor voltage drops are approximated with

$$v_{\text{con},T} = V_{0,T} + R_{0,T} i_T, \qquad v_{\text{con},D} = V_{0,D} + R_{0,D} i_D \tag{6.2}$$

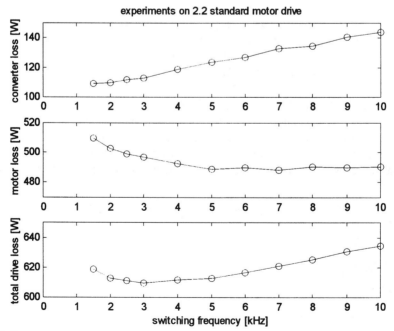

FIGURE 6.7
Measured harmonic losses as a function of switching frequency for a 2.2 kW drive.

where $v_{con,T}$, $v_{con,D}$ are the transistor and diode voltage drops; $V_{0,T}$, $V_{0,D}$ are the transistor and diode constant voltage drops; $R_{0,T}$, $R_{0,D}$ are the transistor and diode dynamic resistances; and i_T, i_D are the transistor and diode currents.

The conduction loss is calculated for sinusoidal modulation with an injected third harmonic, and the result is nearly identical to the loss generated by the most commonly used space-vector modulation:

$$P_{con,T} = \frac{V_{0,T}I_s\sqrt{2}}{\pi} + \frac{V_{0,T}I_s m_i \cos(\varphi)}{\sqrt{6}} + \frac{R_{0,T}I_s^2}{2} + \frac{R_{0,T}I_s^2 m_i}{\sqrt{3}\cos(\varphi)6\pi} - \frac{4R_{0,T}I_s^2 m_i \cos(3\varphi)}{45\pi\sqrt{3}} \quad (6.3)$$

$$P_{con,D} = \frac{V_{0,D}I_s\sqrt{2}}{\pi} - \frac{V_{0,D}I_s m_i \cos(\varphi)}{\sqrt{6}} + \frac{R_{0,D}I_s^2}{2} - \frac{R_{0,D}I_s^2 m_i}{\sqrt{3}\cos(\varphi)6\pi} + \frac{4R_{0,D}I_s^2 m_i \cos(3\varphi)}{45\pi\sqrt{3}} \quad (6.4)$$

where I_s is the RMS stator current, m_i is the modulation index, which varies from 0 to 1, and φ is the phase shift between stator voltage and stator current.

6.2.2.2 Inverter Switching Loss
Although an accurate description of the switching energy is a complex function of the current level [6], it is shown in [4] that the total inverter switching energy can approximately be considered as a linear function of the stator current. This means that the inverter switching loss can be expressed simply as

$$P_{sw} = C_{sw} I_s f_{sw} \quad (6.5)$$

where C_{sw} is an empirically determined constant, and f_{sw} is the switching frequency.

6.2.2.3 Total Inverter Loss
The total inverter loss then becomes

$$P_{loss,inv} = 3(P_{con,T} + P_{con,D}) + P_{sw}. \quad (6.6)$$

The inverter loss is a function of stator current amplitude, phase shift, switching frequency, and modulation index.

6.2.3 Optimization Criteria

Before choosing an energy optimal control strategy it is essential to analyze the behavior of the drive in order to decide how to realize minimum loss. Such an analysis is briefly shown here for a 2.2 kW standard induction motor drive.

It is possible, of course, to make some considerations on an analytical level with the steady-state motor model, as it was also done in Section 6.1.2. The problem is, however, the nonlinearities of the motor, not to mention the complex loss description for the converter. For example, if the motor had no core loss and no saturation it would be easy to show that optimal efficiency is defined by one constant slip frequency, but it will be shown later in this section that this result is totally useless for industrial motors, which are always designed with core saturation in the nominal operating point.

The calculations on the 2.2 kW drive are only shown for 900 rpm (nominal 1500 rpm). All graphs show calculations for four different load torques, and in each case the magnetization is varied, the nominal value of the air-gap flux being 0.66 Wb. On each graph the points of minimum drive loss are indicated by a dot.

Figure 6.8 shows that the input power minima are not well defined—it would be worse at low speed. It means that an input power minimizing search control (see Section 6.3.3) demands a precise power measurement. On the other hand, the same figure shows that the input power

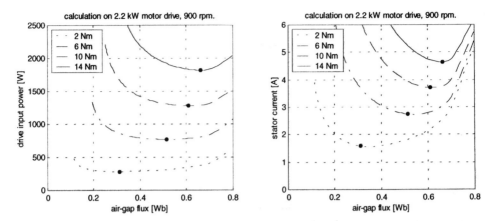

FIGURE 6.8
Calculated drive input power and stator current for a 2.2 kW drive. The dots denote the points of minimum drive loss.

minima almost coincide with the stator current minima and that these are more well defined, so that stator current would be better to use as the variable in search control.

Figure 6.9 shows that the motor and converter losses both have minima very near the drive loss minima. It is natural for the motor because its loss makes up a large part of the total loss. The explanation for the converter is that the loss to a large extent follows the stator current amplitude.

If Fig. 6.10 is compared with Fig. 6.9 it is seen that a constant slip frequency will indeed not ensure a good loss minimization. That will, on the other hand, a constant $\cos(\varphi)$. Although the calculation is only shown for 900 rpm, the picture is not much different at any other speed below nominal speed. If it is wished not to hit the exact loss minimum but to operate with a slightly higher magnetization in order to make the drive more stiff, with the $\cos(\varphi)$ control it is simply a matter of reducing the $\cos(\varphi)$ reference a little bit.

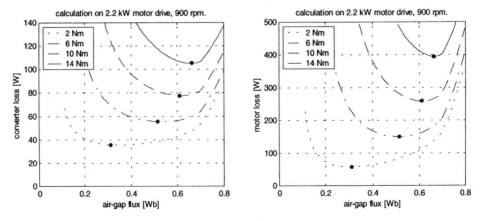

FIGURE 6.9
Calculated converter and motor loss for a 2.2 kW drive. The dots denote the points of minimum drive loss.

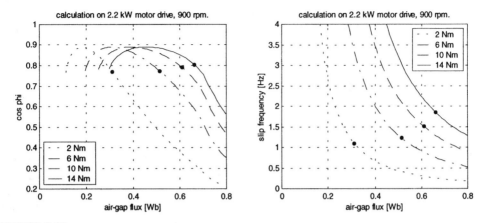

FIGURE 6.10
Calculated cos(φ) and rotor slip frequency for a 2.2 kW drive. The dots denote the points of minimum drive loss.

6.2.4 Loss Minimization in Medium-Size Drives

The literature about loss minimization of induction motor drives almost entirely treats small drives (less than 10 kW) and focuses on the motor loss, probably because small motors with their high losses benefit the most from flux optimization. While the nominal efficiency of a converter varies only between 0.96 and 0.98 for drives up to 100 kW, the motor efficiency may vary between 0.75 and 0.96. So the converter loss becomes more and more important as the drive becomes larger. It is briefly shown here how this affects loss minimization of a 90 kW drive. A more thorough analysis is given in [8].

Figure 6.11 shows calculated drive loss for a 90 kW induction motor drive at loads ranging from low to nominal load torque. Two operating points are indicated on each curve. The filled

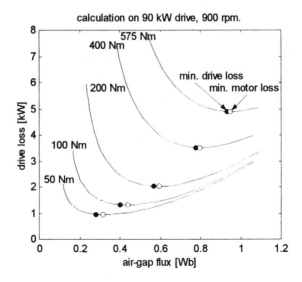

FIGURE 6.11
Calculated drive loss for a 90 kW drive. The filled dots denote the points of minimum drive loss and the circles denote the points of minimum motor loss.

dots denote the points of minimum drive loss and the circles denote the points of minimum motor loss. The difference in air-gap flux between the two criteria is calculated at 900 rpm and will increase for increased speed. In terms of drive loss, however, the flat-bottomed curves mean that even a noticeable difference in air-gap flux has almost no effect on the drive loss. So from the point of view of drive loss minimization, the minimum motor loss criterion is just as good as the minimum drive loss criterion. The only reason to include the converter losses is that it commands a higher flux level and thereby guarantees a higher robustness against load disturbances.

For a grid-connected motor it is good to design the motor as small as possible to ensure that the motor is loaded as much as possible and thereby it operates with a high efficiency. It is interesting to note that the same is not the case for speed controlled motors operated with optimized efficiency. In [4] it was demonstrated that in this case a 3.3 kW motor will always have a higher efficiency than a 2.2 kW motor when they are subjected to the same load. This comparison was done for motors with the same building size, and it was not investigated whether the result is the same if the large motor is constructed with a larger building size.

6.3 PRACTICAL ENERGY OPTIMAL CONTROL STRATEGIES

Five energy optimal control strategies are presented here. In literature there have been presented many more variants than these five, but they represent the most important methods of loss minimum flux level adaptation. The control methods can be combined with a multitude of motor control methods, such as scalar control, vector control, and direct torque control. The experiments are here shown for a scalar drive and a rotor-flux oriented vector controlled drive. All analyses are done only below nominal speed of the motor.

6.3.1 Simple State Control

By simple state control is meant that one parameter of the drive is measured and controlled in a simple way. This includes constant slip frequency control, but as already mentioned, it is poorly performing and is not treated further.

The analysis in Section 6.2.3 showed that a constant displacement power factor, $\cos(\varphi)$, ensures nearly minimum loss. A simple realization of constant $\cos(\varphi)$ control is shown in Fig. 6.12. The best dynamic performance is obtained if the $\cos(\varphi)$ is measured and calculated in every sample of the control loop. Some authors have suggested varying the $\cos(\varphi)$-reference as a function of speed and load, but as this requires estimation of both speed and torque the method then loses its simplicity, and the improvement is only marginal.

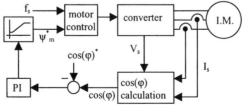

FIGURE 6.12
Scheme for $\cos(\varphi)$ control.

6.3.2 Model-Based Control

The flux level adaptation based on a description of the drive losses can be realized in various ways, and the method chosen to implement is often determined by how the motor control is realized: for example, whether rotor flux oriented or stator flux oriented control is employed. Two examples are shown here: an analytical solution to minimum motor loss and a numerical solution to minimum drive loss, i.e., with converter loss included.

6.3.2.1 Analytical Model Solution

The optimal operating point is solved using the steady-state rotor-flux oriented model in Fig. 6.13. Three loss components are represented: stator copper loss, core loss, and rotor copper loss. The total motor loss is split into the following three parts, the first depending on the d-axis current, the second depending on the q-axis current, and the third depending on both current components:

$$P_{\text{loss,d}} = \left(\frac{(\omega_s L_M)^2}{R_{\text{Fe}}} + R_s + (\omega_s L_M)^2 \frac{R_s}{R_{\text{Fe}}^2} \right) i_{\text{sd}}^2 \tag{6.7}$$

$$P_{\text{loss,q}} = (R_s + R_R) i_{\text{sq}}^2 \tag{6.8}$$

$$P_{\text{loss,dq}} = -2\omega_s L_M \frac{R_s}{R_{\text{Fe}}} i_{\text{sd}} i_{\text{sq}} \tag{6.9}$$

where ω_s is the stator angular velocity, L_M is the rotor-flux magnetizing inductance (rotor inductance), R_{Fe} is the core loss resistance, R_s is the stator resistance, R_R is the rotor resistance in the rotor-flux oriented model, i_{sd} is the d-axis current (field-producing current), and i_{sq} is the q-axis current (torque-producing current).

The developed torque is

$$\tau_{\text{em}} = z_p L_M i_{\text{sd}} i_{\text{sq}} \tag{6.10}$$

where z_p is the pole-pair number.

With the definition of A as

$$A = \frac{i_{\text{sq}}}{i_{\text{sd}}} \tag{6.11}$$

and in combination with Eq. (6.10), the following is obtained:

$$i_{\text{sq}}^2 = A \frac{\tau_{\text{em}}}{z_p L_M}, \quad i_{\text{sd}}^2 = \frac{1}{A} \frac{\tau_{\text{em}}}{z_p L_M}, \quad i_{\text{sd}} i_{\text{sq}} = \frac{\tau_{\text{em}}}{z_p L_M}. \tag{6.12}$$

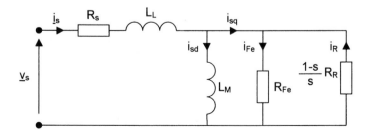

FIGURE 6.13
Steady-state rotor-flux oriented model of the induction motor including core losses.

Using Eqs. (6.7)–(6.9) and (6.12), the total motor loss becomes

$$P_{loss} = P_{loss,d} + P_{loss,q} + P_{loss,dq}$$
$$= \frac{\tau_{em}}{z_p L_M} \left[\left(\frac{(\omega_s L_M)^2}{R_{Fe}} + R_s + (\omega_s L_M)^2 \frac{R_s}{R_{Fe}^2} \right) \frac{1}{A} + (R_s + R_R)A - 2\omega_s L_M \frac{R_s}{R_{Fe}} \right]. \quad (6.13)$$

For a constant torque, the loss minimum is found by differentiating the loss expression with respect to A and assuming that the model parameters are independent of A. This is not entirely true, because the magnetizing inductance and the core loss resistance depend on the flux level, which is contained in A. However, it is assumed initially that these errors can be ignored.

$$\frac{\partial P_{loss}}{\partial A} = 0$$
$$\Updownarrow$$
$$-\left(\frac{(\omega_s L_M)^2}{R_{Fe}} + R_s + (\omega_s L_M)^2 \frac{R_s}{R_{Fe}^2} \right) \frac{1}{A^2} + (R_s + R_R) = 0 \quad (6.14)$$
$$\Updownarrow$$
$$P_{loss,d} = P_{loss,q}.$$

The motor losses thus reach a minimum when the motor loss depending on the current direct with the rotor flux is equal to the loss depending on the current in quadrature to the rotor flux. In [9] it is proposed to solve this equation with a PI-controller, as shown in Fig. 6.14.

The weakness of the method is still that it does not include core saturation, which causes it to command a flux level at high load torque which is too high. Reasonable results can be obtained if the motor flux is limited to its nominal value.

6.3.2.2 Numerical Model Solution The strength of a numerical loss model solution is that it can incorporate nonlinearities of the motor model and also the converter losses. Actually, only the processor power sets the limit. The only requirement is, of course, a good model of the motor and of the drive losses. An implementation is shown in Fig. 6.15. In this case the motor model is used to estimate the speed and load torque, and from that the optimal flux level can be calculated.

A numerical solution normally requires too much time to be executed in every sample of the control loop. One solution is to let it run as a background process in the microcontroller and to update the flux reference every time the optimization is solved.

Another solution is to solve all the optimization off-line and to store values of flux level as a function of load torque and speed in tables. The on-line calculations are then limited to load torque and speed estimation and to make a lookup in a table.

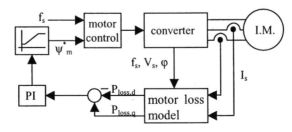

FIGURE 6.14
Scheme for energy optimal analytical model-based control.

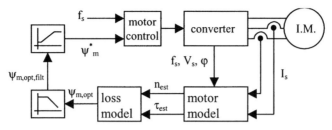

FIGURE 6.15
Scheme for energy optimal numerical model-based control.

The output from the loss model does not necessarily have to be a flux level, but could for example be stator current or stator voltage.

6.3.3 Search Control

The search control method is the only one that requires very good speed information—practically, this implies a speed sensor. The basic principle is to keep the output power of the motor constant, to measure the input power, and then iteratively to step the flux level until the minimum of input power is detected. In practice the motor output power cannot be measured, but it is solved by keeping the speed constant and assuming that the load torque is constant during the optimization period.

Minimum drive loss would require a power measurement at the input of the rectifier and would be too expensive. Another solution is to measure the dc-link power, which only requires one extra current sensor. In both cases the measurement must be very precise because the input power minimum is not well defined (see Section 6.2.3). Here it was also shown that an easier and cheaper solution is to minimize the stator current. One implementation is shown in Fig. 6.16.

An important drawback of the search control is the slow convergence time and the time-consuming trial-and-error process of tuning the searching algorithm—especially if the convergence time should be minimized. The most straightforward method is to decrease the flux level in constant steps, and possibly to reduce the step size near the optimum. Another approach is to use fuzzy logic to determine the step size [10]. The main difference with fuzzy logic is that the step size is determined by the size and the rate of change of speed and the measured variable, for example input power. Although the fuzzy logic approach makes the optimization more systematic it does not take away the need for good knowledge of the dynamics of the drive.

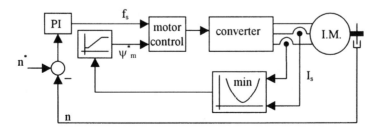

FIGURE 6.16
Scheme for energy optimal search control with stator current minimization.

6.3.4 Comparison of Control Methods

The five energy optimal control strategies have been tested on a 2.2 kW drive operating at low load (2 Nm), to see how fast the transition is from nominal magnetization to a minimum loss operation. The tests are done for a rotor-flux oriented vector controlled drive with speed feedback, and for a scalar drive operating in open loop except for the search control where the speed is controlled.

It is seen from Fig. 6.17 that the optimization is in general faster in a vector controlled drive than in a scalar drive. The best performance is obtained with the off-line numerical model-based control. The convergence times for the search control algorithms are only slightly larger than for the rest of the methods, but their responses are much more noisy. The comparison of the control strategies is summarized in Table 6.1.

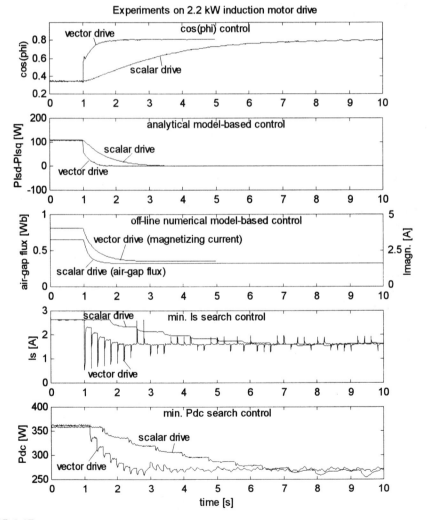

FIGURE 6.17
Experiments with turn-on of energy optimal control in a 2.2 kW drive for both scalar control, and vector control with speed feedback. The load torque is 2 Nm and the speed is 900 rpm.

Table 6.1 Comparison of Energy Optimal Control Strategies

Control strategy	Advantages	Disadvantages
Constant $\cos(\varphi)$ control	Simple, requires very little information about drive	Slow
Analytical model-based control	Relatively simple	Inaccurate—does not include core saturation
Numerical model-based control	High accuracy, fast response, simple to implement	Requires knowledge of motor model and drive loss
Min. P_{in} search control, Min. I_s search control	Loss model not necessary	Slow with perturbations, requires speed sensor and possibly extra sensor for power measurement, time-consuming to tune

The steady-state loss minimizing performances are seen in Fig. 6.18. It shows measured converter and motor loss for a 2.2 kW drive with constant flux and with energy optimal control. It is not possible to see a clear difference between the five control strategies. Either the differences are not present, or they are to some degree hidden in measurement inaccuracies.

6.4 CONCLUSIONS

This chapter showed how energy optimal control can be realized and how well such strategies perform, in terms of both dynamic performance and loss reduction. As the convergence time for the flux adaptation is counted in seconds it is clear that it is applicable only in low-dynamic applications, such as HVAC (heating, ventilation, and air-conditioning). Indeed, large savings can be anticipated in HVAC applications because they typically operate at low load most of the time and operate many hours per year. In most HVAC applications, the drive already operates

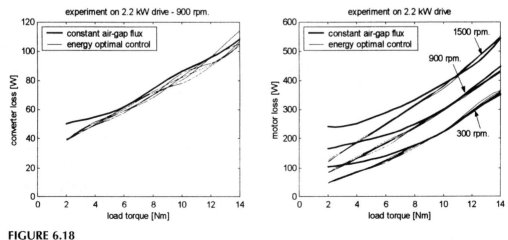

FIGURE 6.18

Measured losses of a 4-pole 2.2 kW drive operating at constant air-gap flux and with five different energy optimal control strategies. Nominal torque is 14 Nm. Switching frequency is 5 kHz.

with a so-called squared volt/hertz characteristic that reduces the magnetization at low speed. Still, the advantage with energy optimal control is that it ensures minimum loss even if the load characteristic is unknown and if it changes from time to time.

It was shown that the energy optimal control can be realized by considering the converter loss and the fundamental motor loss only—and even the converter loss is not very important to include in small drives. The traditional single-phase motor model is sufficient to describe the motor losses, but it is imperative to include core saturation, core losses, and possibly a temperature model of the windings.

If focus is put on HVAC applications, search control is excluded because it requires a speed sensor. A constant and nominal $\cos(\varphi)$ control is simple yet performs well—although it is not very fast. A model-based method is also possible. A numerical model solution where the main calculations are done off-line is simple to realize. The analytical model-based control does not perform so well because it does not include core saturation. The drawback of the model-based control methods is, of course, that they require motor and converter loss models.

Apart from saving energy, flux reduction at low load also has the advantage that it reduces the acoustic noise from both motor and converter. One drawback of the method is that losses are moved from the stator to the rotor when the flux is reduced, and that can be a problem because it is more difficult to remove heat from the rotor than from the stator. Another drawback is that the motor becomes more sensitive to sudden load disturbances. It can be alleviated by having a load surveillance and increasing the flux rapidly in case of a load increase, but the motor is still more sensitive than with nominal flux. It can only be evaluated in each specific case whether or not it will be a real problem.

REFERENCES

[1] F. Abrahamsen, F. Blaabjerg, J. K. Pedersen, P. Grabowski, P. Thøgersen, and E. J. Petersen, On the energy optimized control of standard and high-efficiency induction motor in CT and HVAC applications. Proc. of IAS '97, Oct. 1997, pp. 621–628.

[2] S. Hansen, P. Nielsen, and F. Blaabjerg, Harmonic cancellation by mixing non-linear single-phase and three-phase loads. Proc. of IAS '98, St. Louis, MO, October 1998, Vol. 2, pp. 1261–1268.

[3] S. Nadel, M. Shepard, S. Greenberg, G. Katz, and A. T. de Almeida, *Energy-Efficient Motor Systems*. American Council for an Energy-Efficient Economy, 1001 Connecticut Avenue, N.W., Suite 801, Washington, D.C. 20036, ISBN 0-918249-10-4, 1992.

[4] F. Abrahamsen, Energy optimal control of induction motor drives, Ph.D. thesis, Aalborg University, Denmark, ISBN 87-89179-26-9, Feb. 2000.

[5] D. W. Novotny, S. A. Nasar, B. Jeftenic, and D. Maly, Frequency dependence of time harmonic losses in induction machines. Proc. of ICEM '90, pp. 233–238, 1990.

[6] F. Blaabjerg, U. Jaeger, S. Munk-Nielsen, and J. K. Pedersen, Power losses in PWM-VSI inverter using NPT or PT IGBT devices. *IEEE Trans. Power Electron.*, **10**, 225–232 (1995).

[7] J. W. Kolar, H. Ertl, and F. C. Zach, Calculation of the passive and active component stress of three-phase PWM converter systems with high pulse rate. Proc. of EPE '89, Aachen, Germany, Oct. 9–12, pp. 1303–1311, 1989.

[8] F. Abrahamsen, F. Blaabjerg, J. K. Pedersen, and P.. Thøgersen, Efficiency optimized control of medium-size induction motor drives. Proc. of IAS '2000, Rome, Oct. 2000, Vol. 3, pp. 1483–1496.

[9] K. S. Rasmussen and P. Thøgersen, Model based energy optimizer for vector controlled induction motor drives. Proc. of EPE '97, Trondheim, Norway, pp. 3.711–3.716.

[10] G. C. D. Sousa, B. K. Bose, and J. G. Cleland, Fuzzy logic based on-line efficiency optimization control of an indirect vector controlled induction motor drive. Proc. of IECON, Vol. 2, Nov. 1993, pp. 1168–1174.

CHAPTER 7

Comparison of Torque Control Strategies Based on the Constant Power Loss Control System for PMSM

RAMIN MONAJEMY
Samsung Information Systems America, San Jose, California

R. KRISHNAN
The Bradley Department of Electrical and Computer Engineering, Virginia Tech, Blacksburg, Virginia

7.1 INTRODUCTION

Variable speed permanent magnet synchronous machine (PMSM) drives are being rapidly deployed for a vast range of applications to benefit from their high efficiency and high control accuracy. Vector control of PMSM allows for the implementation of several choices of control strategies while control over torque is retained. The main torque control strategies for the lower than base speed operating region are zero d-axis current, maximum torque per unit current, maximum efficiency, unity power factor, and constant mutual flux linkages. In this chapter, these control strategies are compared based on the constant power loss (CPL) control system for PMSM. The CPL control system allows for maximizing torque at all speeds based on a set power loss for the machine. Comparison of different torque control strategies based on the CPL control system provides a basis for choosing the torque control strategy that optimizes a motor drive for a particular application. The application of the CPL control system for different categories of cyclic loads is also discussed in this chapter.

7.1.1 Background

High-performance control strategies are capable of providing accurate control over torque or speed to within a small percentage error. A high-performance control strategy can also optimize one or more performance indices such as torque, efficiency, and power factor over its operational boundary. The rated current and power usually define the operational boundary of the machine. This operational boundary is only valid at rated speed. However, researchers and practitioners carry the same operational boundary over to variable speed motor drives. Such a step is not necessarily correct, because the true operational boundary of a machine depends on the maximum permissible power loss vs speed profile for the machine.

The main torque control strategies for the lower than base speed operating region for PMSM are the maximum efficiency, maximum torque per unit current, zero d-axis current, unity power factor, and constant mutual flux linkages. The main control strategies for the higher than base speed operating region are constant back emf and six-step voltage. A comprehensive analysis and comparison of the torque control strategies in the operating region with lower than base speed is made in this chapter. Availability of such analysis and comparison is the key to choosing a control strategy that optimizes the operation of a particular motion control system. The torque control strategies are analyzed and compared based on the constant power loss concept that defines the operational boundary in each case. This study lays the foundation for the analysis of truly optimized motor drives for wide speed range motion control systems based on PMSM. Similar techniques can be applied to all types of motor drives.

7.1.2 Literature Review

The number of research papers that directly investigate the subject of operational limits of PMSM motor drives for variable speed applications is limited [1–9]. References [3, 4] deal with choosing motor parameters such that the motor is suitable for a given maximum speed vs torque envelope. References [2, 3] investigate the optimal design of a motor for delivering constant power in the flux-weakening region. Operating limits of PMSM are studied in [5, 6] based on the constant power criterion. Reference [7] studies the CPL-based operation of PMSM and compares the resulting operational boundary to that resulting from limiting current and power to rated values. An implementation strategy for the CPL control system is also provided in [7]. Reference [8] compares the constant back emf and six-step voltage torque control strategies based on the CPL operational boundary in the operating region higher than the base speed for PMSM. A detailed comparison of all torque control strategies for the full range of speed is given in [9].

For the lower than base speed operating region, one performance criterion can be optimized while torque linearity is being maintained at the same time. This degree of freedom can be utilized in implementing different torque control strategies. The main torque control strategies for PMSM for lower than base speed operating region are as follows:

(a) Zero d-axis current (ZDAC)
(b) Maximum torque per unit current (MTPC)
(c) Maximum efficiency (ME)
(d) Unity power factor (UPF)
(e) Constant mutual flux linkages (CMFL)

The ZDAC control strategy [10, 11] is widely used in the industry. It is similar to the armature controlled dc machine in that it forces the torque to be proportional to current magnitude in the

PMSM. The basics behind the MTPC control strategy have been known for several decades. The MTPC control strategy provides maximum torque for a given current. This, in turn, minimizes copper losses for a given torque [12]. However, the MTPC control strategy does not optimize the system for net power loss. The UPF control strategy [10] optimizes the system's apparent power (volt–ampere requirement) by maintaining the power factor at unity. The ME control strategy [13, 14] minimizes the net power loss of the motor at any operating point. The CMFL control strategy [10] limits the air gap flux linkages to any set or desired flux linkages. This control strategy, therefore, leads to a seamless flux-weakening strategy in the PMSM drive and is to be noted.

Each control strategy has its own merits and demerits. Reference [10] provides a comparison of the ZDAC, UPF, and CMFL control strategies from the point of view of torque per unit current ratio and power factor. The UPF control strategy is shown to yield a very low torque per unit current ratio. The ZDAC control strategy results in the lowest power factor. Reference [15] provides a comparison between the MTPC and ZDAC for an interior PMSM. This study shows that the MTPC control strategy is superior in both efficiency and torque per unit current as compared to the ZDAC control strategy. Torque is limited to rated value in all non-CPL-based control schemes for operation lower than base speed. The operating region below base speed is referred to as the constant torque operating region. It is shown in [7] that the maximum torque in the operating region with lower than base speed is not a constant. A thorough comparison of all five control strategies from the point of view of maximum torque vs speed profile provides a sound basis for choosing the optimal control strategy for a particular motor drive application.

Section 7.2 introduces the CPL control system in brief. Comparison of control strategies based on the CPL control system is described in Section 7.3. The application of the CPL control system to cyclic loads is presented in Section 7.4. The conclusions are summarized in Section 7.5. Section 7.6, the Appendix, provides the parameters of the prototype PMSM drive used in Section 7.2.

7.2 CONTROL AND DYNAMICS OF CONSTANT POWER LOSS BASED OPERATION OF PMSM DRIVE SYSTEM

The operational boundary of an electrical machine is limited by the maximum permissible power loss vs speed profile for the machine. The control and dynamics of the PMSM drive operating with constant power loss are presented in this section [7]. This control system is modeled and analyzed. Its comparison to a system that limits current and power, say to rated values, demonstrates the superiority of the CPL control system. The implementation of the CPL control system is given. This has the advantage of retrofitting the present PMSM drives with the least amount of software/hardware effort. The PMSM drives in this case then can use the existing controllers to implement any torque control criterion.

7.2.1 Rationale for Constant Power Loss Control

The maximum torque vs speed envelope for the control strategies for speeds lower than the base speed region is commonly found by limiting the stator current magnitude to the rated (or nominal) value. For speeds higher than base speed operational region, the shaft power is commonly limited to the rated value. Current limiting restricts copper losses but not necessarily the core losses. Similarly, limiting the shaft power does not limit power losses directly. Limiting current and power to rated values ignores the thermal robustness of the machine since that requires the total loss to be constrained to a permissible value. Rated current and power guarantee acceptable power loss only at rated speed. Therefore, these simplistic restrictions are only valid for motion control applications requiring operation at rated speed. Increasingly, at present, single-speed motion control applications are being retrofitted or replaced with variable speed motor drives to increase process efficiency and operational flexibility. Also for cost optimization in manufacturing of the PMSMs, a few standardized lines of machine designs are utilized in vastly different environmental conditions, thus necessitating control methods to maintain the thermal robustness of the machine while extracting the maximum torque over a wide speed range. Only the CPL based operation can provide the maximum torque vs speed envelope from these viewpoints. A comparison of this operational boundary and the operational boundary resulting from limiting current and power to rated values clearly reveals that the CPL control system results in a significant increase in permissible torque at lower than rated speeds. Consequently, the dynamic response is enhanced below the base speed.

7.2.2 PMSM Model with Losses

A dq model for a PMSM in rotor reference frame in steady state with simplified loss representation is given in Fig. 7.1, where I_{qs} and I_{ds} are q and d axis stator currents, respectively, and V_{qs} and V_{ds} are q and d axis stator voltages, respectively. I_q and I_d are q and d axis torque generating currents, respectively, and I_{qc} and I_{dc} are q and d axis core loss currents, respectively. All these variables are in rotor reference frames and therefore, are dc values at steady state. R_s and R_c are stator and core loss resistors, respectively, and L_q and L_d are q and d axis self inductances, respectively. λ_{af} is magnet flux linkages, and ω_r is the rotor's electrical speed.

7.2.2.1 Electrical Equations of PMSM Including Core Losses
Equations (7.1) and (7.2) are derived from the model of Fig. 7.1:

$$\begin{bmatrix} I_{qs} \\ I_{ds} \end{bmatrix} = \begin{bmatrix} 1 & \dfrac{L_d \omega_r}{R_c} \\ -\dfrac{L_q \omega_r}{R_c} & 1 \end{bmatrix} \begin{bmatrix} I_q \\ I_d \end{bmatrix} + \begin{bmatrix} \dfrac{\lambda_{af}\omega_r}{R_c} \\ 0 \end{bmatrix} \qquad (7.1)$$

$$\begin{bmatrix} V_{qs} \\ V_{ds} \end{bmatrix} = \begin{bmatrix} R_s & \omega_r L_d\left(1 + \dfrac{R_s}{R_c}\right) \\ -\omega_r L_q\left(1 + \dfrac{R_s}{R_c}\right) & R_s \end{bmatrix} \begin{bmatrix} I_q \\ I_d \end{bmatrix} + \begin{bmatrix} \omega_r \lambda_{af}\left(1 + \dfrac{R_s}{R_c}\right) \\ 0 \end{bmatrix} \qquad (7.2)$$

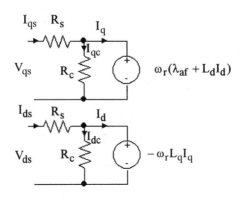

FIGURE 7.1
q and d axis steady-state model in rotor reference frames including stator and core loss resistances.

The air gap torque, T_e, as a function of I_q and I_d is given

$$T_e = 0.75P(\lambda_{af}I_q + (L_d - L_q)I_dI_q) \tag{7.3}$$

where P is the number of rotor poles.

7.2.2.2 Total Power Loss Equation for PMSM
The net core loss, P_{lc}, for the machine is

$$P_{lc} = \frac{1.5\omega_r^2(L_qI_q)^2}{R_c} + \frac{1.5\omega_r^2(\lambda_{af} + L_dI_d)^2}{R_c} = \frac{1.5}{R_c}\omega_r^2\lambda_m^2 \tag{7.4}$$

where λ_m is the air gap flux linkages. Note that in practice a more complex representation of core losses, based on elaborate equations or tables, can be used to increase the accuracy of the core loss estimation at higher speeds. More accurate equations for core losses can be found in [16, 17].

The total machine power loss, P_l, including both copper and core losses, can be described as

$$P_l = 1.5R_s(I_{qs}^2 + I_{ds}^2) + \frac{1.5}{R_c}\omega_r^2[(L_qI_q)^2 + (\lambda_{af} + L_dI_d)^2] \tag{7.5}$$

Equation (7.5) is a simplified representation of the sum of copper and core losses. In practice a more accurate representation of core losses, based on elaborate equations or tables, can be used to increase the accuracy of the total loss estimation. Other types of losses, such as drive losses, friction and windage losses, and stray losses, can also be included. Another major source of power losses is the electronic inverter. However, inverter losses do not affect the operational envelope of the machine. This is due to the fact that the cooling arrangement for the inverter is separate from the cooling arrangement for the machine. For some emerging motor drives where the inverter itself is integrated with the machine at one of the machine end bells, the inverter losses have to be considered in computing the total losses of the machine. In our illustrations, the machine and inverter are not integrated. Inverters limit the maximum current and voltage that can be delivered to a machine. It is presumed that the operating envelope of the inverter satisfies the machine operation.

In the next section the operational envelope resulting from the application of the CPL control system to a PMSM is discussed.

7.2.3 Constant Power Loss Control System and Comparison

The maximum permissible power loss, P_{lm}, depends on the desired temperature rise for the machine. P_{lm} can be chosen to be equal to the net loss at rated torque and speed assuming that the machine is running under exact operating conditions defined in manufacturer's data sheets. At any given speed the current phasor, which is the resultant of I_q and I_d, and its trajectory for maximum power loss is given by (7.5) with P_l replaced by P_{lm}. This trajectory is a circle at zero speed, and a semicircle at nonzero speeds. The operating point of a PMSM must always be on or inside the trajectory defined by (7.5) for that speed so that the net loss does not exceed P_{lm}. At any given speed the operating point on the constant power loss trajectory that also results in maximum torque defines the operational boundary at that speed. At this operating point, maximum torque is generated for the given power loss of P_{lm}. To find the maximum permissible torque at a given speed it is sufficient to move along the trajectory defined by (7.5) for a given P_{lm} and find the operating point that maximizes (7.3). In the flux-weakening region, both voltage and power loss restrictions limit the maximum torque at any given speed. The following relationship is true for any stator current phasor operating point in the flux-weakening region assuming that the voltage drop across the phase resistance is negligible:

$$V_{sm} = [(L_q I_q)^2 + (\lambda_{af} + L_d I_d)^2]^{0.5} \omega_r = \omega_r \lambda_m \qquad (7.6)$$

where V_{sm} is either the maximum desired back emf or the fundamental component of maximum voltage available to the phase. The latter applies to the six-step voltage control strategy. At any given speed in the flux weakening region the stator current phasor that results in a set power loss can be found by solving equations (7.5) and (7.6). This operating point corresponds to the maximum permissible torque at the given speed in the flux weakening region. Figure 7.2 shows the maximum torque possible for the full range of speed for the motor drive described in the Appendix. All variables in Fig. 7.2 are normalized using rated values. The power loss is limited to the rated value of 121 watts at all operating points. The following assumption is used in solving the required equations as discussed earlier:

$$I_{qs}^2 + I_{ds}^2 = I_q^2 + I_d^2. \qquad (7.7)$$

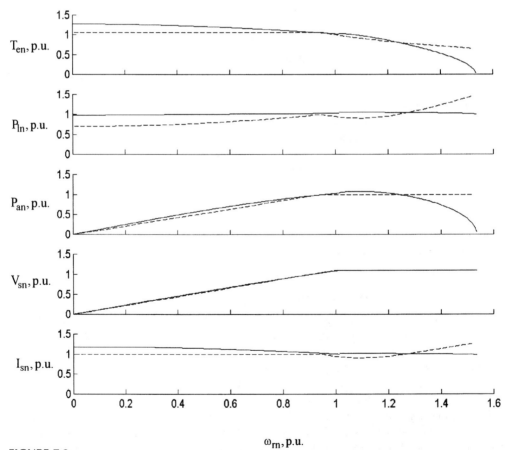

FIGURE 7.2
Normalized maximum torque, power loss, air gap power, voltage, and phase current vs speed for the CPL control system (solid lines) and for the scheme with current and power limited to rated values (dashed lines).

The slight deviation of the power loss from the set value of 1 p.u. in Fig. 7.2 is due to this assumption.

It is to be noted that the operating points along the operational envelope described here are also the most efficient operating points. Therefore, only the maximum efficiency control strategy for torque can lead to an operating point along this envelope. Other torque control strategies, such as maximum torque per current and constant torque angle, will result in operational envelopes that are smaller and narrower than that resulting from the maximum efficiency control strategy.

7.2.3.1 Operational Region with Lower Than Base Speed The base speed, ω_b, is defined as the speed beyond which the applied phase voltage must remain constant along the CPL operational boundary. In the lower than base speed operating region torque is only limited by the power loss, while phase voltage is less than maximum possible value, V_{sm}. This region of operation is shown in Fig. 7.2 between 0 and 1.1 p.u. speed. Power loss, air gap power, phase voltage, and current along the CPL operational boundary are also shown in Fig. 7.2.

7.2.3.2 The Flux-Weakening Operating Region The operation in Fig. 7.2 between 1.1 and 1.55 p.u. speed corresponds to the flux-weakening region of operation, where back emf is limited to 1.1 p.u. It is seen that the CPL operational boundary drops at a faster rate in this region because of voltage restrictions. The maximum possible air gap power continues to rise beyond rated speed up to approximately 1.25 p.u. speed.

The operational boundary resulting from limiting current and power to rated values is also shown in Fig. 7.2. It can be concluded from the example that the application of the constant current and power operational envelope results in:

- Underutilization of the machine at lower than base speed
- Generation of excessive power losses at higher than rated speeds unless both power and current are limited to rated values in the flux weakening region
- Underutilization of the machine in some intervals of the flux weakening region

7.2.4 Secondary Issues Arising Out of the CPL Controller

7.2.4.1 Higher Current Requirement at Lower Than Base Speed It is seen that the CPL control system provides 39% higher torque at zero speed as compared to the maximum torque possible with rated current. This requires 36% more than rated current at zero speed or very low speed and no increase in voltage. Therefore, the power switches have to be upgraded only for current and not for higher voltage.

7.2.4.2 Parameter Dependency The CPL control system is model-based and dependent on machine parameters. Therefore, provision must be made to track the machine parameters, particularly those that vary significantly over the operational region. Let us examine each of the machine parameters. The d axis flux path of the rotor involves a relatively large effective air gap and does not saturate under normal operating conditions. Therefore, the d axis inductance L_d does not vary significantly. The q axis inductance L_q varies as a result of magnetic saturation along the q axis and can be estimated accurately as a function of phase current [18]. An accurate estimation of rotor flux linkages λ_{af} is possible but requires a combination of voltage and current signals in the form of reactive or real power [19]. Stator resistance varies as a function of temperature that is fairly easy to measure inexpensively and instrument. Any implementation

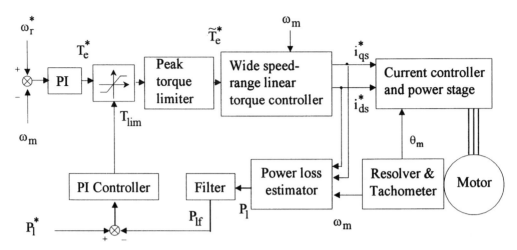

FIGURE 7.3
Constant power loss controller implementation.

strategy for the CPL control system is by nature parameter-dependent. This is the case with all model-based control strategies.

7.2.5 Implementation Scheme for the CPL Controller

Figure 7.3 shows the block diagram for an implementation strategy of the CPL control system. The wide speed-range linear torque controller is assumed to provide torque linearity over the full range of operating speed including the flux weakening region. Any control strategy can be utilized in the torque controller block. The copper and core losses of the machine are estimated using (7.5) utilizing the q and d axis current commands, i_{qs}^* and i_{ds}^*, as well as measured speed, ω_m. All the required variables for power loss estimation are already available within most high-performance control systems. The estimated net power loss, P_l, is always a positive number as seen from (7.5). P_{lf} is the filtered version of P_l. The filtered power loss estimation is compared with the power loss reference, P_l^*. The difference is processed through a proportional and integrator (PI) controller. The output of the power loss controller determines the maximum permissible torque, T_{lim}. The torque command, T_e^*, is the output of a PI controller operating on the difference of the commanded and measured speeds. If the torque command is higher than T_{lim} then the system automatically adjusts the torque to the maximum possible value, T_{lim}. However, if the torque command is less than the maximum possible torque at a given speed, then the torque limit is set at absolute maximum value leaving the torque command unaltered. The same absolute value of the torque limit is applied to both positive and negative torque commands. The transient torque magnitude is limited with the peak torque limiter block. The processed torque command, $\tilde{T}_e^*$, and ω_m are inputs to the wide speed range linear torque controller. The outputs of this block are i_{qs}^* and i_{ds}^*. The current controller and power stage enforce the desired current magnitude and its phase on the motor. The inputs to this stage are i_{qs}^*, i_{ds}^*, and rotor position, θ_m.

The salient features of the implementation strategy for a CPL control system are summarized as follows:

- Off-line calculation of the maximum torque vs speed envelope is not necessary and thereby avoids the highly computational approach adopted by the maximum efficiency implementation
- Maximum power loss can be a variable reference and, therefore, adjusted by an operator or by process demand
- The system is independent of the control strategy employed in the linear torque controller block
- All necessary parameters are usually available in most high performance controllers, thus making the CPL control system dependency on machine parameters not such a serious problem in implementation
- An on-line estimation of L_q, λ_{af}, and R_s can be used by the power loss estimation block to increase accuracy as other parameter does not change significantly
- The scheme lends itself to real-time implementation
- The scheme lends itself to retrofitting in existing PMSM drives with a minimum change in the control software
- Because it is the outermost loop, its execution occurs at sampling intervals that are on the order of seconds rather than the microseconds required for the current and torque control loops, which reduces the computational burden on the processor

Experimental correlation is provided using a prototype motor drive utilizing an interior PMSM with parameters given in the Appendix [7].

7.3 COMPARISON OF CONTROL STRATEGIES BASED ON CPL CONTROL SYSTEM

Each of the torque control strategies discussed in Section 7.1 is described and analyzed in this section. The procedure for deriving the q and d axis current commands as a function of torque and speed is described in each case. Also, the procedure for deriving the maximum possible torque vs speed profile for a given maximum possible power loss is described for each control strategy. Subsequently, a comprehensive comparison of these control strategies is made. The comparison is based on operation along the maximum possible torque vs speed envelope. The maximum possible torque vs speed envelope depends on the maximum possible power loss vs speed profile and is also a function of the chosen control strategy as well as motor drive parameters. Therefore, each control strategy results in a unique operational envelope for a given machine. Consequently, the performance of the system under each control strategy is also unique. Maximum torque, current, power, torque per current, back emf, and power factor vs speed are the key performance indices that are used here to compare different control strategies. It is assumed that the maximum possible power loss is constant for the full range of operating speed. This assumption is made in order to simplify the demonstrations and analytical derivations, and to allow the fundamental concepts to be presented with better clarity. Similar procedures can be used to analyze and compare different control strategies for an arbitrary maximum possible power loss vs speed profile. The procedures provided in this study can be used to choose the optimal control strategy based on the requirements of a particular application and also based on the capabilities of the chosen motor drive.

Section 7.3.1 reviews the analytical details of the five control strategies applicable to a high-performance PMSM control system in the operating region with lower than base speed. The performances of control strategies along the constant power loss operational envelope are compared in Section 7.3.2.

7.3.1 Control Strategies for Lower than Base Speed

The most important objective of high-performance control strategies is to maintain linear control over torque. Therefore, i_q and i_d must be coordinated to satisfy the following equation for a desired torque, T_e:

$$T_e = 0.75P(\lambda_{af}i_q + (L_d - L_q)i_d i_q) \tag{7.8}$$

However, it can be seen from (7.8) that a wide range of i_q and i_d values yield the same torque. Each of the five control strategies discussed in Sections 7.3.1.1 to 7.3.1.5 utilizes the available degree of freedom, seen in (7.8), to meet a particular objective. Equation (7.9) shows the general description of the intended relationship between currents i_q and i_d and torque and speed T_e and ω_r, respectively, for a given control strategy:

$$\begin{bmatrix} i_q \\ i_d \end{bmatrix} = \begin{bmatrix} \Lambda(T_e, \omega_r) \\ \Gamma(T_e, \omega_r) \end{bmatrix} \tag{7.9}$$

where Λ and Γ represent the relationship described by (7.8) in combination with the objective of the specific control strategy. At any given speed the maximum possible torque is limited by the maximum possible power loss, P_{lm}. Therefore, while the system is operating at maximum torque, i_q and i_d must satisfy

$$P_{lm} = 1.5R_s(i_{qs}^2 + i_{ds}^2) + \frac{1.5}{R_c}\omega_r^2[(L_q i_q)^2 + (\lambda_{af} + L_d i_d)^2] \tag{7.10}$$

where i_{qs} and i_{ds} are defined as a function of i_q and i_d in (7.1). In this section the maximum possible torque at any given speed is studied for different control strategies. Generally, the maximum possible torque is a function of speed, maximum possible loss, and the chosen control strategy:

$$T_{em} = Y(\omega_r, P_{lm}) \tag{7.11}$$

where T_{em} is the maximum possible torque, and the function Y depends on the chosen torque control strategy and machine parameters.

The maximum torque under the maximum efficiency, maximum torque per unit current, and zero d-axis current control strategies is only limited by the maximum possible power loss for the motor. However, each of the unity power factor and constant mutual flux linkages control strategies imposes an absolute maximum possible torque, as described in Sections 7.3.1.4 and 7.3.1.5, respectively. Therefore, for the latter two control strategies, the maximum possible torque is the smaller of the respective absolute maximum torque and the maximum torque resulting from the power loss limitation.

7.3.1.1 Maximum Efficiency Control Strategy (ME)
i_q and i_d are coordinated to minimize net power loss, P_l, at any operating torque and speed. The net power loss can be described as

$$P_l = 1.5R_s(i_{qs}^2 + i_{ds}^2) + \frac{1.5}{R_c}\omega_r^2[(L_q i_q)^2 + (\lambda_{af} + L_d i_d)^2] \tag{7.12}$$

7.3 COMPARISON OF CONTROL STRATEGIES BASED ON CPL CONTROL SYSTEM

where

$$\begin{bmatrix} i_{qs} \\ i_{ds} \end{bmatrix} = \begin{bmatrix} 1 & \frac{L_d \omega_r}{R_c} \\ -\frac{L_q \omega_r}{R_c} & 1 \end{bmatrix} \begin{bmatrix} i_q \\ i_d \end{bmatrix} + \begin{bmatrix} \frac{\lambda_{af} \omega_r}{R_c} \\ 0 \end{bmatrix} \quad (7.13)$$

The optimal set of currents, i_q and i_d, that result in minimization of power loss at a given speed and torque can be found using (7.8), (7.12), and (7.13). This operation can be performed using numerical methods.

Minimizing P_l at zero speed results in minimizing copper losses since core losses are zero at zero speed. On the other hand, minimizing copper losses is equivalent to minimizing current. Therefore, the maximum efficiency control strategy results in minimum current for a given torque at zero speed, which means that the ME and MTPC control strategies result in identical performance at zero speed.

7.3.1.2 Zero d-Axis Current Control Strategy (ZDAC)

The torque angle is defined as the angle between the current phasor and the rotor flux linkages in the rotor reference frame. This angle is maintained at 90° in the case of the ZDAC control strategy. The ZDAC control strategy is the control strategy most widely utilized by the industry. The d-axis current is effectively maintained at zero in this control strategy. The main advantage of this control strategy is that it simplifies the torque control mechanism by linearizing the relationship between torque and current. This means that a linear current controller results in linear control over torque as well. The following relationships hold for the ZDAC control strategy:

$$T_e = 0.75 P \lambda_{af} i_s \quad (7.14)$$

where i_s is the phase current magnitude, and

$$i_q = i_s \quad (7.15)$$
$$i_d = 0 \quad (7.16)$$

The current i_s for a given torque T_e can be calculated as

$$i_s = \frac{T_e}{0.75 P \lambda_{af}} \quad (7.17)$$

The air gap flux linkages can be described as

$$\lambda_m = (\lambda_{af}^2 + L_d^2 i_s^2)^{0.5} \quad (7.18)$$

The ZDAC control strategy is the only control strategy that enforces zero d-axis current. This is one of the disadvantages of this control strategy as compared to the other four. A nonzero d-axis current has the advantage of reducing the flux linkages in the d-axis by countering the magnet flux linkages. This serves to generate additional torque for inset and interior PMSM, and also reduces the air gap flux linkages. Lower flux linkages result in lower voltage requirements as well. Therefore, application of the ZDAC control strategy results in higher air gap flux linkages and higher back emf as compared to other control strategies. The maximum possible torque under this control strategy is limited only by the maximum possible power loss.

7.3.1.3 Maximum Torque per Unit Current Control Strategy (MTPC)
This control strategy minimizes current for a given torque. Consequently, copper losses are minimized in the process. The additional constraint imposed on i_q and i_d for motors with magnetic saliency is

$$i_q^2 = i_d\left(i_d + \frac{\lambda_{\text{af}}}{L_d - L_q}\right), \qquad L_d \neq L_q \tag{7.19}$$

For the types of PMSM that do not exhibit magnetic saliency the MTPC and ZDAC control strategies are the same. The MTPC control strategy results in maximum utilization of the drive as far as current is concerned. This is due to the fact that more torque is delivered for unit current as compared to other techniques. The MTPC and ME control strategies result in identical current commands at zero speed. The maximum possible torque under this control strategy is only limited by the maximum possible power loss.

7.3.1.4 Unity Power Factor Control Strategy (UPF)
If the voltage drop across the phase resistance is ignored, power factor can be defined as

$$\text{pf} = \cos(\theta) = \cos(\angle \vec{I} - \angle \vec{E}) \tag{7.20}$$

where $\vec{I}$ and $\vec{E}$ are the current and back emf phasors, respectively, and "$\angle$" denotes the angle of the respective phasor. The angle θ can be described as

$$\theta = \tan^{-1}\left(\frac{\lambda_{\text{af}} + L_d i_d}{-L_q i_q}\right) - \tan^{-1}\left(\frac{i_q}{i_d}\right) \tag{7.21}$$

Unity power factor can be achieved by maintaining the following relationship between i_q and i_d:

$$L_d i_d^2 + L_q i_q^2 + \lambda_{\text{af}} i_d = 0 \tag{7.22}$$

This control strategy imposes an absolute maximum possible torque on the system. This maximum permissible torque is found by inserting i_q as a function of i_d from (7.22) into the torque equation (7.8), and differentiating torque with respect to i_d. By differentiating this equation and equating it to zero, the d-axis current, I_{dm}, for the generation of the maximum possible torque, T_{em}, is derived. The following equation yields I_{dm}:

$$\alpha I_{dm}^2 + \beta I_{dm} + \gamma = 0 \tag{7.23}$$

where

$$\alpha = 4(L_d - L_q)L_d$$
$$\beta = 3(L_d - L_q)\lambda_{\text{af}} + 2\lambda_{\text{af}} L_d$$
$$\gamma = \lambda_{\text{af}}^2$$

Inserting I_{dm} into (7.22) yields the q-axis current, I_{qm}, at absolute maximum torque. Inserting I_{qm} and I_{dm} in (7.8) yields the absolute maximum torque possible under the UPF control strategy.

7.3.1.5 Constant Mutual Flux Linkages Control Strategy (CMFL)
In this control strategy the air gap flux linkage is forced to a set flux linkage. For our illustration, the mutual flux linkage is set to equal the rotor flux linkage. Also, the maximum speed after which flux weakening becomes necessary is extended. In this case Eq. (7.24) must hold:

$$\lambda_{\text{af}}^2 = (\lambda_{\text{af}} + L_d i_d)^2 + (L_q i_q)^2 \tag{7.24}$$

7.3 COMPARISON OF CONTROL STRATEGIES BASED ON CPL CONTROL SYSTEM

This control strategy imposes an absolute maximum torque on the system. This maximum permissible torque is found by inserting i_q as a function of i_d from (7.24) into the torque equation (7.8) and differentiating torque with respect to i_d. By differentiating this equation and equating it to zero, the d-axis current, I_{dm}, that yields the maximum possible torque, T_{em}, is derived. The following equation yields I_{dm} in this case:

$$\alpha I_{dm}^3 + \beta I_{dm}^2 + \gamma I_{dm} + \vartheta = 0 \tag{7.25}$$

where

$$\alpha = 4L_d^2(L_d - L_q)^2$$
$$\beta = 6\lambda_{af}L_d(L_d - L_q)(2L_d - L_q)$$
$$\gamma = 2L_d\lambda_{af}^2(5L_d - 4L_q)$$
$$\vartheta = 2\lambda_{af}^3 L_d$$

The smallest real and negative solution of (7.25) is the right choice. I_{qm} can then be calculated by inserting I_{dm} into (7.24). Inserting I_{dm} and I_{qm} into (7.8) yields the absolute maximum possible torque under the CMFL control strategy. This maximum torque is usually very high for PMSM and requires excessive current:

$$T_{em} = \Phi(\omega_r, P_{lm}) \tag{7.26}$$

7.3.2 Comparison of Torque Control Strategies Based on the CPL Control System

Any of the five control strategies discussed in Section 7.3.1 can be used in the lower than base speed operating region. In the higher than base speed operating region either the constant back emf or six-step voltage control strategy can be implemented. Therefore, 10 different combinations are possible to cover the full range of operating speed. Each combination results in a unique operational envelope. The performance of the system along each envelope is studied in this section with emphasis on the lower than base speed operating region. Detailed comparison of torque control strategies for the higher than base speed operating region is given in [20]. The procedure for comparing the performance of the system under each control strategy is presented here using the parameters of the PMSM drive prototype described in the Appendix. The same procedure can be applied to any motor drive. The five control strategies for the lower than base speed operating region each result in unique performances. The back emf is limited to 1.1 p.u. for the flux weakening region.

Several key performance indices vs speed are evaluated here for each of the five control strategies discussed in Section 7.3.1. Maximum torque, current, power, torque per current, back emf, and power factor are chosen for this purpose. The maximum torque vs speed envelope determines if a motor drive can meet the torque requirements of a particular application. The current vs speed envelope determines part of the requirements imposed on the drive. Higher current requirements translate into a more expensive power stage in the drive. The torque per unit current index is one of the most common performance indices used by researchers. Therefore, the torque per unit current index vs speed is also studied here for each control strategy. Power vs speed shows the maximum possible real power at any given speed for a particular control strategy. The power factor vs speed shows how well a particular control strategy utilizes the apparent power.

Figure 7.4 shows normalized maximum torque, current, power, and power loss vs speed for each of the five control strategies discussed earlier. The net power loss is maintained at rated

238 CHAPTER 7 / COMPARISON OF TORQUE CONTROL STRATEGIES

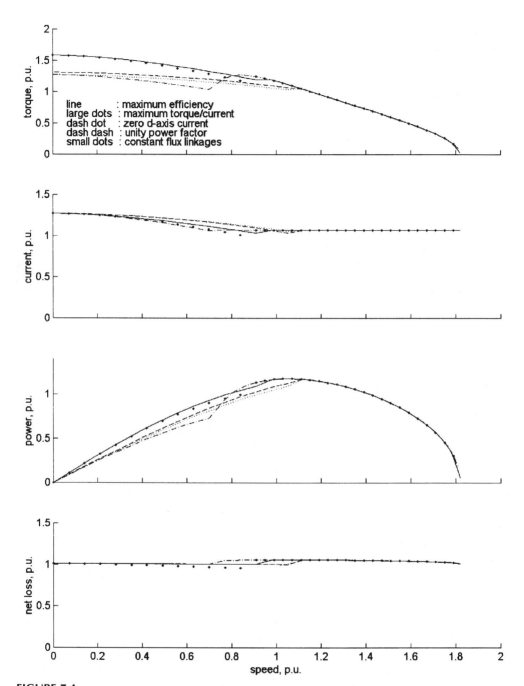

FIGURE 7.4
Maximum torque, current, power, and net loss vs speed at rated power loss for lower than base speed control strategies, and $E_m = 1.1$ p.u. for the flux weakening region.

7.3 COMPARISON OF CONTROL STRATEGIES BASED ON CPL CONTROL SYSTEM 239

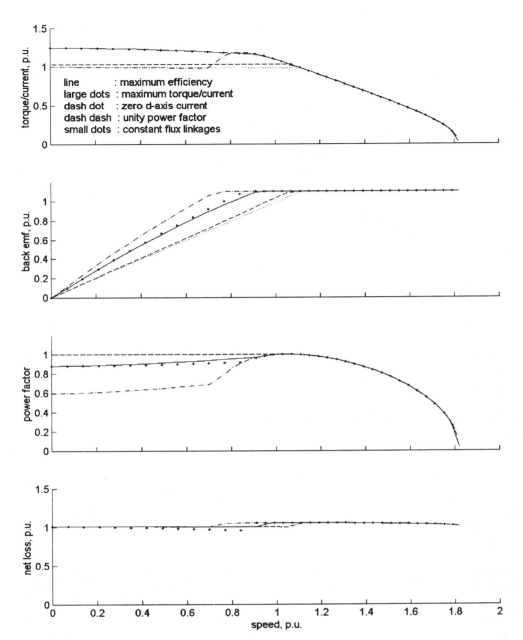

FIGURE 7.5

Torque per current, back emf, power factor, and net loss vs speed at rated power loss for lower than base speed control strategies, and $E_m = 1.1$ p.u. for the flux-weakening region.

level. The CBE control strategy, with E_m of 1.1 p.u., is chosen for the flux weakening region. Figure 7.5 shows torque per current, back emf, power factor, and net loss vs speed trajectories at rated power loss. The motor drive parameters are given in the Appendix. All variables have been normalized using rated torque, current and speed. All operational envelopes are calculated using assumption (7.3) as discussed in Section 7.2. The slight deviation of the power loss vs speed from 1 p.u. is due to this assumption. The performances of the five control strategies are compared next based on Figs. 7.4 and 7.5.

7.3.2.1 Torque vs Speed Envelope The ME control strategy provides more torque at any speed than any other control strategy. The MTPC control strategy provides only slightly less torque. However, the maximum torque for the MTPC control strategy drops at a faster rate as speed increases. The ZDAC control strategy provides the least torque at any given speed. This is mostly due to the fact that the ZDAC control strategy does not utilize the machine's reluctance torque in the case of this example. The maximum torque vs speed envelope for the UPF and CMFL control strategies falls in between those for the ZDAC and ME control strategies. The UPF control strategy produces slightly more torque than the CMFL control strategy at all speeds. Note that whereas the ME control strategy generates more torque than the MTPC control strategy, it also requires more current at any given speed.

The maximum torque for the ZDAC control strategy actually rises between the speeds of 0.7 and 0.9 p.u. speed. This is due to the introduction of nonzero d-axis current in the flux-weakening region, which produces reluctance torque.

7.3.2.2 Current vs Speed All control strategies require the same current at zero speed under the constant power loss criteria. This is due to the fact that core losses are zero at zero speed. Therefore, constant net power loss implies constant copper losses at zero speed. And, constant copper losses result in identical current magnitude for all control strategies at zero speed. As speed increases the current requirements of the five control strategies diverge significantly. The CMFL control strategy has the highest current requirement. The rate of drop of current increases successively for each of the UPF, ME, MTPC, and ZDAC control strategies.

7.3.2.3 Power vs Speed The ME control strategy produces more power at any given speed than any other control strategy. The power levels drop successively for each of the MTPC, UPF, CMFL, and ZDAC control strategies.

7.3.2.4 Torque per Current vs Speed The ME and MTPC control strategies produce near identical torque per current vs speed characteristics. The ZDAC and CMFL control strategies result in roughly 1 p.u. torque per current for the full range of speed. The torque per current envelope for the UPF control strategy is only slightly higher than that for the CMFL control strategy.

7.3.2.5 Back Emf vs Speed The ZDAC control strategy results in the highest back emf among the five control strategies. This, in turn, significantly limits the speed range even before flux weakening is initiated for the ZDAC control strategy. The UPF and CMFL control strategies result in the least back emf among the five control strategies, and lower back emf requirements lead to increased speed range for operation in the non-flux-weakening region for these two control strategies. The ME and MTPC control strategies also have similar back emf requirements, and their back emf requirements are significantly higher than those of the UPF and CMFL control strategies.

7.3.2.6 Air Gap Flux Linkages vs Speed Higher back emf at a given speed indicates higher mutual flux linkages as well. Therefore, mutual flux linkage requirements can be studied using back emf vs speed figures. The ZDAC control strategy requires by far the highest air gap flux linkages of all control strategies at any speed. This may raise concerns regarding saturation of the core for some machines. The ME and MTPC control strategies both require roughly the same air gap flux linkages. The CMFL and UPF control strategies require the least mutual flux linkages among all control strategies. The CMFL and UPF control strategies both require roughly the same back emf. Therefore, the flux linkage of the UPF control strategy is roughly 1 p.u., i.e., almost the same as that of the CMFL control strategy.

7.3.2.7 Power Factor vs Speed The UPF control strategy results in the highest possible power factor of 1 for the full range of speed. The CMFL control strategy results in a nearly unity power factor as well. The ZDAC control strategy results in the worst power factor. The power factor is roughly 0.65 on the average in this case. The ME and MTPC control strategies both result in reasonable power factors ranging from 0.85 at lower speeds to 0.95 at higher speeds. The power factor increases for both of these control strategies as speed increases.

7.3.2.8 Speed Range before Flux Weakening The CMFL control strategy results in the widest speed range before flux weakening among the five control strategies. The UPF control strategy stands with a slightly lower speed range. The ZDAC control strategy yields the narrowest speed range. This is mainly due to the relatively large air gap flux linkages for this control strategy. Note that the ZDAC control strategy is the only control strategy where the magnet flux linkage is never opposed by a countering field in the rotor's d-axis. The speed ranges for the ME and MTPC control strategies fall in between those of the ZDAC and UPF control strategies. The ME control strategy results in a slightly higher speed range than the MTPC control strategy.

7.3.2.9 Base Speed The speed at which the back emf reaches the maximum possible value along the maximum torque vs speed envelope is defined here as base speed. The base speed is a function of maximum permissible power loss, maximum voltage available to the phase, and the choice of control strategies. The specific objective of a control strategy in the lower than base speed operating region cannot be met beyond the base speed along the operational envelope of the system. It is seen that the CMFL control strategy provides the largest base speed among the five control strategies. The base speed reduces successively for each of the UPF, MTPC, ME, and ZDAC control strategies.

7.3.2.10 Complexity of Implementation The ZDAC control strategy is the simplest as far as implementation is concerned. I_d is simply maintained at zero, which makes the torque proportional to the phase current. The MTPC, CMFL, and UPF control strategies all require the implementation of separate functions for each of the d and q axis currents. These currents are functions of torque only. Therefore, implementation of the MTPC, CMFL, and UPF control strategies involves the same level of complexity. However, in the case of the ME control strategy the currents, i_q and i_d, are functions of both torque and speed. The necessary equations, in all cases, can be implemented on-line, or by implementing lookup tables [21].

7.4 APPLICATION OF CONSTANT POWER LOSS CONTROLLER WITH CYCLIC LOADS

7.4.1 Introduction

Many motion control applications require cyclic accelerations, decelerations, stops, and starts. The required movements are usually programmed into the system using a microcontroller. A simple programmed move is the cyclic on–off operation with negligible fall and rise times. The load duty cycle, in this case, is defined as the ratio of on time to the cycle period. In many applications, speed is constant during the on time. Usually several times rated torque is required during transitions from zero to maximum speed and vice versa. Different power losses are generated during different phases of operation of a motor drive with cyclic loads. For applications where the cycle period is significantly smaller than the thermal time constant of the machine the average power loss during one cycle must be limited to the maximum permissible value. This type of application is dealt with in this section. Generally speaking, the maximum possible torque increases as the duty cycle decreases. This is due to the fact that no power loss is generated during the off time. It is to be noted that in some applications the cycle period is large compared to the thermal time constant of the machine. In these applications the average power loss during one cycle in not of significant value, and instead, the instantaneous power loss during on time must be limited to the maximum permissible value.

The objective of this section is to calculate the appropriate power loss command for a constant power loss controller applied to different applications with cyclic loads. It is assumed that the maximum permissible power loss, P_{lm}, during continuous operation of the machine at steady state is known. It is also assumed that the on and off times are significantly shorter than the thermal time constant of the machine. Three different categories of applications with cyclic loads are considered here. In each case the power loss command is calculated such that the average loss over one period is equal to the maximum permissible power loss for the machine. The power loss command and maximum possible torque are calculated in each case as a function of load duty cycle, maximum permissible power loss, and maximum speed during one cycle. The procedures discussed in this section can be applied to any motor drive application with cyclic loads.

In Section 7.4.2, major motor drive applications are classified into three categories as far as cyclic loads are concerned. In each category the required power loss command, average power loss, and maximum torque during on time are calculated.

7.4.2 Power Loss Command in Different Application Categories

A simple and practical power loss estimator for PMSM is developed here. This power loss estimator is used in calculating the appropriate power loss command, average power loss, and maximum possible torque in each of three different categories of applications with cyclic loads. Most applications fall in one of these categories. Applications that are not covered in this section can be treated using similar procedures. All torque and speed profiles are normalized using maximum possible torque and speed in each category.

7.4.2.1 Derivation of a Practical Power Loss Estimator
An instantaneous power loss estimator, applicable to the surface mount PMSM, is developed here. Note that most high-performance motion control applications utilize the surface mount PMSM. All derivations are based on the following assumptions:

- d-Axis current is zero
- The difference between air gap and magnet mutual flux linkages is negligible
- Torque and current are proportional
- Impact of core losses on torque linearity is negligible

The preceeding assumptions are very closely valid for the surface mount PMSM. These assumptions are also valid for the brushless dc motor. Therefore, the results of this section are readily applicable to the brushless dc motor. However, these assumptions are not valid for inset and interior PMSM unless these machines are operated using the zero d-axis control strategy. The inset and interior PMSM are only used in a small percentage of all high-performance PMSM applications. Based on the above assumptions,

$$T_e = K_t i_s \tag{7.27}$$

where T_e is the machine torque, i_s is the stator current magnitude, and

$$K_t = 0.75 P \lambda_{af} \tag{7.28}$$

The equation for the instantaneous power loss, P_l, for a PMSM can be derived using (7.5), (7.27), and (7.28) by applying the assumptions just discussed. P_l is

$$P_l = K_1 T_e^2 + K_2 \omega_r^2 \tag{7.29}$$

where

$$K_1 = \frac{1.5 R_s}{K_t^2}, \quad K_2 = \frac{1.5 \lambda_{af}^2}{R_c} \tag{7.30}$$

Note that the air gap flux linkages and magnet flux linkages are almost equal for the surface mount PMSM. This fact is used in calculating K_2 as given in (7.30).

7.4.2.2 Power Loss Command and Maximum Torque

In this section motion control applications are broadly classified in three categories as far as cyclic loads are concerned. In each category the required power loss command, P_l^*, is calculated. All calculations are based on the implementation scheme described in Fig. 7.3. The average power loss, $\bar{P}_l$, and maximum possible torque, T_{em}, as a function of the maximum permissible power loss, P_{lm}, and for a given maximum speed, ω_{rm}, and load duty cycle, d, are calculated in each category.

(i) On–off operation with negligible rise and fall times. This category applies to applications that run periodically in on–off mode but have negligible rise and fall times compared to the cycle period. Some air conditioning and fan/pump/compressor applications fall in this category. Figure 7.6 shows an example of a torque and speed profile in this category. The operating point of the machine rises to the desired point. Then the operating point stays constant for T_{on} seconds. Subsequently, the operating point drops to zero. The machine remains at this operating point for T_{off} seconds. The net power loss during rise and fall of the operating point is negligible compared to the losses during on time. Applications in this category may require higher than rated torque during the short rise and fall periods. The average power loss in this case is

$$\bar{P}_l = \frac{P_l T_{on}}{T_{on} + T_{off}} = P_l d \tag{7.31}$$

where P_l is the power loss during on time. On the other hand, the maximum average power loss is

$$\bar{P}_l = P_{lm} \tag{7.32}$$

It can be concluded from (7.31) and (7.32) that, while operating at maximum average power loss,

$$P_l d = P_{lm} \tag{7.33}$$

Therefore, the maximum power loss command during on time is

$$P_l^* = \frac{P_{lm}}{d} \tag{7.34}$$

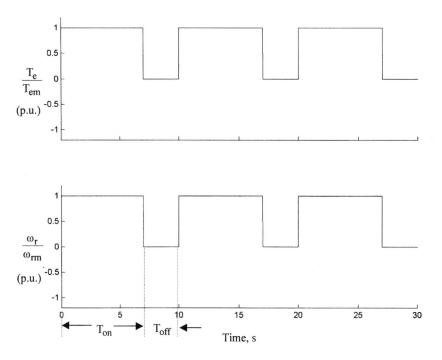

FIGURE 7.6
Normalized torque and speed profiles for on–off operation with small transition times.

The maximum torque during on time, T_{em}, can be calculated by substituting (7.29) into (7.34):

$$(K_1 T_{em}^2 + K_2 \omega_{rm}^2)d = P_{lm} \tag{7.35}$$

$$T_{em} = \left(\frac{P_{lm}/d - K_2 \omega_{rm}^2}{K_1}\right)^{0.5} \tag{7.36}$$

(ii) Speed varying linearly between $\pm \omega_{rm}$. Applications, such as industrial robots and some pick and place machines, where the end effector is always being accelerated in positive or negative directions fall into this category. Figure 7.7 shows an example of the torque and speed profile of an application in this category. The average power loss during the first half of one period is equal to the average power loss during the second half.

The average power loss in the first half of one period can be calculated as

$$\bar{P}_1 = K_1 T_{em}^2 + \frac{\int_0^T K_2 \omega_r^2 dt}{T} \tag{7.37}$$

where

$$\omega_r = \left(\frac{2\omega_{rm}}{T}\right)t - \omega_{rm} \tag{7.38}$$

7.4 APPLICATION OF CONSTANT POWER LOSS CONTROLLER WITH CYCLIC LOADS   245

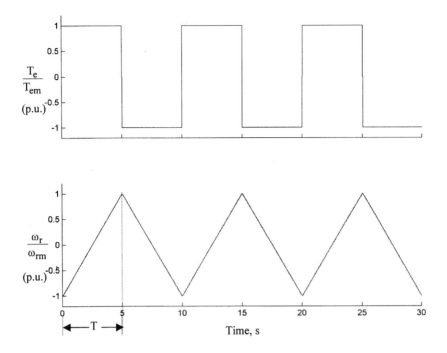

FIGURE 7.7
Normalized torque and speed profiles for operation between $\pm \omega_{rm}$.

and T is one-half of a cycle period. In this case (7.38) can be simplified as

$$\bar{P}_1 = K_1 T_{em}^2 + \frac{1}{3} K_2 \omega_{rm}^2 \tag{7.39}$$

In this category the instantaneous power loss estimation changes with speed throughout every period. If the power loss controller's PI block operates on the difference of the power loss command and the instantaneous power loss estimation, then the resulting torque limit varies linearly within one cycle of operation. However, as seen in Fig. 7.6, the torque needs to be constant during each half-period of one full cycle. The solution to this problem is to use a low-pass filter with a large time constant on the output of the power loss estimator (see Fig. 7.3). Such a filter should effectively output the average power loss of the machine in steady state. Under these conditions the maximum power loss command, which can limit the average power loss to P_{lm}, can be described as

$$P_1^* = P_{lm} \tag{7.40}$$

Since the maximum average power loss must not exceed P_{lm}, it can be concluded from (7.40) that

$$K_1 T_{em}^2 + \frac{1}{3} K_2 \omega_{rm}^2 = P_{lm} \tag{7.41}$$

The maximum possible torque can then be calculated from (7.41) as

$$T_{em} = \left(\frac{3P_{lm} - K_2 \omega_{rm}^2}{K_1}\right)^{0.5} \quad (7.42)$$

(iii) On–off operation with significant rise and fall times. Industrial lifts, elevators, and some pumps, fans, and servo drives are examples that fall in this category. For these applications the speed rises to a target level while several times the rated torque is being applied. Speed and torque remain constant for a period of time. Then the speed is reduced to zero, which again requires several times the rated torque. The operating point stays in this state for a period of time. The power losses during rise and fall times, in this category, constitute a significant portion of the average power loss.

Figure 7.8 shows a typical torque and speed profile in this category. A peak torque of T_{ep} Nm is applied for ΔT_{p1} seconds at the beginning of each cycle to raise the speed to the desired value of ω_{rm}. Then speed is maintained constant for a period of ΔT_m seconds during which time a constant torque of T_{em} Nm is applied to the load. Finally, speed is brought back to zero by applying $-T_{ep}$ Nm for a period of ΔT_{p2} seconds. The speed remains at zero for a period of ΔT_z seconds.

The magnitude of peak torque is set by the peak torque limiter block of Fig. 7.3. The power loss command, P_t^*, is set at maximum value during the peak torque periods in order to saturate

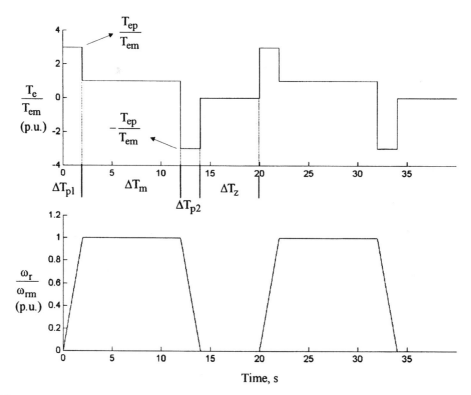

FIGURE 7.8
Normalized torque and speed profiles for operation with significant transition times.

the power loss PI controller. Saturation of this PI controller allows the torque to be limited by the peak torque limiter block. The power loss command during the ΔT_m seconds where nominal torque is being applied must be calculated as follows. The average power loss in this case is calculated by adding the energy losses in each segment, and dividing the result by the time period of one cycle. The net energy loss, W_1, in one cycle can be described as

$$W_1 = \left(K_1 T_{ep}^2 + \frac{1}{3} K_2 \omega_{rm}^2\right) \Delta T_{p1} + P_1^* \Delta T_m + \left(K_1 T_{ep}^2 + \frac{1}{3} K_2 \omega_{rm}^2\right) \Delta T_{p2} \quad (7.43)$$

The average power loss in one cycle is

$$\bar{P}_1 = \frac{W_1}{\Delta T_{p1} + \Delta T_m + \Delta T_{p2} + \Delta T_z} = \frac{W_1}{T} \quad (7.44)$$

where T is the cycle period. The average power loss should not exceed P_{lm}. Therefore,

$$\bar{P}_1 = P_{lm} \quad (7.45)$$

It can be concluded from (7.17), (7.18), and (7.19) that

$$P_{lm} T = \left(K_1 T_{ep}^2 + \frac{1}{3} K_2 \omega_{rm}^2\right) \Delta T_{p1} + P_1^* \Delta T_m + \left(K_1 T_{ep}^2 + \frac{1}{3} K_2 \omega_{rm}^2\right) \Delta T_{p2} \quad (7.46)$$

Therefore, the required P_1^* during the ΔT_m period can be calculated from (7.29) as,

$$P_1^* = \frac{P_{lm} T - \left(K_1 T_{ep}^2 + \frac{1}{3} K_2 \omega_{rm}^2\right) \Delta T_{p1} - \left(K_1 T_{ep}^2 + \frac{1}{3} K_2 \omega_{rm}^2\right) \Delta T_{p2}}{\Delta T_m} \quad (7.47)$$

In brief, the power loss command is set at maximum during the transitional periods and is set at the value calculated in (7.47) for the steady-state period.

The power loss during the steady-state period is

$$P_1 = (K_1 T_{em}^2 + K_2 \omega_{rm}^2) \quad (7.48)$$

where P_1 is calculated from (7.47) and is equal to P_1^* in steady state. Therefore, the maximum possible torque during the ΔT_m period in Fig. 7.8 is

$$T_{em} = \left(\frac{P_1^* - K_2 \omega_{rm}^2}{K_1}\right)^{0.5} \quad (7.49)$$

An alternative to the CPL implementation scheme described in Section 7.2 is to program the torque limiter to limit the torque to maximum desirable values during transition periods, and to limit the torque the value given in (7.49) during nominal operation. In this case the power loss control loop is not required. Similarly, the torque limiter can be utilized in limiting the average power loss in other application categories without resorting to the power loss feedback control loop. However, having a power loss control loop is preferred in the first two categories discussed because of the simplicity of its implementation and the simplicity involved in calculating the power loss command in those categories.

7.5 CONCLUSIONS AND RECOMMENDATIONS FOR FUTURE WORK

The constant power loss control system and its implementation are summarized in this chapter. All major control strategies for linear control of torque for PMSM are analyzed and compared based on the constant power loss operational boundary. The maximum efficiency, maximum torque per unit current, zero d axis current, unity power factor, and constant mutual flux linkage control strategies are the main possible choices for the lower than base speed operating region. Each control strategy results in a unique operational boundary and performance. Therefore, the comparison provided in this chapter allows for choosing the optimal control strategy for a particular application. A procedure to calculate the power loss command and maximum torque for motion control applications with cyclic loads is also presented. This enables the application of constant power loss control system in practice.

The concept presented in this chapter is demonstrated using PMSM drives. However, the concept equally applies to PM brushless dc, induction, switched reluctance, and synchronous reluctance motor drives. Future work is possible along these lines.

7.6 APPENDIX: PROTOTYPE PMSM DRIVE

The interior PMSM parameters are

$$P = 4, \quad R_s = 1.2\,\Omega, \quad R_c = 416\,\Omega, \quad V_{dc} = 118\,\text{V} \text{ (bus voltage)},$$
$$L_q = 12.5\,\text{mH}, \quad L_d = 5.7\,\text{mH}, \quad \lambda_{af} = 123\,\text{mWeber-turns}.$$

Combined motor and load inertia: $0.0019\,\text{Kg.m}^2$
Friction coefficient: $2.7 \times 10^{-4}\,\text{Nm/rad/s}$.
The rated values of the system are

Speed = 3500 rpm Current = 6.6 A
Torque = 2.4 Nm   Power = 890 W
Power loss = 121 W Core losses at rated operating point = 43 W
Copper losses at rated operating point = 78 W.

REFERENCES

[1] Y. Honda, T. Higaki, S. Morimoto, and Y. Takeda, Rotor design optimization of a multi-layer interior permanent-magnet synchronous motor. *IEE Proc.—Electr. Power Appl.*, **145**, 119–124 (1998).
[2] N. Bianchi and S. Bolognani, Parameters and volt–ampere ratings of a synchronous motor drive for flux weakening applications. *IEEE Trans. Power Electron.*, **12**, 895–903 (1997).
[3] S. Morimoto and Y. Takeda, Generalized analysis of operating limits on PM motor and suitable machine parameters for constant-power operation. *Electrical Eng. Japan*, **123**, 55–63 (1988).
[4] S. Morimoto, M. Sanada, Y. Takeda, and K. Taniguchi, Optimum machine parameters and design of inverter-driven synchronous motors for wide constant power operation. IEEE IAS Annual Meeting, 1994, pp. 177–182.
[5] T. Sebastian and G. R. Slemon, Operating limits of inverter-driven permanent magnet motor drives. *IEEE Trans. Indust. Appl.*, **IA-23**, 327–333 (1987).
[6] S. Morimoto, M. Sanada, and Y. Takeda, Wide-speed operation of interior permanent magnet synchronous motors with high-performance current regulators. *IEEE Trans. Indust. Appl.*, **30**, 920–925 (1994).
[7] R. Monajemy and R. Krishnan, Control and dynamics of constant power loss based operation of permanent magnet synchronous motor. *IEEE Trans. Indust. Appl.*, **48**, 839–844 (2001).
[8] R. Monajemy and R. Krishnan, Performance comparison of six-step voltage and constant back emf control strategies for PMSM. Conference Record, IEEE Industry Applications Society Annual Meeting, Oct. 1999, pp. 165–172.
[9] R. Monajemy, Control strategies and parameter compensation for permanent magnet synchronous motor drives. Ph.D. dissertation, Virginia Tech, Blacksburg, VA, Oct. 2000.

[10] S. Morimoto, Y. Takeda, and T. Hirasa, Current phase control methods for PMSM. *IEEE Trans. Power Electron.*, **5**, 133–138 (1990).
[11] P. Pillay and R. Krishnan, Modeling, analysis and simulation of high performance, vector controlled, permanent magnet synchronous motors. Conference Record, IEEE Industry Applications Society Meeting, 1987, pp. 253–261.
[12] T. M. Jahns, J. B. Kliman, and T. W. Neumann, Interior permanent-magnet synchronous motors for adjustable-speed drives. *IEEE Trans. Indust. Appl.*, **IA-22**, 678–690 (1986).
[13] R. S. Colby and D. W. Novotny, Efficient operation of PM synchronous motors. *IEEE Trans. Indust. Appl.*, **IA-23**, 1048–1054 (1987).
[14] S. Morimoto, Y. Tong, Y. Takeda, and T. Hirasa, Loss minimization control of permanent magnet synchronous motor drives. *IEEE Trans. Industr. Appl.*, **41**, 511–517 (1994).
[15] S. Morimoto, K. Hatanaka, Y. Tong, Y. Takeda, and T. Hirasa, Servo drive system and control characteristics of salient pole permanent magnet synchronous motor. *IEEE Trans. Indust. Appl.*, **29**, 338–343 (1993).
[16] G. Slemon and Xiau Liu, Core losses in permanent magnet motors. *IEEE Trans. Magnetics*, **26**, 1653–1656 (1990).
[17] T. J. E. Miller, Back EMF waveform and core losses in brushless DC motors. *IEEE Proc. Electr. Power Appl.*, **141**, 144–154 (1994).
[18] S. Morimoto, M. Sanada, and Y. Takeda, Effects and compensation of magnetic saturation in PMSM drives. *IEEE Trans. Indus. Appl.*, **30**, No. 6, November–December 1994, pp. 1632–1637.
[19] R. Krishnan and P. Vijayraghavan, Fast estimation and compensation of rotor flux linkage in permanent magnet synchronous machines. *Proc. Int. Symp. Indust. Electron.*, **2**, 661–666 (1999).
[20] R. Monajemy and R. Krishnan, Performance comparison of six-step voltage and constant back emf control strategies for PMSM. Conference Record, IEEE Industry Applications Society Annual Meeting, Oct. 1999, pp. 165–172.
[21] R. Monajemy and R. Krishnan, Concurrent mutual flux and torque control for the permanent magnet synchronous motor. Conference Record, IEEE Industry Applications Society Annual Meeting, Oct. 1995, pp. 238–245.

CHAPTER 8

Modeling and Control of Synchronous Reluctance Machines

ROBERT E. BETZ
School of Electrical Engineering and Computer Science, University of Newcastle, Callaghan, Australia

8.1 INTRODUCTION

There has been a revival of interest in reluctance machines over the past 15 years. This revival of interest has been largely focused on the switched reluctance machine (SRM), a trend that continues to this day. Over the same period there has also been an increase of interest in the synchronous reluctance machine (SYNCREL), although it should be said that the commercial application and development of this machine still lags far behind that of the SRM.

SRM development has been motivated by its robust and simple mechanical structure. The machine is capable of high torque density, and this coupled with its fault tolerant nature means that it is considered a serious candidate for aerospace and automotive applications. It has even found some limited application in domestic appliance applications (e.g., washing machines).

However, the SRM in many respects is not an ideal machine. It does not have a sinusoidal winding structure, but instead has concentrated windings on the stator. This means that normal sinusoidal analysis and control techniques cannot be applied to the machine. More importantly, the double salient structure and concentrated windings lead to severe torque pulsations and consequent noise problems. In addition a standard three-leg inverter cannot be used. Much of the current research on SRMs is associated with alleviating the torque pulsation and noise problems.

The SYNCREL on the other hand is a sinusoidally wound machine. In fact most SYNCRELs constructed to date have used an induction machine stator and windings. This means that the ideal machine will naturally produce smooth torque. The traditional three-phase winding structure also means that standard inverter technology and vector based control techniques can be readily applied. On the negative side, the SYNCREL has not received the same research support as the SRM because of the relatively poor performance of traditionally designed machines compared to the SRM and the induction machine. This has been primarily due to the design of the rotor. Rotor designs that have resulted in comparative performance with other machines have, until recently, been impractical to mass produce.

Synchronous reluctance machines are a very old machine. Indeed research papers were being published on the machine in the 1920s [1]. Most of the machines up to the 1970s were designed as direct-on-line start machines [2–4]. This meant that the design of the rotor was a compromise between startup performance and synchronous performance. The main application was in the fiber spinning industry, where the synchronous nature of the machine allowed large numbers of rolls to run at the same speed.

Line start synchronous reluctance machines in general did not have performance comparable with that of induction machines. Their power factor was very poor and torque per unit volume was low. Therefore they did not find use outside the above-mentioned niche application. The poor performance was primarily due to the line start requirement which meant that the rotor had to include an induction machine starting cage. This compromised the inductance ratio of the rotor and hence the synchronous performance. The rotor cage was also required to prevent oscillations of the rotor at synchronous speed.

Developments in power electronics and low-cost computing technology in the early 1980s led to a reexamination of the synchronous reluctance machine as a cageless machine. *It is this machine (without a cage) that is known as the SYNCREL.* The cageless structure required a variable frequency and voltage power supply in order to start. This was now feasible to synthesise using an inverter. In addition the application of a modified form of vector control (being developed at the time for the induction machine) could stabilize the machine and allow high performance. The renewed interest in the SYNCREL was motivated partly by curiosity (as to what performance could be achieved from the machine using modern technology and control) and also the hunch that a SYNCREL may be more efficient than the induction machine [5].

Recent research on the SYNCREL has been in two main areas—control strategies, and the design of the machine itself. The machine design research has mostly concentrated on the SYNCREL rotor. The control strategy research has considered the optimal control of the SYNCREL for a variety of different control objectives.

Although this presentation will concentrate on the control of the SYNCREL, a brief examination of the different rotor designs will be presented for completeness. Figure 8.1 sketches the main rotor design structures for the SYNCREL.

The salient pole rotor is essentially the same type of rotor used in the SRM. It has the advantage that it is simple to manufacture and robust. However, the difference between the inductances of the d-axis and the q-axis is only modest with this rotor. As we shall see in the following sections the performance of a SYNCREL is very dependent on the ratio and difference of these inductances.

The flux barrier design, like the salient pole design, is a very old structure. It was a design still constrained around having a cage in the rotor, and hence could not achieve the inductance ratios that later designs could.

The last two designs are the more modern designs for SYNCREL rotors. The axial laminated rotor in general gives the higher performance of these two. It is constructed of alternate flat steel laminations and insulation layers that run into the page in Fig. 8.1. These are held in place by pole pieces that are bolted into the square section of the shaft. As one can imagine, the precision bending and complex assembly make this rotor very expensive to build. In addition, the bolted structure limits the mechanical strength and hence the maximum angular velocity that can be obtained. If one uses thicker bolts or more bolts then the magnetic performance of the rotor is affected because of the amount of steel lamination material cut out of the rotor. These problems have relegated the axial laminated rotor to university laboratories.

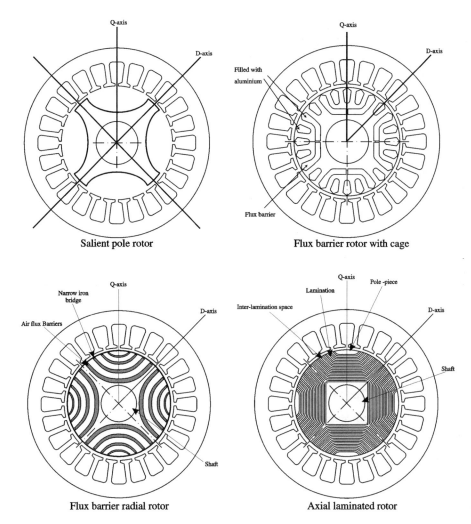

FIGURE 8.1
Various rotor structures used for SYNCRELs.

The radial laminated rotor, on the other hand, offers performance very close to that of the axial laminated design, but it is much easier to manufacture. As the name implies, this rotor is constructed using conventional punched radial laminations. They are punched with cutouts which form the flux barriers, and then assembled on a shaft using conventional techniques. The flux barriers can be filled with plastic or epoxy material for more strength. The steel bridges on the outer periphery of the rotor are designed to saturate under normal operation. Their thickness is a compromise between inductance ratio and mechanical thickness [6].

8.1.1 Scope and Outline

The remainder of this chapter will concentrate on the modeling and control of the SYNCREL. It should be emphasized that the presentation is tutorial in nature, beginning with a look at the basic torque production mechanism of a reluctance machine, and finally ending with the latest research results on the control of the machine. Inevitably, with a presentation of this length some issues will only be briefly introduced, and others omitted. Therefore, where relevant, appropriate references will be supplied for those who require more detail.

Following the torque production discussion, mathematical models suitable for the analysis of the control properties of the machine are developed. The main emphasis is on the development of the *dq* model of the machine, but the space vector model will also be briefly introduced. The models are then used to derived control objectives or set points to achieve a variety of control outcomes from the machine. Finally, a SYNCREL drive system that utilizes some of the theory developed shall be described in detail. Where appropriate practical issues will be highlighted.

8.2 BASIC PRINCIPLES

Reluctance machines are among the oldest electrical machines, since they are based on the physical fact that a magnet attracts a piece of iron. Because a reluctance machine is essentially like a solenoid that has been physically arranged so that it produces rotary motion, one can use the same techniques to obtain quantitative expressions for the torque produced by the machine. This section shall outline the development of the torque expression for a generic reluctance machine using coenergy concepts. For more detail refer to [7] or any other introductory machines textbook.

8.2.1 Coenergy and Torque

The coenergy approach to the calculation of torque is based on the following conservation of energy equation for a machine:

$$\begin{bmatrix} \text{Electrical} \\ \text{energy input} \end{bmatrix} = \begin{bmatrix} \text{Electrical} \\ \text{losses} \end{bmatrix} + \begin{bmatrix} \text{Stored energy} \\ \text{in fields} \end{bmatrix} + \begin{bmatrix} \text{Mechanical} \\ \text{energy} \end{bmatrix} \tag{8.1}$$

which can be written more succinctly as

$$E_e = E_{le} + E_{fe} + E_{me}. \tag{8.2}$$

Therefore if we can calculate the E_e, E_{le}, and E_{fe} components in (8.2) then we can find the amount of energy going into mechanical energy. We shall apply this principle to a very simple reluctance machine as shown in Fig. 8.2. In the following development these assumptions will be made:

1. The iron circuit exhibits saturation—i.e., it has a nonlinear flux versus current relationship.
2. There is negligible leakage flux.
3. Hysteresis and eddy currents are ignored.
4. Mechanical energy storage and losses are ignored.

Applying Kirchhoff's voltage law to Fig. 8.2 one can write:

$$v = Ri + \frac{d\psi}{dt} \tag{8.3}$$

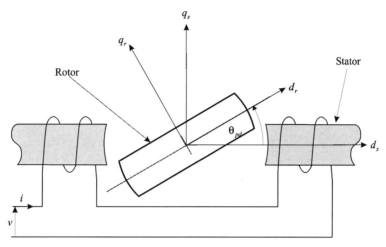

FIGURE 8.2
Simple single excited reluctance machine.

which allows the following expression for the power into the circuit to be written

$$vi = Ri^2 + i\frac{d\psi}{dt}. \tag{8.4}$$

To get the energy absorbed by the circuit over a time t we can integrate this expression as follows:

$$\begin{aligned}\int_0^t vi\,dt &= \int_0^t \left[Ri^2 + i\frac{d\psi}{dt}\right]dt \\ &= \int_0^t Ri^2\,dt + \int_0^t \left[i\frac{d\psi}{dt}\right]dt.\end{aligned} \tag{8.5}$$

Clearly the first term on the right-hand side (RHS) of (8.5) is related to the electrical losses (i.e., term E_{le}), and therefore the second term is related to mechanical and stored energy. We shall consider this term in more detail.

For a nonlinear magnetic structure such as that of Fig. 8.2 the flux is a nonlinear function of the current in the machine. Therefore, the current is also a nonlinear function of the flux—i.e., $i = F(\psi)$, where F denotes a nonlinear function. This relationship allows one to apply a change of variable to the second RHS term of (8.5) since one can write

$$i(F(\psi))\frac{d\psi}{dt}dt = i(\psi)d\psi. \tag{8.6}$$

If we make the further assumption that the resistive losses can be ignored, we can write the energy balance equation as

$$\int_0^t vi\,dt = \int_0^\psi i(\psi)d\psi. \tag{8.7}$$

The following discussion is with reference to Fig. 8.3. This figure shows the flux versus current curve for two extreme positions of the rotor. The unaligned position is when the angle $\theta_{pd} = \pi/2$ rad in Fig. 8.2—this corresponds to the position where the mmf of the stator sees the maximum reluctance to flux formation. The aligned position is when $\theta = 0$ rad, and therefore the

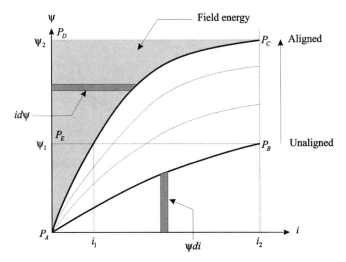

FIGURE 8.3
Flux plots for static movement of the simple reluctance machine.

mmf sees the minimum reluctance and the flux is maximized. Clearly there are an infinite number of intermediate positions between these two extremes.

Remark 8.1 *Note from Fig. 8.3 that the flux versus current curve for the aligned position has a more significant bend in it, indicating that the machine is more heavily saturated when the rotor is in the aligned position. In the unaligned position the flux path is dominated by air.*

Let us assume that the rotor of Fig. 8.2 is held stationary at the aligned position. Because the rotor is stationary and is being prevented from moving we know that any energy put into the system cannot be going into mechanical energy, and therefore must be going into stored field energy. Referring to Fig. 8.3, if the current is increased from 0 to i_2 A, then the flux goes from 0 to ψ_2 W along the line P_A to P_B. Therefore, according to (8.7) the energy input into the machine corresponds to the shaded area in Fig. 8.3. If the rotor is in the unaligned position, and if the same test is repeated, the flux moves to ψ_1, along the P_A to P_B contour. One can immediately see that there is a substantial difference in the stored field energy in the system when the rotor is in the unaligned and aligned positions.

The situation when the machine is stationary does not tell us much about what happens when the rotor is moved. In order to make this easy to understand the normal approach taken in most books is to consider two thought experiments—one is to imagine that the rotor can be moved so slowly that there is insignificant voltage induced in the windings due to the rate of change of flux in the system and consequently the current does not change. The other is to assume that the rotor can be moved instantaneously from the unaligned to the aligned position, and the flux remains constant throughout this movement. To keep the discussion brief we shall only consider one of these movements, the slow movement, since the results of both lead to similar results.

Consider a slow movement from P_B to P_C in Fig. 8.3. The current is constant at i_2 A. In order to calculate the energy supplied to the system we use (8.7) with the current value at i_2. Therefore the *total* energy is

$$\int_{\psi_1}^{\psi_2} i_2 \, d\psi = \int_{\psi_1}^{\psi_2} d\psi = i_2(\psi_2 - \psi_1). \tag{8.8}$$

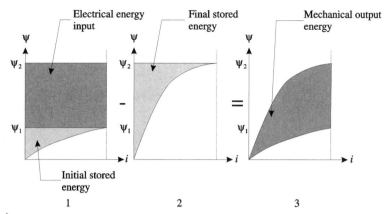

FIGURE 8.4
Diagrammatic representation of the mechanical output energy in the slow rotor movement scenario.

This is the area of a rectangular region of dimensions $(\psi_2 - \psi_1)$ and i_2.

Using conservation of energy one can write[1]:

$$E_{me} = \underbrace{i_2(\psi_2 - \psi_1)}_{\text{Elec input energy}} - \underbrace{\left[\int_0^{\psi_2} i\,d\psi - \int_0^{\psi_1} i\,d\psi\right]}_{\text{Change in stored field energy}} \tag{8.9}$$

$$= \underbrace{i_2(\psi_2 - \psi_1) + \int_0^{\psi_1} i\,d\psi}_{\text{Total energy}} - \underbrace{\int_0^{\psi_2} i\,d\psi}_{\text{Final stored energy}} \tag{8.10}$$

The various areas represented by this equation can be seen in Fig. 8.4. One can see that in the leftmost figure of the diagram that the total shaded area is the total energy in the system—i.e., the initial stored energy in the field (light gray area) plus the added electrical energy from the supply. The shaded area in the middle diagram is the final stored energy. If we subtract the final stored energy from the total energy then we should be left with the amount of energy that has not gone into stored energy. Because we are assuming that there are no electrical losses this energy must be going into mechanical energy. The third diagram shows the difference between the first and second diagrams, where the shaded area is equal to the mechanical output energy.

As we have seen in (8.7) the energy in stored in the field of a stationary system is obtained by integrating along the ψ axis. However, we could also integrate along the i axis and find another area. In general this area is not equal to the energy area (although in the special case of a linear magnetic system it is). The area obtained by integrating along the i axis is known as the *coenergy*. The reason for coenergy importance can be seen in diagram 3 of Fig. 8.4. The shaded area is the difference between the coenergy of the final position and the coenergy of the initial position. Therefore the mechanical energy output can be written as

$$E_{me} = E'_{fe_2} - E'_{fe_1} = \delta E'_{fe} \tag{8.11}$$

---

[1] Note that the sign convention is that energy out of the machine is positive.

where

$$E'_{fe_1} = \int_0^{i_2} \psi_{\text{unaligned}}\, di \triangleq \text{the coenergy in the unaligned position}$$

$$E'_{fe_2} = \int_0^{i_2} \psi_{\text{aligned}}\, di \triangleq \text{the coenergy in the aligned position.}$$

Now that we have an expression for the mechanical energy output, we require an expression for the torque. This is easily constructed by a straight application of the mathematical definition of average torque realizing that the slow movement of the rotor has been through an angle of $\Delta\theta_{pd}$:

$$T_{\text{ave}} = \frac{\text{mechanical energy}}{\text{angular rotor movement}} = \frac{E_{me}}{\Delta\theta_{pd}}. \tag{8.12}$$

If one considers that the angular movement approaches zero then the average torque becomes the instantaneous torque:

$$\begin{aligned} T_e &= \lim_{\delta\theta_{pd}\to 0}\left(\frac{\delta E'_{fe}}{\delta\theta_{pd}}\right)_{i\text{ constant}} \\ &= \left.\frac{\partial E'_{fe}}{\partial\theta_{pd}}\right|_{i\text{ constant}} \end{aligned} \tag{8.13}$$

Equation (8.13) has been derived under the assumption that the movement of the rotor was slow. Another extreme is to assume that the movement is instantaneous. If this is followed through then a very similar expression can be found in terms of the rate of change of energy (as opposed to coenergy). The two cases are then combined by considering a real movement that is neither slow nor instantaneous, and it can be shown that the two different expressions for energy give the same torque expression for infinitesimally small movements of the rotor [7]. Therefore (8.13) can be considered to be the general expression for instantaneous torque.

8.2.2 Coenergy and Inductances

The previous section established the connection between the rate of change of coenergy with rotor angular movement. The next step is to connect the coenergy concept to a machine's inductance. This connection is particularly useful when dealing with machines that are modeled with a *linear iron circuit*. In order to develop the basic expressions we shall consider a singly excited system such as that shown in Fig. 8.2.

If a magnetic system is linear then it can be characterized by

$$\psi = Li \tag{8.14}$$

where $L \triangleq$ the inductance of the system. For any particular position of the rotor in Fig. 8.2 one will then get a straight line for the flux versus current relationship, as shown for two positions in Fig. 8.5.[2]

---
[2] This diagram is the linear equivalent to Fig. 8.3.

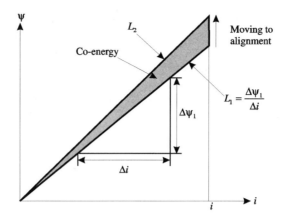

FIGURE 8.5
Flux versus current for a linear magnetic system.

It is clear because of the linearity that $E_{fe} = E'_{fe}$. Therefore the coenergy and energy can be calculated as follows:

$$E'_{fe} = \int_0^i \psi \, di$$
$$= \int_0^i Li \, di \qquad (8.15)$$
$$= \frac{1}{2}Li^2.$$

It can be seen from the diagram and reasoned from the physics of the machine of Fig. 8.2 that E'_{fe} is a function of i and θ_{pd}. Therefore, substituting for E'_{fe} in (8.13) we can write

$$T_e = \left.\frac{\partial E'_{fe}(i, \theta_{pd})}{\partial \theta_{pd}}\right|_{i \text{ constant}}$$
$$= \left.\frac{\partial}{\partial \theta_{pd}}\left[\frac{1}{2}Li^2\right]\right|_{i \text{ constant}} \qquad (8.16)$$
$$\therefore T_e = \frac{1}{2}i^2 \frac{dL}{d\theta_{pd}}.$$

Remark 8.2 *Equation (8.16) is a fundamental relationship that is used extensively to analyze many different types of machines. If one can find the expressions for the relationship between the inductances of a machine and the rotor position then (8.16) allows one to quickly evaluate the torque.*

Remark 8.3 *Equations (8.3) and (8.16) are essentially the dynamic equations for a simple reluctance machine. In order to complete the equations we would require the relationship between the inductance and the angle of the rotor. The mechanical load equation would also be required.*

8.2.2.1 Doubly Excited System.

Consider the doubly excited reluctance machine system sketched in Fig. 8.6. One may be tempted to ask what such a machine has to do with the SYNCREL, since the SYNCREL does not have a winding on the rotor. It turns out that the torque expression for a machine such as that of Fig. 8.6 can be applied directly to the evaluation of the torque expression for the SYNCREL.

We shall not go through the full derivation of the torque expression for this machine (for space reasons), but instead the procedure will be briefly outlined and the final expression for the torque stated.

The key difference between the machine of Fig. 8.6 and that of Fig. 8.2 is the second winding on the rotor. This leads to a difference in the flux expressions as there is now a mutual flux component. The flux expressions for the machine are

$$\psi_1 = L_1 i_1 + M i_2 \tag{8.17}$$
$$\psi_2 = L_2 i_2 + M i_1 \tag{8.18}$$

where

$L_1 \triangleq$ the self-inductance of the stator winding

$L_2 \triangleq$ the self-inductance of the rotor winding

$M \triangleq$ the mutual inductance between the stator and the rotor.

Remark 8.4 *It is very important to realise that all of the inductances in this machine are functions of θ_{pd}, the angle of the rotor.*

As with the singly excited system we can write the voltage equations for the doubly excited system as:

$$v_1 = R_1 i_1 + \frac{d\psi_1}{dt} \tag{8.19}$$

$$v_2 = R_2 i_2 + \frac{d\psi_2}{dt}. \tag{8.20}$$

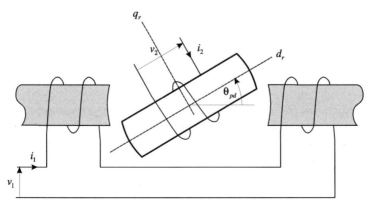

FIGURE 8.6
Doubly excited reluctance machine.

The preceding flux expressions can be substituted into these voltage equations and the appropriate derivatives taken (assuming that the inductances are not a function of the current) to give

$$v_1 = R_1 i_1 + L_1 \frac{di_1}{dt} + i_1 \frac{dL_1}{dt} + M \frac{di_2}{dt} + i_2 \frac{dM}{dt} \tag{8.21}$$

$$v_2 = R_2 i_2 + L_2 \frac{di_2}{dt} + i_2 \frac{dL_2}{dt} + M \frac{di_1}{dt} + i_1 \frac{dM}{dt}. \tag{8.22}$$

Using conservation of energy arguments similar to those used for the singly excited system it is now possible to develop the expression for the coenergy of the system [7]:

$$E'_{fe} = \frac{1}{2} L_1 i_1^2 + \frac{1}{2} L_2 i_2^2 + i_1 i_2 M. \tag{8.23}$$

Once we have the coenergy then the torque for this machine can be derived using (8.13):

$$\begin{aligned} T_e &= \left. \frac{\partial E'_{fe}}{\partial \theta_{pd}} \right|_{i \text{ constant}} \\ &= \frac{1}{2} i_1^2 \frac{dL_1}{d\theta_{pd}} + \frac{1}{2} i_2^2 \frac{dL_2}{d\theta_{pd}} + i_1 i_2 \frac{dM}{d\theta_{pd}}. \end{aligned} \tag{8.24}$$

8.3 MATHEMATICAL MODELING

In this section we shall develop two main models of the SYNCREL. The main emphasis will be on the development of the *dq* model since it is the model that is mostly used in the following sections.

8.3.1 SYNCREL Inductances

In the previous section we found that a knowledge of the inductances of a machine is useful as a technique for finding its torque expression. Therefore, we need the inductance expressions for the SYNCREL.

The following assumptions are made in the derivation of the inductances:

1. The stator windings are sinusoidally distributed. When excited with current a sinusoidal spatial distribution of mmf is produced.
2. The machine does not exhibit any stator or rotor slotting effects.
3. The machine iron is a linear material—i.e., it is not subject to magnetic saturation effects. The permeability of the material is very large in comparison to that of air. Therefore the permeance of the magnetic paths is dominated by the air gaps.
4. The air gap flux density waveforms can be adequately represented by their fundamental component.
5. The stator turns are all full pitched (i.e., they cover π electrical radians).
6. There is no leakage flux—all the windings are perfectly coupled.

There are two inductance values to be evaluated—the self-inductances of the three-phase windings, and the mutual inductances of the windings.

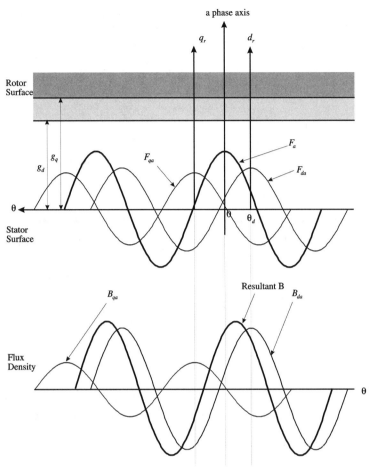

FIGURE 8.7
Developed diagram of a SYNCREL showing the "a" phase mmf, its d_r and q_r components, and the resultant flux density waveforms.

Two different techniques are usually employed to work out inductances—the twin air gap technique, or the winding function technique. The trick in all these techniques is the modeling of the irregular air gap. We shall briefly outline the first technique and then present the inductance expressions.

The twin air gap technique [7] for determining the inductances of the SYNCREL is based on the approximation that the mmf produced by a single winding (say for example the "a" phase) can be broken into two components, one aligned with the d_r axis of the rotor, and the other with the q_r axis. The d_r mmf component is assumed to be operating on a constant air gap of g_d, and the q_r mmf component on a constant air gap of g_q. Using this assumption the flux density can be calculated for each of the air gaps, and then the total resultant air gap flux can be found by the use of superposition. These concepts are illustrated in Fig. 8.7 for the "a" phase.[3]

---

[3] F_a, F_{da}, and F_{qa} denote the "a" phase mmf and its two components.

Remark 8.5 *Note that the process of taking components in the twin air gap technique automatically takes into account the variation of the mmf with angular position around the machine.*

Using these flux density relationships, the flux linkage of d_r and q_r axes to the "a" phase can be found by integrating the flux density over half the coil span of the "a" phase, taking into account the sinusoidal turns distribution of the winding. This will then give the self flux linkage of the winding. The self-inductance of the winding can then be calculated using the relationship

$$L = \frac{\psi_{aa}}{i_a}. \tag{8.25}$$

The other inductance that occurs in the SYNCREL is mutual inductance. The phase windings of the machine are spatially separated by $2\pi/3$ electrical radians. Therefore there is coupling between the windings, and this coupling is also a function of the rotor position. The technique for calculating the mutual inductance is almost identical to the self inductance calculation, except that the flux density is now being generated by another winding.

If the foregoing processes are carried out for all the windings in the machine, then we end up with the following expressions for the inductances of the SYNCREL:

Self-inductances:

$$L_{aa} = L_1 + L_2 \cos 2\theta_{pd} \tag{8.26}$$

$$L_{bb} = L_1 + L_2 \cos 2\left(\theta_{pd} - \frac{2\pi}{3}\right) \tag{8.27}$$

$$L_{cc} = L_1 + L_2 \cos 2\left(\theta_{pd} + \frac{2\pi}{3}\right) \tag{8.28}$$

Mutual inductances:

$$L_{ba} = L_{ab} = -\frac{L_1}{2} + L_2 \cos 2\left(\theta_{pd} - \frac{\pi}{3}\right) \tag{8.29}$$

$$L_{cb} = L_{bc} = -\frac{L_1}{2} + L_2 \cos 2\theta_{pd} \tag{8.30}$$

$$L_{ca} = L_{ac} = -\frac{L_1}{2} + L_2 \cos 2\left(\theta_{pd} + \frac{\pi}{3}\right) \tag{8.31}$$

where

$$L_1 = \frac{N^2}{8}(P_d + P_q)$$

$$L_2 = \frac{N^2}{8}(P_d - P_q)$$

$N \triangleq$ total number of turns in a sinusoidal winding

$P_d, P_q \triangleq$ the axis permeances.

Table 8.1 Summary of Transformations between the Three-Phase Stationary and Two-Phase Rotating Reference Frames

To $dq\gamma$	To abc
$\mathbf{F}^r_{dq\gamma} = \frac{2}{3}\mathbf{C}\mathbf{F}_{abc}$	$\mathbf{F}_{abc} = \mathbf{C}^T \mathbf{F}^r_{dq\gamma}$
$\mathbf{i}^r_{dq\gamma} = \frac{2}{3}\mathbf{C}\mathbf{i}_{abc}$	$\mathbf{i}_{abc} = \mathbf{C}^T \mathbf{i}^r_{dq\gamma}$
$\mathbf{v}^r_{dq\gamma} = \frac{2}{3}\mathbf{C}\mathbf{v}_{abc}$	$\mathbf{v}_{abc} = \mathbf{C}^T \mathbf{v}^r_{dq\gamma}$
$\mathbf{\Psi}^r_{dq\gamma} = \frac{2}{3}\mathbf{C}\mathbf{\Psi}_{abc}$	$\mathbf{\Psi}_{abc} = \mathbf{C}^T \mathbf{\Psi}^r_{dq\gamma}$
$\mathbf{L}^r_{dq\gamma} = \frac{2}{3}\mathbf{C}\mathbf{L}_{abc}\mathbf{C}^T$	$\mathbf{L}_{abc} = \frac{2}{3}\mathbf{C}^T \mathbf{L}^r_{dq\gamma}\mathbf{C}$
$\mathbf{R}^r_{dq\gamma} = \frac{2}{3}\mathbf{C}\mathbf{R}_{abc}\mathbf{C}^T$	$\mathbf{R}_{abc} = \frac{2}{3}\mathbf{C}^T \mathbf{R}^r_{dq\gamma}\mathbf{C}$
$\mathbf{Z}^r_{dq\gamma} = \frac{2}{3}\mathbf{C}\mathbf{Z}_{abc}\mathbf{C}^T$	$\mathbf{Z}_{abc} = \frac{2}{3}\mathbf{C}^T \mathbf{Z}^r_{dq\gamma}\mathbf{C}$

$$\mathbf{C} = \begin{bmatrix} \cos\theta_{pd} & \cos\left(\theta_{pd} - \frac{2\pi}{3}\right) & \cos\left(\theta_{pd} + \frac{2\pi}{3}\right) \\ -\sin\theta_{pd} & -\sin\left(\theta_{pd} - \frac{2\pi}{3}\right) & -\sin\left(\theta_{pd} + \frac{2\pi}{3}\right) \\ \frac{1}{\sqrt{2}} & \frac{1}{\sqrt{2}} & \frac{1}{\sqrt{2}} \end{bmatrix} \quad \mathbf{i}^r_{dq\gamma} = \begin{bmatrix} i^r_d \\ i^r_q \\ i^r_\gamma \end{bmatrix} \quad \mathbf{i}_{abc} = \begin{bmatrix} i_a \\ i_b \\ i_c \end{bmatrix}$$

8.3.2 The SYNCREL dq Model

Most electrical machines with sinusoidally distributed windings are modeled using a technique called dq modeling. This technique essentially converts a three-phase machine into an equivalent two-phase machine. There are a variety of different transformations that can be used to carry out this process, and the transformation to a synchronously rotating reference frame has the property that the currents and voltages in steady state become dc values.

It is outside the scope of this presentation to develop the theory behind the development of dq models. A complete development of this can be found in any standard textbook on generalized machine theory (e.g., [7]). A summary of the transformations appears in Table 8.1.[4]

In order to derive the dq equations for the SYNCREL we simply apply these transformations to the voltage equations for the SYNCREL. These equations are (using Faraday's law once again)[5]:

$$v_a = Ri_a + \frac{d\psi_a}{dt} \tag{8.32}$$

$$v_b = Ri_b + \frac{d\psi_b}{dt} \tag{8.33}$$

$$v_c = Ri_c + \frac{d\psi_c}{dt} \tag{8.34}$$

which can be written more succinctly in matrix form as

$$\mathbf{v}_{abc} = R\mathbf{i}_{abc} + \frac{d\mathbf{\Psi}_{abc}}{dt}. \tag{8.35}$$

---
[4] Note that these transformations are known as power variant transformations, since the two phase machine only produces 2/3 of the power and torque of the three-phase machine.
[5] Note that $\psi_{a,b,c}$ is the total flux linkage with the respective phase—i.e., it includes the self and mutual flux.

Substituting into this using the transformations in Table 8.1 we can write this equation as (expanding the derivatives appropriately)[6]

$$\mathbf{v}_{abc} = R\mathbf{C}^T \mathbf{i}^r_{dq\gamma} + \{p\mathbf{C}^T\}\mathbf{\Psi}^r_{dq\gamma} + \mathbf{C}^T\{p\mathbf{\Psi}_{dq\gamma}\} \tag{8.36}$$

where $p \triangleq$ the derivative operator d/dt.

If we further expand the derivatives and rearrange we can write

$$\mathbf{v}_{abc} = \mathbf{C}^T R \mathbf{i}^r_{dq\gamma} + \mathbf{C}^T \left\{ p \begin{bmatrix} \psi^r_d \\ \psi^r_q \\ \psi^r_\gamma \end{bmatrix} + \omega_{pd} \begin{bmatrix} -\psi^r_q \\ \psi^r_d \\ 0 \end{bmatrix} \right\} = \mathbf{C}^T \mathbf{v}^r_{dq\gamma}. \tag{8.37}$$

Therefore we can see that

$$\begin{bmatrix} \psi^r_d \\ \psi^r_q \\ \psi^r_\gamma \end{bmatrix} = R \begin{bmatrix} i^r_d \\ i^r_q \\ i^r_\gamma \end{bmatrix} + p \begin{bmatrix} \psi^r_d \\ \psi^r_q \\ \psi^r_\gamma \end{bmatrix} + \omega_{pd} \begin{bmatrix} -\psi^r_q \\ \psi^r_d \\ 0 \end{bmatrix}. \tag{8.38}$$

The flux expressions in the preceding equation are of the form $\mathbf{L}^r_{dq\gamma}\mathbf{i}^r_{dq\gamma}$; therefore we need to apply the inductance transformation from Table 8.1 to the three-phase inductance expressions in order to get the $L_{dq\gamma}$ expressions. The resultant inductance matrix is

$$\mathbf{L}^r_{dq\gamma} = \begin{bmatrix} \frac{3}{2}(L_1 + L_2) & 0 & 0 \\ 0 & \frac{3}{2}(L_1 + L_2) & 0 \\ 0 & 0 & 0 \end{bmatrix} \tag{8.39}$$

If we assume that the machine is balanced and is star connected then there cannot be any zero sequence components. This allows us to drop the γ related rows and columns from the equations. The resultant model for the electrical dynamics of the machine is

$$\mathbf{v}^r_{dq} = \begin{bmatrix} R & 0 \\ 0 & R \end{bmatrix} \mathbf{i}^r_{dq} + \begin{bmatrix} L^r_d & 0 \\ 0 & L^r_q \end{bmatrix} p\mathbf{i}^r_{dq} + \omega_{pd} \begin{bmatrix} -L^r_q & 0 \\ 0 & L^r_d \end{bmatrix} \mathbf{i}^r_{dq}, \tag{8.40}$$

which can be written in scalar form as

$$v^r_d = Ri^r_d + L^r_d \frac{di^r_d}{dt} - \omega_{pd} L^r_q i^r_q \tag{8.41}$$

$$v^r_q = Ri^r_q + L^r_q \frac{di^r_q}{dt} + \omega_{pd} L^r_d i^r_d \tag{8.42}$$

where

$$L^r_d = \frac{3}{2}(L_1 + L_2) \tag{8.43}$$

$$L^r_q = \frac{3}{2}(L_1 - L_2). \tag{8.44}$$

This model is shown diagrammatically in Fig. 8.8.

The final piece to the SYNCREL model is the torque expression for the machine. In order to use previous results we require the dq inductances in the correct form to apply (8.24). This implies that we need to transform the three-phase inductances to their two-phase *stationary*

---

[6] Note that the r superscript denotes that these values are expressed in a rotating reference frame. An s superscript is used to denote the stationary reference frame.

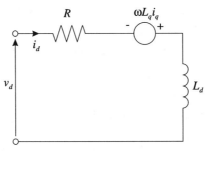

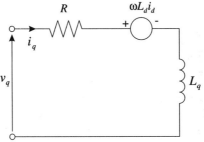

FIGURE 8.8
Ideal *dq* model of the SYNCREL.

frame values (so that the two *dq* windings still contain their dependence on θ_{pd}). This transformation is implemented as

$$\mathbf{L}_{dq\gamma}^s = \frac{2}{3}\mathbf{S}\mathbf{L}_{abc}\mathbf{S}^T \tag{8.45}$$

where

$$\mathbf{S} = \begin{bmatrix} 1 & -\frac{1}{2} & -\frac{1}{2} \\ 0 & \frac{\sqrt{3}}{2} & -\frac{\sqrt{3}}{2} \\ \frac{1}{\sqrt{2}} & \frac{1}{\sqrt{2}} & \frac{1}{\sqrt{2}} \end{bmatrix}. \tag{8.46}$$

If this transformation is applied to

$$\mathbf{L}_{abc} = \begin{bmatrix} L_{aa} & L_{ab} & L_{ac} \\ L_{ba} & L_{bb} & L_{bc} \\ L_{ca} & L_{cb} & L_{cc} \end{bmatrix} \tag{8.47}$$

then we get

$$\mathbf{L}_{dq\gamma}^s = \frac{3}{2}\begin{bmatrix} L_1 + L_2\cos 2\theta_{pd} & L_2\sin 2\theta_{pd} & 0 \\ L_2\sin 2\theta_{pd} & L_1 - L_2\cos 2\theta_{pd} & 0 \\ 0 & 0 & 0 \end{bmatrix}. \tag{8.48}$$

Because we have two windings (for the d and q axes), Eq. (8.24) is applicable, even though the winding configuration is not exactly the same as that in Fig. 8.6. Therefore (8.24) can be rewritten as

$$T_e = \frac{1}{2}(i_d^s)^2 \frac{dL_d^s}{d\theta_{pd}} + \frac{1}{2}(i_q^s)^2 \frac{dL_q^s}{d\theta_{pd}} + i_d^s i_q^s \frac{dL_{dq}^s}{d\theta_{pd}}. \tag{8.49}$$

Using (8.48) we can write this equation as

$$T_e = \frac{3}{2}[((i_q^s)^2 - (i_d^s)^2)L_2 \sin 2\theta_{pd} + 2L_2 i_d^s i_q^s \cos 2\theta_{pd}]. \tag{8.50}$$

This is the torque in a stationary reference frame. Notice that it still contains θ_{pd} terms. The next step is to convert this stationary frame expression to a rotating frame expression. In order to do this we need the transformation matrix from a stationary frame to a rotating frame. This matrix can be shown to be[7]

$$\mathbf{B} = \begin{bmatrix} \cos\theta_{pd} & \sin\theta_{pd} & 0 \\ -\sin\theta_{pd} & \cos\theta_{pd} & 0 \\ 0 & 0 & 1 \end{bmatrix}. \tag{8.51}$$

Using

$$\mathbf{i}_{dq\gamma}^s = \mathbf{B}^T \mathbf{i}_{dq\gamma}^r \tag{8.52}$$

we can substitute for i_d^s and i_q^s in (8.50) to obtain

$$T_e = 3L_2 i_d^r i_q^r \tag{8.53}$$
$$= (L_d^r - L_q^r) i_d^r i_q^r. \tag{8.54}$$

This equation is the torque for a two-phase single-pole pair machine. To account for multiple poles and a three-phase machine we have to adjust the two phase machine torque so that it becomes[8]

$$T_e = \frac{3}{2} p_p (L_d^r - L_q^r) i_d^r i_q^r \tag{8.55}$$

Figure 8.9 shows the space vector diagram for a SYNCREL. Some of the crucial angles for the SYNCREL are defined on this figure.

Summary 8.1 To summarise, the dq equations for the SYNCREL are

$$v_d^r = R i_d^r + L_d^r \frac{di_d^r}{dt} - \omega_{pd} L_q^r i_q^r \tag{8.56}$$

$$v_q^r = R i_q^r + L_q^r \frac{di_q^r}{dt} + \omega_{pd} L_d^r i_d^r \tag{8.57}$$

$$T_e = \frac{3}{2} p_p (L_d^r - L_q^r) i_d^r i_q^r \tag{8.58}$$

---

[7] The **C** matrix in Table 8.1 is actually **BS**.
[8] These adjustments are a consequence of the power variant transformation from the three-phase machine to the two-phase machine. These transformations are such that the two-phase machine produces 2/3 the power and torque of the three-phase machine.

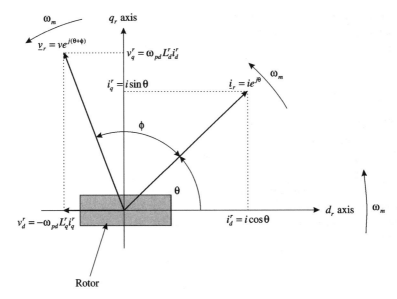

FIGURE 8.9
Vector diagram for the *dq* model of the SYNCREL.

where

$$L_d^r = \frac{3}{2}(L_1 + L_2) \tag{8.59}$$

$$L_q^r = \frac{3}{2}(L_1 - L_2) \tag{8.60}$$

$$p_p \triangleq \text{the pole pairs of the machine.} \tag{8.61}$$

8.3.3 The SYNCREL Space Vector Model

Space vector modeling is the other form of modeling commonly used to model AC machines. The main advantage of this modeling technique is the notational simplicity. Of course the *dq* model and the space vector model are equivalent, and it is relatively simple to convert between them.

The space vector model will only be stated for the SYNCREL, for completeness reasons. It is not very common to use the space vector model with the SYNCREL, and the future work on the control properties does not depend on it. If the reader wishes to learn more about space vector modeling, [8] is an excellent reference.

The stationary frame space vector model is

$$\underline{v}_s = R\underline{i}_s + \frac{d\underline{\psi}_s}{dt} \tag{8.62}$$

where the $\underline{x}$ quantities denote space vectors (the $\underline{x}$ is a generic vector).

It is difficult to evaluate (8.62) because of the complex nature of the flux linkage space vector, $\underline{\psi}_s$. Therefore, as was the case with the *dq* model we transform the stationary frame model to a rotating frame model. This can be achieved using the following reference frame transformation relationships:

$$\left.\begin{array}{l}\underline{v}_s = \underline{v}_r e^{j\theta_{pd}} \\ \underline{i}_s = \underline{i}_r e^{j\theta_{pd}} \\ \underline{\psi}_s = \underline{\psi}_r e^{j\theta_{pd}}\end{array}\right\}. \tag{8.63}$$

Substituting these into (8.62) and taking the appropriate derivatives the following rotating frame space vector equation can be written:

$$\underline{v}_r = R\underline{i}_r + \frac{d|\underline{\psi}_r|}{dt} + j\omega_{pd}\underline{\psi}_r \tag{8.64}$$

where

$$\omega_{pd} = \frac{d\theta_{pd}}{dt} \triangleq \text{rotor angular velocity}. \tag{8.65}$$

The following current and flux linkage relationships can be derived:

$$\underline{i}_r = I_{pk} e^{j\theta} \tag{8.66}$$

$$\underline{\psi}_r = \frac{3}{2} I_{pk}[L_1 e^{j\theta} + L_2 e^{-j\theta}] \tag{8.67}$$

when the three-phase currents supplying the machine are of the form:

$$\left.\begin{array}{l}i_a = I_{pk}\cos(\theta_{pd} + \theta) \\ i_b = I_{pk}\cos\left(\theta_{pd} + \theta - \frac{2\pi}{3}\right) \\ i_c = I_{pk}\cos\left(\theta_{pd} + \theta + \frac{2\pi}{3}\right)\end{array}\right\}. \tag{8.68}$$

Remark 8.6 *Equation (8.66) states that the current vector is at a constant angle of θ radians and has a magnitude equal to the magnitude of the three-phase currents supplying the machine. One can clearly see the the temporal and spatial currents are directly related.*

Remark 8.7 *As noted previously, one of the main advantages of the space vector modeling approach is the succinct nature of the modeling equations. The electrical dynamics of the machine are now represented by (8.64), instead of the two equations of the dq model case.*

The torque expression for the machine is

$$T_e = \frac{9}{4} I_{pk}^2 L_2 \sin 2\theta \tag{8.69}$$

which can be shown to be equivalent to (8.55).

8.3.4 Practical Issues

The analysis carried out so far has been for the ideal model of the SYNCREL, which neglected the following practical effects:

1. Saturation of the stator and rotor iron
2. Leakage inductance
3. Iron losses
4. Stator and rotor slotting effects

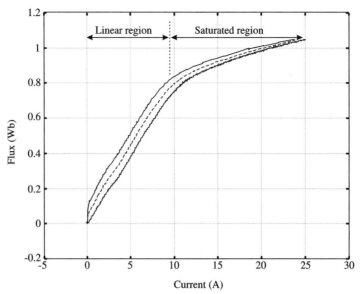

FIGURE 8.10
Flux linkage versus current plot for the d-axis of a 6-kW axial laminated SYNCREL.

The relevance of including these practical effects in a model depends on the use of the model. For example, if one is carrying out the initial design of a controller a simple dq model is usually sufficient to represent the main control properties of the machine. A more detailed control design may include saturation and iron loss effects.

Let us briefly look at the four listed practical effects.

8.3.4.1 Saturation. Saturation refers to the nonlinear relationship between the magnetizing force and the resultant flux density for a ferromagnetic material. The d-axis of the SYNCREL, because it is dominated by iron, does show saturation in normal operation. Figure 8.10 shows the flux versus current relationship for the d-axis of an axial laminated SYNCREL.[9] Note that for currents above 10 A there is a marked departure of the characteristic from a linear one, this indicating that the iron in the machine is now starting to saturate.

The inclusion of saturation in dynamic models of machines has traditionally been a difficult process. Moreover, the resultant models cannot be solved analytically. The advent of digital computers, however, has made it fairly straightforward to generate numerical solutions for machine equations containing saturation.

The process of including saturation into the models involves getting the flux linkage versus current characteristic of the particular machine being modeled (this can be done experimentally, or perhaps via finite element modeling). This data is then stored as a lookup table in a computer. The differential equations of the machine are then written in flux linkage form:

$$v_d^r = R i_d^r + \frac{d\psi_d^r}{dt} - \omega_{pd}\psi_q^r \qquad (8.70)$$

$$v_q^r = R i_q^r + \frac{d\psi_q^r}{dt} + \omega_{pd}\psi_d^r \qquad (8.71)$$

$$T_e = \frac{3}{2} p_p (\psi_d^r i_q^r - \psi_q^r i_d^r). \qquad (8.72)$$

---

[9] The two plots in this diagram are for rising currents and falling currents. The dashed line is the average value. The difference between the rising and falling current flux linkage indicates that the iron exhibits hysteresis.

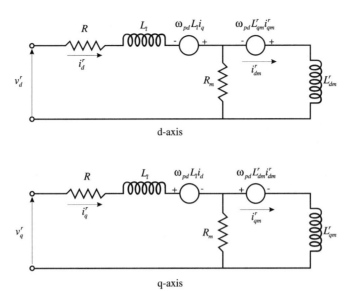

FIGURE 8.11
dq circuit including the stator iron loss resistor.

These equations are solved numerically by using the lookup table at each step of the numeric solver to calculate the value of i_d^r that corresponds to the present value of the *d*-axis flux linkage.

Remark 8.8 *The nonlinear differential equations, though allowing precise performance for the machine to be found in simulation, are not a great aid to intuitive understanding. The linear equations are far more useful for this purpose.*

Remark 8.9 *Other effects such as cross-magnetization[10] also occur due to the presence of saturation. The practical effects of cross-magnetization on the performance of the machine are relatively minor, so this will not be pursued here.*

8.3.4.2 Leakage Inductance.
Leakage inductance is present in every machine. In the particular case of the SYNCREL these inductances are related to the flux linking the stator but not linking the rotor. It is mainly due to end turn leakage and stator slot leakage. In terms of the model we can consider the leakage to be included in the L_d and L_q inductances. It does not influence the torque because it is a function of $(L_d - L_q)$, and hence the leakage terms tend to cancel out (assuming that the *d*- and *q*-axis leakages are the same). Leakage inductance must be more formally included in the model if iron losses are present, since this prevents the leakage from being subsumed into the $L_{d,q}$ inductances (see Fig. 8.11).

Remark 8.10 *In general, leakage inductance has only a minor influence on the performance of the machine, and for control purposes it can be safely ignored.*

8.3.4.3 Iron Losses.
Iron losses are from two sources in electrical machines—eddy currents and hysteresis losses. Depending on the design of the SYNCREL, iron losses may have a significant effect on the control performance.

---
[10] Cross-magnetization is an effect where saturation of the iron in one axis of the machine can affect the electrical parameters in an orthogonal axis.

Hysteresis losses are a consequence of the energy required to change the direction of the domains in a ferromagnetic material. An empirical expression for the power loss per unit volume in a ferromagnetic material is:

$$p_h = k_h f B^n \text{ W/m}^3 \qquad (8.73)$$

where k_h and n are empirically derived constants for a given material,[11] B is the maximum flux density, and f is the frequency that the hysteresis is traversed.

Eddy current losses occur in conducting materials subject to time-varying magnetic fields. Lenz's law results in a current in the material that produces a flux density to oppose the impinging flux density. For a material subject to sinusoidally varying flux density, the average power dissipated due to eddy currents is

$$p_{e_{ave}} = k_e \omega^2 B^2 \text{ W/m}^3 \qquad (8.74)$$

where

$$k_e = \frac{\delta^2}{24\rho}$$

$\delta \triangleq$ thickness of the material

$\rho \triangleq$ resistivity of the material.

Remark 8.11 *Notice that eddy current losses are proportional to the applied frequency squared, whereas the hysteresis losses are only proportional to the frequency. Also note the effect of having thin laminations (i.e., δ is small)—the eddy current losses are proportional to the square of the lamination thickness. From a rotor design viewpoint, it is therefore important to keep the laminations in the rotor thin.*

Remark 8.12 *In an axial laminated machine the laminations in the rotor are edge on to the d-axis, but the flux density produced in the q-axis is orthogonal to the laminations. Therefore the q-axis eddy currents can be substantial, especially taking into account the high-frequency flux density oscillations in this axis from the stator slotting.*

The task of deriving the modified model of the SYNCREL with iron losses is a rather lengthy process and too detailed to present here. The stator power loss due to the iron is modeled as a resistor across the magnetizing branch of the machine. To a first approximation, a constant value of resistance models the eddy current losses, since the power dissipated in a resistor in parallel with an inductor is proportional to $\omega^2 B^2$. However, the resistor does not model the hysteresis losses.

Figure 8.11 shows the dq model of the SYNCREL including the resistor to model the eddy currents in the stator. Notice that the leakage inductance has been separated out from the magnetizing inductance.

Remark 8.13 *From Figure 8.11 one can see that one consequence of iron losses is that the terminal input currents are no longer equal to the flux producing current.*

Remark 8.14 *One can also include a resistor across the transient inductance section of Fig. 8.11 (i.e., the L^r_{dm} and L^r_{qm} elements) to account for the losses in the rotor. There also can be a difference between the placement of the resistor in the d- and q-axes [9].*

---

[11] n typically has a value of 1.5 to 2.5.

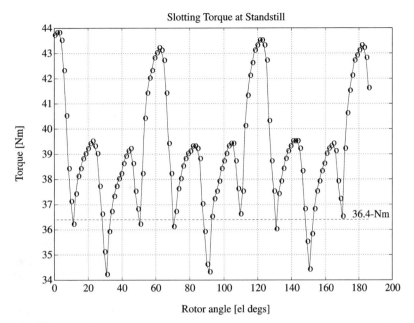

FIGURE 8.12
Slotting torque ripple in an axial laminated SYNCREL with 60° current angle and 20 A current.

8.3.4.4 Rotor and Stator Slot Effects. Interaction between the laminations of the rotor and the slots in the stator of the SYNCREL result in torque ripple. This effect can be measured by fixing the current angle at some angle and then rotating the rotor and measuring the torque. Ideally the torque should be constant, its value defined by the current angle. Figure 8.12 shows the result of this test for a 6-kW axial laminated SYNCREL using a 36-slot induction machine stator. The peak-to-peak torque ripple is approximately 10 Nm! The 36.4 Nm is the torque calculated using the nonlinear model of the machine.

This is probably not surprising in retrospect, since the rotor is not skewed and the stator is not skewed. In the case of an axial laminated rotor it is not practical to skew the rotor; therefore the stator would need to be skewed. If a flux barrier design is used for the rotor, then it is feasible to skew the rotor to minimize the torque ripple. With careful design of a flux barrier rotor it is possible to get the torque ripple low enough for a SYNCREL to be suitable for high-performance servo applications.

8.4 CONTROL PROPERTIES

In this section we use the models derived in the previous section to derive a number of control properties for the SYNCREL. One important aspect of this analysis is that it is done in a normalized fashion so that it is machine independent.

A point of clarification at this stage. The term "control properties" when referring to a machine means the following: given a control objective, what are the set points of the relevant control variables to achieve this in an optimal fashion? Secondly how do machine parameters affect performance and controllability?

274 CHAPTER 8 / SYNCHRONOUS RELUCTANCE MACHINES

The following control objectives will be investigated:

1. Maximum torque per ampere
2. Maximum efficiency
3. Maximum rate of change of torque
4. Maximum power factor
5. Field weakening

Before considering these we first have to develop the normalized form of the machine model.

8.4.1 Normalizations

The choice of the base values used in a per-unit or normalized system are to some degree arbitrary. In the particular case of the SYNCREL several different schemes have been used in the literature. One should realize though, that the conclusions drawn from using different normalizations have to be the same, as they are simply looking at the system in a slightly different way. It should also be realised that some normalizations are more suited for the analysis of particular control strategies—they produce simpler expressions that are more easily analyzed.

Because the following analysis is based on the models derived in the previous section, they are subject to the same assumptions. In addition most of the analysis also assumes that the stator resistance can be ignored. This assumption creates expressions simple enough that the basic properties of the machine can be gleaned from them.

One normalization that can be used for the SYNCREL is based on the maximum torque per ampere and rated voltage and current of the machine [10]. When this normalization is used the inductances disappear, since they are represented in the model as the ratio L_d/L_q, which is denoted by the symbol ξ.

In order to determine the maximum torque per ampere we need to ascertain the angle of the current vector relative to the d-axis. Consider expression (8.54), repeated here for convenience:

$$T_e = \frac{3}{2}p_p(L_d^r - L_q^r)i_d^r i_q^r \tag{8.75}$$

This expression can also be written as

$$T_e = \frac{3}{2}p_p(L_d^r - L_q^r)(i\cos\theta)(i\sin\theta)$$
$$= \frac{3}{4}p_p(L_d^r - L_q^r)i^2 \sin 2\theta \tag{8.76}$$

where $\theta \triangleq$ the angle of the current space vector with respect to the d-axis of the machine, and $i \triangleq$ the current vector magnitude (as defined in Fig. 8.9).

One can see by inspection from (8.76) that for a given current vector magnitude the torque is maximized if $\theta = \pi/4$ radians. Therefore the maximum torque for the SYNCREL is

$$T_{e_{\max}} = \frac{3}{4}p_p(L_d^r - L_q^r)i_0^2 \tag{8.77}$$

where $i_0 \triangleq$ the rated current for the SYNCREL.

8.4 CONTROL PROPERTIES

For convenience reasons we shall defined the base torque for the machine in terms of the two-phase machine. Therefore

$$T_0 = \frac{1}{2} p_p (L_d^r - L_q^r) i_0^2. \tag{8.78}$$

The base frequency is defined as the frequency at which the machine runs out of voltage at base torque and current. This is the normal "break point" in the torque characteristic of the machine. Therefore the base frequency is

$$\omega_0 \triangleq p_p \omega_{\text{brk}}. \tag{8.79}$$

The rated voltage of the machine (i.e., the voltage at the break frequency) is denoted as v_0.[12]

The base flux for the machine can be derived as follows:

$$\psi_0 = \sqrt{(L_d^r i_{d_0}^r)^2 + (L_q^r i_{q_0}^r)^2} \tag{8.80}$$

where $i_{d_0}^r \triangleq$ the d-axis current, and $i_{q_0}^r \triangleq$ the q-axis current, both when the current magnitude is i_0. As can be seen from Fig. 8.9 one can write these currents as

$$i_{d_0}^r = i_0 \cos\theta = \frac{1}{\sqrt{2}} i_0 \quad \text{for} \quad \theta = \pi/4 \tag{8.81}$$

$$i_{q_0}^r = i_0 \sin\theta = \frac{1}{\sqrt{2}} i_0 \quad \text{for} \quad \theta = \pi/4. \tag{8.82}$$

Therefore using these expressions the base flux can be written as

$$\psi_0 = \frac{i_0}{\sqrt{2}} \sqrt{(L_d^r)^2 + (L_q^r)^2}. \tag{8.83}$$

The other bases can now defined in terms of those defined already. The base voltage is

$$v_0 = \omega_0 \psi_0. \tag{8.84}$$

The base power can now be defined:

$$\begin{aligned} P_0 &= v_0 i_0 \\ &= \omega_0 \psi_0 i_0 \\ &= \frac{\omega_0 i_0^2}{\sqrt{2}} \sqrt{(L_d^r)^2 + (L_q^r)^2}. \end{aligned} \tag{8.85}$$

The base resistance and inductance can also now be defined:

$$R_0 = \frac{v_0}{i_0} \tag{8.86}$$

$$L_0 = \frac{\psi_0}{i_0}. \tag{8.87}$$

Let use now summarize the normalized values using the foregoing bases for the major parameters for the machine.

---

[12] The base voltage definition implicitly assumes that the stator resistance of the machine is zero, since as we shall see later it is defined in terms of base frequency and flux linkage.

Summary 8.2

$$\left.\begin{aligned} T_n &= \frac{T_e}{T_0} & P_n &= \frac{P}{P_0} & \psi_n &= \frac{\psi_n}{\psi_0} \\ \omega_n &= \frac{\omega}{\omega_0} & i_n &= \frac{i}{i_n} & v_n &= \frac{v}{v_0} \\ R_n &= \frac{R}{R_0} & L_n &= \frac{L}{L_0} & \end{aligned}\right\} . \quad (8.88)$$

Using the normalizations in Summary 8.2 and assuming that the stator resistance can be neglected,[13] we can derive the following normalized electrical equations from those in Summary 8.1:

$$v_{dn} = \frac{\sqrt{2}\xi}{\sqrt{\xi^2+1}} \left(\frac{1}{\omega_0} p i_{dn} - \frac{\omega_n}{\xi} i_{qn} \right) \quad (8.89)$$

$$v_{qn} = \frac{\sqrt{2}\xi}{\sqrt{\xi^2+1}} \left(\frac{1}{\xi\omega_0} p i_{qn} + \omega_n i_{dn} \right) \quad (8.90)$$

$$T_n = i_n^2 \sin 2\theta = 2 i_n^2 \frac{\tan\theta}{1+\tan^2\theta} \quad (8.91)$$

where $p \triangleq$ the derivative operator d/dt and

$$\xi = \frac{L_d^r}{L_q^r} \quad \text{(which is known as the saliency ratio)}. \quad (8.92)$$

Using these basic expressions one can generate a number of other auxiliary expressions. The steady-state voltages of the SYNCREL can be written as (by letting the p terms in (8.89) and (8.90) equal zero)

$$v_{dn} = \frac{-\sqrt{2}\omega_n i_{qn}}{\sqrt{\xi^2+1}} \quad (8.93)$$

$$v_{qn} = \frac{\sqrt{2}\xi\omega_n i_{dn}}{\sqrt{\xi^2+1}}. \quad (8.94)$$

Using the fact that $\tan\theta = i_{qn}/i_{dn}$ and $i_n = \sqrt{i_{dn}^2 + i_{qn}^2}$ one can write the currents into the machine as

$$i_{dn} = \frac{i_n}{\sqrt{1+\tan^2\theta}} \quad (8.95)$$

$$i_{qn} = \frac{i_n \tan\theta}{\sqrt{1+\tan^2\theta}} \quad (8.96)$$

---

[13] The stator resistance is neglected so that the expressions derived are simplified. This approximation only affects the accuracy of the results at very low speeds.

which can be substituted into (8.93) and (8.94) to give

$$v_{dn} = \frac{-\sqrt{2}\omega_n(\tan\theta)i_n}{\sqrt{(\xi^2+1)(1+\tan^2\theta)}} \qquad (8.97)$$

$$v_{qn} = \frac{\sqrt{2}\xi\omega_n i_n}{\sqrt{(\xi^2+1)(1+\tan^2\theta)}}. \qquad (8.98)$$

These voltage expressions can be substituted into $v_n^2 = v_{dn}^2 + v_{qn}^2$ and rearranged to give the following expression for the normalized current amplitude into the machine:

$$i_n^2 = \frac{(\xi^2+1)(1+\tan^2\theta)v_n^2}{2\omega_n^2(\tan^2\theta + \xi^2)}. \qquad (8.99)$$

This can then be substituted into (8.91) to give:

$$T_n = \frac{(\xi^2+1)(\tan\theta)v_n^2}{\omega_n^2(\tan^2\theta + \xi^2)}. \qquad (8.100)$$

Remark 8.15 *This expression for the torque of the machine implicitly assumes that the current angle is constant. This occurs as a consequence of the steady-state assumption.*

Another very useful expression can be obtained if we get the voltage magnitude in terms of the torque *under transient conditions*. If we utilize the fact that $i_n^2 = i_{dn}^2 + i_{qn}^2$ together with (8.91) one can write

$$i_{dn} = \sqrt{\frac{T_n}{2}\cot\theta} \qquad (8.101)$$

$$i_{qn} = \sqrt{\frac{T_n}{2}\tan\theta} \qquad (8.102)$$

which when substituted into (8.89) and (8.90) gives the normalized voltages in terms of the torque and current angle:

$$v_{dn} = \frac{\xi}{\sqrt{\xi^2+1}}\left[\frac{\sqrt{\cot\theta}}{\omega_0}p\sqrt{T_n} - \frac{\omega_n}{\xi}\sqrt{T_n\tan\theta}\right] \qquad (8.103)$$

$$v_{qn} = \frac{\xi}{\sqrt{\xi^2+1}}\left[\frac{\sqrt{\tan\theta}}{\xi\omega_0}p\sqrt{T_n} + \omega_n\sqrt{T_n\cot\theta}\right]. \qquad (8.104)$$

Remark 8.16 *Note that these voltage expressions assume that θ is constant—i.e., it is not changing with respect to time. This has allowed the θ-based terms to be moved outside the p operator. Consequently these equations and the following equation derived from it are restricted to constant angle control (CAC) control strategies. This implies that the i_{dn} and i_{qn} currents are not independent, but are related by $\tan\theta$.*

Using $v_n^2 = v_{dn}^2 + v_{qn}^2$ and substituting (8.103) and (8.104) we can write

$$v_n^2 = \frac{\tan\theta + \xi^2\cot\theta}{\xi^2+1}\left[\frac{1}{4T_n\omega_0^2}(pT_n)^2 + \omega_n^2 T_n\right]. \qquad (8.105)$$

Finally, another useful normalization is the normalized rate of change of normalized torque—i.e., pT_n. This can be normalized to the angular velocity as follows:

$$p'T_n = \frac{pT_n}{\omega_0} \quad (8.106)$$

which has the units of pu/radian.

Remark 8.17 *One can interpret $p'T_n$ as how much the torque in pu rises for one radian of an electrical cycle at ω_0 frequency. For example, if $p'T_n = 5/2\pi$ then the torque is rising 5 pu in 2π radians, or 1 pu in $2\pi/5$ radians, which is 1/5th of the base electrical cycle.*

8.4.2 Maximum Torque per Ampere and Maximum Efficiency Control

Section 8.4.1 implicitly worked out the correct control strategy for the maximum torque per ampere control (MTPAC) of the SYNCREL—i.e., the current angle should be at $\pi/4$.

Another interesting and very important property of a machine is its maximum efficiency. Given the assumption that the stator resistance is zero then the efficiency of the machine must be unity at any current angle (since there are no losses). In order to get something other than this trivial solution we must reintroduce the stator resistance into the model.

The steady-state voltage equations (8.93) and (8.94) with stator resistance become

$$v_{dn} = R_n i_{dn} - \frac{\sqrt{2}\omega_n i_{qn}}{\sqrt{\xi^2+1}} \quad (8.107)$$

$$v_{qn} = R_n i_{qn} - \frac{\sqrt{2}\xi\omega_n i_{dn}}{\sqrt{\xi^2+1}}. \quad (8.108)$$

The normalized power into the machine can be written as

$$P_n^{in} = R_n i_n^2 + \frac{(\xi-1)\omega_n i_n^2 \sin 2\theta}{\sqrt{2(\xi^2+1)}}. \quad (8.109)$$

The output power of the machine is given by

$$P_{out} = \frac{1}{2}\omega(L_d^r - L_q^r)i^2 \sin 2\theta \quad (8.110)$$

where ω is the angular velocity in electrical radians/sec (i.e., $\omega = p_p\omega_m$). This allows the normalized output power to be written as

$$P_n^{out} = \frac{P_{out}}{P_0}$$

$$= \frac{(\xi-1)\omega_n i_n^2 \sin 2\theta}{\sqrt{2(\xi^2+1)}}. \quad (8.111)$$

Using these two expressions the efficiency of the machine is defined as

$$\eta = \frac{P_n^{out}}{P_n^{in}}$$

$$= \frac{1}{\left[\dfrac{R_n i_n^2 \sqrt{2(\xi^2 + 1)}}{(\xi - 1)\omega_n i_n^2 \sin 2\theta}\right] + 1}. \tag{8.112}$$

In order to calculate the maximum efficiency one should formally take the derivative of this expression. However, the maximum value can be seen by inspection of (8.112) to occur at $\theta = \pi/4$.

Remark 8.18 *The maximum efficiency control (MEC) current angle of the SYNCREL with stator resistance included occurs at exactly the same angle as the MTPA angle of $\pi/4$. Therefore this angle not only maximizes the torque per ampere, it also minimizes the losses in the machine for a given output power.*

Another quantity of importance is the break frequency for this control mode. This can be found using (8.105) and setting $pT_n = 0$, $v_n = 1$ and $T_n = \sin 2\theta$ (i.e., $i_n = 1$), and then rearranging so that ω_n is the subject of the expression:

$$\omega_{n_{max}} = \sqrt{\frac{\xi^2 + 1}{\sin 2\theta(\tan \theta + \xi^2 \cot \theta)}}. \tag{8.113}$$

In the particular case of MTPAC and MEC, $\tan \theta = 1$, and therefore

$$\omega_{n_{max}} = 1. \tag{8.114}$$

Remark 8.19 *This value of $\omega_{n_{max}}$ should be expected, since the MTPA normalization is defined using the break frequency as the base frequency.*

8.4.3 Maximum Power Factor Control

One of the fundamental quantities for any machine is the power factor. Consider the dq phasor diagram for a steady state machine in Fig. 8.9 (which assumes that the stator resistance is zero). The projection of $\underline{v}_r$ onto $\underline{i}_r$ can be seen to be

$$v_{r_i} = \omega_{pd} L_d^r i_d^r \sin \theta - \omega_{pd} L_q^r i_q^r \cos \theta. \tag{8.115}$$

The power factor is defined as the cosine of the angle between the current and voltage vectors. Therefore, from Fig. 8.9, it can be seen that the power factor is $\cos \phi$. Using trigonometry one can write the following expression:

$$\cos \phi = \frac{v_{r_i}}{|\underline{v}_r|}$$

$$= \frac{\omega_{pd} L_d^r i_d^r \sin \theta - \omega_{pd} L_q^r i_q^r \cos \theta}{\sqrt{(\omega_{pd} L_d^r i_d^r)^2 + (\omega_{pd} L_q^r i_q^r)^2}}. \tag{8.116}$$

This expression can be manipulated to give

$$p_{f_i} = \cos\phi = \frac{(\xi - 1)\cos\theta}{\sqrt{\xi^2 \cot^2\theta + 1}}$$

$$= \frac{\xi - 1}{\sqrt{2}} \sqrt{\frac{\sin 2\theta}{\xi^2 \cot\theta + \tan\theta}}. \tag{8.117}$$

In order to determine the angle for the maximum power factor we differentiate (8.117) with respect to θ,

$$\frac{dp_{f_i}}{d\theta} = \frac{\xi - 1}{\sqrt{2}}\left[\frac{\cos 2\theta}{\sqrt{\sin 2\theta(\tan\theta + \xi^2\cot\theta)}} + \frac{(\tan\theta - \xi^2\cot\theta)}{(\tan\theta + \xi^2\cot\theta)}\sqrt{\frac{(\tan\theta + \xi^2\cot\theta)}{\sin 2\theta}}\right], \tag{8.118}$$

and equate this expression with zero to give

$$\cos 2\theta - \left(\frac{\tan\theta - \dfrac{\xi^2}{\tan\theta}}{\tan\theta + \dfrac{\xi^2}{\tan\theta}}\right) = 0. \tag{8.119}$$

Using the substitution $\tan\theta = \sigma$ and $\cos 2\theta = (1 - \sigma^2)/(1 + \sigma^2)$ one can solve for $\tan\theta$ to give

$$\tan\theta = \sqrt{\xi}. \tag{8.120}$$

If (8.120) is substituted into (8.117) then one obtains the expression for the maximum power factor of the SYNCREL:

$$p_{f_i} = \frac{\xi - 1}{\xi + 1}. \tag{8.121}$$

Remark 8.20 *An important attribute of (8.121) is the dependence on the saliency ratio of the machine. This emphasizes the fact that the performance of the machine is critically dependent on this ratio.*

Remark 8.21 *Operating at the maximum power factor angle of (8.120) makes a significant difference to the power factor of the machine. This is shown in Fig. 8.13, where the power factor for maximum power factor control (MPFC) is plotted along with the power factor for MTPAC. Note that as ξ increases the difference between the two control strategies increases. Also note that $\xi \geq 8$ allows the power factor of the SYNCREL to be competitive with that of the induction machine.*

Remark 8.22 *The better power factor obtained under MPFC is not obtained without cost. The angle required for maximum power factor is much larger than the optimal angle for MTPAC; therefore we are sacrificing the torque output of the machine to obtain maximum power factor.*

If the machine is ideal the maximum power factor angle is also the *minimum volt–amps (VA) angle*. This can be deduced by considering the situation where the machine is producing a constant torque at constant speed (i.e., constant output power and the machine is in steady state).

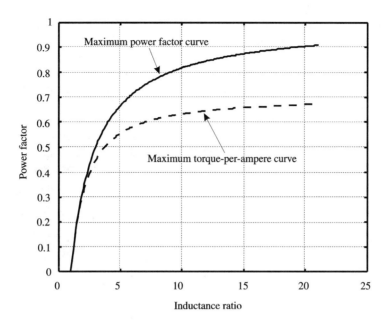

FIGURE 8.13
Power factor versus inductance ratio for MTPAC and MPFC.

If the machine is lossless, then the input power must be equal to the output power. Since the real input power is

$$P_{in} = VI \cos \phi \\ = P_{out}, \quad (8.122)$$

if the power factor increases the VI term must decrease to maintain P_{in} constant. Therefore VI is a minimum when $\cos \phi$ is a maximum.

As mentioned in the last remark, the increased power factor is being traded off against lower torque output (assuming the same maximum current). The precise value of the torque can be calculated by substituting (8.120) into (8.91) to give

$$T_n = \frac{2\sqrt{\xi}}{\xi + 1}. \quad (8.123)$$

The break frequency for this control strategy can be evaluated by substituting (8.120) into (8.113)[14] to give

$$(\omega_{n_{max}})_{\tan \theta = \sqrt{\xi}} = \sqrt{\frac{\xi^2 + 1}{2\xi}}. \quad (8.124)$$

8.4.4 Maximum Rate of Change of Torque Control

One of the important properties of any machine that is to be used in high-performance applications is how fast it can change its torque. Therefore it is of interest to know whether

---
[14] Remember that $\sin 2\theta = 2 \tan \theta / (1 + \tan^2 \theta)$.

there is an optimal angle which will maximize the pT_n of the SYNCREL. Intuitively one would think that such an angle would exist. This can be deduced by considering the rate of change of torque if the current angle is small—i.e., the current vector is close to the d-axis of the machine. In this situation, the current is virtually lying in the most inductive axis of the machine; therefore for some applied voltage the rate of change of current, and therefore torque, will be relatively slow. The converse occurs as the current vector approaches $90°$.

Maximization of the rate of change of torque implies that we wish to choose the optimal angle to obtain maximum pT_n *for a given voltage*. Therefore one can view maximum rate of change of torque control (MRCTC) as a strategy that maximizes the usage of the available voltage.

Equation (8.105) can be rearranged to make pT_n the subject of the expression:

$$(pT_n)^2 = 4T_n\omega_0^2\left[\frac{v_n^2(\xi+1)}{\xi^2 \cot\theta + \tan\theta} - \omega_n^2 T_n\right]. \tag{8.125}$$

In order to maximize (8.125) with respect to θ we take the following derivative:

$$\frac{d(pT_n)^2}{d\theta} = \frac{-K\left(\dfrac{1}{\cos^2\theta} - \dfrac{\xi^2}{\sin^2\theta}\right)}{\xi^2 \cot\theta + \tan\theta} \tag{8.126}$$

where $K = 4T_n\omega_0(\xi^2+1)v_n^2$.

If (8.126) is equated to zero and then rearranged one can obtain the following:

$$\sin\theta = \frac{\xi}{\sqrt{\xi^2+1}} \tag{8.127}$$

which implies (using trigonometry) that

$$\tan\theta = \xi \tag{8.128}$$

$$\text{or} \quad \theta = \tan^{-1}\xi. \tag{8.129}$$

Remark 8.23 *Note that the MRCTC optimal angle has been defined under the constraint that the current angle is held at a constant angle. This is implicit in the voltage expression used to derive the angle. However, this does not mean that the maximum rate of change of torque (MRCT) derived under this condition is the optimal value if the current vector angle is allowed to vary.*

Remark 8.24 *Investigation of (8.125) indicates that pT_n will increase as ξ increases.*

Another interesting parameter to look at is the angle of the flux vector in the machine. The angle of the flux in the machine is defined as

$$\theta_f = \tan^{-1}\frac{\psi_q^r}{\psi_d^r} = \tan^{-1}\frac{L_q^r i_q^r}{L_d^r i_d^r}. \tag{8.130}$$

If we are controlling the machine for MRCT then we have

$$\tan\theta = \xi = \frac{i_q^r}{i_d^r} = \frac{L_d^r}{L_q^r}. \tag{8.131}$$

Substituting for i_q^r into (8.130) one can write

$$\theta_f = \tan^{-1} \frac{L_q^r \frac{L_d^r}{L_q^r} i_d^r}{L_d^r i_d^r} = \tan^{-1} 1 = \frac{\pi}{4} \text{ rad.} \quad (8.132)$$

Remark 8.25 *It is interesting that under this condition it is the flux vector, and not the current vector, that has the angle of $\pi/4$ radians.*

As in the previous control strategies we can calculate the normalized torque for MRCTC by substituting (8.128) into (8.91) to give

$$T_n = \frac{2\xi}{\xi^2 + 1}. \quad (8.133)$$

Similarly the break frequency can be found by substituting (8.128) into (8.113) to give

$$\omega_{n_{\max}} = \frac{\xi^2 + 1}{2\xi}. \quad (8.134)$$

Remark 8.26 *From (8.133) and (8.134) it is clear that T_n falls and $\omega_{n_{\max}}$ increases with increasing ξ.*

8.4.5 Constant Current in *d*-Axis Control

The control strategies described thus far have all been constant angle control (CAC) strategies. However, obviously there is another family of controllers that are variable angle control (VAC) strategies, where the current vector angle is allowed to vary transiently.

One such control strategy that falls into the VAC category is constant current in *d*-axis control (CCDAC). This control approach is based on the realization that the total flux in the machine is largely due to the flux in the *d*-axis as this is the axis with the least reluctance (note that we are assuming that there is a reasonable amount of current in this axis). Therefore, to some degree if we keep the current in the *d*-axis constant, we are setting a major component of the flux. The *q*-axis, because of its low inductance, contributes a much smaller component of flux per ampere as compared to the *d*-axis. However, by controlling the *q*-axis flux we are effectively controlling the current angle, and therefore the torque produced by the machine. The other motivation for this approach is that the dynamic performance of the *q*-axis is faster than the *d*-axis due to the low *q*-axis inductance.

We shall concentrate on the rate of change of torque for CCDAC, and compare its performance with MRCTC. Consider (8.89) and (8.90) with $di_{dn}/dt = 0$ and $i_{dn} = I_{dn}$ (where $I_{dn} \stackrel{\Delta}{=}$ some constant value of current in the *d*-axis):

$$v_{dn} = \frac{-\sqrt{2}\omega_n}{\sqrt{\xi^2 + 1}} i_{qn} \quad (8.135)$$

$$v_{qn} = \frac{\sqrt{2}\xi\omega_n}{\sqrt{\xi^2 + 1}} \left(\frac{1}{\xi\omega_0} p i_{qn} + I_{dn} \right). \quad (8.136)$$

Using the relationship $\sin 2\theta = 2I_{dn}i_{qn}/i_n^2$, (8.91) can be written as

$$i_{qn} = \frac{T_n}{2I_{dn}}. \quad (8.137)$$

Since I_{dn} is constant, the derivative of this expression is

$$p'i_{qn} = \frac{1}{2I_{dn}} p'T_n \qquad (8.138)$$

where $p' \triangleq d/(\omega_0 dt)$.

Rearranging (8.136) and substituting for v_{qn} and using $v_n^2 = v_{dn}^2 + v_{qn}^2$ and (8.135), allows (8.138) to be written as

$$p'i_{qn} = \pm \sqrt{\frac{\xi^2+1}{2} v_n^2 - \frac{\omega_n^2 T_n^2}{4I_{dn}^2}} - \omega_n \xi I_{dn}. \qquad (8.139)$$

By substituting (8.135) and (8.138) into (8.139) the expression for the rate of change of torque can be obtained:

$$p'T_n = \pm \sqrt{2I_{dn}^2(\xi+1)v_n^2 - \omega_n^2 T_n^2} - 2\omega_n \xi I_{dn}^2. \qquad (8.140)$$

The break frequency and maximum torque output from this control strategy are limited to the same values as the other CAC strategies, depending on the value of the current angle.

Remark 8.27 *The CCDAC strategy considered in this section is but one VAC control approach. Another obvious control strategy that superficially appears to offer fast transient response is the constant flux strategy. As the name implies this strategy maintains the same flux magnitude in the machine and achieves different torques by varying the angle of the flux vector. This strategy is closely related to MRCTC, since for any constant flux value the maximum torque is achieved at the MRCTC optimal angle.*

The constant flux control strategy is generally inferior in all performance measures and will not be investigated any further. Its performance is shown in some later examples for comparison purposes.

8.4.6 Comparative Performance of Control Strategies

This section will look at the comparative performance of the various control strategies and establish the interrelationships between them.

8.4.6.1 Torque Interrelationships.
Let use first consider the relationships between torque output from the different control strategies. Because all the strategies that we have considered have been derived using the maximum-torque-per-ampere normalization then they can be directly compared without any conversion factors. Table 8.2 lists the torques for the CAC strategies.

Table 8.2 Comparison of Normalized Torques

Control strategy	Rated normalized torque
MTPAC/MEC	1
MPFC	$\dfrac{2\sqrt{\xi}}{\xi+1}$
MRCTC	$\dfrac{2\xi}{\xi^2+1}$

An examination of the values in Table 8.2 allows the following ordering of the torques:

$$(T_{n_{\max}})_{\text{MRCTC}} \leq (T_{n_{\max}})_{\text{MPFC}} \leq (T_{n_{\max}})_{\text{MTPAC/MEC}}. \tag{8.141}$$

Remark 8.28 *Note that as $\xi \to \infty$, $T_{n_{\max}}$ for MPFC and MRCTC approaches zero.*

8.4.6.2 Break Frequency Interrelationships. In a manner similar to the torque interrelationships we can catalog the break frequency interrelationships. This appears in Table 8.3. We can also form the following expression for the relationships between the break frequencies:

$$(\omega_{n_{\max}})_{\text{MTPAC/MEC}} \leq (\omega_{n_{\max}})_{\text{MPFC}} \leq (\omega_{\max})_{\text{MRCTC}} \tag{8.142}$$

Remark 8.29 *One trend that is clear from Table 8.3 is that the MPFC and MRCTC strategies trade off lower torque output for a higher break frequency.*

Remark 8.30 *By examining Tables 8.2 and 8.3 one can see that there is an interesting relationship between MTPAC and MRCTC. For MTPAC clearly $T_n \omega_{n_{\max}} = 1$. The same expression for the MRCTC strategy is*

$$T_{n_{\max}} \omega_{n_{\max}} = \left(\frac{2\xi}{\xi^2 + 1}\right)\left(\frac{\xi^2 + 1}{2\xi}\right) = 1 \tag{8.143}$$

$$\therefore \quad (T_{n_{\max}} \omega_{n_{\max}})_{\text{MTPAC}} = (T_{n_{\max}} \omega_{n_{\max}})_{\text{MRCTC}}. \tag{8.144}$$

Therefore these two control strategies produce the same maximum output power under rated angular velocity and output torque conditions. It can be shown that this relationship is unique between these two CAC strategies.

The torque and break frequency relationships are captured in Fig. 8.14. This figure also illustrates the gain × bandwidth trade-off that occurs with the different control strategies.

Remark 8.31 *Notice that for MPFC, $T_{n_{\max}} \approx 0.58 \, pu$ and $\omega_{\max} \approx 2.3$. This means that $T_{n_{\max}} \omega_{n_{\max}} \approx 0.58 \times 2.3 = 1.334$, which is larger than the same figure for either MTPAC or MRCTC. Therefore MPFC can produce more power than either MTPAC or MRCTC at $v_n = 1$ and $i_n = 1$.*

Remark 8.32 *Above the break frequency, with the angle still held constant at the particular value for a control strategy, the torque can be shown to fall $\propto 1/\omega_n^2$.*

Table 8.3 Comparison of Normalized Break Frequencies

Control strategy	Normalized break frequency
MTPAC/MEC	1
MPFC	$\sqrt{\dfrac{\xi^2 + 1}{2\xi}}$
MRCTC	$\dfrac{\xi^2 + 1}{2\xi}$

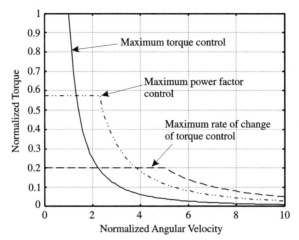

FIGURE 8.14
Comparison of the torque speed characteristics for various constant angle control strategies for the SYNCREL (assuming $\xi = 10$).

8.4.6.3 Rate of Change of Torque Interrelationships.
The relevant rate of change of torque equations for each of the CAC strategies can be obtained by rearranging (8.105) to give

$$p'T_n = \pm 2\sqrt{T_n(v_n^2 - \omega_n^2 T_n)} \quad \text{for MTPAC} \tag{8.145}$$

$$p'T_n = \pm 2\sqrt{T_n\left(\frac{\xi^2 + 1}{2\xi}v_n^2 - \omega_n^2 T_n\right)} \quad \text{for MRCTC} \tag{8.146}$$

$$p'T_n = \pm 2\sqrt{T_n\left(\frac{\xi^2 + 1}{\sqrt{\xi}(\xi + 1)}v_n^2 - \omega_n^2 T_n\right)} \quad \text{for MPFC} \tag{8.147}$$

The expression for the rate of change of torque for the CCDAC strategy was given in (8.140).

Given these expressions there are several different plots one can use to carry out a comparison. For example, a plot of the voltage required on the machine to achieve a particular value of $p'T_n$ at a particular torque level. Alternatively, one can fix the voltage and plot the rate of change of torque versus normalized torque.

Remark 8.33 *Interpretation of the different strategies in the plots just suggested is not as simple as it may seem. For example, for MTPAC and MRCTC if we plot over a torque range of 0 to 1 pu, then this means that MRCTC will be operating at far larger torques than the normal rated torque for this strategy (which implies that the currents will be larger than the rated currents).*

The plot of the voltage required for each strategy to obtain a normalized rate of change of torque of $p'T_n = 5/2\pi$ with $\omega_n = 1$ and $\xi = 10$ is shown in Fig. 8.15.

As can be seen the MRCTC requires the least voltage to achieve the required $p'T_n$ over the entire torque range, closely followed by MPFC. However, at very low load torques all the CAC strategies required an increasing voltage to maintain the $p'T_n$. This is due to the smaller back-emf in the d-axis of the machine which normally aids an increase in the d-axis current. Since the current angles are constrained to be constant, the rate of change of torque is dominated by the rate at which current can be changed in the d-axis of the machine.

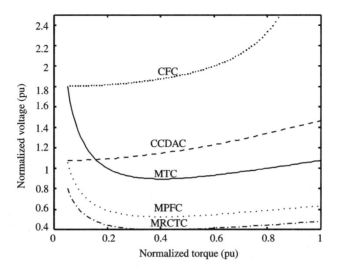

FIGURE 8.15
Voltage required for the control strategies when $p'T_n = 5/2\pi$ and $\omega_n = 1$, $\xi = 10$.

When $T_n = 0$ the voltage required goes to infinity. Because the current in the machine is now zero there is no flux in the machine. Consequently, when a voltage is applied the flux experiences a rate of change, but instantaneously the flux is still zero. The reason for the infinite voltage is a little harder to explain heuristically. The key to this behavior is the relationship between the currents in the machine (the currents are related by $\tan\theta = i^r_{qn}/i^r_{dn}$). This means that the torque is related in a quadratic sense to either of the currents, and therefore the derivative of the torque has to be zero at zero current (as is the derivative of a general quadratic function at zero). The torque effectively reacts to the second derivative of the current.

Remark 8.34 *The reader will note that the constant flux control (CFC) strategy previously mentioned in Remark 8.27 is also plotted on this diagram. Note the very poor performance of this strategy in relation to all the others.*

The CCDAC control strategy does not appear to work very well under the conditions of Fig. 8.15, except at very low torques where it obviously will require less voltage for a given rate of change of torque.[15]

As mentioned previously, the other way to look at the rate of change of torque is to fix the voltage and frequency in magnitude and vary the torque level. The result of this analysis is shown in Figs. 8.16 and 8.17.

The main observations from Fig. 8.16:

1. At very low torques all the strategies have very poor rates of change of torque.
2. For high angular velocities the CAC strategies generally have higher rates of change of torque.
3. The MRCTC strategy has the fastest $dT_n/d(\omega_0 t)$ as expected from the previous analysis.
4. The CFC strategy again shows very poor performance.

---

[15] The CCDAC strategy has a constant current of $I_{dn} = 1/\sqrt{2}$ for these plots.

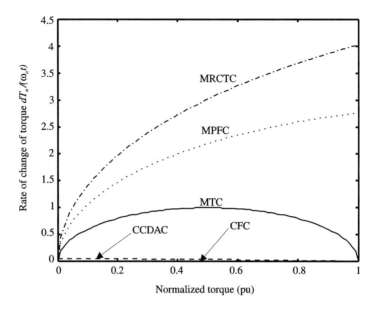

FIGURE 8.16
Rate of change of torque with $v_n = 1$, $\omega_n = 1$, $\xi = 10$.

5. Under this condition the CCDAC performs poorly because this strategy maintains a fairly constant high level of flux in the machine. Therefore much of the available voltage is absorbed to counter the back-emf produced by this high flux. The CAC strategies on the other hand have a flux level that varies with the torque level in the machine.

6. At $T_n = 1$ MTPAC has zero $p'T_n$ since there is no voltage available to change the current.

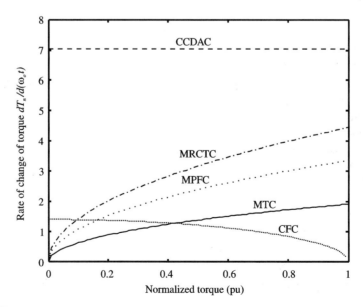

FIGURE 8.17
Rate of change of torque with $v_n = 1$, $\omega_n = 0.3$, $\xi = 10$.

Figure 8.17 is a similar plot to Fig. 8.16 except that the angular velocity is lower. The most significant change is that CCDAC now has significantly better performance than all the other CAC strategies. In fact the voltage required for a certain rate of change of torque is virtually independent of the torque level. The CFC strategy also has improved performance under this condition, but again its performance is only average and less than many of the CAC strategies.

The overall conclusions that one can draw are as follows:

1. At high speeds the MRCTC gives superior $p'T_n$ performance for most torque levels compared to other strategies.
2. At medium to low speed the CCDAC strategy gives superior performance in terms of $p'T_n$.
3. In order to optimize the rate of change of torque performance one needs to switch between the CCDAC and MRCTC strategy depending on the speed and torque level at which the machine is operating [10].

8.4.7 Field Weakening

Field weakening refers to the ability to push the machine to speeds higher than the break frequency by weakening the flux in the machine. This process allows the machine to produce torque higher than that obtainable if the flux is not weakened.

The term field weakening itself comes from the separately excited dc machine, where the field is a separately controlled entity. Therefore the field can be explicitly weakened. In the case of the SYNCREL field weakening is a little more difficult to see physically, since independent torque and field producing currents are not present.

8.4.7.1 Classical Field Weakening. Classical field weakening is field weakening that produces *constant power* from the machine above the break frequency (this is the field weakening normally associated with dc machines). If constant power is produced from the machine then the torque versus angular velocity relationship has to be

$$T \propto \frac{1}{\omega}. \tag{8.148}$$

If one simply pursues a constant angle control strategy above the break frequency, then (8.100) shows that

$$T_n \propto \frac{1}{\omega_n^2}, \tag{8.149}$$

whereas what we require is

$$T_n = \frac{K}{\omega_n}. \tag{8.150}$$

Using (8.150) and (8.100) one can write

$$\frac{(\xi^2 + 1)v_n^2}{\tan\theta + \xi^2 \cot\theta} = K\omega_n \tag{8.151}$$

where $K = 1$ in (8.150) from $T_n = 1$ and $\omega_n = 1$.

Rearranging, we can write

$$\tan\theta + \xi^2 \cot\theta - \frac{\xi^2 + 1}{\omega_n} = 0 \tag{8.152}$$

which can be solved to give

$$\theta = \tan^{-1}\left\{\frac{\xi^2+1}{2\omega_n} \pm \frac{1}{2}\sqrt{\left(\frac{\xi^2+1}{\omega_n}\right)^2 - 4\xi^2}\right\}. \tag{8.153}$$

This equation gives the value of the current angle to maintain constant power for angular velocities $\omega_n > 1$.

Remark 8.35 *The first point to note about (8.153) is that the angle θ is no longer constant but is a function of the angular velocity of the rotor.*

Remark 8.36 *One can see from (8.153) that for real solutions we must have*

$$\left(\frac{\xi^2+1}{\omega_n}\right) \geq 4\xi^2, \tag{8.154}$$

which implies that

$$\omega_n \leq \frac{\xi^2+1}{2\xi} \tag{8.155}$$

This is the frequency limit for classical field weakening. Note that this is also the break frequency for the MRCTC strategy.

Therefore one can see that the classical field weakening curve links the break frequency of the MTPAC and MRCTC strategies. This is shown in Fig. 8.18, which accurately plots the torque speed characteristics of the CAC strategies along with the classical field weakening characteristic. Note how the MPFC strategy is able to move to the right of the classical curve.

Remark 8.37 *It can be shown that for reasonably large values of ξ (say $\xi \geq 8$), the normalized field weakening range is approximately $\xi/2$ using the MTPA normalization. Therefore, in the case shown in Fig. 8.18 the field weakening range is $\omega_n = \xi/2 = 10/2 = 5$.*

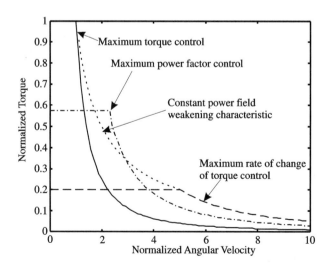

FIGURE 8.18
Torque speed characteristic of MTPAC, MPFC, and MRCTC together with the classical field weakening characteristic ($\xi = 10$).

8.4.7.2 Maximum Power Field Weakening.
As alluded to in Remark 8.36, MPFC appears to produce points on the torque–speed characteristic of the machine that are outside those produced by classical field weakening. This observation leads to a new type of field weakening that exploits this phenomenon to produce more power during field weakening than is possible with the classical approach.

The best way to visualize maximum power field weakening (MPFW) is to use a circle diagram [11, 12], which is a plot of the voltage, current, torque, and angular velocity characteristics on a set of axes corresponding to the dq currents. This diagram allows a graphical constrained optimization to be carried out, so that the maximum torque can be visually seen subject to current and voltage magnitude constraints.

In order to construct the circle diagram it is necessary to get the key variables in terms of the dq currents. For example, consider the voltage equation

$$v_{dn}^2 + v_{qn}^2 = 1. \tag{8.156}$$

Substituting (8.93) and (8.94) into this we can write

$$(\xi i_{dn})^2 + i_{qn}^2 = \frac{\xi^2 + 1}{2\omega_n^2}. \tag{8.157}$$

This equation is an ellipse centered at the origin of the dq plane and with an ellipticity equal to the saliency ratio. As ω_n increases, the dimensions of the ellipse decrease.

One can also plot the constant T_n curves on the circle diagram using $i_n^2 = i_{dn}^2 + i_{qn}^2$, $\tan \theta = i_{qn}/i_{dn}$, and (8.91):

$$i_{qn} = \frac{0.5 T_n}{i_{dn}}, \tag{8.158}$$

which is a hyperbola for each value of T_n.

The final expression on the circle diagram is for the current. Similarly to the voltage constraint we can write

$$i_n = \sqrt{i_{dn}^2 + i_{qn}^2} = 1. \tag{8.159}$$

This is clearly the equation of a circle.

The foregoing circle equations are plotted for one quadrant in Fig. 8.19. In order to understand this diagram a few points should be noted:

1. The line contours for the maximum voltage magnitude (i.e., $v_n = 1$) are drawn for different angular velocities. The outermost line is for the lowest angular velocity ($\omega_n = 1$). As angular velocity increases, the contour for maximum voltage approaches the center of the diagram.
2. The constant torque hyperbola increase in torque as one moves away from the origin of the diagram.
3. The field weakening portion of the diagram is from point B to C and then back to A.

The line drawn from point A to B is the line that is traversed if the maximum torque per ampere strategy is being applied. Notice that this line is orthogonal to the tangents of the constant torque hyperbolae. This is the normal 45° current angle line for maximum torque. For any point along this line $v_n \leq 1$ if $T_n \leq 1$ and $\omega_n \leq 1$.

At point B one hits the current constraint. If $\omega_n \leq 1$ then v_n would be less than or equal to 1 at this point. Notice however, that the $v_n = 1$, $\omega_n = 1$ contour goes through this point. This means

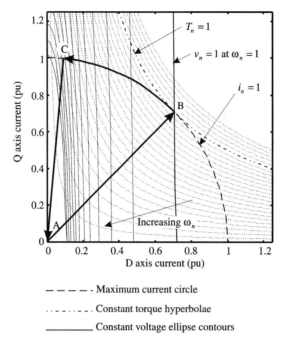

FIGURE 8.19
Circle diagram for maximum power field weakening.

that if $\omega_n = 1$ then the machine will have maximum voltage applied to it and field weakening would have to begin to increase ω_n without the $1/\omega_n^2$ loss of torque. The valid operating points for $\omega_n \geq 1$ are confined to the curve from point A to B. This can be seen from the diagram by moving along to a new voltage contour for each new value of ω_n, at the same time trying to maximize the torque subject to the current constraint. The net result is that the current constraint curve is traversed. Therefore during this phase of field weakening the machine is current and voltage constrained, and the current constraint trajectory is traversed on the circle diagram. Notice that the current angle is changing from the 45° value at point A to higher angles during this movement.

Remark 8.38 *While traversing from point B to C the current magnitude is equal to 1. This is different from classical field weakening strategy where the current drops below 1 as field weakening progresses.*

At point C another change of mode occurs. It is here that the machine operation moves from a current-limit-based trajectory to a voltage-limit-based trajectory. At point C the constant torque hyperbolae are tangential to the constant voltage ellipses. If one uses the same process for determining the operating point (i.e., traverse the constant voltage contour trying to maximize the torque subject to the current and voltage constraints) then as the angular frequency rises above that corresponding to point C one finds that the point of maximum torque lies along the line C to A. As can be seen from Fig. 8.19 this means that the current limit is no longer defining the trajectory of the current angle, but the voltage limit is.

The trajectory from point C to A is a trajectory that is effectively trying to maximize the torque produced from the machine for a given voltage. In other words it is a maximum torque per volt trajectory (as opposed to a maximum torque per ampere trajectory). Although we have drawn the line from point C to A as a straight line, this is very difficult to see from the diagram. The following analysis will prove that the current angle is fixed during this field weakening mode.

To find the maximum torque per volt we take $\partial T_n/\partial\theta$ of (8.100), which gives

$$\frac{\partial T_n}{\partial \theta} = \frac{(\xi^2+1)v_n^2}{\omega_n^2}\left\{\frac{\tan^2\theta+1}{\tan^2\theta+\xi^2} - \frac{2\tan^2\theta(\tan^2\theta+1)}{(\tan^2\theta+\xi^2)^2}\right\}. \quad (8.160)$$

This expression equals zero when

$$\tan\theta = \xi. \quad (8.161)$$

This is the angle that corresponds to the line from point C to point A. Note that this is the same angle as the MRCTC angle. In hindsight this makes sense, because if the rate of change of torque is to be maximized then the best possible usage of the available voltage must occur.

8.4.8 Practical Considerations

The analyses of the control properties of the SYNCREL have thus far been largely based on the ideal linear, lossless model of the machine. This has allowed us to develop a large number of expressions that succinctly capture the main properties of the machine. However, in reality most machines are subject to saturation, iron losses, and slotting effects, and these usually make the performance vary from that of the ideal machine.

Unfortunately the nonlinear nature of many of these practical effects makes it impossible to generate easily understood expressions describing their influence on the performance. In almost all cases very complex models result which have to be numerically solved to generate plots of the vital parameters. A comprehensive analysis of the influence of a variety of nonlinearities on the performance of the machine appears in [13, 14].

It turns out that saturation for most practical SYNCRELs is by far the major influence on the variation of the control strategies from their ideal performance. The control strategy that is influenced most is MTPAC since saturation is higher in this case (because of the smaller current angle). The larger current angle strategies can afford to ignore saturation.

A whole chapter could be devoted to the study of the difference that various nonlinearities and model simplifications cause between real machines and their models. In order to give the reader a feel for the influence of saturation on the performance of MTPAC consider Fig. 8.20, which shows the variation of the optimal maximum torque per ampere angle. Note that this figure was derived using the expressions in [13, 14] with the actual L_d versus i_d data for a real axial laminated SYNCREL. Only saturation is considered—iron losses are not included.

One can clearly see that the inclusion of saturation into the evaluation of the optimal MTPA angle causes a major deviation of this angle away from the ideal one, especially as the current increases. At 20 A input current, for example, one would lose approximately 5 Nm of torque by operating at the ideal current angle as compared to the optimal one (which is about a 13% drop on maximum torque output).

Compensation for saturation (or iron losses) in a real controller involves storing lookup tables that tabulate the appropriate characteristic. An appropriate on-line algorithm is then required to allow the calculation of the required angle, based on the current operating point, in real time. These issues are considered in the next section on controller implementation.

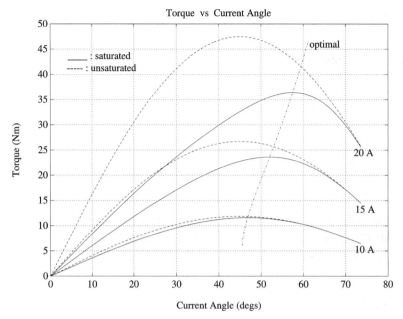

FIGURE 8.20
Torque output with constant input current magnitude for a SYNCREL with and without saturation.

8.4.9 A SYNCREL Drive System

In this section we shall briefly describe the major components required to implement a variable speed drive using a SYNCREL. It will not be possible to discuss in detail all the components of the drive system, but we shall instead concentrate on the major components, with particular emphasis on those that are unique to a SYNCREL drive system.

Remark 8.39 *It should be noted that this section will concentrate in a vector based drive system, because this follows naturally from the analysis carried out in the previous sections. However, one can also apply direct torque control (DTC) techniques to this machine [9, 15]. This control technique is fundamentally different from the vector-based techniques and will not be considered in this presentation. However, DTC of synchronous machines is considered in another chapter of this book.*

Figure 8.21 shows the block diagram of a SYNCREL based variable speed drive. Many of the blocks in this figure would be familiar to a reader with knowledge of induction machine vector drives. However, in some cases the contents of the blocks are quite different.

8.4.9.1 L_d Lookup Table. One of the major differences between an induction machine drive system and the SYNCREL drive system is the presence of the L_d lookup block. This block is essentially a table storing the SYNCREL saturation characteristic in the form of L_d and L'_d versus i^r_d values. This table is formed from measurements of L^r_d versus i^r_d compiled off-line, which has a curve is fitted to it. This curve is then differentiated to give $L'_d = dL_d/di^r_d$. The resultant L_d and L'_d equations are then used to generate the table that is actually stored. This table can be quickly indexed in software, with interpolation techniques being used for points between the stored values. The output of the L_d lookup table is used in several other blocks in the controller.

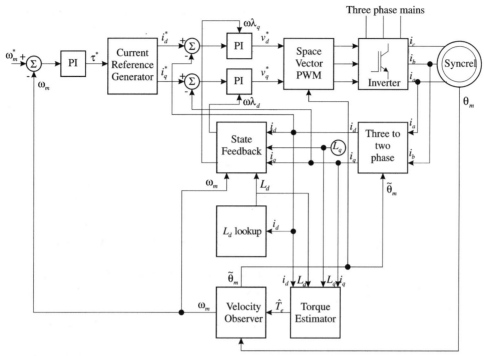

FIGURE 8.21
Block diagram of a typical vector-based SYNCREL drive system.

8.4.9.2 The Torque Estimator. The purpose of this block is to generate an estimate of the electromagnetic torque being produced by the machine. This is required in the velocity observer block, but would also be required if one wanted to implement torque control.

The block simply implements the torque equation of the SYNCREL with the appropriate modifications at account for saturation:

$$\hat{T}_e = \frac{3}{2} p_p [L_d(i_d^r) - L_q] i_d^r i_q^r \qquad (8.162)$$

where $L_d(i_d^r)$ denotes that the d-axis inductance is a function of i_d^r.

Remark 8.40 *The values of i_d^r and i_q^r are effectively the average values of these currents over a control interval. The currents in a PWM fed machine (of any type) have a lot of ripple in them, but if symmetrical PWM is implemented and the current is sampled in the middle of the control interval, then the value of current is the average value of the current over the control interval. Therefore the estimated torque is effectively the average torque over the control interval.*

8.4.9.3 Velocity Observer. The purpose of the velocity observer is to produce relatively noise-free values of the rotor angular velocity from rotor position measurements. This is achieved by using a classical observer structure based in the load dynamics of the machine [16]. The reason for using the observer is that it provides a delay-free filter of the discrete position measurements one obtains from an incremental encoder or digital absolute position encoder.

The observer is based on a rearrangement of the normal rotating load equation. The rearranged rotating load equation is

$$\dot{\omega}_m = \frac{1}{J}[\hat{T}_e - T_f - f\omega_m]. \tag{8.163}$$

Therefore the estimates of the angular velocity and position are

$$\hat{\omega}_m = \frac{1}{J}\int_0^t [\hat{T}_e - T_f - f\omega_m]\, dt \tag{8.164}$$

$$\hat{\theta}_m = \int_0^t \omega_m\, dt. \tag{8.165}$$

The $\hat{\theta}_m$ estimate can be compared with the measured θ_m and the error used to ensure convergence of the observer.

The preceding expressions are all in continuous time. To implement the observer we need to convert the continuous time expressions into a discrete implementation. The result is the discrete observer shown in Fig. 8.22. Notice that the observer is driver by the feedback error $\theta_{enc} - \hat{\theta}_m$, and the feedback gains K_1, K_2, and K_3 are adjusted depending on the noise present in this feedback signal (more noise → lower gains). This feedback helps to overcome cumulative errors that could build up in an open-loop observer, as well as to compensate for errors in the estimated torques being fed into the observer.

Remark 8.41 *The K_2 gain is not standard and is there to improve the dynamics of the observer with respect to errors. This gain needs to be carefully chosen as it feeds the noise error directly into the output.*

8.4.9.4 State Feedback. The state feedback block is required for three reasons:

1. To prevent the integrators in the current PI regulators from having very large values on them (the value would be to compensate for the back-emf).
2. The back-emfs effectively act as unknown disturbances to the current PI regulators. The state feedback eliminates these disturbances.
3. The back-emfs introduce cross coupling between the d- and q-axes.

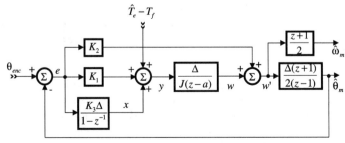

FIGURE 8.22
Block diagram of the digital implementation of a velocity–position observer.

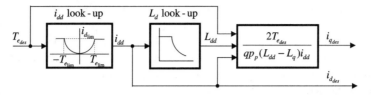

FIGURE 8.23
Current reference generator for maximum-torque-per-ampere control.

The basic equations of the SYNCREL can be written in integral form as

$$i_d^r(\tau) = \int_0^\tau \frac{v_d^r - R i_d^r + \omega_{pd} L_q^r i_q^r}{L_d^r(i_d^r) + (L_d^r)' i_d^r} \, dt \qquad (8.166)$$

$$i_q^r(\tau) = \int_0^\tau \frac{v_q^r - R i_q^r - \omega_{pd} L_d^r(i_d^r) i_d^r}{L_q^r} \, dt. \qquad (8.167)$$

As can be seen from these equations there is a lot of cross-coupling between them via the $\omega_{pd} L i$ terms. The PI regulators controlling the equations have to "fight" these terms via the v_d^r and v_q^r applied voltages. Ideally we want the d-axis voltage to influence only the d-axis current, and similarly with the q-axis voltage and current.

If we add $R i_r^r - \omega_{pd} L_q^r i_q^r$ to (8.166) and $R i_q^r + \omega_{pd} L_d^r(i_d^r) i_d^r$ to (8.167), we then get

$$i_d^r(\tau) = \int_0^\tau \frac{v_d^r}{L_d^r(i_d^r) + (L_d^r)' i_d^r} \, dt \qquad (8.168)$$

$$i_q^r(\tau) = \int_0^\tau \frac{v_q^r}{L_q^r} \, dt. \qquad (8.169)$$

As can be seen the cross-coupling is removed, and the PI controllers no longer have to combat the back-emf disturbance.

8.4.9.5 Current Reference Generator. The current reference generator is one of the main differences between the SYNCREL drive system and an induction machine vector control drive system.

The purpose of the block is to accept a desired torque and then produce the desired currents in the machine that would produce the desired torque. The basic layout of the current reference generator is shown in Fig. 8.23.

As can be seen from Fig. 8.21 the desired torque $T_{e_{des}}$ is generated by the speed control loop PI controller. This torque is passed through a nonlinear function that produces an $i_{d_{des}}^r$ which will produce the maximum torque from the machine. The nonlinear relationship between torque and i_d^r is a lookup table determined off-line by a numerical procedure using the saturation characteristic of the machine together with (8.162). The table is multidimensional as values have to be stored under different ω and T_e conditions. The d-axis current generated by this lookup table is then passed to the L_d lookup table mentioned previously and the value of L_d is determined. Finally a rearranged version of the torque equation with i_q^r as the subject of the expression is used to generate the desired i_q^r current. The two axis currents are then passed to the control loops.[16]

---

[16] This particular structure for the reference generator arises because of the complexity of the expressions required to generate the lookup tables.

The remainder of the SYNCREL drive system is conventional. The current loops are controlled by PI regulators (with the state feedback mentioned previously), and the desired dq voltages are then passed to a space vector PWM algorithm which generates the firing signals for the legs of the inverter. Note that this algorithm also performs the rotating frame to stationary frame transformation required. The inverter (as mentioned previously) is a conventional three-leg inverter used for induction machine drives.

Overall a SYNCREL drive system of the type described here has performance very similar to that of a high-performance induction machine vector based drive system. The hardware required for the two systems is virtually identical, the main differences being in the software and control algorithms, and of course the machine being driven.

One nice property of the SYNCREL is that it is a very amenable to sensorless control—i.e., to control without a shaft position sensor. This can be achieved by sensing the shaft position via current and/or voltage measurements on the terminals of the machine. This is simpler to do for the SYNCREL, as compared to, say, the induction machine, because the saliency of the rotor allows easy sensing via position dependent inductance variation.

8.5 CONCLUSION

This chapter has attempted, in a short dissertation, to develop the models and control principles for the SYNCREL, starting with basic operation right through to advanced control strategies. Inevitably one has to omit or skim through certain aspects, but references have been included for the reader who requires more detail.

The future of the SYNCREL as a competitive machine with the induction machine is not clear at this stage. It would appear that in general the performance of the SYNCREL is very close to that of the induction machine. At low speeds and high torques (i.e., both of these conditions together) the efficiency of the SYNCREL should be significantly better than that of the induction machine because of the very low rotor losses in the SYNCREL compared to the relatively high rotor bar losses in the induction machine. Under these conditions the iron losses in the SYNCREL should be low.

One form of the SYNCREL that has not been discussed in this chapter is the magnet-assisted SYNCREL [12, 17]. This is essentially a special form of interior permanent magnet machine in which a significant part of the torque arises from the saliency, the magnets performing a secondary role. For an axial laminated machine the magnet material replaces the normal interlamination material. One only needs to use low-quality ferromagnetics—the case of the machine designed in [17] rubberized refrigerator magnet material was used. The effects on performance are dramatic, with the machine clearly outperforming the induction machine on virtually every measure. This is due to the fact that the field produced by the introduced magnets effectively cancels the q-axis flux and therefore creates the illusion that $L_q = 0$. Hence the saliency ratio $\xi \to \infty$, and as has been noted in this chapter all the properties of the machine improve with increased saliency ratio.

REFERENCES

[1] J. Kostko, Polyphase reaction synchronous motors. *J. Am. Inst. Elec. Eng.* **42**, 1162–1168 (1923).
[2] V. Honsinger, The inductances l_d and l_q of reluctance machines. *IEEE Trans. Power Apparatus Syst.* **PAS-90**, 298–304 (1971).
[3] V. Honsinger, Steady-state performance of reluctance machines. *IEEE Trans. Power Apparatus Syst.* **PAS-90**, 305–317 (1971).

REFERENCES

[4] V. Honsinger, Inherently stable reluctance motors having improved performance. *IEEE Trans. Power Apparatus Syst.* **PAS-91**, 1544–1554 (1972).

[5] A. Vagati and T. Lipo, eds., *Synchronous Reluctance Motors and Drives—a New Alternative*. IEEE Industry Applications Society, Oct 1994. Tutorial course presented at the IEEE-IAS Annual Meeting, Denver, Colorado.

[6] A. Vagati, A. Canova, M. Chiampi, M. Pastorelli, and M. Repetto, Improvement of synchronous reluctance motor design through finite-element analysis. *Conference Record, 34th IEEE Industry Applications Society Annual Meeting*, Vol. 2, pp. 862–871, Oct. 1999.

[7] D. O'Kelly and S. Simmons, *Introduction to Generalized Electrical Machine Theory*. McGraw-Hill, U.K., 1968.

[8] P. Vas, *Vector Control of AC Machines*. Oxford University Press, 1990.

[9] I. Boldea, *Reluctance Synchronous Machines and Drives*. Oxford University Press, 1996.

[10] R. Betz, Theoretical aspects of control of synchronous reluctance machines. *IEE Proc-B* **139**, 355–364 (1992).

[11] T. Miller, *Brushless Permanent-Magnet and Reluctance Motor Drives*. Oxford University Press, 1989.

[12] W. Soong, Design and modelling of axially-laminated interior permanent magnet motor drives for field-weakeing applications. Ph.D. thesis, Department of Electronics and Electrical Engineering, University of Glasgow, 1993.

[13] M. Jovanović, Sensorless control of synchronous reluctance machines. Ph.D. thesis, University of Newcastle, Australia, 1997.

[14] M. Jovanović and R. Betz, Theoretical aspects of the control of synchronous reluctance machines including saturation and iron losses. Technical Report EE9305, Department of Electrical and Computer Engineering, University of Newcastle, Australia, 1993.

[15] R. Lagerquist, I. Boldea, and T. Miller, Sensorless control of the synchronous reluctance motor. *IEEE Trans. Indust. Appl.* **IA-30**, 673–682 (1994).

[16] R. Lorenz and K. Patten, High-resolution velocity estimation for all-digital, ac servo drives. *IEEE Trans. Indust. Appl.* **IA-27**, 701–705 (1991).

[17] W. Soong, D. Staton, and T. Miller, Design of a new axially-laminated interior permanent magnet motor. *IEEE Trans. Indust. Appl.* **IA-31**, 358–367 (1995).

CHAPTER 9

Direct Torque and Flux Control (DTFC) of ac Drives

ION BOLDEA
University Politehnica, Timisoara, Romania

9.1 INTRODUCTION

Ac motors are by now predominant in variable speed drives at about 75% of all markets. Essentially the absence of the mechanical commutator makes the difference between the dc and ac motors for adjustable speed.

In general, ac motor drives may be used without restriction in chemically aggressive and volatile environments.

The advantage of the rather simpler power electronics control of dc motors in two-quadrant applications is practically lost in four-quadrant operation.

The superior torque density, speed range, and ruggedness of ac motors is paid for, in variable speed drives, by more sophisticated control systems. Energy-efficient wide speed range control with ac motors may be performed only through frequency and voltage coordinated changes. Moreover, for fast torque control, required in high-performance ac drives, decoupled control of flux current and torque current virtual components of stator current has to be performed. As flux variation tends to be slow, the flux current is, in general, maintained constant. We end up with only torque current variation for torque change, much as in a dc motor with separate excitation.

This decoupled flux and torque currents control is called vector (or field orientation) control [1]. Vector control in ac drives has by now become a mature technology with sizable markets.

Equivalently fast and robust transient response in torque may be obtained by other nonlinear transformations of motor variables so as to obtain again decoupled flux and torque currents control. This new breed of methods, known as feedback linearization control or input–output decoupling control [2], is still in the laboratory stage.

Both vector (or field orientation) control and feedback linearization control tend to require a large amount of on-line computation if torque response quickness, robustness, and precision are to be secured.

302 CHAPTER 9 / DIRECT TORQUE AND FLUX CONTROL (DTFC) OF AC DRIVES

In search of a simpler and more robust control system capable of preserving high performance, the direct torque and flux control (DTFC) method was born. The DTFC principle for induction motors was introduced in 1985–1986 [3, 4] and generalized for all ac drives in 1988 [5]. By 1995 DTFC (as DTC) for induction motors reached markets [6] and is now used from 2 kW to 2 MW, with the controller, basically implemented in the same hardware and using the same software.

As expected, the DTFC literature is growing by the day while continuous improvements in field-oriented control are also produced [7–28].

As known today, ac drives are manufactured menu-driven, both with and without motion (position or speed) sensors. To shorten the presentation we will deal directly with those lacking motion sensors, as drives with motion sensors are a particular case of the former.

At first we introduce the DTFC principles for both induction and synchronous motor drives. Further on the implementation of DTFC for induction, PM synchronous, reluctance synchronous, and large power (electromagnetically excited) synchronous motor drives are all treated separately.

Basic and refined (with space vector modulation added) solutions are presented. Mathematical derivations are kept to a minimum, while concepts, flow signal diagrams, and results are given extensively.

The high pace of laboratory developments in DTFC for synchronous motors suggests its imminent implementation in industry.

Also, DTFC induction motors, now produced by only a few manufacturers, are likely to spread to multiple manufacturers, given the now-proven assets of this new technology. Field orientation control and DTFC are expected to be neck-and-neck competitors in the future market for high-performance ac drives.

9.2 DTFC PRINCIPLES FOR INDUCTION AND SYNCHRONOUS MOTORS

9.2.1 The Induction Motor

It is well known that, in electric motors, motion is produced by the electromagnetic torque T_e. On the other hand the magnetic flux $\bar{\Psi}_i$ in the motor shows the degree of iron utilization and is related to core losses.

What flux linkage $\bar{\Psi}_i$ are we talking about? In induction motors main (air gap) flux $\bar{\Psi}_m$, "in the rotor" flux $\bar{\Psi}_r$, and stator flux $\bar{\Psi}_s$ are well defined:

$$\begin{aligned}\bar{\Psi}_s &= L_s \bar{i}_s + L_m \bar{i}_m; & L_s &= L_{sl} + L_m \\ \bar{\Psi}_m &= L_m(\bar{i}_s + \bar{i}_r); & L_{sc} &= L_{sl} + L_{sc} \\ \bar{\Psi}_r &= L_r \bar{i}_r + L_m \bar{i}_m; & L_r &= L_{rl} + L_m.\end{aligned} \quad (9.1)$$

Space vector models for the induction motor are defined in general orthogonal reference systems rotating at ω_b with respect to the stator:

$$\bar{V}_s = R_s \bar{i}_s + \frac{d\bar{\Psi}_s}{dt} + j\omega_b \bar{\Psi}_s \quad (9.2)$$

$$0 = R_r \bar{i}_r + \frac{d\bar{\Psi}_r}{dt} + j(\omega_b - \omega_r)\bar{\Psi}_r \quad (9.3)$$

$$T_e = \frac{3}{2} p_1 \mathrm{Re}[j\bar{\Psi}_s \bar{i}_s^*] = -\frac{3}{2} p_1 \mathrm{Re}[j\bar{\Psi}_r \bar{i}_r^*]. \quad (9.4)$$

9.2 DTFC PRINCIPLES FOR INDUCTION AND SYNCHRONOUS MOTORS

Also from (9.1),

$$\bar{\Psi}_r = \frac{L_m}{L_r}\bar{\Psi}_s + L_{sc}\bar{i}_s; \qquad \bar{i}_s = i_d + ji_q. \tag{9.5}$$

In rotor coordinates $\bar{\Psi}_r = \Psi_r$ we obtain

$$i_d = \left(1 + S\frac{L_r}{R_r}\right)\frac{\Psi_r}{L_m}. \tag{9.6}$$

From (9.6) it is clear that the rotor flux transients are slow because $L_r/R_r \sim (0.1$–$0.3)$ seconds in general.

For millisecond time intervals we may thus consider it constant $(d\Psi_r/dt = 0)$, when

$$i_q = i_d(\omega_{\Psi_i} - \omega_r)\frac{L_r}{R_r}; \qquad i_d = \frac{\Psi_r}{L_m} \tag{9.7}$$

and

$$\Psi_s = L_s i_d + jL_{sc}i_q. \tag{9.8}$$

The vector diagram in Fig. 9.1 illustrates Eqs. (9.7) and (9.8).
Now from (9.4) and (9.8) the torque T_e is

$$T_e = \frac{3}{2}p_1(L_s - L_{sc})i_d i_q. \tag{9.9}$$

Further on (from Fig. 9.1):

$$T_e = \frac{3}{2}p_1(L_s - L_{sc})\frac{\Psi_s^2 \sin 2\delta}{2L_s L_{sc}}. \tag{9.10}$$

Note: For constant rotor flux—steady state and transient—the induction motor behaves as a reluctance synchronous motor with rather high saliency as $L_s \gg L_{sc}$. The "virtual" saliency is produced by the rotor currents.

Modifying the torque is possible (see (9.10)) by changing the stator flux level Ψ_s or the "torque" angle δ.

Moving the stator flux vector $\bar{\Psi}_s$ ahead (the acceleration) or slowing it down would make δ increase in the positive direction and produce more positive torque, or decrease and become negative to yield negative torque.

The torque T_e versus angle δ for two stator flux levels shows a limited zone of stability (Fig. 9.2) between A_m and A_g, that is $|\delta| < \pi/4$.

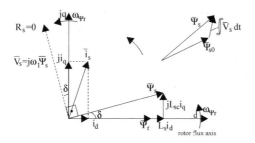

FIGURE 9.1
IM vector diagram.

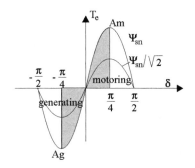

FIGURE 9.2
IM torque/angle curve for constant rotor flux.

Now according to Eq. (9.1) in stator coordinates ($\omega_b = 0$), neglecting R_s,

$$\bar{\Psi}_s \approx \frac{\bar{V}_s}{j\omega_1}; \qquad \Psi_s \approx \frac{V_s}{\omega_1}; \qquad \omega_1 = S\omega_1 + \omega_r \tag{9.11}$$

where ω_1 is the stator flux speed (voltage frequency at steady state). The flux level Ψ_s is limited by the voltage ceiling in the inverter V_{sn} and by the frequency ω_1 (indirectly by the motor speed as $S\omega_1$ is limited).

So the torque reference T_e^* should be limited to the case when $|\delta| = \pi/4$ to retain stability, for given stator flux level Ψ_s^*:

$$|T_e^*| = \frac{3}{2}p_1\left(\frac{1}{L_{sc}} - \frac{1}{L_s}\right)\frac{\Psi_s^2}{2} \tag{9.12}$$

$$|\Psi_s^*| \leq V_{sn}/\omega_1^*; \qquad \omega_1^* \leq \omega_r + (\omega_1)_{max}^* \tag{9.13}$$

For $\delta_{max} = \pi/4$:

$$(S\omega_1)_{max}^* = \left(\frac{L_{sc}i_q}{L_s i_d}\right)_{\delta=\pi/4} \cdot \frac{R_r}{L_r}\frac{L_s}{L_{sc}} = \frac{R_r}{L_r}\frac{L_s}{L_{sc}}. \tag{9.14}$$

Now the stator flux equation (9.1) for stator coordinates (9.1) with $R_s i_s \sim 0$ leads to

$$\Delta\bar{\Psi}_s = \bar{\Psi}_s - \bar{\Psi}_{s0} = \int \bar{V}_s dt, \tag{9.15}$$

as illustrated in Fig. 9.1.

Equation (9.15) implies that the variation of stator flux falls along the applied voltage vector direction.

Changing the voltage vector direction and timing of its application should then lead to acceleration or deceleration of stator flux vector and the rise or fall of its amplitude. It is now time to remember that a voltage-source single-level PWM inverter produces six nonzero and two zero voltage vectors only (Fig. 9.3).

The stator voltage space vector $\bar{V}_s$ may thus be written as

$$\bar{V}_s(v) = \begin{cases} \frac{2}{3}V_{dc}e^{j(v-1)\pi/3} & \text{for } v = 1,\ldots,6 \\ 0 & \text{for } v = 0, 7. \end{cases} \tag{9.16}$$

The principle of direct torque and flux control—the original version—consists of triggering directly a certain voltage vector $\bar{V}_s(v)$ in the inverter based on stator flux error $\varepsilon_{\lambda_s} = \Psi_s^* - \Psi_s$,

9.2 DTFC PRINCIPLES FOR INDUCTION AND SYNCHRONOUS MOTORS

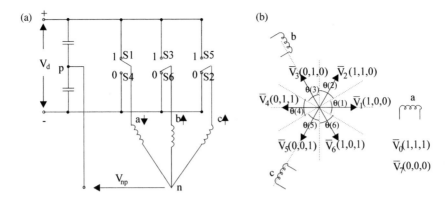

FIGURE 9.3
The voltage source PWM inverter. (a) Schematics; (b) voltage vectors.

torque error $\varepsilon_{T_e} = T_e^* - T_e$, and the stator flux vector location in one of the six 60° wide sectors $\theta(v) - v = 1, \ldots, 6$, shown in Fig. 9.3.

The zero voltage (V_0, V_7) which corresponds to IM short time short circuit at the terminals or the stoppage of voltage vector motion, may be used to trigger a zero flux error, while for the torque error mainly nonzero voltage vectors are triggered to secure fast torque response.

Hysteresis flux and torque controllers have been initially used. Proper nonzero and zero voltage timings may also produce limited torque pulsations. Notable on-line computation efforts are needed to do so [25, 26].

There are two main items to clarify in Fig. 9.4: the table of optimal switchings (TOS) and the state estimator. The TOS will be dealt with here.

Let us first suppose that the flux vector $\bar{\Psi}_s$ is located in the first 60° wide sector (Fig. 9.5) spanning from $-30°$ to $+30°$.

In essence $\bar{V}_2$ and $\bar{V}_3$ move the flux ahead (flux accelerating, that is, more torque) while $\bar{V}_5$ and $\bar{V}_6$ do flux decelerating (that is less—negative—torque). Also, $\bar{V}_3$ and $\bar{V}_5$ lead to flux amplitude reduction while $\bar{V}_2$ and $\bar{V}_6$ lead to flux amplitude increase. As the flux hysteresis

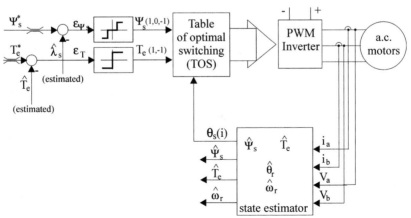

FIGURE 9.4
Principle of direct torque and flux control.

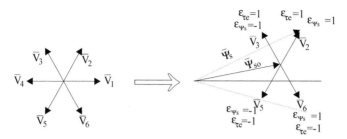

FIGURE 9.5
From flux and torque errors ε_{λ_s}, ε_{T_e}, to optimal voltage vectors.

controller is tripositional, $|\varepsilon_{\lambda_s}| < h$, a zero voltage vector is triggered also. Constant frequency triggering has been also proposed [13, 15].

For the other five sectors the adequate voltage vectors are obtained by adding one digit to the previous voltage vector (V_i becomes V_{i+1}) along the counterclockwise direction.

Typical flux hodograph and torque pulsations in DTFC are shown in Fig. 9.6

When six-pulse operation is reached the flux hodograph becomes, during steady state, a hexagon. Zero voltage vectors stop the flux motion and represent dots on the hodograph.

Now if the stator estimators are fast, robust, and precise for a wide range of speeds, DTFC should work fine. The absence of vector rotators, corroborated by the direct control of flux and torque, is the main asset of DTFC.

Reference flux/reference torque coordination may be used to optimize performance for steady state. For fast transient response, full flux is required. Above base speed, flux weakening is mandatory, as the voltage ceiling V_{sn} was reached (9.13).

9.2.2 Flux/Torque Coordination in IM

When the stator resistance is neglected, for steady state (implicitly constant rotor flux for constant load), the flux and torque expressions to be used (from (9.8), (9.9), and (9.13)) are

$$V_s = \Psi_s \omega_1; \qquad \omega_1 = \omega_s + S\omega_1; \qquad S\omega_1 = \frac{R_r}{L_r} \frac{i_q}{i_d} \qquad (9.17)$$

$$\Psi_s = \sqrt{(L_s i_d)^2 + (L_{sc} i_q)^2}; \qquad T_e = \frac{3}{2} p_1 (L_s - L_{sc}) i_d i_q. \qquad (9.18)$$

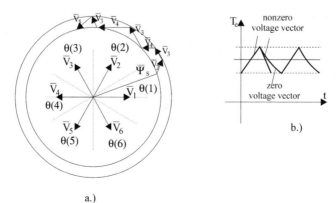

FIGURE 9.6
The flux hodograph (a) and torque pulsation (b).

9.2 DTFC PRINCIPLES FOR INDUCTION AND SYNCHRONOUS MOTORS

Neglecting the hysteresis losses we might claim that the core loss P_{core} is

$$P_{core} = \frac{3}{2} \frac{\Psi_s^2 \omega_1^2}{R_{core}} \qquad (9.19)$$

and the winding losses are

$$P_{copper} = \frac{3}{2} R_s (i_d^2 + i_q^2) + \frac{3}{2} R_r \cdot i_q^2 \cdot \left(\frac{L_r}{L_m}\right)^2. \qquad (9.20)$$

The goal is to establish a relationship between Ψ_s^*, T_e^*, eventually for a certain ω_1, according to some objective function such as

- Maximum torque/stator current, below base speed
- Maximum torque/flux, above base speed
- Maximum ideal power factor, below base speed
- Maximum torque/losses, below and above base speed

Here we will treat the first three criteria as the last one implies somewhat more cumbersome mathematics.

For the maximum torque/current criterion, the stator current i_s is given by

$$i_s = \sqrt{i_d^2 + i_q^2} = \text{const.} \qquad (9.21)$$

With this condition, from the torque expression (for $\partial T_e / \partial (i_q/i_d) = 0$) we obtain

$$\frac{i_q}{i_d} = 1; \quad \Psi_s = \sqrt{\frac{2T_e^*(L_s^2 + L_{sc}^2)}{3p_1(L_s - L_{sc})}}; \quad \Psi_s^* \leq \frac{V_s}{\omega_1}. \qquad (9.22)$$

Now as the rotor flux $\Psi_r = L_m i_d$ it is evident that the maximum value of i_d before heavy saturation occurs should be equal to the rated no-load current $i_{dmax} \leq I_{0n}\sqrt{2}$. As $I_{0n}/I_n < 0.5$, in general, the level of maximum torque to be obtained is maximum 50% of rated torque.

Only for low torque does this method seem adequate. For maximum torque per flux we have the flux as given and use the flux and torque equations (9.18) to obtain ($\partial T_e / \partial (i_q/i_d) = 0$):

$$i_{dk} = \frac{\Psi_s}{L_s \sqrt{2}}; \quad i_{qk} = \frac{\Psi_s}{L_{sc}\sqrt{2}}; \quad \delta_{ik} = \pi/4 \qquad (9.23)$$

$$\Psi_s^* = \sqrt{\frac{4T_e^* L_{sc}}{3p_1(1 - L_{sc}/L_s)}}. \qquad (9.24)$$

The ideal power factor angle φ_1 (Fig. 9.1) is

$$\varphi_1 = \tan^{-1}\frac{i_d}{i_q} + \tan^{-1}\delta = \tan^{-1}\frac{L_{sc}}{L_s} + \tan^{-1}\frac{L_{sc} i_{qk}}{L_s i_{dk}} = \tan^{-1}\frac{L_{sc}}{L_s} + \frac{\pi}{4}. \qquad (9.25)$$

This approach is producing more torque per given flux level but at a rather low power factor. It may be used above base speed, beyond the constant flux range.

In between stands the maximum power factor criterion. From (9.25), for $\partial \varphi_1 / \partial (i_d/i_q) = 0$ we obtain

$$\tan \delta_m = i_{dm}/i_{qm} = \sqrt{L_{sc}/L_s}; \quad (\cos \varphi_1)_{max} = \frac{1 - L_{sc}/L_s}{1 + L_{sc}/L_s}. \qquad (9.26)$$

Finally for this case,

$$\Psi_s^* = \sqrt{\frac{2T_e^*(L_s^2 + L_{sc}L_s)}{3p_1(L_s - L_{sc})}} \sqrt{\frac{L_{sc}}{L_s}}. \tag{9.27}$$

It may prove practical to choose the maximum power factor criterion for base speed rated torque design in order to secure enough "room" for a sizable constant power speed range above it.

As inferred earlier, the flux/torque reference coordination (management) seems an important "battlefield" for improving steady-state performance of DTFC IM drives.

9.2.3 The Synchronous Motors

Synchronous motors are built with PM excitation, with nonexcited high magnetic saliency or with electromagnetically excited rotors (for large powers). When used for variable-speed drives, in general, with voltage source converters (cycloconverters, matrix converters, or single- or double-level PWM inverters) no cage in the rotor is placed. Only for current source machine commutated inverter synchronous motor drives is a strong rotor cage required to reduce the machine commutation inductance $L_c = (L_d'' + L_q'')/2$ such that up to 150% rated current natural commutation is secured.

For the time being let us neglect the rotor damper cage presence, as is the case for most low and medium power drives.

The space vector equations in rotor coordinates are

$$\bar{V}_s = R_s \bar{i}_s + \frac{d\bar{\Psi}_s}{dt} + j\omega_r \bar{\Psi}_s; \qquad \bar{i}_s = i_d + ji_q \tag{9.28}$$

$$\bar{\Psi}_s = L_d i_d + L_{dm} i_F + jL_q i_q \tag{9.29}$$

$$V_F = R_F i_F + \frac{d\Psi_F}{dt}; \qquad \Psi_F = L_F i_F + L_{dm}(i_d + i_F) \tag{9.30}$$

$$T_e = \frac{3}{2}p_1 R_e(j\bar{\Psi}_s \bar{i}_s^*) = \frac{3}{2}p_1[L_{dm} i_q i_F + (L_d - L_q)i_d i_q]. \tag{9.31}$$

Now for $i_F = i_{F0}$ = constant ($L_{dm} i_{F0} = \Psi_{PMd}$) we have the case of PM rotor if $L_q \geq L_d$. On the other hand for $i_{F0} = 0$ and $L_d \gg L_q$ we obtain the high saliency passive rotor.

For this latter case low remanent flux density ($B_r < 0.4\,\text{T}$) low-cost PMs may be located in axis q, and thus

$$\Psi_q = L_q i_q - \Psi_{PMq}. \tag{9.32}$$

The torque T_e is

$$T_e = \tfrac{3}{2}p_1[\Psi_{PMq} i_d + (L_d - L_q)i_d i_q]. \tag{9.33}$$

Thus (9.28)–(9.33) represent the space phasor model of practically all cageless rotor synchronous motors.

The space phasor diagrams for the three distinct cases are shown in Fig. 9.7.

The PMs in axis q in Fig. 9.7c serves evidently to increase torque production (9.33) and to improve the power factor. This motor is called a PM assisted reluctance synchronous motor. Sometimes it may also be called the IPM reluctance synchronous motor.

9.2 DTFC PRINCIPLES FOR INDUCTION AND SYNCHRONOUS MOTORS

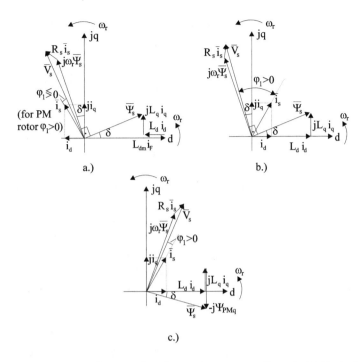

FIGURE 9.7
Synchronous motor space vector diagrams (steady state in rotor coordinates). (a) Excited ($L_d \geq L_q$) or d axis PM ($L_d \leq L_q$); (b) variable reluctance rotor $L_d \gg L_q$; (c) variable reluctance rotor with PMs in axis q.

Based on the space vector diagram we alter the torque expressions. We may express the torque (9.31) in terms of stator flux Ψ_s and the torque angle δ, instead of i_d, i_q:

$$L_q i_q = \Psi_s \sin \delta \tag{9.34}$$

$$L_{dm} i_F + L_d i_d = \Psi_s \cos \delta. \tag{9.35}$$

Finally (9.31) yields:

$$T_e = \frac{3}{2} p_1 \left[L_{dm} i_F \frac{\Psi_s}{L_d} \sin \delta + \frac{1}{2}(L_d - L_q) \frac{\Psi_s^2}{L_d L_q} \sin 2\delta \right]. \tag{9.36}$$

Note again that for the d axis PM rotor $L_d \leq L_q$ while for the d axis excited rotor $L_d \geq L_q$. However, the expression (9.36) of torque remains the same in both cases.

Finally for the q axis PM reluctance rotor ($L_d \geq L_q$) synchronous motor (IPM-RSM) the torque expression (9.33) becomes

$$T_e = \frac{3}{2} p_1 \left[(\pm) \Psi_{PMq} \frac{\Psi_s}{L_q} \cos \delta + \frac{1}{2}(L_d - L_q) \frac{\Psi_s^2}{L_d L_q} \sin 2\delta \right]. \tag{9.37}$$

In (9.37), the sign + is for positive δ and − for negative δ.

Let us notice that in both cases, to modify the torque we have to change either the stator flux amplitude Ψ_s or the torque angle δ, much as in the case of the IM, with the difference that now the coordinates are fixed to the rotor rather than to the rotor (stator) flux (for the IM). Also for the IM only the second terms in (9.36) and (9.37) are visible as no source of dc magnetization on the rotor was considered for the IM.

The stator voltage equation of SMs in stator coordinates is identical to that of the IM:

$$\bar{V}_s = R_s \bar{i}_s + \frac{d\bar{\Psi}_s}{dt}. \tag{9.38}$$

It follows that all the rationale for the DTFC of IMs is valid also for the SMs—with a few pecularities:

- The state (flux, torque, speed) observers are to be adapted to SM model
- The table of optimal switchings (TOS) of d axis PM and q axis PM high saliency rotor SMs is identical to that for the IM (Table 9.1)
- For the electromagnetically excited SM, where the power factor is unity or leading, the TOS is to be slightly changed [5]
- For the electromagnetically excited rotor SM, the excitation (field) current (voltage) control is introduced (additionally) to keep the stator flux under control and the power factor angle constant either at zero ($\varphi_1 = 0$) or negative ($\varphi_1 = -(6-8)°$); the case of DTFC for excited rotor SM will be dealt with in a separate paragraph
- The flux–torque coordination is to be treated in what follows

9.2.4 Flux/Torque Coordination for the SMs

As for the IM, the stator flux amplitude is limited by the voltage V_s and speed $\omega_r = \omega_1$:

$$\Psi_s \leq \frac{V_{sn}}{\omega_r}. \tag{9.39}$$

Also, to maintain stability, the torque angle δ should be less than δ_{max}:

$$|\delta| < |\delta_{max}|. \tag{9.40}$$

The maximum δ is to be calculated from (9.36) and (9.37) by making $\partial T_e / \partial \delta = 0$. So

$$T_e(\delta_{max}) = T_{emax}(\Psi_s, \delta_{max}). \tag{9.41}$$

From (9.41) we may extract, for the maximum available flux Ψ_{smax}, the maximum reference torque $T_{emax} = T_e(\Psi^*_{smax}, \delta^*_{max})$ in the torque limit. The actual value of field current should be used to secure stability of the response if the field current loop is slow.

Then, the expression of torque limitation T_{emax} is somewhat complicated but still straightforward.

Table 9.1 The Table of Optimal Switchings

Ψ_s	τ_e	$\theta(1)$	$\theta(2)$	$\theta(3)$	$\theta(4)$	$\theta(5)$	$\theta(6)$
1	1	V_2	V_3	V_4	V_5	V_6	V_1
1	−1	V_6	V_1	V_2	V_3	V_4	V_5
0	1	V_0	V_7	V_0	V_7	V_0	V_7
0	−1	V_0	V_7	V_0	V_7	V_0	V_7
−1	1	V_3	V_4	V_5	V_6	V_1	V_2
−1	−1	V_5	V_6	V_1	V_2	V_3	V_4

9.2 DTFC PRINCIPLES FOR INDUCTION AND SYNCHRONOUS MOTORS

Now within the preceeding limits we adopt flux/torque relationships to produce:

- Maximum torque/current (for d axis PM rotors or variable reluctance rotors without or with PMs in axis q)
- Maximum torque/flux (for the same as before, but above base speed or for fast transients)
- Unity power factor ($\varphi_1 = 0$): voltage source inverter-fed excited rotor SM
- Leading power factor ($\varphi_1 = -(6-8)°$): for current source inverter-fed excited rotor SMs

To save space let us treat here only the first two criteria for the d axis PM rotor IPM ($L_d < L_q$) SM and the third one for the excited rotor SM.

For the IPM-SM (with $L_d < L_q$) and maximum torque/current we have available

$$i_s^2 = i_d^2 + i_q^2 = \text{given} \tag{9.42}$$

and the torque expression (9.31). For $(\partial T_e/\partial(i_d/i_q))_{i_s} = 0$, and i_s constant, we get:

$$2i_{di}^2 - i_{di}\frac{\Psi_{PMd}}{(L_d - L_q)} - i_s^2 = 0. \tag{9.43}$$

From (9.43) we retain the $i_{di} < 0$ solution for given i_s and then introduce it into the torque expression:

$$T_e^*(i_s^*) = \frac{3}{2}p_1[\Psi_{PMd} + (L_d - L_q)i_{di}]\sqrt{i_s^2 - i_{di}^2}. \tag{9.44}$$

The stator flux expression (9.29) is

$$\Psi_s^2 = (L_d i_{di} + \Psi_{PMd})^2 + L_q^2(i_s^2 - i_{di}^2). \tag{9.45}$$

Finally we obtain a table $\Psi_s^*(T_e^*)$ for maximum torque per current. For $L_d = L_q$ (surface PM rotor) $i_{di} = 0$, $i_{qi} = i_s$, as expected.

This criterion should be used below base speed where the reluctance torque component is worth producing (with $i_{di} < 0$).

Above base speed, for the same motor, the voltage is constant and thus $\Psi_{smax} = V_{sn}/\omega_r$ decreases with speed.

Now the flux level (decreasing with speed) is given by

$$\Psi_s^2 = (\Psi_{PM} + L_d i_d)^2 + (L_q i_q)^2. \tag{9.46}$$

And again from torque expression (9.31), for $(\partial T_e/\partial(i_d/i_q))_{\Psi_s} = 0$, we obtain finally

$$(L_d i_{d\Psi} + \Psi_{PMd})^2(2L_d - L_q) - L_d L_q i_{d\Psi}(L_d i_{d\Psi} + \Psi_{PMd}) + (L_d - L_q)\Psi_s^2 = 0. \tag{9.47}$$

Apparently only for $2L_d > L_q$ do we have a solution $i_{d\Psi} < 0$. Then from (9.46), we calculate $i_{q\Psi}$ and finally from (9.44) the torque $T_e(i_{d\Psi}, i_{q\Psi})$. Finally we may build a table $\Psi_s^*(T_e^*)$.

This kind of feedforward $\Psi_s^*(T_e^*)$ relationship depends heavily on SM parameters Ψ_{PMd}, L_d, L_q, and in some cases it may be preferable to find some other on-line methods to optimize performance during steady state, mainly.

As for the excited SM, let us calculate the required field current for unity power factor and given stator flux Ψ_s^* and torque T_e^*.

From the space vector diagram of Fig. 9.7a we obtain (for $\varphi_1 = 0$)

$$\frac{L_q i_q^*}{\Psi_s^*} = \sin \delta^*; \qquad \cos \delta^* = \frac{L_{dm} i_F^* - L_d i_d^*}{\Psi_s^*} \qquad (9.48)$$

$$i_q^*/i_s^* = \cos \delta^*; \qquad T_e^* = \frac{3}{2} p_1 \Psi_s^* i_s^*. \qquad (9.49)$$

First from (9.48) we find δ^*:

$$\delta^* = \tan^{-1} \frac{L_q 2 T_e^*}{3 p_1 (\Psi_s^*)^2} \qquad (9.50)$$

Then with

$$|i_d^*| = i_s^* \sin \delta^* = \frac{2 T_e^*}{3 p_1 \Psi_s^*} \sin \delta^* \qquad (9.51)$$

we may calculate, from (9.48), the field current i_F^*:

$$i_F^* = \left(\Psi_s^* \cos \delta^* + L_d \frac{2 T_e^*}{3 p_1 \Psi_s^*} \sin \delta^* \right) / L_{dm}. \qquad (9.52)$$

As expected $i_F^*(T_e^*, \Psi_s^*)$ is a rising function because more field current is required to preserve unity power factor with increasing torque (load).

Again (9.52) is "plagued" by the need to know the machine parameters which are saturation dependent.

It might thus prove practical to add a power factor corrector loop for the field current reference based on the reactive power in the machine (or reactive torque):

$$T_{\text{reactive}} = p_1 \frac{Q_1}{\omega_1} = \frac{3}{2} p_1 \text{Re}(\bar{\Psi}_s, \bar{i}_s^*). \qquad (9.53)$$

T_{reactive} is frequency (speed) independent. We might use only a T_{reactive} loop on the field winding voltage (avoiding field current measurement) to control the reactive power flow from (into) the SM drive. The procedure may also be used for generator (or synchronous condenser) reactive and active separate power control through DTFC.

The principles of DTFC as described earlier should lead to remarks such as the following:

- DTFC is a direct torque and flux close loop control for ac drives in both motoring and generating modes.
- DTFC uses direct triggering of appropiate voltage sequences in the voltage source converter and respectively appropiate current vectors in current source converters (to be shown in the section on large high speed and power SM drives).
- DTFC does not use ac current controllers with all the robustness problems related to them. Neither does it use, in general, vector rotators.
- DTFC uses stator flux, rather than rotor flux as in vector control. In general more robustness in the controller is inherent this way.
- DTFC presupposes flux, torque, and (for sensorless drives) speed observers. The DTFC performance depends esssentially on these observers' performance. In general, for IM, DTFC has been implemented and, in sensorless configurations, proved to produce servo-like torque response time at zero speed of 1–5 ms, 0.1% rated speed static precision in speed response, rotor temperature independent torque response, etc.
- DTFC may be implemented in the same hardware and software on a menu basis to serve small and high powers alike much more easily than vector control with similar performance

where more scaling and adaptation are required.
- The autocommisioning for DTFC is similar to that for vector control, indeed slightly simpler.
- DTFC paves the way for a true ac universal drive.

As the state observer—for flux, torque, and speed—bears, to some extent, the pecularities of various ac machines under study, we will deal with it in the following section dedicated to IMs, PM-SMs, reluctance SMs, and large power SMs.

9.3 DTFC OF INDUCTION MOTORS

Despite its apparent simplicity DTFC is able to produce fast torque and stator flux control. Provided the torque and flux observers produce pertinent results, DTFC is rather robust to motor parameters and external perturbation.

However, during steady state, notable torque and flux pulsations occur, especially with hysteresis torque and flux loops at low speeds. They are reflected in the speed estimator (observer) and also increased noise is radiated.

Closed-loop stator flux predictive control with open-loop torque control using space vector modulation [29] has been proposed to reduce torque pulsations. Torque ripple reduction has been tackled also through predictive nonzero and zero voltage vector timing computation [25, 26]. Notable on-line computation effort is inherent in such methods.

Here we present for comparison both the conventional DTFC solution and the so-called improved DTFC with space vector modulation (SVM) for a sensorless drive [27]. Let us call it DTFC-SVM.

9.3.1 The DTFC-SVM System

The DTFC-SVM sensorless induction motor drive block diagram is shown in Fig. 9.8. It operates with constant rotor flux, direct stator flux, and torque control. The speed controller is a

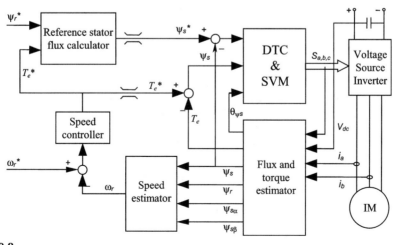

FIGURE 9.8
The DTFC-SVM sensorless IM drive.

classical PID regulator which produces the reference torque. Only the dc-link voltage and two line currents are measured.

The stator flux and torque close loop control is achieved by the DTFC-SVM unit. In order to reduce the torque and flux pulsations and, implicitly, the current harmonics content, in contrast to the standard DTFC, we do use decoupled PI flux and torque controllers and space vector modulation. Also the flux, torque, and speed estimators are visible on Fig. 9.8.

9.3.2 The Flux, Torque, and Speed Estimator

The estimator calculates the stator flux $\underline{\Psi}_s$, the rotor flux $\underline{\Psi}_r$, the electromagnetic torque T_e, and the rotor speed ω_r. It is based on the induction motor equations (9.1) to (9.4). The inputs of the state estimator are the stator voltage $\underline{V}_s$ and current $\underline{i}_s$ space vectors. They are referred to a stationary reference frame.

The flux estimator is a full-order, wide speed range stator and rotor flux observer (Fig. 9.9). It contains two models—the open-loop current model, which is supposed to produce an accurate value, especially for low-speed operation, and the adaptive voltage model for wide speed range operation.

The rotor flux current model estimator is derived from (9.3) and (9.1) in a rotor flux reference frame ($\omega_e = \omega_{\Psi r}$, subscript "$dq$") using the measured stator current (9.54):

$$\bar{\Psi}_{rdq} = \frac{L_m}{1+sT_r}\bar{i}_{sdq} - j\frac{\omega_{\Psi_r} - \omega_r}{1+sT_r}\bar{\Psi}_{rdq} \qquad (9.54)$$

where $T_r = L_r/R_r$ is the rotor time constant.

For rotor flux coordinates, the dq rotor flux components are

$$\Psi_{rd} = \frac{L_m}{1+sT_r}i_{sd} \qquad (9.55)$$

$$\Psi_{rq} = 0. \qquad (9.56)$$

The output of the open-loop current model (superscript "i") is the stator flux $\underline{\Psi}_s^i$ calculated in stator coordinates (9.57):

$$\bar{\Psi}_s^i = \frac{L_m}{L_r}\bar{\Psi}_r^i + \frac{L_s L_r - L_m^2}{L_r}\bar{i}_s \qquad (9.57)$$

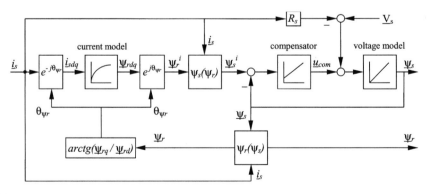

FIGURE 9.9
The flux estimator for the DTC-SVM drive.

where $\bar{\Psi}_r^i$ is the estimated rotor flux from (9.55) and (9.56) in a stationary reference frame (Fig. 9.9).

The voltage model is based on (9.2) and uses the stator voltage vector and current measurement. For the stator reference frame, the stator flux vector $\underline{\Psi}_s$ is simply

$$\bar{\Psi}_s = \frac{1}{s}(\bar{V}_s - R_s \bar{i}_s - \bar{V}_{\text{comp}}). \quad (9.58)$$

In order to correct the value of estimated stator flux, to compensate for the errors associated with pure integrator and stator resistance R_s measurement (estimation) at low speed, and to provide a wide speed range operation for the entire observer, the voltage model is adapted through a PI compensator:

$$\bar{V}_{\text{comp}} = \left(K_{\text{P}} + K_{\text{I}}\frac{1}{s}\right)(\bar{\Psi}_s - \bar{\Psi}_s^i). \quad (9.59)$$

The coefficients K_{P} and K_{I} may be calculated such that, at zero frequency, the current model stands alone, while at high frequency the voltage model prevails:

$$K_{\text{P}} = \omega_1 + \omega_2, \qquad K_{\text{I}} = \omega_1 \cdot \omega_2. \quad (9.60)$$

Values such as $\omega_1 = 2\text{--}5\,\text{rad/s}$ and $\omega_2 = 20\text{--}30\,\text{rad/s}$ are practical for a smooth transition between the two models.

The rotor flux $\bar{\Psi}_r$ is calculated in a stator reference frame:

$$\bar{\Psi}_r = \frac{L_r}{L_m}\bar{\Psi}_s - \frac{L_s L_r - L_m^2}{L_m}\bar{i}_s. \quad (9.61)$$

A detailed parameter sensitivity analysis of this observer can be found in [30].

The speed estimator has the structure of a model reference adaptive controller (MRAC) [31]. In order to achieve a wide speed range, an improved solution which uses the full order flux estimator is proposed (Fig. 9.10).

The reference model is the rotor flux estimator presented so far (9.61). It is supposed to operate accurately for a wide frequency band (1 to 100 Hz). The adaptive model is a current model based on (9.3) for a stationary reference frame ($\omega_e = 0$, superscript "a"):

$$\bar{\Psi}_r^a = \frac{L_m}{1 + sT_r}\bar{i}_s + j\frac{\omega_r}{1 + sT_r}\bar{\Psi}_r^a. \quad (9.62)$$

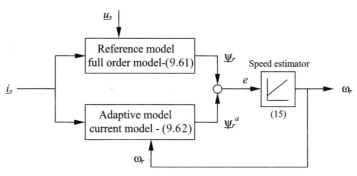

FIGURE 9.10
The MRAC speed estimator.

The rotor speed ω_r is calculated and corrected by a PI adaptation mechanism,

$$\omega_r = \left(K_{P\omega} + K_{I\omega}\frac{1}{s}\right)e, \qquad e = \Psi_{r\alpha}^a \Psi_{r\beta} - \Psi_{r\beta}^a \Psi_{r\alpha} \qquad (9.63)$$

applied on the error between the two models (9.61) and (9.62).

The dynamic analysis of a rather similar speed estimator [31] proved that the achievable bandwidth with which the actual speed can be tracked is limited only by noise considerations. However, very low speed and fast dynamic operation remain an incompletely solved problem.

9.3.3 The Torque and Flux Controllers

The proposed topology for the direct flux and torque controllers (DTFC-SVM) is shown in Fig. 9.11. The controller contains two PI regulators—one for flux and one for torque—and a space vector modulation unit. It receives as inputs the stator flux and torque errors and generates the inverter's command signals. The *dq* components of the reference voltage vector in a stator flux reference frame are

$$V_{sd}^* = (K_{P\Psi} + K_{I\Psi}/s)(\Psi_s^* - \Psi_s) \qquad (9.64)$$

$$V_{sq}^* = (K_{Pm} + K_{Im}/s)(T_e^* - T_e) + \Psi_s \omega_{\Psi_s}. \qquad (9.65)$$

From (9.2), for a stator flux reference frame ($\omega_e = \omega_{\Psi_s}$—the stator flux speed, $\Psi_s = \Psi_{sd}$), the voltage vector components are:

$$u_{sd} = R_s i_{sd} + s\Psi_s \qquad (9.66)$$

$$u_{sq} = R_s i_{sq} + \omega_{\Psi_s} \Psi_s \qquad (9.67)$$

and the electromagnetic torque is

$$T_e = 1.5 p_1 \Psi_s i_{sq}. \qquad (9.68)$$

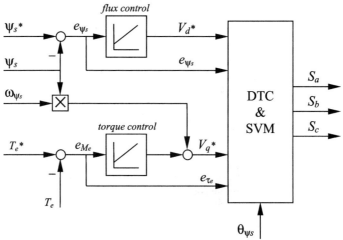

FIGURE 9.11
The DTFC-SVM controller: emf compensation.

9.3 DTFC OF INDUCTION MOTORS

If the stator flux is constant, it is evident that the torque can be controlled by the imaginary component V_{sq}—the torque component—of the voltage vector (9.22):

$$V_{sq} = \frac{2R_s M_e}{3p_1 \Psi_s} + \Psi_s \omega_{\Psi_s}. \tag{9.69}$$

The stator flux speed ω_{Ψ_s} is calculated in a stationary reference frame from two successive estimations of the stator flux $\underline{\Psi}_{s(k)}$ and $\underline{\Psi}_{s(k+1)}$ as

$$\omega_{\Psi_s} = (\Psi_{sd(k)} \Psi_{sq(k+1)} - \Psi_{sq(k)} \Psi_{sd(k+1)})/(\Psi_{s(k+1)}^2 T_s). \tag{9.70}$$

The precision of the calculation is not so important since a PI regulator is present on the torque channel. It corrects the torque even if the last term in (9.69) is somewhat erroneously estimated.

The flux control is accomplished by modifying the real component V_{sd}—the flux component—of the voltage vector.

For each sampling period T_s, one can approximate the V_{sd} voltage as

$$V_{sd} = R_s i_{sd} + \Delta \Psi_s / T_s. \tag{9.71}$$

At high speed the $R_s i_{sd}$ voltage drop can be neglected and the voltage becomes proportional to the flux change $\Delta \Psi_s$ and to the switching frequency $1/T_s$. At low speed, the $R_s i_{sd}$ term is not negligible. The current–flux relations are rather complicated (in stator flux coordinates):

$$(1 + sT_r)\Psi_s = L_s(1 + s\sigma T_r)i_{sd} - (\omega_{\Psi_s} - \omega_r)L_s \sigma T_r i_{sq} \tag{9.72}$$

$$(\omega_{\Psi_s} - \omega_r)T_r \Psi_s = L_s(1 + s\sigma T_r)i_{sq} - (\omega_{\Psi_s} - \omega_r)L_s \sigma T_r i_{sd} \tag{9.73}$$

where

$$\sigma = (L_s L_r - L_m^2)/L_m^2. \tag{9.74}$$

It is evident that a cross-coupling is present in terms of i_{sd} and i_{sq} currents. The simplest way to realize the decoupling is to add the $R_s i_{sd}$ term at the output of the flux regulator in the same manner as the speed-dependent term was added to the torque controller output. However, the computation of the voltage drop term requires a time-consuming stator flux coordinate transformation. Instead of it, a PI controller was used on the flux channel.

The inverter control signals are produced by the SVM unit. It receives the reference voltages (9.64) and (9.65) in a stator flux reference frame. The SVM principle is based on the switching between two adjacent active vectors and a zero vector during one switching period. The reference voltage vector $\bar{V}^*$ defined by its length V (9.75) and angle α (9.76) in a stator reference frame can be produced by adding two adjacent active vectors $\bar{V}_a$ (V_a, α_a) and $\bar{V}_b$ (V_b, α_b) ($\alpha_b = \alpha_a + \pi/3$) and, if necessary, a zero vector $V_0(000)$ or $V_7(111)$.

$$V = \sqrt{V_d^{*2} + V_q^{*2}} \tag{9.75}$$

$$\alpha = \operatorname{arctg} \frac{V_q^*}{V_d^*} + \theta_{\Psi_s} \tag{9.76}$$

where θ_{Ψ_s} is the stator flux position.

The duty cycles D_a and D_b for each active vector are the solutions of the complex equation (9.77):

$$V(\cos \alpha + j \sin \alpha) = D_a V_a (\cos \alpha_a + j \sin \alpha_a) + D_b V_b (\cos \alpha_b + j \sin \alpha_b) \quad (9.77)$$

$$D_a = \frac{\sqrt{3}V}{V_{dc}} \sin \alpha \quad (9.78)$$

$$D_b = \frac{3V}{2V_{dc}} \cos \alpha - \frac{1}{2} D_a \quad (9.79)$$

where V_{dc} is the dc link voltage.

The duty cycle for the zero vector is the remaining time inside the switching period T_s

$$D_0 = 1 - D_a - D_b. \quad (9.80)$$

The vector sequence and the timing during one switching period is

$$\begin{array}{ccccccc} \bar{V}_0 & \bar{V}_a & \bar{V}_b & \bar{V}_7 & \bar{V}_b & \bar{V}_a & \bar{V}_0 \\ \tfrac{1}{4}D_0 & \tfrac{1}{2}D_a & \tfrac{1}{2}D_b & \tfrac{1}{2}D_0 & \tfrac{1}{2}D_b & \tfrac{1}{2}D_a & \tfrac{1}{4}D_0 \end{array}$$

The sequence guarantees that each transistor inside the inverter switches once and only once during the SVM switching period. A strict control of the switching frequency can be achieved by this approach. Figure 9.12 shows the command signals for the inverter when the vectors $V_1(100)$ and $V_2(110)$ and zero vectors $V_0(000)$ and $V_7(111)$ are applied.

A situation which must be considered appears when the control requirements surpass the voltage capability of the inverter—the reference voltage is too high. The PI control method does not guarantee for six-pulse operation. The adopted solution is to switch to the classical DTFC when the PI controllers saturate. If the torque or flux is "far from target," the respective error is large positive or negative and the forward–backward DTFC strategy is applied. A single voltage vector is applied the whole switching period. It ensures that the target will be reached quickly. If the torque and flux are "close to target" the errors are small and now the SVM strategy based on PI controllers is enabled instead of applying a zero vector as a classical DTFC would imply.

The "saturation point" for PI regulators is considered at $1.5V_{dc}$. Normally, the voltage amplitude control becomes ineffective for a reference voltage higher than V_{dc}, but the voltage angle control is still effective. This observation permits to choose the switching point from SVM to DTFC at a relatively higher voltage—up to $2V_{dc}$ where the PI antiwindup becomes active.

The classical DTFC topology is presented in Fig. 9.13.

The DTFC strategy can be simply defined as follows: Each sampling period, the adequate voltage vector is selected in order to decrease rapidly, in the same time, the torque and flux errors. The voltage vector selection is done in accordance with the signals produced by two hysteresis comparators and the stator flux vector position, as explained in Section 9.1.

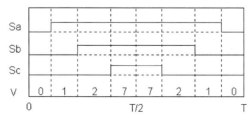

FIGURE 9.12
The SVM voltage vector timing.

9.3 DTFC OF INDUCTION MOTORS 319

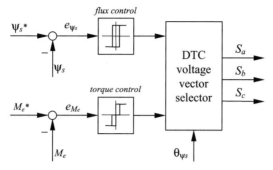

FIGURE 9.13
The classical DTC controller.

The digital simulation results with DTC-SVM strategy are presented in Fig. 9.14—speed and torque—and Fig. 9.15—stator and rotor flux amplitude. Both figures show the starting transients and the speed transients for a 4 kW IM. A smooth operation can be observed.

The simulation was used to perform a parameter sensitivity analysis. Figures 9.14 and 9.15 show operation with the values $K_{P\psi} = 100$, $K_{I\psi} = 300$, $K_{Pm} = 2$, $K_{Im} = 200$ for the PI torque

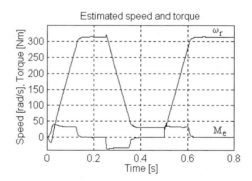

FIGURE 9.14
The simulated speed and torque transients. The parameters of the DTFC-SVM controller are $K_{P\psi} = 100$, $K_{I\psi} = 300$, $K_{Pm} = 2$, $K_{Im} = 200$.

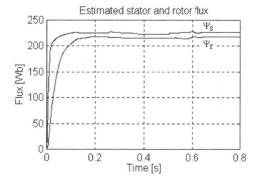

FIGURE 9.15
The simulated stator and rotor flux transients. The parameters of the DTFC-SVM controller are $K_{P\psi} = 100$, $K_{I\psi} = 300$, $K_{Pm} = 2$, $K_{Im} = 200$.

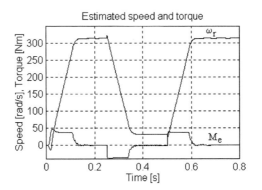

FIGURE 9.16
The simulated speed and torque transients. The parameters of the DTC-SVM controller are $K_{P\psi} = 200$, $K_{I\psi} = 600$, $K_{Pm} = 4$, $K_{Im} = 400$.

and flux controllers, and Figs. 9.16 and 9.17 show the operation with doubled values: $K_{P\psi} = 200$, $K_{I\psi} = 600$, $K_{Pm} = 4$, $K_{Im} = 400$. Further increase is possible, but a too-high gain for torque controller will cause oscillations. A very high gain will produce operation similar to that of the DTFC. This way, the robustness of the DTFC-SVM controller is partly proved.

The design of the two PI controllers is based on (9.69) and (9.71). The torque controller gain should be equal to $2R_s/3p_1\Psi_s$ from the first term in (9.69). The values $K_{Pm} = 2$ to 4 denote a high-gain torque controller but are necessary to produce a fast torque response. For the flux controller, the gain $K_{P\psi} = 100$ is smaller than the switching frequency $1/T_s = 8$ kHz, but the overall system's stability is improved even if the flux controller is not a very fast one. The integrator term in both controllers introduces a unitary discrete pole and compensates for the cross-coupling errors.

9.3.4 The Exprimental Results

The experimental setup of the DTC-SVM system is shown in Fig. 9.18. The induction motor has the rated values $P_N = 4$ kW, $f_N = 50$ Hz, $V_N = 400$ V, $M_N = 27$ Nm, $p_1 = 2$ pole pairs and the

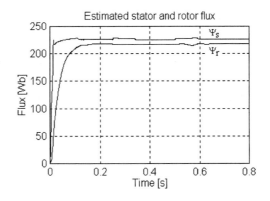

FIGURE 9.17
The simulated stator and rotor flux transients. The parameters of the DTC-SVM controller are $K_{P\psi} = 200$, $K_{I\psi} = 600$, $K_{Pm=4}$, $K_{Im} = 400$.

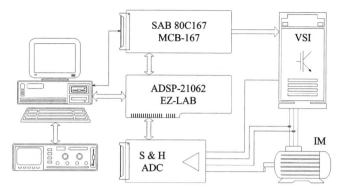

FIGURE 9.18
The experimental setup.

parameters are $R_s = 1.55\,\Omega$, $R_r = 1.25\,\Omega$, $L_s = 0.172\,\text{H}$, $L_r = 0.172\,\text{H}$, $L_m = 0.166\,\text{H}$. The inverter is a 7 kVA industrial voltage source inverter with dc voltage $V_{dc} = 566\,\text{V}$. The digital control system contains a DSP (ADSP-21062) and a microcontroller (SAB 80C167). The DSP is responsible for all the calculations and the microcontroller produces the PWM signals. The sampling time was 125 μs and the switching frequency 8 kHz. Because the phase voltage is calculated, a dead time compensation was included. Both DTFC-SVM and classical DTFC strategies were implemented.

The following controller parameters are used for experiments:

- The PI compensator for the flux estimator in Fig. 9.9 uses the values $K_P = 18$ and $K_I = 80$ calculated for a transition between the two models around 10 Hz.
- The PI speed estimator in Fig. 9.10: $K_{P\omega} = 100$ and $K_{I\omega} = 22000$ determined as in [31]. Higher gains produce instability.
- The PI torque and flux controllers in Fig. 9.11: $K_{Pm} = 2$, $K_{Im} = 200$, and $K_{P\psi} = 100$, $K_{I\psi} = 300$.

Comparative experimental results for low speed no load operation are presented first. Figure 9.19 shows the estimated speed, torque, stator and rotor flux, and the measured current for steady-state 1 Hz DTFC-SVM operation. Figure 9.20 shows the estimated speed, torque, and stator and rotor flux for steady-state 1 Hz DTFC operation. An improved operation in terms of high-frequency ripple can be noticed with DTFC-SVM.

The no-load starting transient performances are presented in Fig. 9.21—estimated speed and torque—for DTFC-SVM and in Fig. 9.22—the same quantities—for DTFC. Again the torque ripple is drastically reduced while the fast response is preserved.

The same conclusions are evident in the no-load speed transients—from 5 to 50Hz—presented in Fig. 9.23 for DTFC-SVM and in Fig. 9.24 for DTFC. A zoom of torque proves the fast torque response of the new strategy.

Figure 9.25 shows the speed reversal from 25 Hz to −25 Hz—speed, flux, and current—for DTFC-SVM. Some small flux oscillations can be observed when the flux changes because of the absence of the decoupling term in the flux controller.

The system's stability is influenced by the precision and the speed of convergence of the flux and speed estimation. The speed estimator is not a very fast one, and this can be seen on Fig. 9.25 where some speed oscillations occur. The DTFC-SVM controller does not depend on motor parameters and is relatively robust as was partly proved by simulation.

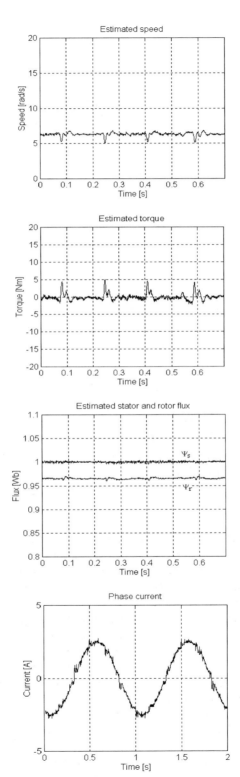

FIGURE 9.19
DTFC-SVM: 1 Hz (30 rpm) no-load steady state.

9.3 DTFC OF INDUCTION MOTORS **323**

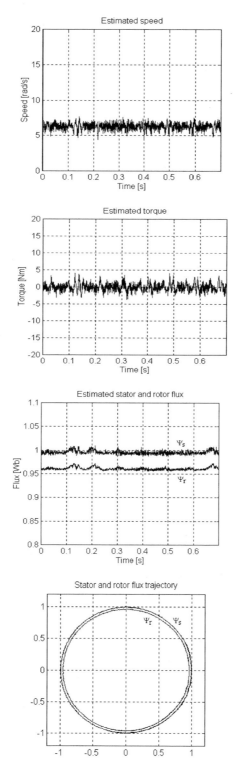

FIGURE 9.20
Classical DTFC: 1 Hz (30 rpm) no-load steady state.

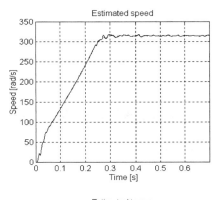

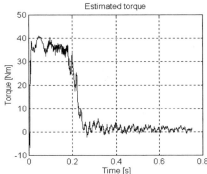

FIGURE 9.21
DTFC-SVM no-load starting transients.

9.3.5 Discussion

We presented a direct torque and flux control strategy based on two PI controllers and a voltage space vector modulator. The complete sensorless solution is developed.

The main conclusions are as follows:

- The DTFC-SVM strategy realizes almost ripple free operation for the entire speed range. Consequently, the flux, torque, and speed estimation are improved.
- The fast response and robustness merits of the classical DTFC are entirely preserved.
- The switching frequency is constant and controllable. In fact, the better results are due to the increasing of the switching frequency. While for DTFC a single voltage vector is applied during one sampling time, for DTFC- SVM, a sequence of six vectors is applied during same time. It is the merit of the SVM strategy.
- An improved MRAC speed estimator based on a full-order rotor flux estimator as reference model was proposed and tested at high and low speed.

It can be stated that using the DTFC-SVM topology, the overall system performance is increased.

9.3.6 Stator Resistance Estimation

Though the DTFC-SVM approach proved to provide good performance and some digital simulations demonstrated some robustness of the controller, the flux estimator precision through

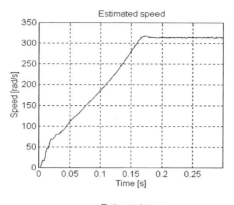
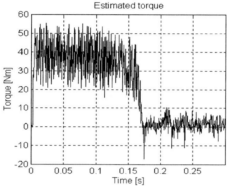

FIGURE 9.22
Classical DTC no-load starting transients.

the voltage model (Fig. 9.9), at least at low speeds, requires the correct value of the stator resistance. Consequently, at least for low speeds, stator resistance correction should be highly beneficial.

Fortunately the stator and torque estimation paves the way for an easy estimation of stator resistance based on stator current error $i_s^* - I_s$ with i_s (stator current) measured (Fig. 9.26) [24].

As Fig. 9.26 is self-explanatory we have to insist only about how to calculate the reference stator current i_s^* for given stator flux and torque reference values Ψ_s^* and T_e^*.

To find i_s^* we have to go back to IM equations (9.1)–(9.4), but this time in stator flux coordinates ($\bar{\Psi}_s = \Psi_s = \Psi_{ds}$, $\Psi_{qs} = 0$, $d\Psi_{ds}/dt = d\Psi_{qs}/dt = 0$). With $\omega_{s\Psi_s}$ as slip frequency of stator flux $\omega_{s\Psi_s} = \omega_{\Psi_s} - \omega_r$ we finally obtain in a rather straightforward manner [24] (for steady state)

$$i_q^{e*} = \frac{2T_e^*}{3p_1 \Psi_s^*} \tag{9.81}$$

$$L_s(i_d^{e*})^2 - \Psi_s^*\left(1 - \frac{1}{\sigma}\right)i_d^{e*} + L_s(i_q^{e*})^2 - \frac{(\Psi_s^*)^2}{L_s} = 0 \tag{9.82}$$

$$\omega_{s\Psi_s} = \frac{R_r(\Psi_s^* - L_s i_d^{e*})}{i_q^{e*}(L_s L_r - L_m^2)}. \tag{9.83}$$

Finally the stator reference current i_s^* is

$$i_s^* = \sqrt{(i_d^{e*})^2 + (i_q^{e*})^2}. \tag{9.84}$$

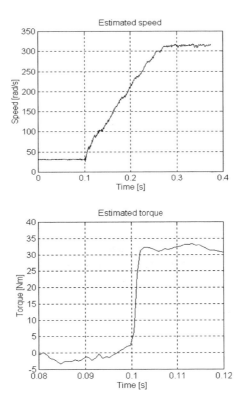

FIGURE 9.23
DTC-SVM speed and torque transients zoom during no-load acceleration from 5 to 50 Hz.

As a bonus we have obtained the slip frequency of stator flux. Unfortunately (9.83) is very sensitive to any variation in rotor resistance so we should not rely heavily on it for other purposes in the control system.

As L_s occurs in (9.82) the value i_d^{e*} is still dependent on saturation. However, $i_d^{e*} < i_q^{e*}$, at least above a certain load level. Instability in the torque and current response with the wrong value of R_s, in a purely voltage model flux observer, have been eliminated through a correction scheme as before [24].

A similar procedure may be applied to SM.

9.4 DTFC OF PMSM DRIVES

The DTFC of PMSM drives is very similar to the solution applied to IMs. A rather conventional DTFC system is shown in Fig. 9.27.

Notice the comprehensive state observer including stator flux, torque, rotor position and speed: $\hat{\Psi}_s$, $\hat{T}_e$, $\hat{\theta}_r$, $\hat{\omega}_r$. When position feedback θ_r is available, the speed estimator is rather straightforward.

The TOS is identical to the one in Table 9.1. The flux torque $\Psi_s^*(T_e^*)$ relationships have already been discussed in Section 9.2.4.

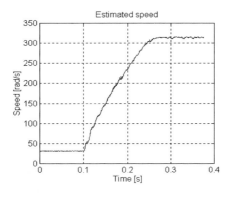
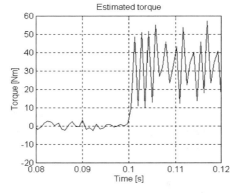

FIGURE 9.24
Classical DTFC speed and torque transients zoom during no-load acceleration from 5 to 50 Hz (IM).

9.4.1 The Flux and Torque Observer

For the case of the sensorless drive the state observer is very important. We describe here a combined voltage–current model flux estimator very similar to the one adopted for the IM. The voltage model in stator coordinates is the same as for the IM:

$$\frac{d\bar{\Psi}_{sV}}{dt} = \bar{V}_s - R_s \bar{i}_s, \tag{9.85}$$

while the current model in rotor coordinates is

$$\bar{\Psi}_{si}^r = L_d i_d + \Psi_{PM} + jL_q i_q; \qquad \bar{i}_{d,q} = \bar{i}_s^r = i_d + ji_q \tag{9.86}$$

$$\bar{i}_s^r = \bar{i}_s^e e^{-j\theta_{er}}; \qquad \theta_{er} = p_1 \theta_r. \tag{9.87}$$

Now the flux current model is transformed back to stator coordinates to become $\bar{\Psi}_{si}^s$:

$$\bar{\Psi}_{si}^s = \bar{\Psi}_{si}^r e^{j\theta_{er}}. \tag{9.88}$$

The difference between $\bar{\Psi}_{sV}^s$ and $\bar{\Psi}_{si}^s$ is flowed through a PI loop compensator (Fig. 9.28). For low frequency the current model is predominant while for higher frequencies (speed) the voltage model takes over.

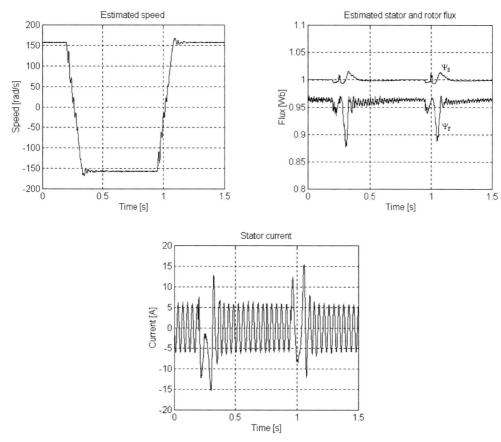

FIGURE 9.25
DTFC-SVM speed reversal transients (from 25 Hz to −25 Hz) (IM).

The values of K_i, τ_i are such that

$$K_i = -(\omega_1 + \omega_2); \qquad \tau_i = \frac{K_i}{\omega_1 \cdot \omega_2} \tag{9.89}$$

with $\omega_1 = -(3-10)\,\text{rad/s}$ and $\omega_2 = -(3-10)\omega_1$ for a rather smooth transition from current to voltage model at low speeds.

Still the rotor position $\hat{\theta}_{er}$, if not measured, should be estimated. So an additional position and speed observer is required.

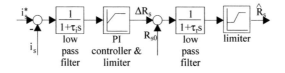

FIGURE 9.26
Stator resistance estimation.

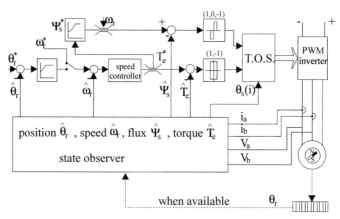

FIGURE 9.27
DTFC of PMSM.

9.4.2 The Rotor Position and Speed Observer

An extended Luenberger observer, which uses a current error $(\tilde{i}_s - i_s)$ compensator K, is used for position and speed estimation while a nonlinear estimator calculates the stator current vector $\hat{i}_s$. The extended Luenberger observer has the equations [32]

$$\begin{vmatrix} \hat{\theta}_{er} \\ \hat{\omega}_r \\ \hat{T}_{load} \end{vmatrix} = \begin{vmatrix} 0 & 1 & 0 \\ 0 & 0 & -p_1/J_0 \end{vmatrix} \times \begin{vmatrix} \hat{\theta}_{er} \\ \hat{\omega}_r \\ \hat{T}_e \end{vmatrix} + \begin{vmatrix} 0 \\ p_1/J_0 \\ 0 \end{vmatrix} \tilde{T}_e + K(i_s - \tilde{i}_s). \qquad (9.90)$$

The load torque $\hat{T}_{load}$ has the same step signal class as an exogenous model. The gain matrix K provides for a predictive correction for desired speed convergence and better robustness.

Still the nonlinear stator current vector $\hat{i}_s$ estimator is to be obtained. To estimate the current vector $\hat{i}_s$ we simply use the dq model of the machine with the speed already estimated in the previous time step. The position error is considered zero:

$$s\begin{bmatrix} L_d \hat{i}_d \\ L_q \hat{i}_q \end{bmatrix} = -\begin{bmatrix} R_s & \hat{\omega}_r L_q \\ -\hat{\omega}_r L_d & -R_s \end{bmatrix}\begin{bmatrix} i_d \\ i_q \end{bmatrix} + \begin{bmatrix} V_d \\ V_q \end{bmatrix} + \begin{bmatrix} 0 \\ -\hat{\omega}_r \Psi_{PMd} \end{bmatrix}. \qquad (9.91)$$

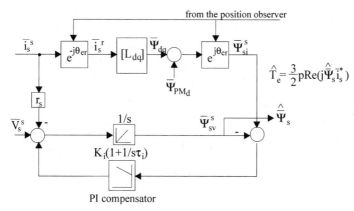

FIGURE 9.28
Combined voltage–current flux and torque observer for PMSMs.

As we measure V_a, V_b, V_c, the stator voltage vector has to be transformed to rotor coordinates:

$$V_{d,q} = \bar{V}_s^s e^{-j\hat{\theta}_{er}}; \qquad \bar{i}_{d,q} = \bar{i}_s e^{-j\hat{\theta}_{er}} \qquad (9.92)$$

where $\hat{\theta}_{er}$ is the estimated value of the rotor position in the previous time step.

For the right values of parameters and a $\Delta\theta_{er}$ position error it may be shown from (9.91) that

$$s\begin{bmatrix} L_d & \Delta i_d \\ L_q & \Delta i_q \end{bmatrix} = \begin{bmatrix} \hat{\omega}_r \Psi_{PMd} & \Delta\theta_{er} \\ -\Psi_{PMd} & \Delta\omega_r \end{bmatrix}; \qquad \begin{bmatrix} \Delta i_d \\ \Delta i_q \end{bmatrix} = \begin{bmatrix} i_d^* - i_d \\ i_q^* - i_q \end{bmatrix}. \qquad (9.93)$$

Thus the compensator K takes the decoupled form

$$K(\bar{i}_s - \bar{i}_s^*) = \begin{bmatrix} K\hat{\omega}_r & 0 \\ 0 & K_2 \\ 0 & K_3 \end{bmatrix} \begin{bmatrix} \Delta i_d \\ \Delta i_q \end{bmatrix}. \qquad (9.94)$$

Now Eqs. (9.90), (9.91), and (9.94) may be assembled into the position and speed observer with current vector error compensation.

When the motor parameters are tuned in [32, 33] angle errors of less than 6° and speed errors less than 5 rpm at 20 rpm speed have been reported (Fig. 9.29) [33].

As expected the position error gets larger at larger speeds, but in DTFC the estimated position $\hat{\theta}_{er}$ plays a role only within the flux observer, and here, fortunately, its influence at high speeds becomes negligible.

9.4.3 Initial Rotor Position Detection

Most position and speed observers, including the one presented earlier in this section, are not capable of detecting the initial rotor position. Special starting methods or initial position detection before starting is required for a safe start.

The drive itself could, for example, send short voltage vector signals $\bar{V}_1$, $\bar{V}_3$, $\bar{V}_5$ and, respectively, V_2, V_4, V_6 and measure the current levels reached after a given few microseconds.

Based on the sinusoidal inductance variation (for IPM) the initial position can be calculated univoquely. Thus a nonhesitant start is obtained.

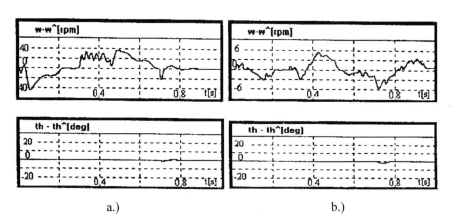

FIGURE 9.29
Speed and position error at (a) 1000 rpm; (b) 20 rpm.

A simple approach would be to replace the integrator in the flux observer (Fig. 9.28) by a first-order delay with a time constant of tens of miliseconds. A good start with this method was reported for light loads [19].

Finally a strong autocommisioning methodology to estimate the machine parameters and to tune the controllers by the drive itself is still due for DTFC of PMSM.

9.5 DTFC OF RELUCTANCE SYNCHRONOUS MOTORS

The DTFC of RSM is evidently based first on typical stator flux and torque expressions:

$$\bar{\Psi}_s^r = L_d i_d + j(L_q i_q - \Psi_{PMq}) \tag{9.95}$$

$$T_e^* = \frac{3}{2} p_1 [\Psi_{PMq} + (L_d - L_q) i_q] i_d; \qquad L_d \gg L_q. \tag{9.96}$$

Notice the presence of the PMs along axis q to improve power factor and torque production, and finally increase the constant power speed range up to more than 5 to 1 [34]. The flux/torque limitations due to current or voltage limitations, and flux/torque coordination have been discussed in principle in Section 9.2. We may present directly the block diagram for the DTFC of RSM, the flux–torque observer and the position and speed observer. These are quite similar to those of PMSM but without PMs in axis d and eventually with weak PMs along axis q in the rotor. This is why we decide here to present for comparisions a DTFC sensorless RSM drive implementation both in the conventional form and with space vector modulation (DTFC-SVM) as we did for the IM.

In order to reduce the torque and current pulsations, in steady state a mixed DTFC-SVM control method seems more suitable. SVM techniques offer better dc link utilization and decreased torque ripple. Lower THD in the motor current is also obtained. It is different from standard PWM techniques, such as sinusoidal PWM or third harmonic PWM methods. The proposed control strategy, its implementation, and test results constitute the core of what follows.

9.5.1 The Proposed RSM Sensorless Drive

The proposed DTFC-SVM sensorless RSM drive block diagram is presented in Fig. 9.30.

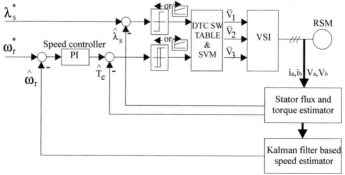

FIGURE 9.30
DTFC–SVM sensorless RSM drive.

Its components are described as follows:

- RSM and VSI (voltage source inverter—VLT 3008—Danfoss) in the present implementation with current and dc link voltage measurements—necessary for flux, torque, and speed estimators (together with the known inverter switch states)
- Stator flux and torque estimator
- Kalman filter based speed estimator
- PI speed controller
- DTFC TOS and SVM switching and voltage selection table
- Hysteresis controller on stator flux and estimated torque errors
- Reference speed and stator flux as inputs

The RSM model used in d–q coordinates is

$$\begin{cases} V_d = R_s i_d + \dfrac{d\Psi_d}{dt} - \omega_r \Psi_q \\ V_q = R_s i_q + \dfrac{d\Psi_q}{dt} + \omega_r \Psi_d \end{cases} \tag{9.97}$$

$$T_e = \frac{3}{2} p_1 (L_d - L_q) i_d i_q \tag{9.98}$$

where ω_r is the rotor electrical angular speed (rad/s), Ψ_d is the d axis stator flux ($L_d i_d$), Ψ_q is the q axis stator flux ($L_q i_q$), i_d, i_q are the d, q axis stator currents, and p_1 is the pole pair number.

$$\begin{cases} \Psi_d = L_{s\sigma} i_d + L_{md} i_d \\ \Psi_q = L_{s\sigma} i_q + L_{mq} i_q \end{cases} \tag{9.99}$$

$L_{s\sigma}$, L_{md}, L_{mq} are the leakage and d–q axis magnetizing inductances.

For a good estimation accurate current and voltage measurements are necessary. In this case the available variables are dc link voltage and two phase currents (star connection of stator winding). The phase voltages are calculated from dc link measurement and the inverter switching state logic.

9.5.2 Estimators

9.5.2.1 Flux Estimator. The flux estimator (Fig. 9.31) is based on the combined voltage–current model of stator flux in RSM. The block diagram of this estimator is presented in the figure. The voltage model is supposed to be used for higher speeds and the current model for the

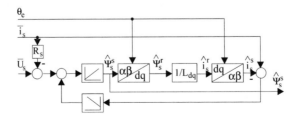

FIGURE 9.31
The stator flux estimator.

low-speed region. A PI regulator makes the smooth transition between these two models, at medium speed, on the current error. For the current model we need the rotor position angle θ_e based on the stator flux position.

The equations describing these flux estimators are

$$\hat{\bar{\Psi}}_s^s = \int (\bar{V}_s - R_s \cdot \bar{i}_s + \bar{V}_{comp}) dt \tag{9.100}$$

$$\bar{V}_{comp} = (K_p + K_i/s)(\hat{\bar{i}}_s^s - \bar{i}_s). \tag{9.101}$$

The coefficients K_p and K_i may be calculated such that, at zero frequency, the current model stands alone, while at high frequency the voltage model prevails. Again:

$$K_p = \omega_1 + \omega_2, \quad K_i = \omega_1 \cdot \omega_2. \tag{9.102}$$

Values such as $\omega_1 = 2\text{--}3 \text{ rad/s}$ and $\omega_2 = 20\text{--}30 \text{ rad/s}$ are practical for a smooth transition between the two models. The values $K_p = 50$ and $K_i = 30$ were used in experiments.

9.5.2.2 The Torque Estimator. The torque estimation uses the estimated stator flux (9.100) and the measured currents:

$$\hat{T}_e = \frac{3}{2} p_1 (\Psi_{s\alpha} i_{s\beta} - \Psi_{s\beta} i_{s\alpha}). \tag{9.103}$$

9.5.2.3 The Speed Estimator. The speed estimator is based on a Kalman filter approach [35] with rotor position as input. This way a position derivative calculation is eliminated.

The equations are

$$\begin{cases} \varepsilon_k = \sin\theta_e \cdot \cos\theta_k - \sin\theta_k \cdot \cos\theta_e \\ \theta_k = \theta_k + 2T_s \cdot \omega_r + K_1 \cdot \varepsilon_k \\ \text{if } \theta_k > \pi \Rightarrow \theta_k = -\pi \\ \text{if } \theta_k < -\pi \Rightarrow \theta_k = \pi \\ \hat{\omega}_r = \hat{\omega}_r + w_k + K_2 \cdot \varepsilon_k \\ w_k = w_k + K_3 \cdot \varepsilon_k \end{cases} \tag{9.104}$$

where K_1, K_2, K_3 are Kalman filter parameters; ω is the estimated rotor speed; θ_e is the estimated rotor position, and T_s is sampling time.

The rotor position necessary for the speed estimator has been calculated from the stator flux position.

9.5.3 DTFC and SVM

The speed controller is a digital PI one with equation

$$T_e^* = (K_{ps} + 1/sT_{is}) \cdot \varepsilon_\omega \tag{9.105}$$

where $\varepsilon_\omega = \omega_r^* - \hat{\omega}_r$. The output of the speed controller is the reference torque. PI parameters are variable depending on speed error in three steps: $K_{ps} = 10, 3, 1$ and $T_{is} = 0.01, 0.1, 0.5$ (for $\text{abs}(\varepsilon_\omega) \geq 1$; $\text{abs}(\varepsilon_\omega) < 1$ and $\text{abs}(\varepsilon_\omega) > 0.5$; $\text{abs}(\varepsilon_\omega) \leq 0.5$). The PI controller has output limits at $\pm 15 \text{ Nm}$.

Now that we have the reference and estimated stator flux and torque values, their difference is known as $\varepsilon_{\lambda s}$ and ε_{Te}:

$$\varepsilon_{\Psi_s} = \Psi_s^* - \hat{\Psi}_s \quad (9.106)$$

$$\varepsilon_{Te} = T_e^* - \hat{T}_e. \quad (9.107)$$

A bipositional hysteresis comparator for flux error and a three- positional one for torque error have been implemented. The two comparators produce the Ψ_s and T_e command laws for torque and flux paths:

$$\begin{array}{lll} \lambda_s = +1 & \text{for} & \varepsilon_{\Psi s} > 0 \\ \lambda_s = -1 & \text{for} & \varepsilon_{\Psi s} < 0 \\ T_e = +1 & \text{for} & \varepsilon_{Te} > h_m \\ T_e = 0 & \text{for} & |\varepsilon_{Te}| < h_m \\ T_e = -1 & \text{for} & \varepsilon_{Te} < -h_m \end{array} \quad (9.108)$$

where h_m is a relative torque value of a few percent.

The discrete values of flux and torque commands of $+1$ means flux (torque) increases, -1 means flux (torque) decreases, while zero means zero voltage vector applied to the inverter.

To detect the right voltage vector in the inverter, we also need the position of the stator flux vector in six 60° extension sectors, starting with the axis of phase "a" in the stator.

In each switching period only one voltage vector is applied to the motor and a new vector is calculated for the following period. This is the standard DTFC.

In the addition here a combined DTC-SVM strategy has been chosen in order to improve the steady-state operation of the drive by reducing the torque–current ripples caused by DTFC.

DTFC-SVM is based on the fact that a SVM unit modulates each voltage vector selected from the DTFC TOS before applying it to the inverter.

The SVM principle is based on switching between two adjacent active vectors and a zero vector during a switching period. An arbitrary voltage vector V defined by its length $\bar{V}$ and angle α can be produced adding two adjacent active vectors $\bar{V}_a$ (V_a, α_a) and $\bar{V}_b$ (V_b, α_b) and, if necessary, a zero vector $\bar{V}_0$ or $\bar{V}_7$. The duty cycles D_a and D_b for each active vector are calculated as a solution of the complex equations:

$$V(\cos \alpha + j \sin \alpha) = D_a V_a (\cos \alpha_a + j \sin \alpha_a) + D_b V_b (\cos \alpha_b + j \sin \alpha_b) \quad (9.109)$$

$$D_a = M_{\text{index}} \sin \alpha \quad (9.110)$$

$$D_b = \frac{\sqrt{3}}{2} M_{\text{index}} \cos \alpha - \frac{1}{2} D_a \quad (9.111)$$

where the modulation index M_{index} is

$$M_{\text{index}} = \frac{\sqrt{3} V}{V_{\text{dc}}}. \quad (9.112)$$

Because $\bar{V}_a$ and $\bar{V}_b$ are adjacent voltage vectors, $\alpha_b = \alpha_a + \pi/3$.

The duty cycle for the zero vectors is the remaining time inside the switching period:

$$D_0 = -D_a - D_b + 1. \quad (9.113)$$

9.5 DTFC OF RELUCTANCE SYNCHRONOUS MOTORS

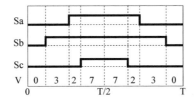

FIGURE 9.32
A switching pattern example.

The vector sequence and the timing during one switching period is (Fig. 9.32)

$$\begin{array}{ccccccc}
\bar{V}_0 & \bar{V}_a & \bar{V}_b & \bar{V}_7 & \bar{V}_b & \bar{V}_a & \bar{V}_0 \\
1/4D_0 & 1/2D_a & 1/2D_b & 1/2D_0 & 1/2D_b & 1/2D_a & 1/4D_0
\end{array}$$

The sequence, identical to that for IMs, guarantees each transistor IGBT in the inverter switches once and only once during the SVM switching period. Strict control of the inverter's switching frequency can be achieved this way.

The duty cycles for each arm can be easily calculated. For SVM control the reference value of voltage vector comes from its components as PI controller outputs on the flux error and the torque error:

$$\bar{V}_r^* = V^*(\cos\alpha + j\sin\alpha) \tag{9.114}$$

$$V^* = \sqrt{V_d^{*2} + V_q^{*2}} \tag{9.115}$$

$$\alpha = (V_q^*/V_d^*). \tag{9.116}$$

If the torque and/or flux error is large the DTFC strategy is applied with a single voltage vector through a switching period. In case of a small torque and/or flux error the SVM strategy is enabled based on the two PI controllers discussed before.

The DTFC strategy is suitable for transients because its fast-response; SVM control is suitable for steady state and small transients for its low torque ripple. The switching frequency is being increased in this latter case, but the drive noise is also reduced.

It is time to note that the entire solution is similar to that described for the IM.

9.5.4 The Experimental Setup

The voltage source inverter used for tests was a Danfoss VLT 3008 7KVA one working at 9 KHz switching frequency.

The reluctance synchronous motor has the following parameters: $P_N = 2.2\,\text{kW}$, $V_N = 380/220\,\text{V}\ (Y/\Delta)$, $I_N = 4.9/8.5\,\text{A}$, $f_N = 50\,\text{Hz}$; $R_s = 3\,\Omega$, $L_d = 0.32\,\text{H}$, $L_q = 0.1\,\text{H}$.

In what follows, test results are presented with sensorless control of RSM at different speeds and during transients (speed step response and speed reversing) using DTFC and DTFC-SVM control strategies.

Though simulation studies were made before tests, only test results are given in Figs. 9.34–9.39. Notable reduction of torque pulsations, speed, flux, and current ripple is produced by the DTFC-SVM.

The experimental setup of the RSM sensorless drive with DTFC-SVM is presented in Fig. 9.33.

The control system contains a 32-bit floating point DSP and a 16-bit microcontroller. The digital control system is supervised by a personal computer. The calculations for state estimation

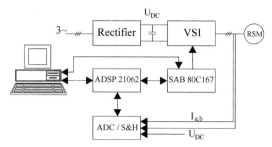

FIGURE 9.33
The experimental setup.

and control strategy are performed by ADSP-21062 DSP. The inverter command signals (space vector modulation) are produced by a SAB 80C167 microcontroller.

The estimation and control algorithm is implemented using standard ANSI-C language. All differential equations are transformed to the digital form using the bilinear approximation

$$s = \frac{2}{T_s}\left(\frac{z-1}{z+1}\right) \quad (9.117)$$

with the sampling time $T_s = 111\,\mu s$ for a 9 KHz switching frequency. All the calculations are done by DSP.

For current and voltage measurement one 12-bit, 3.4-μs analog to digital converter (ADC) and an 8-channel simultaneous sampling and hold circuits are used. The sampling time is the same as the switching time. To avoid problems related to frequency aliasing, analog filters with 300 μs time constant are used for the current measurement path. The current and voltage measurement is done at the beginning of each switching period. This technique reduces the measurement noise because of the symmetry of the switching pattern. Each current is measured with a Hall sensor current transformer.

For speed information (for comparisons), the speed was measured using an incremental encoder with a resolution of 1024 increments/revolution. The measured speed is filtered with an analog filter with 300 μs time constant.

To eliminate the nonlinear effects produced by the inverter, the dead time compensation was introduced. The inverter's dead time is 2.0 μs.

9.5.5 Test Results

Steady-state performance of RSM drive with DTFC and DTFC-SVM control strategy at 15 rpm is presented in Figs. 9.34 and 9.35. The represented waveforms are: phase current, estimated torque, measured and estimated speed.

Comparative results are also presented in Figs. 9.36 and 9.37 for DTFC and DTFC-SVM control during transients (reference speed step from 450 to 30 rpm). A faster torque response has been obtained with DTFC control strategy, as expected.

Test results are compared in Fig. 9.38 with simulation results in Fig. 9.39 during speed reversal at 150 rpm using DTFC control strategy.

9.5.6 Discussion

A direct torque and stator flux control strategy combined with space vector modulation was implemented for RSMs. The complete sensorless solution was given.

9.5 DTFC OF RELUCTANCE SYNCHRONOUS MOTORS

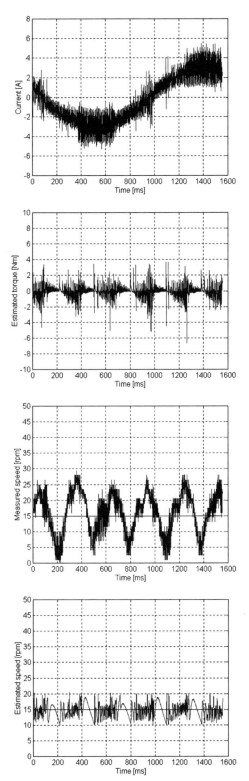

FIGURE 9.34
DTFC control of RSM: 15 rpm steady-state no-load operation.

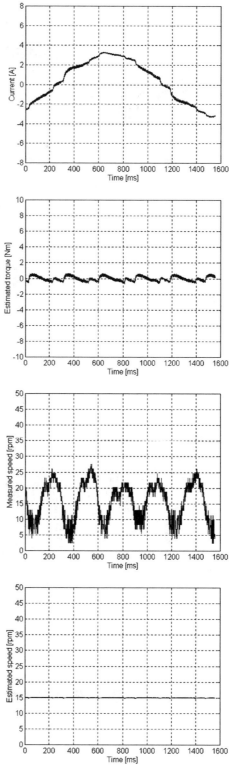

FIGURE 9.35
DTFC–SVM control of RSM: 15 rpm steady-state no-load operation.

9.5 DTFC OF RELUCTANCE SYNCHRONOUS MOTORS

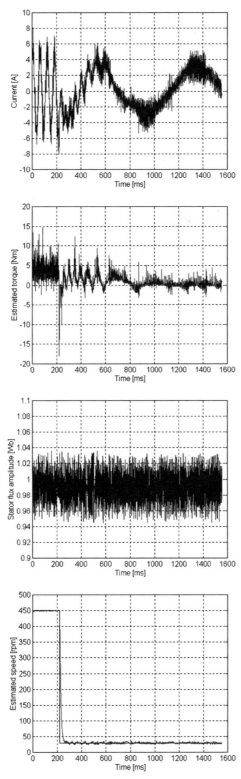

FIGURE 9.36
Speed, torque, stator flux, and current transients during no-load deceleration from 450 to 30 rpm with DTFC control.

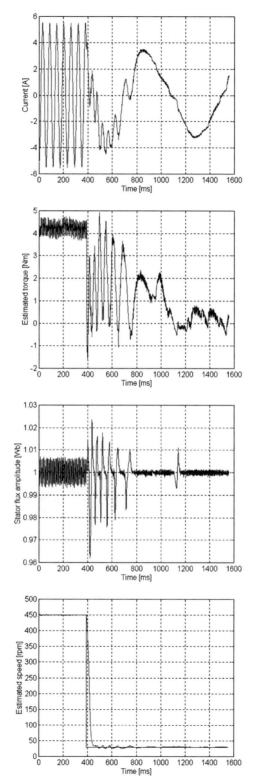

FIGURE 9.37
Speed, torque, stator flux, and current transients during no-load deceleration from 450 to 30 rpm with DTFC-SVM control.

9.5 DTFC OF RELUCTANCE SYNCHRONOUS MOTORS

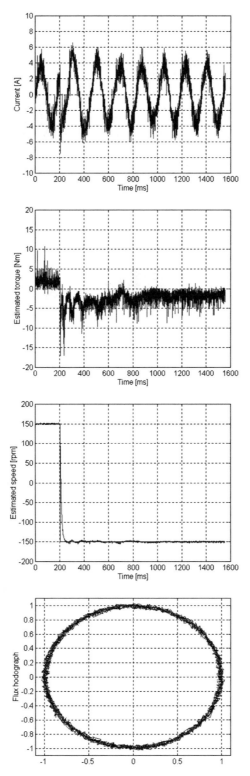

FIGURE 9.38
DTFC speed reversal transients at 150 rpm: test results.

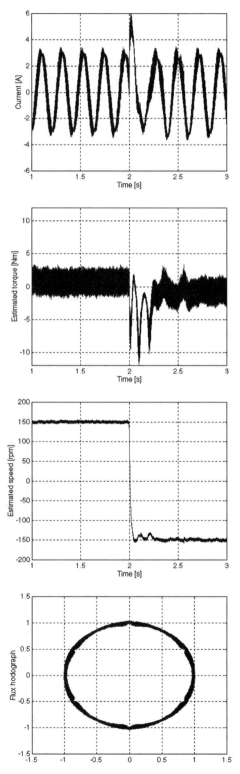

FIGURE 9.39
DTFC speed reversal transients at 150 rpm: simulation results.

With the combined DTFC-SVM strategy, low-torque-ripple operation has been obtained with RSM.

Further improvement of the rotor position estimation is necessary in order to obtain very fast transient response with this machine.

While the maximum switching frequency with DTFC control strategy was 9 KHz, with DTFC-SVM the real commutation frequency is much higher, and the torque pulsations and current harmonics are notably reduced. The stator resistance correction may be performed as for the IM (Section 9.3.6).

9.6 DTFC OF LARGE SYNCHRONOUS MOTOR DRIVES

Large synchronous motor drives [22, Chapter 14] are now dominated by low-speed, high-power cycloconverter SM drives; high power, high-speed three-level voltage source PWM SM drives; and controlled—rectifier current—source inverter high-power SM drives.

Doubly fed IM drives with the wound rotor fed through a bidirectional power flow converter, working as a motor and generator in pump storage power plants or in limited variable speed applications, have been introduced. In an ultimate analysis these cascade IMs operate essentially as synchronous motors.

For the sake of brevity we will deal here only with the voltage source converter (cycloconverter, three-level PWM inverter, matrix converter, etc.) and, respectively, controlled rectifier current—source inverter SM drives with DTFC.

9.6.1 Voltage Source Converter SM Drives

In general such high-power drives, especially with cycloconverters for low speeds (for cement and orr mills, ship propulsion), work at unity power factor.

The flux/torque limitations and relationships for unity power factor and steady state have been discussed in Section 9.2.4.

Also we have already introduced the concept of reactive torque (9.53), T_{reactive}. This concept will allow us to operate at given reactive power (for given speed), leading or lagging, if the static power converter allows for. We might introduce a reactive torque loop in addition to the stator flux loop, and the current controller limiter, to control the field current in the machine.

With the TOS and the state estimators to be detailed later we may directly develop the DTFC block diagram in Fig. 9.40. For a large motor a speed sensor (even a position sensor) may be available as its relative costs are low. A sensorless solution is not to be ruled out either.

Notice that the field control channel has three loops. The T_{reactive} correction loop makes for various detuning effects, to provide unity (or slightly lagging) power factor when required.

The stator flux error loop, however, is the basic loop for field current control.

For flux/torque/speed coordination there are two main options. One is the flux/speed option for applications when the load torque is only a function of speed (pump applications) while the flux/torque option is suitable for various load torque/time perturbations (ship propulsion, etc.).

Concerning the state observer, there are many competing solutions. The combined voltage–current model observer for flux is similar to that for the PMSM where the PM flux $\Psi_{\text{PM}d}$ is replaced by $L_{dm}i_F$ (with i_F measured) (Fig. 9.41). The rotor position $\hat{\theta}_{er}$ may be either measured or estimated. Also the speed may be measured or estimated.

Here we introduce a simpler configuration suitable for a light start (pumplike applications) (Fig. 9.41).

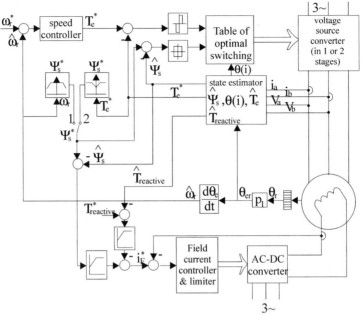

FIGURE 9.40
DTFC of voltage converter SM drives.

Figure 9.41 reveals basically a voltage model enhanced with information from the reference flux transformed into rotor coordinates to reduce the stator flux estimation transients. A similar method has been used succesfully for IM drives. The stator resistance correction may be approached as in Fig. 9.26 for the IM.

The table of optimal switchings (TOS) may be similar to the one used before for the IM, PMSM, and RSM. Vector control in such cases is known to need measures to preserve machine stability while the field current changes rather slowly according to the need to yield unity power factor.

The reactive torque concept with DTFC seems to produce fast transient response avoiding such problems as implicitly the TOS compensates for the upcoming i_F current contribution to stator flux.

Some implementation schemes on small power prototypes [18, 19, 37] showed promising results. A large power implementation is still due.

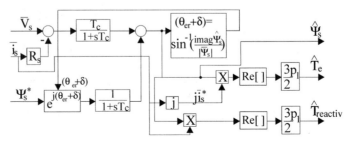

FIGURE 9.41
Flux $\hat{\Psi}_s$, torque $\hat{T}_e$, and reactive torque $\hat{T}_{reactive}$, observer for light starting.

9.6 DTFC OF LARGE SYNCHRONOUS MOTOR DRIVES 345

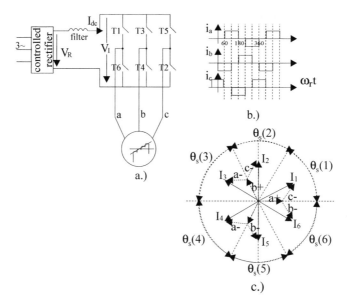

FIGURE 9.42
The current source inverter ac drive. (a) The converter: motor block diagram. (b) Ideal stator phase current waveforms. (c) The 6 nonzero current vectors.

9.6.2 The Current Source Inverter SM Drive

Traditionally thyristor current source inverters with emf commutation, paired with a source side controlled rectifier and an induction filter, are being used for many high-speed high-power SM drives (Fig. 9.42a). The main reason for its success is the converter costs, which are rather low. Essentially a six-leg current source inverter produces ideal 120° wide rectangular constant currents of alternate polarity in the machine stator winding (Fig. 9.42b).

The dc link current is conveniently flowed through two successive phases to produce six nonzero current vectors 60° apart (Fig. 9.42c).

It is also possible to deviate the dc current for a short interval through one of the three legs T1T6, T3T4, T5T2 when the phase currents are all zero. Current notches are thus created. The level of the dc line current is changed through the phase delay angle in the rectifier.

In reality the phase current half-waves are trapezoidal with some overlapping angle u, when all three phases are in conduction (Fig. 9.43).

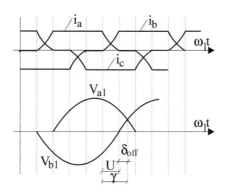

FIGURE 9.43
The commutation process.

It is evident from Fig. 9.43 that in order to provide negative line voltage on the just turned off thyristor T1 (i_a off) and positive along the turning on thyristor T3 (i_b on) with phase C in conduction (T2 on), the current i_{a1} fundamental should lead the phase voltage V_a. Thus leading power factor is required for machine commutation of thyristors. Moreover, the overlapping angle u increases with the dc line current and thus, for safe commutation, the safety angle δ_{off} in Fig. 9.43 should be above a certain value $\delta_{off} \geq (5-6°)$ for the highest value of commutated current.

As shown in [22] (p. 391), the following relationship is valid during commutation:

$$V_1\sqrt{6}[\cos(\delta - u) - \cos\delta] = 2L_c I_{dc}\omega_r \qquad (9.118)$$

where L_c is the commutation inductance: $L_c = (L_d'' + L_q'')/2$. The smaller L_c, the larger the maximum commutable current I_{dc} with $\delta_{off} < (5-6°)$.

In general, a constant leading power factor angle $\varphi_1 = (6-9)°$ suffices to provide safe commutation up to 150% rated current if a strong damper cage on the rotor provides for a low commutation inductance:

$$\varphi_1 \approx \gamma - u/2. \qquad (9.119)$$

The average inverter voltage V_{Iav} (along the dc link side) V_{Iav} is ([22], p. 392):

$$V_{Iav} \approx \frac{3}{\pi}(V_1\sqrt{6}\cos\gamma - L_c I_{dc}\omega_r). \qquad (9.120)$$

Neglecting the inverter and motor losses we may write

$$V_{Iav} I_{dc} = T_{eav}\omega_r \qquad (9.121)$$

So we may calculate the average ideal torque T_{eav} from (9.121). Finally between the rectifier V_R and inverter voltage V_I we have the relationship:

$$V_R - (R_F + sL_F)I_{dc} = V_F. \qquad (9.122)$$

For ideal no-load $V_1 = E_1$, $u = 0$, $\gamma = \gamma_0$ and

$$V_1 = L_{dm} i_F \omega_{r0}/\sqrt{2} \qquad (9.123)$$

From (9.120)–(9.123) and $I_{dc} = 0$,

$$\omega_{r0} = \frac{V_{Iav}\pi}{3\sqrt{3}L_{dm}I_F}. \qquad (9.124)$$

The rectifier voltage V_R is

$$V_R = \frac{3V_L\sqrt{2}}{\pi}\cos\alpha; \qquad V_R = V_1 + R_F I_{dc}. \qquad (9.125)$$

α is the phase delay angle in the rectifier.

The ideal no-load speed depends thus on the field current i_F and on the rectifier voltage $V_{Rav} = V_{Iav}$.

As the current vector jumps 60° we have to rely on kind of traveling flux, as the stator flux is "bumpy" because of the trapezoidal current shape.

The subtransient flux Ψ'' is defined as:

$$\bar{\Psi}'' = \bar{\Psi}_s - L_c \bar{i}_s \qquad (9.126)$$

9.6 DTFC OF LARGE SYNCHRONOUS MOTOR DRIVES

and fulfills this condition. It stems from the voltage behind subtransient inductance. The torque T_e is

$$T_e = \frac{3}{2}p_1 \text{Re}[j\bar{\Psi}''\bar{i}_s^*] = \frac{3}{2}p_1\bar{\Psi}'' i_s \sin\theta_i. \qquad (9.127)$$

Now we may notice that for the subtransient flux vector in the first sector we may use the current vector $\bar{I}_2$ for torque increasing in the positive (motion) direction. Also a combination of $\bar{I}_1$, $\bar{I}_2$, $\bar{I}_0$ ($\bar{I}_0'$ or $\bar{I}_0''$) may be used for the scope, as long as the flux/current angle θ_i increases. In general we use here the TOS based only on the flux position and torque error.

The flux error will be used to trigger the rectifier voltage modification. However, for negative torque we have to use values of α (phase delay angle in the rectifier) greater than 90° to send the energy back to the grid.

Finally the concept of reactive $T_{\text{reactive}} < 0$ loop can also be used to control the field current for a rather constant power factor angle.

The basic block diagram for the DTFC is shown in Fig. 9.44. Notice that the rectifier voltage reference is proportional to speed to secure rather fast dc link current response at all speeds.

Also, the reactive torque loop solely controls the field current for constant power factor angle. The rotor position (speed) may be measured or estimated (in sensorless configurations). The state observer may be similar to those shown in Fig. 9.28 or Fig. 9.41 if Eq. (9.126) is added. Other solutions are to be tried.

The presence of current notches ($\bar{I}_0$, $\bar{I}_0'$, $\bar{I}_0''$) leads to controlled torque pulsations. For starting there is not enough emf to secure machine commutation. A separate capacitor commutator in the dc link may be provided below 5% of rated speed. Alternatively, supply commutation for start can be done through switching the rectifier to inverter mode (from $\alpha < 90°$ to $\alpha > 90°$ and back) after which current vector is turned off completely. The new current vector is then turned on

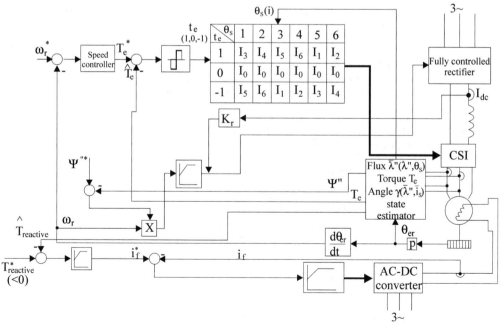

FIGURE 9.44
Basic DTFC for CSI-SM drive.

through triggering both thyristors. When a rather strong power grid is available such a simple starting commutation proves very efficient in terms of both performance and costs.

9.7 CONCLUSION

In this chapter we have introduced in some detail the DTFC of various ac motor drives. Direct torque and stator flux control (through hysteresis controllers and direct triggering of voltage source inverter voltage vectors, for given stator flux position) shows fast response in torque and notable robustness to machine parameter detuning. At low speeds, more elaborate flux and torque SVM controllers provide low torque pulsations and lower current and flux ripples shown through two rather complete case studies on the IM and RSM, respectively. The case studies revealed very good performance over a wide range of speeds and loads. Flux/torque/speed coordination is discussed in some depth according to various optimization criteria both for IMs and various SMs. Notable attention to flux/torque/speed/position observers led to some unitary configuration similar for both IMs and SMs.

A special section is dedicated to large SMs. The DTFC for the CSI-SM drives poses special problems as it handles current vectors instead of voltage vectors and the subtransient flux rather than stator flux variable is controlled.

As of now DTFC IM drives are on the market from 2 to 2000 kW units. The DTFC SM drives are still in the laboratory stages, but they might reach the market soon. Given the DTFC definite merits, even when compared with very advanced vector control ac drives, and their high degree of generality, real universal DTFC ac drives are expected in the near future for wide ranges of speed and power.

REFERENCES

[1] F. Blaschke, The method of field orientation for induction machine control, *Siemens Forsch. Entwicklungsber.* **1** (German), 184–193 (1972).
[2] Z. Krzeminski, Nonlinear control of induction motors. Proc. 10th IFAC World Congress, Munich 1987, pp. 349–354.
[3] I. Takahashi and T. Noguchi, A new quick response and high efficiency control strategy of an induction motor. Rec. IEEE—IAS, 1995; *IEEE Trans.* **IA-22**, 820–827 (1986).
[4] M. Depenbrock, Direct selfcontrol (DSC) of inverter-fed induction machine. *IEEE Trans.* **PE-3**, 420–429 (1988).
[5] I. Boldea and S. A. Nasar, Torque vector control. A class of fast and robust torque, speed and position digital controllers for electric drives. *EMPS* **15**, 135–147 (1988).
[6] P. Tiiten, P. Pohjalainen, and J. Lalu, Next generation motion control method: Direct torque control (DTC). *EPE J.* **5**, (1995).
[7] I. Boldea and A. Trica, Torque vector controlled (TVC) voltage-fed induction motor drives—very low speed performance via sliding mode control. Rec. ICEM 1990, Vol. III, pp. 1212–1217.
[8] T. G. Habetler and D. M. Divan, Control strategies for direct torque control using discrete pulse modulation. *IEEE Trans.* **IA-27**, 893–901 (1991).
[9] H. Y. Zhong, H. P. Messinger, and M. H. Rashid, A new microcomputer-based direct torque control system for three-phase induction motor. *IEEE Trans.* **IA-27**, 294–298 (1991).
[10] I. Boldea, Z. X. Fu, and S. A. Nasar, Torque vector control (TVC) of axially laminated anisotropic (ALA) rotor reluctance synchronous motors. *EMPS* **19**, 381–398 (1991).
[11] I. Boldea, Torque vector control of ac drives. Rec. of PCIM 1992, Europe, April 1992, Vol. IM, pp. 20–25.
[12] M. P. Kazmierkowski and A. Kasprowicz, Improved direct torque flux control of PWM inverter-fed induction motor drive. *IEEE Trans.* **IE- 42**, 344–350 (1995).
[13] D. Casadei, G. Serra, and A. Tani, Constant frequency operation of DTC induction motor drive for electric vehicle. Proc. of ICEM 1996, Vol. 3, pp. 224–229.

[14] M. P. Kazmierkowski, Control philosophies of PWM inverter-fed induction motors. Proc. of IECON 1997, pp. 16–26.
[15] Ch. Lochot, X. Roboam, and P. Maussion, A new direct torque control strategy with constant switching frequency operation. Rec. of EPE 1995, pp. 2431–2436.
[16] M. F. Rahman, L. Zhong, M. A. Rahman, and K. Q. Liu, Voltage switching strategies for the direct torque control of interior magnet synchronous motor drives. Rec. of ICEM 1998, pp. 1385–1389.
[17] S. K. Jackson and W. J. Kamper, Position sensorless control of medium power traction reluctance synchronous machine. Rec. of ICEM 1998, pp. 2208–2211.
[18] J. Pyrhonen, J. Kaukonen, M. Nionela, J. Jyrhonen, and J. Lunkko, Salient pole synchronous motor excitation and stability control in direct torque control driver. Rec. of ICEM 1998, pp. 83–87.
[19] M. R. Zolghadri and D. Roye, Sensorless direct torque control of synchronous motor drives. Rec. of ICEM, 1998, pp. 1385–1389.
[20] W. Leonhard, *Control of Electric Drives*, 2nd ed. Springer Verlag, New York, 1995.
[21] P. Vas, *Sensorless Vector and Direct Torque Control*. Oxford University Press, Oxford, 1998.
[22] I. Boldea and S. A. Nasar, *Electric Drives*. CRC Press, Boca Raton, FL, 1998.
[23] J. Maes and J. Melkebeek, Discrete direct torque control of induction motors using back e.m.f. measurements. Rec. of IEEE—IAS, 1998, Vol. 1., pp. 407–414.
[24] B. S. Lee and R. Krishnan, Adaptive stator resistance compensation for high performance direct torque controlled induction motor drives. Rec. of IEEE—IAS, 1998, Vol. 1, pp. 423–430.
[25] J. K. Kong and S. K. Sul, Torque ripple minimization strategy for direct torque control of induction motor. Rec. of IEEE—IAS, 1998, Vol. 1, pp. 438–443.
[26] P. Mutschler and E. Flach, Digital implementation of predictive direct control algorithms for induction motors. Rec. of IEEE—IAS, 1998, Vol. 1, 444–451.
[27] C. Lascu, I. Boldea, and F. Blaabjerg, A modified direct torque control (DTC) for induction motor sensorless drive. Rec. of IEEE—IAS, 1998, Vol. 1, pp. 415–422.
[28] I. Boldea, L. Janosi, and F. Blaabjerg, A modified DTC of reluctance synchronous motor sensorless drive. *EMPS J.* **28**, 115–128 (2000).
[29] D. Casadei, G. Sera, and A. Tani, Stator flux vector control for high performance induction motor drives using space vector modulation. Proc. of OPTIM '96, Brasov, Romania, pp. 1413–1420.
[30] P. C. Jansen and R. D. Lorenz, A physically insightful approach to the design and accuracy assessment of flux observers for field oriented IM drive. *IEEE Trans.* **IA-30**, 101–110 (1994).
[31] H. Taima and Y. Hori, Speed sensorless field oriented control of the IM. *IEEE Trans.* **29**, 175–180 (1993).
[32] G. D. Andreescu, Nonlinear observer for position and speed sensorless control of PM-SM drives. Rec. of OPTIM, 1998, Vol. II, pp. 473–482.
[33] G. D. Andreescu, Robust direct torque vector control system with stator flux observer for PM-SM drives. Rec. of OPTIM, 1996.
[34] I. Boldea, *Reluctance Synchronous Machines and Drives*. Oxford University Press, Oxford, 1996.
[35] L. Harnefors, Speed estimation from noisy resolver signals. *Power Electronics and Variable Speed Drives*. IEE Conference Publication, 1996, pp. 279–282.
[36] H. Stemmler, High power industrial drives. *Proc. IEEE* **82**, 1266–1286 (1994).
[37] J. Kaukonen *et al.*, Salient pole synchronous motor saturation in a direct torque controlled drive. Rec. of ICEM, 1998, Istanbul, Turkey, Vol. 3., pp. 1397–1401.

CHAPTER 10

Neural Networks and Fuzzy Logic Control in Power Electronics

MARIAN P. KAZMIERKOWSKI
Warsaw University of Technology, Warsaw, Poland

10.1 OVERVIEW

Similarly to the Industrial Revolution (around 1760), which replaced human muscle power with the machine, artificial intelligence (AI) aims to replace human intelligence with the computer (machine). In spite of the fact that the term AI was introduced around 1956, there is no standard definition. From the Oxford English Dictionary we can conclude:

artificial: "produced by human art or effect rather than occurring naturally"
intelligence: "the faculty of reasoning, knowing and thinking of a person or an animal"
artificial intelligence: "the application of computers to areas normally regarded as requiring human intelligence"

A broader definition, "the study of making computers do things that the human needs intelligence to do," includes mimicking human thought processes and also the technologies that make computers accomplish intelligent tasks even if they do not necessarily simulate human thought processes.

Basically, there are three fundamentally different groups of AI [1]:

- *Classical symbolic AI*: knowledge-based (expert) system, logical reasoning, search techniques, natural language processing
- *Biological model-based AI*: neural networks, genetic algorithms (also known as evolutionary computing)
- *Modern AI*: fuzzy and rough sets theory, chaotic systems

352 CHAPTER 10 / NEURAL NETWORKS AND FUZZY LOGIC CONTROL

Sometimes the areas of neural networks, genetic algorithms, fuzzy systems, rough sets, and chaotic systems are commonly refered to as *soft computing*, to stress an approximate computation (in contrast to precise computation, which is called *hard computing*) [1].

In the area of power electronics and drives, mostly the new AI tools such as neural networks, fuzzy logic, and neuro-fuzzy systems are widely studied. The main groups of applications can be listed as follows:

- *State variable estimation*: stator and rotor flux vector of induction motor [2–12], mechanical speed [13–19]
- *Parameter estimation*: stator and rotor resistance [5, 18, 20, 21], leakage inductance [12, 22, 23]
- *PWM strategies*: open-loop [24, 25] and closed-loop current control [4, 12, 22, 25–35]
- *Closed-loop ac motor control*: field oriented control [4, 8, 15, 18, 36, 27], direct torque control [4, 21, 38, 39]
- *Closed-loop PWM boost rectifier control*: voltage oriented control [40], instantaneous active and reactive power control [42]
- *Fault detection/prediction (not discussed in this chapter)*: converter [23, 43], motor [41, 44–46]

In the first part of this chapter (Sections 10.2–10.4) the basics of neural networks, fuzzy, and neuro-fuzzy systems will be reviewed. Afterwards, in Sections 10.5–10.9, we will present the number of examples illustrating application of AI for estimation and control in power electronics and drives.

10.2 BASICS OF ARTIFICIAL NEURAL NETWORKS

Artificial neural networks (ANN) have several important characteristics that are of interest to control and power electronics engineers:

- *Modeling*: Because of their ability to be trained using data records for the particular system of interest
- *Nonlinear systems*: The nonlinear networks have the ability to learn nonlinear relationships
- *Multivariable systems*: Artificial neural networks, by their nature, have many inputs and many outputs and so can be easily applied to multivariable systems
- *Parallel structure*: This feature implies very fast parallel processing, fault tolerance, and robustness

10.2.1 Artificial Neuron Model

The elementary computational elements that create neural networks have many inputs and only one output (Fig. 10.1). These elements are inspired by biological neuron systems and, therefore, are called *neurons* (or by analogy with directed graphs, *nodes*).

The individual inputs x_j weighted by elements w_j are summed to form the weighted output signal:

$$e = \sum_{j=0}^{N} w_j \cdot x_j \qquad (10.1)$$

10.2 BASICS OF ARTIFICIAL NEURAL NETWORKS

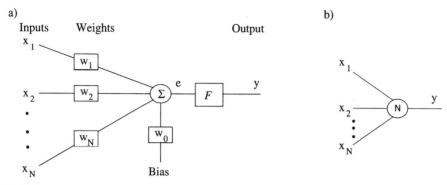

FIGURE 10.1
Neuron model: (a) full diagram; (b) symbol.

and

$$x_0 = 1 \tag{10.2}$$

where elements w_j are called *synapse weights* and can be modified during the learning process. The output of the neuron unit is defined as follows:

$$y = F(e). \tag{10.3}$$

Note that w_0 is adjustable *bias* and F is the *activation function* (also called transfer function). Thus, the output, y, is obtained by summing the weighted inputs and passing the results through a nonlinear (or linear) activation function F. The activation function F maps, a weighted sum's e (possibly) infinite domain into a specified range. Although the number of F functions is possibly infinite, five types are regularly applied in the majority of ANN: linear, step, bipolar, sigmoid, hyperbolic tangent. With the exception of the linear F function, all of these functions introduce a nonlinearity in the network by bounding the output within a fixed range. In the next section some examples of commonly used activation functions are briefly presented.

10.2.1.1 Activation Functions. The linear F function (Fig. 10.2) produces a linearly modulated output from the input e as described by

$$F(e) = \xi e \tag{10.4}$$

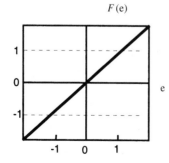 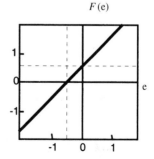

FIGURE 10.2
Linear activation function.

where e ranges over the real numbers and ξ is a positive scalar. If $\xi = 1$, it is equivalent to removing the F function completely. In this case,

$$y = \sum_{j=0}^{N} w_j \cdot x_j. \tag{10.5}$$

The step F function (Fig. 10.3a) produces only two outputs—typically, a binary value in response to the sign of the input, emitting $+1$ if e is positive and 0 if it is not. This function can be described as

$$F(e) = \begin{cases} 1 & \text{if } e \geq 0 \\ 0 & \text{otherwise} \end{cases} \tag{10.6}$$

One small variation of Eq. (10.6) is the bipolar F function (see Fig. 10.3b)

$$F(e) = \begin{cases} 1 & \text{if } e \geq 0 \\ -1 & \text{otherwise} \end{cases} \tag{10.7}$$

which replaces the 0 output value with $a = -1$.

The sigmoid F function is a continuous, bounded, monotonic, nondecreasing function that provides a graded, nonlinear response within a prespecified range. The most common function is the logistic function:

$$F(e) = \frac{1}{1 + \exp(-\beta e)} \tag{10.8}$$

where $\beta > 0$ (usually $\beta = 1$), which provides an output value from 0 to 1.

The alternative to the logistic sigmoid function is the *hyperbolic tangent*,

$$F(e) = \tanh(\beta e), \tag{10.9}$$

which ranges from -1 to 1.

10.2.2 ANN Topologies

In the biological brain, a large number of neurons are interconnected to form the network and perform advanced intelligent activities. An artificial neural network is built by neuron models and in most cases consists of neuron layers interconnected by weighted connections. The arrangement of the neurons, connections, and patterns into a neural network is referred to as a *topology* (or *architecture*).

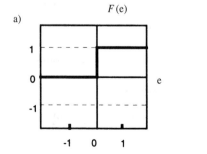
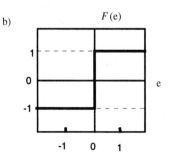

FIGURE 10.3
Step activation function.

10.2.2.1 The Layer of Neurons. Neural networks are organized into *layers* of neurons. Within a layer, neurons are similar in two respects:

- The connections that feed the layer of neurons are from the same source.
- The neurons in each layer utilize the same type of connections and activation F function.

A one-layer network with N inputs and M neurons is shown in Fig. 10.4. In this topology, each element of the input vector X is connected to each neuron input through the weight matrix W. The sum of the appropriate weighted network inputs $W*X$ is the argument of the activation F function. Finally, the neuron layer outputs form a column vector Y. Note that it is common for the number of inputs to be different from the number of neurons, i.e., $N \neq M$.

10.2.2.2 Linear Filter. For the linear activation F function, with $\xi = 1$, the output of the neuron layer can be described by the matrix equation

$$Y = W_k X \tag{10.10}$$

where

$$W_k = \begin{bmatrix} w_{11} & w_{21} & \cdots & w_{N1} \\ \vdots & \vdots & & \vdots \\ w_{1M} & w_{2M} & \cdots & w_{NM} \end{bmatrix} \tag{10.11}$$

is the weight matrix.

Such a simple network can recognize M different classes of patterns. The matrix W_k defines linear transformation of the input signals $X \in \Re^N$ into output signals $Y \in \Re^M$. This linear transformation can have an arbitrary form (for example, Fourier transform). Therefore, such a network can be viewed as a *linear filter*.

10.2.2.3 Multilayer Neural Networks (MNN). A neural network can have several layers. There are two types of connections applied in MNN:

- *Intralayer connections* are connections between neurons in the same layer.
- *Interlayer connections* are connections between neurons in different layers.

It is possible to build ANN that consist of one or both types of connections.

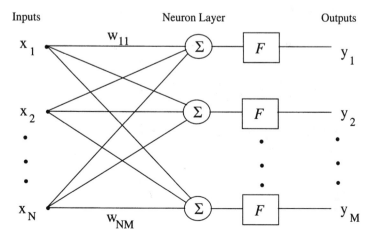

FIGURE 10.4
One-layer network.

Organization of the MNN is classified largely into two types:

- Feedforward networks
- Feedback (also called recurrent) networks.

When the MNN has connections that feed information in only one direction (e.g., input to output) without any feedback pathways in the network, it is a *feedforward MNN*. But if the network has any feedback paths, where feedback is defined as any path through the network that would allow the same neurons to be visited twice, then it is called a *feedback MNN*.

An example of a multilayer feedforward network is shown in Fig. 10.5. Each layer has a weight matrix $W_k^{(l)}$, a weighted input $E^{(l)}$, and an output vector $Y^{(l)}$, where l is the layer number. The layers of a multilayer ANN play different roles. Layers whose output is the network output are called *output layers*. All other layers are called *hidden layers*. In many publications an additional layer called the *input layer* is introduced. This layer consists of an input vector to the whole MNN (in this layer the input vector is equal to the output vector). Note that the output of each layer is the input of the next one.

Feedback ANN has all possible connections between neurons. Some of the weight can be set to zero to create layers within the feedback network if that is desired. The feedback networks are quite powerful because they are sequential rather than combinational like the feedforward networks. The output of such networks, because of the existing feedback, can either oscillate or converge.

Finally, note that the multilayer linear neural network is equivalent to a neural network with one layer. So, it is senseless to use linear ANN with multiple layers.

10.2.3 Learning and Training of Feedforward ANN

10.2.3.1 Introduction. One of the most important qualities of ANN is their ability to learn. Learning is defined as a change of connection weight values that result in the capture of information that can later be recalled. Several algorithms are available for a learning process. Generally, the learning methods can be classified into two categories:

- *Supervised learning:* a process that incorporates an external teacher and (or) global information (Fig. 10.6). The supervised learning algorithms include error correction learning, reinforcement learning, and stochastic learning.

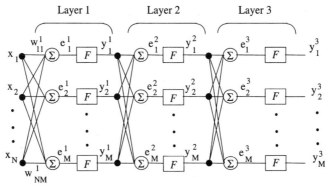

FIGURE 10.5
Multilayer feedforward ANN.

10.2 BASICS OF ARTIFICIAL NEURAL NETWORKS

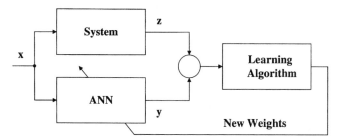

FIGURE 10.6
Supervised learning. x, input vector; z, system output (teacher) vector; y, ANN output vector.

- *Unsupervised learning* (also referred to as *self-organization*): a process that incorporates no external teacher and relies upon only local information during the entire learning process. Examples of unsupervised learning include Hebbian learning, principal component learning, differential Hebbian learning, min-max learning, and competitive learning.

Most learning techniques utilize off-line learning. When the entire pattern set is used to condition the connections prior to the use of the network, it is called *off-line learning*. For example, the backpropagation training algorithm is used to adjust connections in multilayer feedforward ANN, but it requires thousands of cycles through all the pattern pairs until the desired performance of the network is achieved. Once the network is performing adequately, the weights are stored and the resulting network is used in *recall mode* thereafter. Off-line learning systems have the inherent requirement that all the patterns have to be resident for training.

Not all networks perform off-line learning. Some networks can add new information "on the fly" nondestructively. If a new pattern needs to be incorporated into the network's connections, it can be done immediately without any loss of prior stored information. The advantage of off-line learning networks is that they usually provide superior solutions in difficult problems such as nonlinear classification, but on-line learning allows ANN to learn during the system operation.

In control and identification systems mostly the feedforward ANN are applied. Therefore, in this section only the supervised learning algorithms based on error correction for feedforward ANN will be described.

10.2.3.2 The Widrow–Hoff (Standard Delta) Learning Rule

Learning rule for one linear neuron. Let us consider the simplest case of ANN. It means that the ANN consists of one linear neuron with N inputs. We will study the supervised learning process of this network. So, it is convenient to introduce a so-called *teaching sequence*. We can define this sequence as follows:

$$T = \{(\mathbf{X}^{(1)}, z^{(1)}\}, \{\mathbf{X}^{(2)}, z^{(2)}\}, \ldots, \{\mathbf{X}^{(P)}, z^{(P)}\}\} \quad (10.12)$$

where each element $\{\mathbf{X}^{(j)}, z^{(j)}\}$ consists of the input vector X in the jth step of the learning process and the appropriate desired output signal z.

In order to show the learning algorithm, we define the error function as

$$Q = \frac{1}{2}\sum_{j=1}^{P}(z^{(j)} - y^{(j)})^2. \quad (10.13)$$

We can rewrite this equation in the form

$$Q = \sum_{j=1}^{P} Q^{(j)} \tag{10.14}$$

where

$$Q^{(j)} = \frac{1}{2}(z^{(j)} - y^{(j)})^2. \tag{10.15}$$

Since Q is a function of W, the minimum of Q can be found by using the gradient descent method:

$$\Delta w_i = -\eta \frac{\partial Q}{\partial w_i} \tag{10.16}$$

where η is a proportionality constant called the *learning rate*.

For the jth step of the learning process we can obtain

$$w_i^{(j+1)} - w_i^{(j)} = \Delta w_i = -\eta \frac{\partial Q^{(j)}}{\partial w_i} \tag{10.17}$$

and, using the chain rule,

$$\frac{\partial Q^{(j)}}{\partial w_i} = \frac{\partial Q^{(j)}}{\partial y^{(j)}} \frac{\partial y^{(j)}}{\partial w_i}. \tag{10.18}$$

The first part shows the error changes in the jth step of the learning process with the output of the neuron and the second part shows how much changing w_i changes that output. From Eq. (10.15)

$$\frac{\partial Q^{(j)}}{\partial y^{(j)}} = -(z^{(j)} - y^{(j)}) = -\delta^{(j)}. \tag{10.19}$$

Since the linear network output is defined as

$$y^{(j)} = \sum_{k-1}^{N} w_k^{(j)} \cdot x_k^{(j)} \tag{10.20}$$

then

$$\frac{\partial y^{(j)}}{\partial w_i} = x^{(j)}. \tag{10.21}$$

Substituting Eq. (10.19) and Eq. (10.21) back into Eq. (10.18), one obtains

$$-\frac{\partial Q^{(j)}}{\partial w_i} = \delta^{(j)} \cdot x^{(j)}. \tag{10.22}$$

Thus, the rule for changing weights in Eq. (10.17) is given by

$$\Delta w_i^{(j)} = \eta \delta^{(j)} \cdot x_i^{(j)}, \tag{10.23}$$

or in vector form,

$$\Delta \mathbf{W}^{(j)} = \eta \delta^{(j)} \cdot \mathbf{X}^{(j)}. \tag{10.24}$$

Finally, the algorithm for new values of the weight vector $\mathbf{W}$ can be written as

$$\mathbf{W}^{(j+1)} = \mathbf{W}^{(j)} + \eta \delta^{(j)} \cdot \mathbf{X}^{(j)}. \tag{10.25}$$

Learning rule for linear ANN. In the case of the linear ANN with M outputs and N inputs, we can introduce the learning algorithm by results generalization of the previous section. In this section the teaching sequence is defined as

$$T = \{\{\mathbf{X}^{(1)}, \mathbf{Z}^{(1)}\}, \{\mathbf{X}^{(2)}, \mathbf{Z}^{(2)}\}, \ldots, \{\mathbf{X}^{(P)}, \mathbf{Z}^{(P)}\}\} \tag{10.26}$$

where each element $\{\mathbf{X}^{(j)}, \mathbf{Z}^{(j)}\}$ consists of input vector $\mathbf{X}$ in the jth step of learning process and the appropriate desired output vector $\mathbf{Z}$.

One can obtain in analogy to Eq. (10.25) the following learning algorithm:

$$\mathbf{W}_k^{(j+1)} = \mathbf{W}_k^{(j)} + \eta(\mathbf{Z}^{(j)} - \mathbf{Y}^{(j)}) \cdot (\mathbf{X}^{(j)})^T. \tag{10.27}$$

One can define the error vector

$$\mathbf{D}^{(j)} = \mathbf{Z}^{(j)} - \mathbf{Y}^{(j)}. \tag{10.28}$$

This vector consists of the elements

$$\mathbf{D}^{(j)} = [\delta_1^{(j)}, \delta_2^{(j)}, \ldots, \delta_M^{(j)}]^T \tag{10.29}$$

where

$$\delta_i^{(j)} = z_i^{(j)} - y_i^{(j)} \tag{10.30}$$

is the difference between the desired and actual ith output in the jth step of the learning process.

Finally substituting Eq. (10.28) in Eq. (10.27), the algorithm for new values of the weight vector $\mathbf{W}$ can be rewritten as

$$\mathbf{W}_k^{(j+1)} = \mathbf{W}_k^{(j)} + \eta \mathbf{D}^{(j)} \cdot (\mathbf{X}^{(j)})^T \tag{10.31}$$

where $\mathbf{Y}^{(j)}$ is the network output vector $M \times 1$ in the jth step of the learning process, $\mathbf{Z}^{(j)}$ is the target vector $M \times 1$ in the jth step of the learning process, $\mathbf{X}^{(j)}$ is the input vector $N \times 1$ in the jth step of the learning process, $\mathbf{D}^{(j)}$ is the error vector $M \times 1$ in the jth step of the learning process, $\mathbf{W}_k^{(j+1)}$ and $\mathbf{W}_k^{(j)}$ are weights matrices $M \times N$ in the $j + l$th and jth steps of the learning process, and N is the number of neurons, M the number of outputs, P the number of learning steps, and η the learning rate. The algorithm given by Eq. (10.31) is called the *Widrow–Hoff learning rule*. Also, because the amount of learning is proportional to the difference $\mathbf{D}^{(j)}$—or delta—between the target and actual network output, the algorithm is often called the *standard delta rule* [47].

The delta rule is the basis for most applied learning algorithms. As shown, the standard delta rule essentially implements gradient descent in a sum-squared error for linear functions. In this case, *without hidden layers*, the error surface is shaped like a bowl with only one minimum, so that the gradient descent is guaranteed to find the best set of weights with hidden layers. However, it is not so obvious how to compute the derivatives, and the error surface is not concave upward, so there is the danger of getting stuck in local minima. Note that the same algorithm (Eq. (10.33)) can be used to adapt weights in a *single-layer perceptron* with nonlinearity described by Eq. (10.7) or (10.6).

Acceleration of the learning process. The algorithm described by Eq. (10.31) is a generalization of the least mean squares (LMS) algorithm using the gradient search technique. In this case the learning process is convergent, but this convergence is very slow. However, we can have an effect on some features of the learning algorithm to improve learning convergence:

Selection of the initial value of the weight matrix $\mathbf{W}_k^{(1)}$: In most cases, one can assume that the initial values of this matrix should be selected at random and be rather small. A very important condition is to *avoid the same values* of any pairs of elements.

Selection of the learning rate coefficient η: It is possible to change this value during the time of the learning process. But in the many applications, it is enough to assume a typical constant value of $\eta = 0.6$.

Modification of the standard delta rule algorithm: One can improve the learning process by adding the additional term to Eq. (10.31) proportional to momentum.

Momentum is defined by

$$\mathbf{M}^{(j)} = \mathbf{W}_k^{(j)} - \mathbf{W}_k^{(j-1)}. \tag{10.32}$$

Thus, the algorithm for new values of the weight vector can be written as

$$\mathbf{W}_k^{(j+1)} = \mathbf{W}_k^{(j)} + \eta \mathbf{D}^{(j)} \cdot (\mathbf{X}^{(j)})^T + \mu \mathbf{M}^{(j)}. \tag{10.33}$$

Good results of the learning process are obtained with $\eta = 0.9$ and $\mu = 0.6$.

10.2.3.3 The Generalized Delta Learning Rule: Backpropagation.
The generalized delta rule has been developed by Rumelheart *et al.* [47] for *learning the layered feedforward ANN with hidden layers*. First, we will derive a learning formula for one nonlinear neuron using the consideration described in the previous section. Second, the actual generalized delta learning rule will be derived.

Learning rule for one nonlinear neuron. In the case of one nonlinear neuron the output in the *j*th step can be expressed as

$$y^{(j)} = F(e^{(j)}) \tag{10.34}$$

where

$$e^{(j)} = \sum_{i=0}^{N} w_i^{(j)} \cdot x_i^{(j)} \tag{10.35}$$

and $F(\cdot)$ is the activation function.

We can use the sigmoidal activation F function, which is continuous, nondecreasing and differentiable. The derivative in Eq. (10.17) is a product of two parts:

$$\frac{\partial Q^{(j)}}{\partial w_i} = \frac{\partial Q^{(j)}}{\partial e^{(j)}} \frac{\partial e^{(j)}}{\partial w_i}. \tag{10.36}$$

From Eq. (10.35) we see that the second factor is

$$\frac{\partial e^{(j)}}{\partial w_i} = \frac{\partial}{\partial w_i} \sum_{k=1}^{N} w_k x_k = x_i. \tag{10.37}$$

The first part of Eq. (10.36) can be written as

$$\delta^{(j)} = -\frac{\partial Q^{(j)}}{\partial e^{(j)}} = -\frac{\partial Q^{(j)}}{\partial y^{(j)}} \frac{\partial y^{(j)}}{\partial e^{(j)}}. \tag{10.38}$$

By Eq. (10.34) we see that

$$\frac{\partial y^{(j)}}{\partial e^{(j)}} = \frac{d}{de} F(e) \tag{10.39}$$

is the derivative of the activation function. In the case of the sigmoid function this derivative is very easy to calculate:

$$\frac{d}{de} F(e) = y^{(j)} (1 - y^{(j)}). \tag{10.40}$$

The first factor in Eq. (10.38) we can calculate as

$$\frac{\partial Q^{(j)}}{\partial y^{(j)}} = -(z^{(j)} - y^{(j)}) \tag{10.41}$$

and substituting for the two factors in Eq. (10.38) we get

$$\delta^{(j)} = (z^{(j)} - y^{(j)}) \cdot \frac{d}{de} F(e). \tag{10.42}$$

Finally Eqs. (10.42) and (10.23) give us the description of the algorithm in the form

$$w_i^{(j+1)} = w_i^{(j)} + \eta (z^{(j)} - y^{(j)}) \cdot \frac{dF(e)}{de} x_i^{(j)}. \tag{10.43}$$

For the sigmoid activation function, we can rewrite Eq. (10.43) using Eq. (10.40) as

$$w_i^{(j+1)} = w_i^{(j)} + \eta (z^{(j)} - y^{(j)})(1 - y^{(j)}) y^{(j)} x_i^{(j)}. \tag{10.44}$$

Learning rule for nonlinear multilayer ANN. In the case of the nonlinear ANN with M outputs and N inputs, we can introduce the learning algorithm by results generalization of the previous paragraph. For the output layer this generalization is very simple:

$$w_{im}^{(j+1)(L)} = w_{im}^{(j)(L)} + \eta (z_m^{(j)} - y_m^{(j)}) \frac{dF(e)}{de} x_i^{(j)(L)} \tag{10.45}$$

where L is the output layer number, $z_m^{(j)}$, $y_m^{(j)}$ are the mth components in the jth step of the desired and actual output, respectively.

For the hidden layer l we use the chain rule to write

$$\begin{aligned}
\frac{\partial Q^{(j)}}{\partial y_m^{(j)(l)}} &= \sum_k \frac{\partial Q^{(j)}}{\partial e_k^{(j)(l)}} \frac{\partial e_k^{(j)(l)}}{\partial y_m^{(j)(l)}} \\
&= \sum_k \frac{\partial Q^{(j)}}{\partial e_k^{(j)(l)}} \frac{\partial}{\partial y_m^{(j)(l)}} \sum_i w_{ik}^{(j)(l+1)} y_i^{(j)(l+1)} \\
&= \sum_k \frac{\partial Q^{(j)}}{\partial e_k^{(j)(l)}} w_{mk}^{(j)(l+1)} = -\sum_k \delta_k^{(j)(l+1)} w_{mk}^{(j)(l)}.
\end{aligned} \tag{10.46}$$

In this case, substituting for the two parts in Eq. (10.38) yields

$$\delta_m^{(j)(l)} = \frac{dF(e)}{de} \sum_k \delta_k^{(j)(l+1)} w_{mk}^{(j)(l+1)}. \tag{10.47}$$

Using Eq. (10.47), one can obtain the algorithm to adapt weights in the hidden layer l in the jth step:

$$w_{im}^{(j+1)(l)} = w_{im}^{(j)(l)} + \eta \delta_m^{(j)(l)} x_{im}^{(j)(l)} \tag{10.48}$$

where $\delta_m^{(j)(l)}$ can be calculated from Eq. (10.46).

Equation (10.45) and Eqs. (10.47) and (10.48) give a recursive algorithm for computing the weights of the MNN. This algorithm is known as the *generalized delta rule* [47, 48].

Note that in this case we can use the same improvements as in linear ANN (for instance, learning with momentum).

10.2.3.4 The Backpropagation Training Algorithm.
The application of the generalized delta rule involves four phases:

1. The *presentation phase*: Present an input training vector and calculate each layer's output until the last layer's output is found.
2. The *check phase*: Calculate the network error vector and the sum squared error $Q^{(j)}$ for the input vector. Stop if the sum of the squared error for all P training vectors (Eqs. (10.13), (10.14), (10.15)) is less than the specified value or your specified maximum number of epochs has been reached. Otherwise continue calculations.
3. The *backpropagation phase*: Calculate delta vectors for the output layer using the target vector, then backpropagate the delta vector to preceding layers (Eq. (10.47)).
4. The *learning phase*: Calculate each layer's new weight matrix, then return to the first phase.

This training algorithm is commonly known as *error backpropagation* [47, 48].

10.2.3.5 Backpropagation ANN.
Backpropagation ANN is the common name given to multilayer feedforward ANN which are trained by the backpropagation learning algorithm described in Section 10.2.3.4. Currently, over 90% of ANN applications are BP-ANN. This popularity of BP-ANN is due to its simple topology and well-known (tested) learning algorithm.

Capabilities of BP-ANN

- Based on the nonlinearities used within neurons
- Multilayer feedforward ANN with only one hidden layer, with sufficient number of neurons; can learn to approximate any continuous (nonlinear) function [33].

Therefore, ANN is a universal nonlinear approximator. Note that it makes no sense to use linear multilayer feedforward ANN, because it can be replaced by single-layer topology.

Limitations of BP-ANN

- The topology design (number of layers and neurons) is carried out in a fairly heuristic way—based on designer experience
- The local minimum problems and slow convergence lead to very time-consuming learning (sometimes requiring thousands of epochs)

- The parallelism of ANN is not fully utilized, because the majority of ANN are simulated/implemented on sequential processors, giving rise to a very rapid increase in computation time requirements as the size of the problem expands

Some design advice
- How many hidden neurons?

    Too few, meaning that the ANN cannot solve the task?
    Too many, so it overfits the data and gives insufficient generalization?

Start small and then increase the number of hidden neurons if the ANN is unable to learn (but not before trying to change the initial conditions).

- How much data?

    Use enough data to force generalization.
    Too few data points leads to overfitting, insufficient generalization.
    High concept complexity requires more weights and data points.

- Note:

    High initial weight leads to high activation levels, then to low gradients, then to slow learning.
    Zero initial weights mean no learning.
    All equal weights in a fully connected BP-ANN mean no learning.
    Very low interference (ANN is trained to learn problem A, then B; it forgets the solution to problem A).

10.3 STRUCTURES OF NEUROMORPHIC CONTROLLERS

10.3.1 Introduction

Efforts in applying ANN to control and identification of dynamic processes have resulted in the new field of *neurocontrol*, which can be considered as a nonconventional branch of adaptive control theory. The attractiveness of neurocontrol for engineers can be explained by the following reasons:

1. In the same way that transfer functions provide a generic representation for linear black-box models, ANNs potentially provide a generic representation for nonlinear black-box models.
2. Biological nervous systems are living examples of intelligent adaptive controllers.
3. ANN are essentially adaptive systems able to learn how to perform complex tasks.
4. Neurocontrol techniques are believed to be able to overcome many difficulties that conventional adaptive techniques suffer when dealing with nonlinear plants or plants with unknown structure.

Although several ANN architectures have been applied to process control, most of the actual neurocontrol literature concentrates on multilayer feedforward neural networks (MNN). This is because of the following basic reasons:

- MNN are essentially feedforward structures in which the information flows forward, from the inputs to the outputs, through hidden layers. This characteristic is very convenient for

control engineers, used to work with systems represented by blocks with inputs and outputs clearly defined and separated.
- MNN with a minimum of one hidden layer using arbitrary sigmoidal activation functions are able to perform any nonlinear mapping between two finite-dimensional spaces to any desired degree of accuracy, provided there are enough hidden neurons. In other words, MNN are versatile mappings of arbitrary precision. In control, usually many of the blocks involved in the control system can be viewed as mappings and, therefore, can be emulated by MNN with inputs and outputs properly defined.
- The basic algorithm for learning in MNN, the backpropagation algorithm, belongs to the class of gradient methods largely applied in optimal control, and is, therefore, familiar to control engineers.

In this chapter the main structures of *neural controllers* (also called *neuromorphic controllers* [49, 65]) (NC), i.e., controllers based on an ANN structure, will be described.

10.3.2 Inverse Control and Direct Inverse Control (Off-Line)

The principle of the inverse neurocontroller is shown in Fig. 10.7. The neural network MNN learn the inverse dynamic of the plant $f^{-1}(u)$ by using an appropriate training signal (Fig. 10.7a). When training is performed the ANN's weights are fixed and the network is used as a feedforward neural controller before the plant (Fig. 10.7b) to compensate for the plant nonlinearity $f(u)$.

The controller's training signal in Fig. 10.7a provides information needed for the NC to learn the inverse dynamics of the plant in such a way that an error function J of the plant output error $e = y_r - y$ is minimized. Next, three controller training configurations are briefly presented.

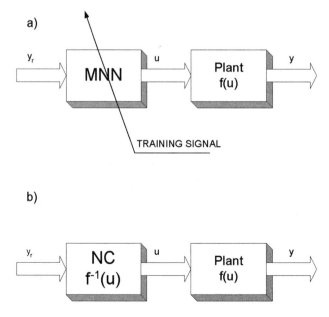

FIGURE 10.7
Inverse control: (a) training phase; (b) feedforward neural controller.

10.3 STRUCTURES OF NEUROMORPHIC CONTROLLERS

The inverse model is built up as shown in Fig. 10.8, and the plant output y for the known input u is used for input to the MNN to obtain network output u_c. The learning process of MNN is carried out to minimize the overall error e^2 between u and u_c. Therefore, this method is also called *general learning architecture* [48, 49].

The success of this method depends largely on the ability of the ANN to generalize or learn to respond correctly to inputs that were not specifically used in the training phase. In this architecture it is not possible to train the system to respond correctly in regions of interest because it is normally not known which plant inputs correspond to the desired outputs. Some improvement can be achieved by using closed loop training architecture (Fig. 10.8b) [50].

10.3.3 Direct Adaptive Control (On-Line)

To overcome the general training problems, the ANN learns during on-line feedforward control (Fig. 10.9) [49, 51]. In this method, the NC can be trained in regions of interest only since the reference value is the input signal for the ANN. The ANN is trained to find out the plant output that derives the system output y to the reference value y_r. The weights of the ANN are adjusted so that the error between the actual system output and the reference value is maximally decreased in every iteration step.

In this method, the dynamic model of the plant can be regarded as an additional layer. Consequently, it is necessary to use some prior information such as the sensitive derivatives or the Jacobian of the system in order to apply the backpropagation algorithm. The problem of

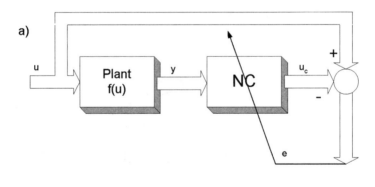

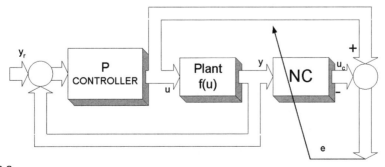

FIGURE 10.8
Direct inverse control architectures: (a) open-loop training; (b) closed-loop training.

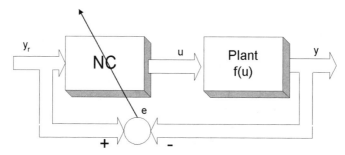

FIGURE 10.9
Direct adaptive control.

direct inverse control is that since the ANN controls the system directly by itself, the controlled system may be unstable at the first stage of learning. Therefore, it is necessary to prepare the initial value of the weights for the ANN in the controller, which may be acquired by prior off-line learning as supervised control, in order to avoid instability.

10.3.4 Feedback Error Training

In this method (Fig. 10.10), the ANN is used as a feedforward controller NC and trained by using the output of a feedback controller P as error signal. The problem with feedback error learning as indirect inverse control is that a priori knowledge must be used as input to the ANN to handle dynamics. There are problems with such knowledge in that the assumption is that plant dynamics are unknown.

10.3.5 Indirect Adaptive Control

Although in many practical cases the sensitive derivatives or the Jacobian of the system can be easily estimated or replaced by $+1$ or -1, that is, the signum of the derivative, this is not the general case. In the indirect adaptive control scheme shown in Fig. 10.11, an ANN based plant emulator NE is used to compute the sensitivity of the error function J with respect to the controller's output. Since NE is an MNN, the desired sensitivity can be easily calculated by using the backpropagation algorithm.

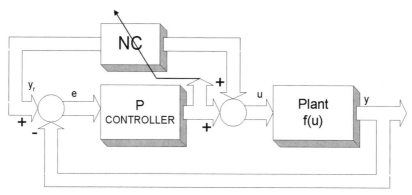

FIGURE 10.10
Feedback error learning configuration.

10.3 STRUCTURES OF NEUROMORPHIC CONTROLLERS 367

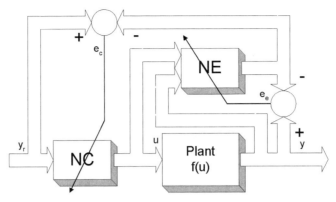

FIGURE 10.11
Indirect adaptive control.

Furthermore, the configuration in Fig. 10.11 is useful when the inverse of the plant is ill defined, i.e., the function f does not admit a true inverse.

The NE should be off-line trained with a data set sufficiently rich to allow plant identification, and then both the NC and NE are on-line trained. In a sense, the NE performs system identification and, therefore, for rapidly changing systems, it is preferred to update the NE more often than the NC.

10.3.6 Adaptive Neurocontrol

In this subsection two configurations of adaptive controllers based on ANN will be presented. In Fig. 10.12 the structure of a neural self-tuning regulator (STR) is shown. The ANN is used to identify system parameters (NI) and tune the conventional controller. In Fig. 10.13, in turn, a neural model reference adaptive control (MRAC) system is presented. In this structure both the controller NC and identifier NI use neural networks. The overall system error e_c between model

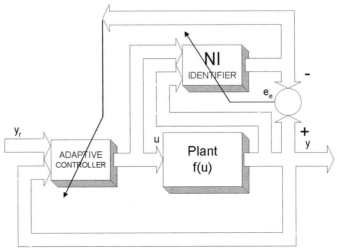

FIGURE 10.12
Neuro self-tuning regulator (STR).

368 CHAPTER 10 / NEURAL NETWORKS AND FUZZY LOGIC CONTROL

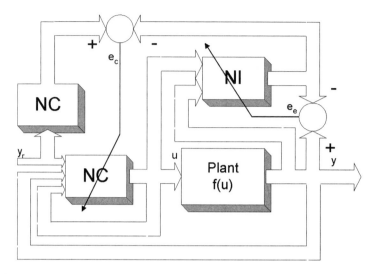

FIGURE 10.13
Example of neural MRAC system.

reference y_m and system outputs y is applied for NC tuning, while error e_e adjusts the neural network identifier NI.

10.4 BASICS OF FUZZY AND NEURO-FUZZY CONTROL

10.4.1 Fuzzy Logic Control System

In the fuzzy logic system the design is based directly on expert knowledge and is formulated in easy human language definitions, such as "*if... then...*" rules [52–54]. There are many different types of fuzzy logic controllers, but generally all of them are based on a model like that in Fig. 10.14 [55–58].

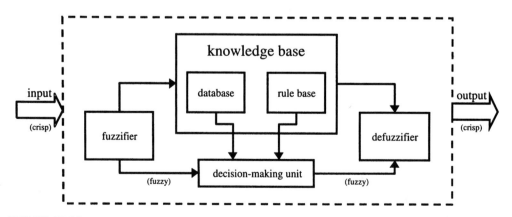

FIGURE 10.14
Fuzzy inference system.

The fuzzy system is composed of five functional blocks: fuzzifier, defuzzifier, database, rule base, and decision-making unit. The fuzzifier performs measurement of the input variable, scale mapping and fuzzification. As a result of operations in this block, degrees of matching are expressed in linguistic values.

The fuzzy set A, in not-empty universe X, can be characterized by the function μ_A, whose values are in the $[0, 1]$ partition. The function μ_A is called the fuzzy membership function. There are three most used shapes of the membership function are the following

1. Triangular:

$$\begin{aligned} &\text{If } x < c - b & &\text{then } \mu_A(x) = 0 \\ &\text{If } c - b < x < c & &\text{then } \mu_A(x) = \frac{1}{b}x + 1 \\ &\text{If } c < x < c + b & &\text{then } \mu_A(x) = -\frac{1}{b}x + 1 \\ &\text{If } x > c + b & &\text{then } \mu_A(x) = 0. \end{aligned} \qquad (10.49)$$

2. Exponential:

$$\mu_A(x) = \exp\left\{-\left[\left(\frac{x-c}{a}\right)^2\right]^b\right\}. \qquad (10.50)$$

3. Gaussian:

$$\mu_A(x) = \frac{1}{1 + \left[\left(\frac{x-c}{a}\right)^2\right]^b}, \qquad (10.51)$$

where a, b, c are the membership function parameters.

Each of the functions in the universe has its own linguistic value, i.e., NEGATIVE SMALL or POSITIVE ZERO. The user initially determines the number of the membership function and the shape. The membership functions of the fuzzy sets used in the fuzzy rules are a defined database.

The rule base contains linguistic control *if–then* rules. The rules can be set by using the experience and knowledge of an expert for the application and the control goals and next modeling the process manually or automatically.

The decision-making unit is the most important part of the fuzzy logic controller. It performs the inference operation on the rules. In general for the controllers, the linguistic rules are in the form IF–THEN. For instance, for the PI controller it takes the form: IF (e is A and ce is B) THEN (cu is C), where A, B, C are fuzzy subsets for the universe of discourse of the error (e), change of the error (ce), and change of the output (cu), respectively. The defuzzifier transforms the fuzzy results of the inference into a crisp output. There are many defuzzification methods. The most used and known is the center of gravity method. The commonly used fuzzy *if–then* rules and fuzzy reasoning mechanism are shown in Fig. 10.15.

10.4.2 Adaptive Neuro-Fuzzy Inference System

As mentioned in Section 10.4.1, fuzzy logic is well suited for dealing with ill-defined and uncertain systems. Fuzzy inference systems employ fuzzy *if–then* rules, which are very familiar

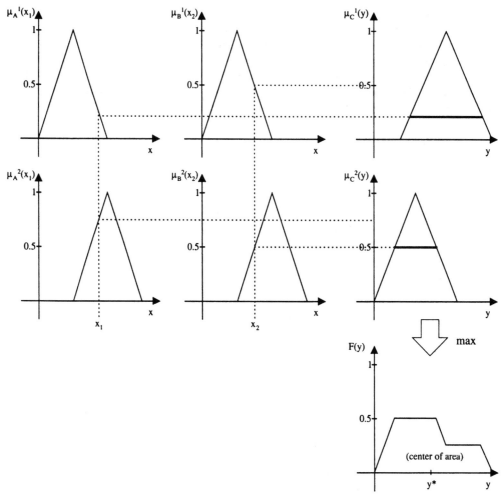

FIGURE 10.15
Commonly used fuzzy *if–then* rules and fuzzy reasoning mechanism.

to human thinking methods. It is possible to build a complete control system without using any precise quantitative analyses. However, to conceive a fuzzy controller, it is necessary to choose a lot of parameters, such as the number of membership functions in each of input and output, the shape of this function and fuzzy rules.

On the other hand, neural networks have proved theoretically and experimentally their capacity for modeling large classes of nonlinear structures. Nevertheless, it is often necessary to run quite a long learning procedure, which can be an obstacle to gaining *on-line* control of the process.

Combining both fuzzy logic and neural networks gives good advantages. Human expert knowledge can be used to build the initial structure of the regulator. Underdone parts of the structure can be improved by *on-* or *off-line* learning processes.

An adaptive neuro-fuzzy inference system (ANFIS) has been proposed for the first time in [59–60]. For simplicity in Fig. 10.16 the ANFIS is reduced to two inputs with two membership

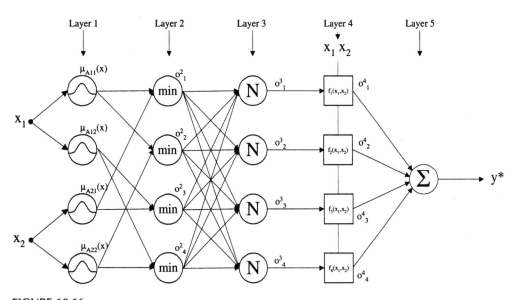

FIGURE 10.16
Two-input adaptive neuro-fuzzy inference system (ANFIS) scheme.

functions for each input and one output. For the presented structure the rule base contains four fuzzy *if–then* rules of Takagi and Sugeno's type [23], which are as follows:

$$
\begin{aligned}
&\text{Rule 1:} \quad \text{If } x_1 \text{ is } A_{11} \text{ and } x_2 \text{ is } A_{21}, \text{ then } f_1 = p_1 x_1 + q_1 x_2 + r_1 \\
&\text{Rule 2:} \quad \text{If } x_1 \text{ is } A_{11} \text{ and } x_2 \text{ is } A_{22}, \text{ then } f_2 = p_2 x_1 + q_2 x_2 + r_2 \\
&\text{Rule 3:} \quad \text{If } x_1 \text{ is } A_{12} \text{ and } x_2 \text{ is } A_{21}, \text{ then } f_3 = p_3 x_1 + q_3 x_2 + r_3 \\
&\text{Rule 4:} \quad \text{If } x_1 \text{ is } A_{12} \text{ and } x_2 \text{ is } A_{22}, \text{ then } f_4 = p_4 x_1 + q_4 x_2 + r_4
\end{aligned}
\tag{10.52}
$$

where x_1, x_2 are input values, A_{11}, A_{12}, A_{21}, A_{22} are linguistic labels, and p, q, r are consequent output function f parameters.

The ANFIS structure contains five network layers:

Layer 1: Every node in this layer contains a membership function. Usually, triangular or bell-shaped functions as in Eqs. (10.49) and (10.51) are used. The number of membership functions depends on the control object. The parameters of the functions, called premise parameters, can be tuned by the backpropagation algorithm.

The first phase generally can be written as

$$O_k^1 = \mu_{A_{ij}}(x_i), \tag{10.53}$$

where i is the input number, j is the membership function number in the ith input, k is the node number in the present layer, x_i is the input signal, O_k^1 is the first-layer output, and $\mu_{A_{ij}}(x_i)$ is the membership function. Generally the node number (K) is

$$K = IJ, \tag{10.54}$$

where I is the number of inputs, and J is the number of membership functions.

Layer 2: The second part in the ANFIS corresponds to the MIN calculation in a classical fuzzy logic system. It can be written as

$$O_k^2 = \min[\mu_{A_{ij}}(x_i), \mu_{A_{i'j'}}(x_2)], \quad (10.55)$$

where O_k^2 is the second-layer output, with condition $i \neq i'$. Not all nodes are connected together, as in the neural network classical structure. The connections are between outputs of membership functions with different inputs.

Layer 3: Every node of this layer calculates the weight, which is normalized firing strengths. The output results are in the range $[0, 1]$. It can be written as

$$O_k^3 = \frac{O_k^2}{\sum_{k=1}^{K} O_k^2}, \quad (10.56)$$

where O_k^3 is the third-layer output.

Layer 4: The fourth phase can be called the decision layer. Every node in this layer is a connection point with the node function

$$O_k^4 = O_k^3 f_k(x_1, x_2) = O_k^3 \sum_{i=1}^{I} p_{ik} x_i, \quad (10.57)$$

where O_k^4 is the fourth-layer output and p_{ik} are the consequent parameters. The linear class of functions has been chosen to simplify the learning process. The consequent parameters of the functions can be tuned by a backpropagation algorithm. Also, thanks to the linear functions, the parameters can be identified by the least square estimate. For the ANFIS of Fig. 10.16, the decision layer can be presented in a graphical example as shown in Fig. 10.17, where the numbers inside the $X_1 X_2$ surface are the decision numbers (consequent function numbers).

Layer 5: The last phase of ANFIS is the summation of all incoming signals. The result of this node creates control signal. The calculation can be written as

$$O^5 = \sum_{k=1}^{K} O_k^4, \quad (10.58)$$

where O^5 is the fifth-layer output.

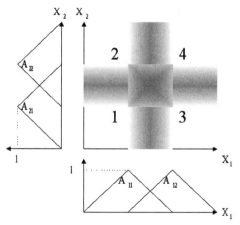

FIGURE 10.17
Graphical example of decision layer.

The presented neuro-fuzzy structure was initially tested and employed to model nonlinear functions, identify nonlinear components, predict a chaotic time series [59], and stabilize the inverted pendulum [60].

It has been shown in [61] that the adaptive neuro-fuzzy inference system can be used successfully instead of almost all neural networks or fuzzy logic-based systems. The advantages of the ANFIS structure are these:

- The human expert knowledge can be used to build the initial structure of the regulator (faster design than a pure neural network)
- The underdone parts of the structure can be improved by *on-line* or *off-line* learning processes (impossible in classical fuzzy logic-based systems).

10.5 PWM CONTROL

10.5.1 Open-Loop Space Vector Modulation

Space vector modulation (SVM) for three-phase voltage sourced converters was discussed in Chapter 4, Section 4.2. Although the DSP software implementation of the conventional SVM is simple, the application of an ANN can considerably reduce the required computation times. Thus, higher switching frequencies, higher bandwidth of the control loops, and reduced harmonics can be achieved. The BP-ANN has been successfully applied for SVM pattern generation [24, 62] in both undermodulation and overmodulation regions. In this application the ANN maps a nonlinear relation between the reference voltage vector magnitude U^* and angle α (see Fig. 4.7) and the three-phase PWM pattern generated at the output.

10.5.1.1 Direct Method. In this approach the feedforward backpropagation ANN directly replace the conventional SVM algorithm, without using any timer (Fig. 10.18). Keeping in mind that feedforward ANN can map for one input pattern onto only one output pattern, the sampling time T_s is subdivided into k intervals. Thus each subinterval T_s/k includes only one output pattern for every input pattern. The main advantages of this method, namely simple structure and fast response, are offset by the large size of training sets and very long off-line learning time.

For example, for BP-ANN topology, 2-25-3 operated in the full range of induction motor speed control (0–50 Hz), the required training data size is (for 1 µs resolution) above 1 million input/output pairs.

10.5.1.2 Indirect Method. In this approach two separate feedforward BP-ANN for amplitude and position of reference voltage vector are used. Additionally, for PWM pulse pattern generation a timer section is applied (Fig. 10.19). The amplitude network approximates the characteristic $f(U^*)$ which in overmodulation range is nonlinear (see Fig. 4.15). The digital words corresponding to ON time of power transistors w_A, w_B, w_C are calculated as follows:

$$w_A = f(U^*) \cdot g_A(\alpha) + \frac{wT_s}{4} \tag{10.59}$$

$$w_B = f(U^*) \cdot g_B(\alpha) + \frac{wT_s}{4} \tag{10.59a}$$

$$w_C = f(U^*) \cdot g_C(\alpha) + \frac{wT_s}{4} \tag{10.59b}$$

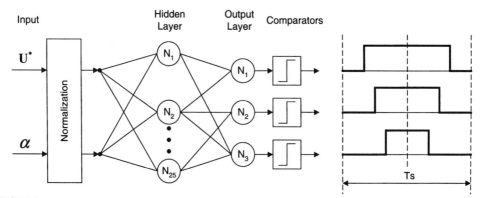

FIGURE 10.18
Direct method of ANN-based PWM signal generation.

where $f(U^*)$ is the output of the amplitude network, $g_A(\alpha)$, $g_B(\alpha)$, $g_C(\alpha)$ are the outputs of the position angle network, and $wT_s/4$ is the bias signal.

Finally, the PWM pattern S_A, S_B, S_C is generated by comparing the signal from an up/down counter with w_A, w_B, and w_C, respectively.

Typical values of training data are:

Amplitude network: 1 V increment in the range $0-\sqrt{2}V_{SN}$
Position network: 2° increment in the range 0–36°

The off-line training usually takes fewer than 5000 steps for an error below 1%.

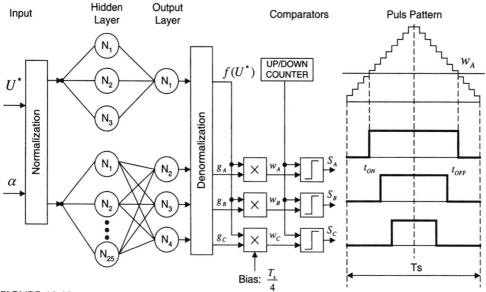

FIGURE 10.19
Indirect method of ANN-based PWM signals generation.

10.5.2 Closed-Loop PWM Current Control

10.5.2.1 On-Line Trained ANN Current Controller.

The block scheme of a digital ANN-based current controller for three-phase PWM converter is shown in Fig. 10.20. The controller operates with components defined in stator oriented coordinates α–β. Thus, the coordinate transformation is not required. The output voltages $u_{\alpha c}$, $u_{\beta c}$ are delivered to the space vector modulator, which generates control pulse S_A, S_B, S_C for power transistors of the PWM converters.

To ensure a fast response and high performance of current control, the configuration of ANN is based on linear adaptive filter topology [29]. Figure 10.21 shows the ANN controller for one component (phase A). The input of the controller is the reference current $i_{Ac}(n)$, which is sampled by delay blocks Z^{-1}, and the output is the sampled voltage command $u_{Ac}(n)$.

There are L units in the input layer and the number L is set to be the same as the sampling number in a period of the reference current so that the information on harmonics in the reference current is known to the network.

The relationship between the output and input is

$$u_{Ac}(n) = \sum i_{Ac}(n-i+1)w- = i_{Ac}(n)w_i \tag{10.60}$$

where $i = 1, 2, \ldots, L$ is the units number, $i_{Ac}(n) = \{i_{Ac}(n), \ldots, I_{Ac}(n-L+1)\}^T$ is the command current vector, and $w_i = \{w_1, \ldots, w_L\}^T$ is the weight vector.

The ANN consists of two layers:

Input layer: In this layer there are L units $V_{11}, V_{12}, \ldots, V_{1L}$. The outputs of these units are $OUT_1, OUT_2, \ldots, OUT_L$, which are connected with the output layer through the weights $w_1, w_2, \ldots, w_L$.

Output layer: This layer consists of one unit only. The inputs to this layer are outputs OUT_L from the input layer. This layer acts as a fan-out layer and hence the output of this layer is reference voltage $u_{Ac(n)}$.

The error signal user for learning of ANN can be expressed as

$$e(n) = (\chi + \delta z^{-1})(i_{Ac}(n) - i_A(n)) \tag{10.61}$$

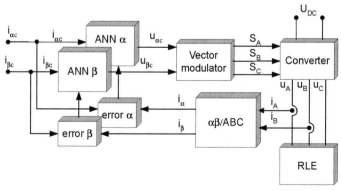

FIGURE 10.20
On-line trained ANN current controller of PWM converter.

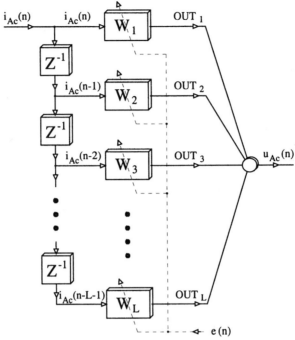

FIGURE 10.21
ANN topology for one component (phase).

where

$$\chi = \frac{\frac{R}{K}}{1-\varepsilon}, \qquad \delta = -\varepsilon, \qquad \varepsilon = \exp\left(\frac{-RT_s}{L}\right).$$

R and L are load parameters, and K is the gain of the PWM inverter.
The weight vector $w_i(n)$ are modified by the rule

$$w_i(n) = w_i(n-1) + \mu e(n) i_A(n). \tag{10.62}$$

Example 10.1: Simulink simulation of on-line ANN based current controller for PWM rectifier

ANN CR simulation results for PWM rectifier (Fig. 10.22) are shown in Fig. 10.23.
ANN parameters: 10 kHz sampling frequency, 100 levels
$U_{DC} = 600\,V$, $R_L = 0.1\,\Omega$, $L_L = 10\,mH$, $f_t = 10\,kHz$
reference current step change: 10 A to 30 A.

Line voltage (per phase)	$220\,V_{rms}$
Line frequency	50 Hz
Line inductor	$L_L = 10\,mH$, $R_L = 0.1\,\Omega$
Dc link voltage	600 V
Dc link capacitor	470 µF

10.5 PWM CONTROL 377

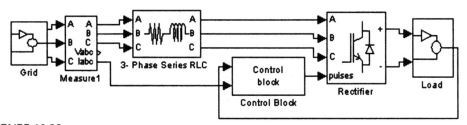

FIGURE 10.22
Simulink: simulation panel.

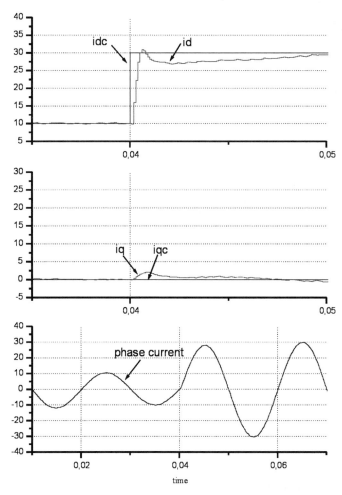

FIGURE 10.23
Simulation results.

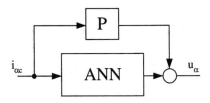

FIGURE 10.24
ANN with parallel P controller.

ANN parameters:
Sampling frequency 10 kHz
Number of levels 100 (for half period training)

The ANN (Fig. 10.21) is an on-line trained controller, and it needs time to learn the reference signal waveform. To improve transient response a proportional controller P with a control gain K_P is connected in parallel with the ANN, as shown in Fig. 10.24. This combination provides faster learning (Figs. 10.25a and 10.25b) and improves dynamic response of the controller (Figs. 10.25c and 10.25d). Learning and adaption (the fact that it can learn the reference shape) abilities

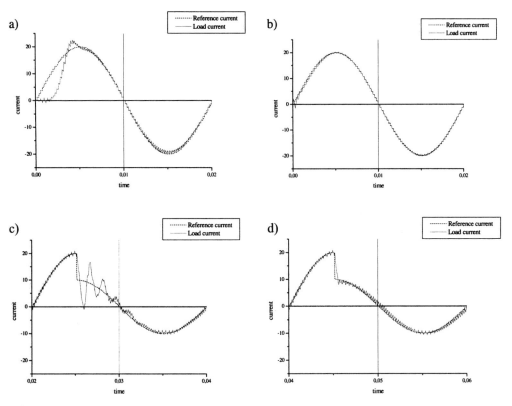

FIGURE 10.25
Learning process and response of the amplitude change of the ANN current controller ($f_{sw} = 5$ kHz) (a) and (c) ANN without proportional gain; (b) and (d) ANN with proportional gain $K_P = 7$ (learning rate $\mu = 0.01$).

are the main advantages of the on-line trained ANN current controller. However, high sampling frequency (for good reference tracking) and a time-consuming design procedure are required to ensure high-performance current control [40].

10.5.2.2 Off-Line Trained ANN Current Controller for Resonant Dc Link Converters (RDCL).
In soft-switched RDCL three-phase converters with zero voltage switching (ZVS), the commutation process is restricted to the discrete time instance, when the dc-link voltage pulses are zero [31]. Therefore, a special technique called delta modulation (DM) or pulse density modulation (PDM) is used (see Section 4.3.3). In order to develop ANN topology and generate training data, for the current controller an optimal mode discrete modulation algorithm will be discussed.

Optimal mode current control algorithm. The algorithm in every sampling interval T_s selects voltage vectors that minimize the RMS current error (see Section 4.3. as well). This is equivalent to selecting the available inverter voltage vector that lies nearest to command vector $u_{Sc}(t)$. The command vector is calculated by assuming that the inverter voltage $u_{Sc}(t)$ and EMF voltage $e(T)$ of the load are constant over the sampling interval T_s. Based on the calculated $u_{Sc}(t)$ value the voltage vector selector chooses the nearest available inverter voltage vector (Fig. 10.26). The reference voltage vector can be calculated using motor parameters r_s, l_σ, and electromotive force EMF according to

$$u_{s\alpha c} = \frac{l_\sigma}{\Delta t}(i_{s\alpha c} - i_{s\alpha}) + r_s i_{s\alpha c} + e_\alpha \tag{10.63}$$

$$u_{s\beta c} = \frac{l_\sigma}{\Delta t}(i_{s\beta c} - i_{s\beta}) + r_s i_{s\beta c} + e_\beta \tag{10.63a}$$

and

$$|u_{sc}(n)| = \sqrt{u_{s\alpha c}^2 + u_{s\beta c}^2} \tag{10.64}$$

$$\gamma_u = \arctg \frac{u_{s\alpha c}}{u_{s\beta c}}. \tag{10.64a}$$

For each inverter voltage vector (seven possibilities) the quality index J,

$$J(n) = \sqrt{(u_{s\alpha c}(n) - u_{s\alpha}(n))^2 + (u_{s\beta c}(n) - u_{s\beta}(n))^2} \tag{10.65}$$

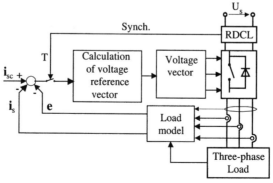

FIGURE 10.26
Optimal mode discrete modulation controller.

where n is the voltage vector number, $u_{s\alpha c}$ is the command voltage, and $u_{s\alpha}$ is the actual voltage computed, and this vector is selected to minimize the quality index.

The described procedure is repeated in each sampling time, and therefore requires very fast processors. The performance of an optimal regulator is much better than in the case of a delta regulator (see Fig. 10.28).

ANN controllers trained by optimal PWM pattern. The ANN controller, which allows for the elimination of the on-line calculations needed to implement the optimal discrete CC of Fig. 10.26, is shown in Fig. 10.27 [4, 18, 31]. The three-layer feedforward BP-ANN with sigmoidal nonlinearity—before using as a controller—were trained using a backpropagation algorithm with randomly selected data from the output pattern of the optimal controller of Fig. 10.26. In order to reduce the noise generated by PWM in signal processing instead of emf voltage, the rotor flux ψ_r components and synchronous frequency ω_s signals are applied to the ANN inputs. These signals are used in conventional vector control schemes of ac motors (see Chapter 5). After training, the performance of the three-layer (architecture: 5-10-10-3) ANN-based controllers only slightly differs from that of the optimal regulator (Fig. 10.28). Thus, the ANN-based controller can be used to regulate PWM converter output current without the need for on-line calculation required for an optimal controller.

With this approach, however, no further training of the ANN is possible during controller operation. Therefore, the performance of such an *off-line* trained ANN controller depends upon the amount and quality of training data used and is also sensitive to parameter variations. For systems where parameter variations have to be compensated, an *on-line* trained ANN controller can be applied [12, 22, 29]. In [22] an ANN induction motor CC with parameter identification was proposed. To achieve very fast on-line training (8 μs for one training cycle) a new algorithm called random weight change (RWC) is applied. This algorithm allows the motor currents to be identified and controlled within a few milliseconds.

10.5.2.3 Current Controller Based on Off-Line Trained Neural Comparator.

The instantaneous phase current error signals (ε_A, ε_B, ε_C), after scaling by factor K, are delivered to the feedforward (3-5-3 architecture) BP-ANN (Fig. 10.29). The signs of the ANN outputs are detected by the comparators, the outputs of which are sampled at a fixed rate, so that the inverter switching signals S_A, S_B, S_C are kept constant during each sampling time T_s.

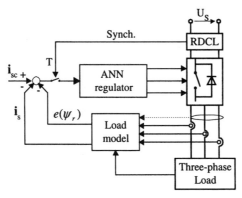

FIGURE 10.27
ANN-based discrete modulation current controller.

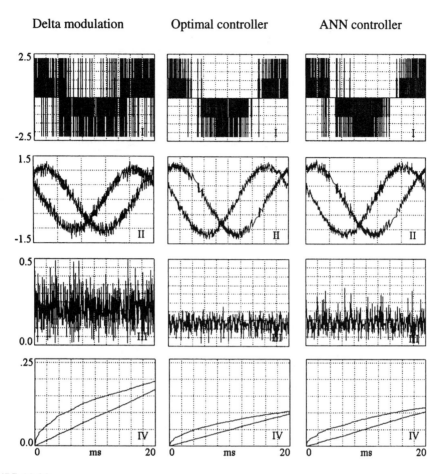

FIGURE 10.28
Current control in RDCL based on discrete modulation. I, line to line voltage u_β; II, current vector components i_α, i_β; III, current error $(\varepsilon_\alpha^2 + \varepsilon_\beta^2)^{1/2}$, IV, RMS and J of current error $\varepsilon(t)$.

ANN is trained according to the rules given in Table 10.1 using back propagation algorithm. If, for example, $\delta = 0.05$ A, the output in the phase $A(S_A)$ will be 0 for $\varepsilon_A < -0.05$ A and 1 for $\varepsilon_A > 0.05$ A. For the range $|\varepsilon_A| < 0.05$ A the output signal S_A will be unchanged.

After the learning process, when the errors between the desired output and the actual ANN output are less than a specified value (e.g., 1%), the ANN is applied as current controller. However, for the three-phase balanced load without zero leader the sum of the instantaneous current errors is always zero:

$$\varepsilon_A + \varepsilon_B + \varepsilon_C = 0. \tag{10.66}$$

Therefore, to satisfy (10.66) the current error signals must be different in polarity. As a consequence, states 1 and 8 in Table 10.1 are not valid for training and the ANN controller cannot select zero voltage vectors $V_0(000)$ or $V_7(111)$. This is when the S&H blocks in all phases are activated at the same instant. However, if the sampling is shifted $T_s/3$ (see Fig. 10.29b) the zero states can be applied (10.66). This leads to better harmonic quality of the output current generation, reduced device voltage stress by avoiding ±1 transition of the dc voltage, and instantaneous current reversal in the dc link.

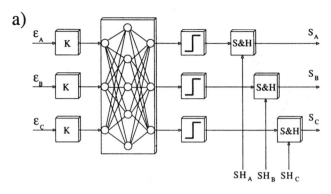

FIGURE 10.29
Neural comparator for three-phase current control: (a) block scheme, (b) modified sampling technique.

10.5.3 Fuzzy Logic-Based Current Controller

In most applications the fuzzy logic controller (FLC) is used as a substitute for the conventional PI compensator [25, 27, 32]. The block scheme of FL current control (FL-CC) (Fig. 10.30a) is similar to the system of Fig. 4.32, where instead of PI, FL self-tuned PI controllers are used. The internal block scheme of an FL-tuned discrete PI controller, including the fuzzy inference mechanism, is shown in Fig. 10.30b. The current error ε and its increment $\Delta\varepsilon$ are FL controller

Table 10.1 Learning Table for Three Input–Three Output ANN

	Input			Output vector
	ε_A	ε_B	ε_C	$\underline{V}(S_A, S_B, S_C)$
1	δ	δ	δ	$\underline{V}_7(1, 1, 1)^*$
2	$-\delta$	δ	δ	$\underline{V}_4(0, 1, 1)$
3	δ	$-\delta$	δ	$\underline{V}_6(1, 0, 1)$
4	δ	δ	$-\delta$	$\underline{V}_2(1, 1, 0)$
5	δ	$-\delta$	$-\delta$	$\underline{V}_1(1, 0, 0)$
6	$-\delta$	δ	$-\delta$	$\underline{V}_3(0, 1, 0)$
7	$-\delta$	$-\delta$	δ	$\underline{V}_5(0, 0, 1)$
8	$-\delta$	$-\delta$	$-\delta$	$\underline{V}_0(0, 0, 0)^*$

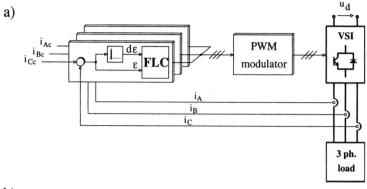

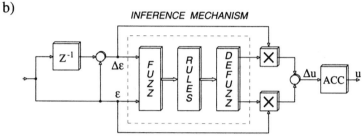

FIGURE 10.30
Fuzzy current controller. (a) Basic scheme, (b) internal structure of fuzzy logic controller for one phase.

input crisp values. The reference voltages for PWM modulator are the FL-CC crisp output commands u.

Example 10.2: Design of fuzzy logic tuned discrete current controller

An incremental form of the PI transfer function can be written as

$$\Delta u = K_P \cdot \Delta \varepsilon + \frac{1}{\tau_I} \cdot \varepsilon \tag{10.67}$$

where $\varepsilon(t)$, $u(t)$, K_P, τ_I are the input signal, the output command, the proportional gain, and the integral time constant, respectively, and

$$\Delta u = u(t_n) - u(t_{n-1})$$
$$\Delta \varepsilon = \varepsilon(t_n) - \varepsilon(t_{n-1})$$
$$\varepsilon = (t_n)$$

are given for the successive sampling instants.

The FL-based structure of Fig. 10.30b is applied for determining two components of expression (10.67), according to the operating point, which is defined by the error ε and the error change $\Delta \varepsilon$. It results in the use of four linguistic variables—two for the inputs and two for the outputs.

The fuzzy sets for linguistic input variables have been defined as follows (Figs. 10.31a and 10.31b): NB, negative big; NS, negative small; N, negative; PB, positive big; PS, positive small; P, positive; Z, zero. Similarly, for each of the two output universes the three primary output fuzzy

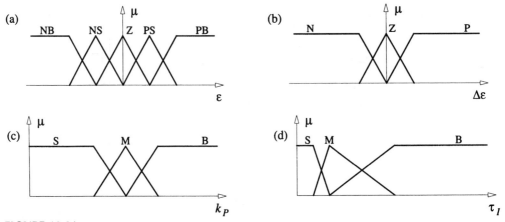

FIGURE 10.31
Definitions of input (a, b) and output (c, d) fuzzy set membership functions for the error space (a) for the error change space (b) the k_P space (c) and the τ_I space (d).

sets are defined: S, small; M, medium; B, big. All fuzzy sets are triangular. The shape of membership functions, the fuzzy sets, and their locations were chosen arbitrarily.

The rule base (Table 10.2) describing the inference mechanism consists of two groups of fuzzy control rules having the form

$$R_i: \text{ if } \varepsilon \text{ is } A_i \text{ and } \Delta\varepsilon \text{ is } B_i \text{ then } k_P \text{ is } C_i \qquad (10.68)$$

$$R_n: \text{ if } \varepsilon \text{ is } A_n \text{ and } \Delta\varepsilon \text{ is } B_n \text{ then } \tau_I \text{ is } D_n \qquad (10.68a)$$

where A, B and C, D are linguistic values of the linguistic variables ε, $\Delta\varepsilon$ and k_P, τ_I in the universe of *Error*, *Change of Error*, *Proportional Gain*, and *Integral Time*, respectively.

Table 10.2 Rule Base of Fuzzy Logic Controller Described by Eqs. (10.68) and (10.68a)

		NB	NS	Z	PS	PB
P		B	B	S	M	M
		S	S	B	M	M
Z		B	B	M	S	S
		S	S	M	S	S
N		M	M	S	B	B
		M	M	B	S	S
		$\Delta\varepsilon$	ε	k_P	τ_I	

The inference process, based on the min–max method, is described by

$$\alpha_i = \mu_{A_i}(\varepsilon) \wedge \mu_{B_i}(\Delta\varepsilon) = \min[\mu_{A_i}(\varepsilon); \mu_{B_i}(\Delta\varepsilon)] \tag{10.69}$$

$$\mu_{C'_i}(k_P) = \alpha_i \wedge \mu_{C_i}(k_P) = \min[\alpha_i; \mu_{C_i}(k_P)] \tag{10.69a}$$

$$(k_P)_i = \mathrm{COG}(C'_i) \tag{10.69b}$$

$$\alpha_n = \mu_{A_n}(\varepsilon) \wedge \mu_{B_n}(\Delta\varepsilon) \tag{10.69c}$$

$$\mu_{D'_n}(\tau_I) = \alpha_n \wedge \mu_{D_n}(\tau_I) \tag{10.69d}$$

$$(\tau_I)_n = \mathrm{COG}(D'_n) \tag{10.69e}$$

where α_i is the firing string, and COG is the center of gravity.

The output crisp values are obtained by using the COG method [54]. The final equation, determining the control action of the FL controller, is

$$\Delta u = \left\{ \sum_{i=1}^{I} \alpha_i \cdot (k_P)_i \Big/ \sum_{i=1}^{I} \alpha_i \right\} \cdot \Delta\varepsilon + \left\{ \sum_{n=1}^{N} \alpha_n \Big/ \sum_{n=1}^{N} \alpha_n \cdot (\tau_I)_n \right\} \cdot \varepsilon. \tag{10.70}$$

The FLC control surface—a graphical representation of Eqs. (10.69)—is shown in Fig. 10.32c.

When instead of conventional discrete PI, an FL controller is used as a current controller, the tracking error and transients overshoots of PWM current control can be considerably reduced (Figs. 10.32a and 10.32b). This is because—in contrast to conventional PI compensator—the control surface of the FL controller can be shaped to define appropriate sensitivity for each operating point (Fig. 10.32c and d). The FL tuned PI controller can easy be implemented as an *off-line* precalculated 3D look-up table consisting of control surface [27]. However, the properties of FL controller are very sensitive to any change of fuzzy sets shapes and overlapping.

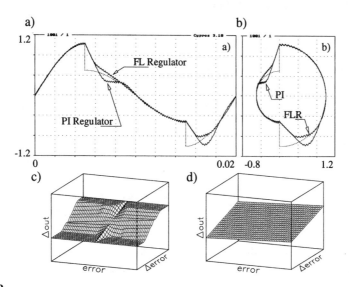

FIGURE 10.32
Comparison of current tracking performance with PI and FL controller: (a) current waveform, (b) current vector loci; (c) control surface of FL controller according to Eq. (10.70); (d) control surface of classical PI controller.

Therefore, the design procedure and achieved results depend strongly on the knowledge and expertise of the designer.

10.6 ANN-BASED INDUCTION MOTOR SPEED ESTIMATION

10.6.1 Introduction

Artificial neural networks (ANN) offer an alternative way to handle the problem of mechanical speed estimation [17]. Two kinds of approaches can be used for ANN-based speed estimation of induction motor:

- Method based on neural modeling—*on-line* trained estimation [4, 13, 14, 18, 19]
- Method based on neural identification—*off-line* trained estimation [3, 6, 7, 16, 60]

In the system, which operates on-line, the ANN is used as a model reference adaptive system (MRAS) and mechanical speed is proportional to one of the ANN weights. As the adaptive model the one-layer linear ANN is used and the estimated speed is one of the weight factors of this network. The block scheme is shown in Fig. 10.33a.

In the off-line approach a multilayer ANN is used and the motor speed is obtained as the output of the neural network. The input vector to the ANN consists of voltage u_s, current i_s, and vectors of the induction motor:

$$\mathbf{x}_{in} = [\mathbf{u}_s(k), \mathbf{u}_s(k-1), \mathbf{i}_s(k), \mathbf{i}_s(k-1), \ldots]^T. \tag{10.71}$$

The block diagram of this approach is presented in Fig. 10.33b. This approach needs an off-line training procedure of ANN. In the further part of this chapter both approaches are described in detail, evaluated, and tested in an experimental system.

10.6.2 On-Line Trained Speed Estimation Based on Neural Modeling

10.6.2.1 Neural Modeling Method Based on Rotor Flux Error.
The neural modeling method used for IM speed estimation is based on an analogy between mathematical description of a simple linear perceptron and the differential equation for rotor flux simulator of the

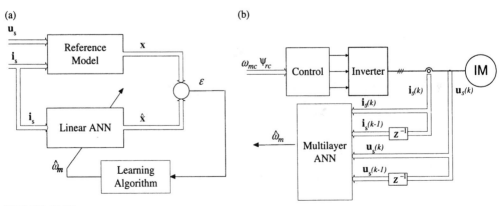

FIGURE 10.33
Block scheme of ANN-based speed estimators: (a) *on-line* estimator, (b) *off-line* estimator.

10.6 ANN-BASED INDUCTION MOTOR SPEED ESTIMATION

induction motor, written in a discrete form. This kind of speed estimation, based on the parallel neural model, was described first by Ben-Brahim and Kurosawa [13] and is shown in Fig. 10.34a.

In Fig. 10.34b a modification is presented, based on a series–parallel neural model. Both MRAS methods are similar from the point of view of estimator sensitivity to motor parameter changes, but the system of Fig. 10.33b works faster in transients [18, 19].

The concept of the neural speed estimator is based on the comparison of two rotor flux models: the first based on the voltage model, obtained from Eqs. (5.1), (5.2), and (5.4):

$$\frac{d}{dt}\underline{\Psi}_{ru} = \frac{x_r}{x_M}\left(\mathbf{u}_s - r_s\mathbf{i}_s - x_s\sigma T_N \frac{d\mathbf{i}_s}{dt}\right)\frac{1}{T_N}, \tag{10.72}$$

and the second model based on the current model, obtained from Eqs. (5.2), (5.3) and (5.4):

$$\frac{d}{dt}\underline{\Psi}_{ri} = \left[\frac{T_N}{T_r}(x_M\mathbf{i} - \underline{\Psi}_{ri}) + j\omega_m\underline{\Psi}_{ri}\right]\frac{1}{T_N} \tag{10.73}$$

where $\sigma = 1 - x_M^2/x_s x_r$, $T_r = x_r T_N/r_r$ is the rotor time constant.

For neural modeling purposes the current model can be written

$$\hat{\underline{\Psi}}_{ri}(k) = (w_i\mathbf{I} + w_2\mathbf{J})\hat{\underline{\Psi}}_{ri}(k-1) + w_3 i_s(k-1), \tag{10.73a}$$

where

$$\mathbf{I} = \begin{bmatrix} 1 & 0 \\ 0 & 1 \end{bmatrix}, \quad \mathbf{J} = \begin{bmatrix} 0 & -1 \\ 1 & 0 \end{bmatrix}.$$

This model, transformed to the stationary (α, β) coordinate system (see Chapter 5), can be treated as a simple connection of two neurons with linear activation functions, as illustrated in Fig. 10.35. One of the weights of these neurons, w_2, is the estimated rotor speed. The BP algorithm for ω_m modification is the following:

$$\hat{\omega}_m[k] = \hat{\omega}_m[k-1] - \eta(-e_\alpha\psi_{ri\beta} + e_\beta\psi_{ri\alpha})[k-1] + \mu\Delta\omega_m[k-1], \tag{10.74}$$

where

$$e_\alpha = \psi_{ri\alpha} - \psi_{ri\alpha}$$
$$e_beta = \psi_{ru\beta} - \psi_{ri\beta}.$$

η is the learning rate and μ is the momentum factor.

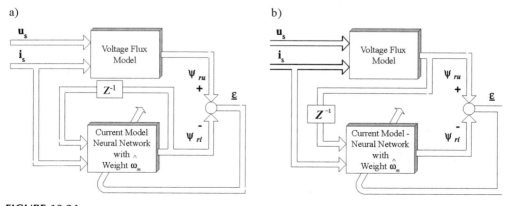

FIGURE 10.34
Speed estimator based on neural modelling method: (a) proposed in [13], (b) modified structure [19].

388   CHAPTER 10 / NEURAL NETWORKS AND FUZZY LOGIC CONTROL

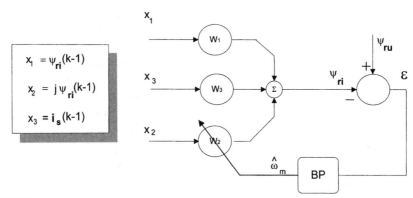

FIGURE 10.35
Structure of linear ANN used in on-line speed estimation.

This ANN identifier works very well for nominal parameters of the induction motor used in both models. In the case when rotor resistance is varying, estimation errors occur, especially in the low-speed region. Typical speed estimation errors in function of induction motor parameter changes are shown in Fig. 10.36.

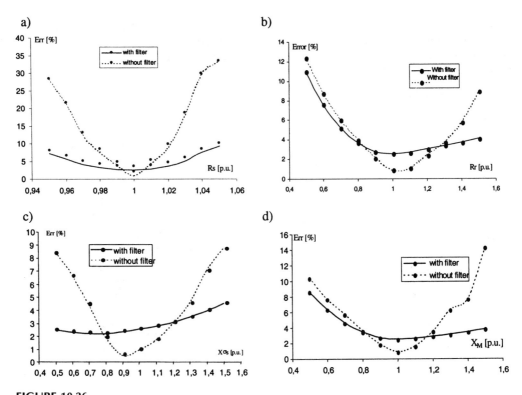

FIGURE 10.36
Average speed estimation error versus changes of (a) stator resistance, (b) rotor resistance, (c) stator leakage reactance, (d) magnetizing reactance.

The average speed estimation errors are calculated as

$$\text{Err}[\%] = \sum_{k=1}^{n} \frac{e_{\omega_k}[\%]}{n} \tag{10.75}$$

where

$$e_{\omega_k} = \frac{|\omega_{mk} - \hat{\omega}_{mk}|}{\omega_{mk}} \cdot 100\%, \tag{10.75a}$$

ω_m is the rotor speed, $\hat{\omega}_m$ is the estimated rotor speed, and n is the number of speed samples during the transient process.

It can be seen that speed estimation error increases significantly when incorrect motor parameters occur:

- Too low a rotor resistance value
- Too high a stator resistance and stator leakage reactance value
- Too low a magnetizing reactance value

Identification errors of stator leakage reactance and stator resistance strongly influence speed estimation errors, especially in the case when low-pass filters are not used (see Fig. 10.36). The best speed reconstruction was obtained for the second-order Butterworth filter with cutoff angular frequency of 300 rad/s.

The speed estimation errors depend strongly on the learning rate used in the backpropagation algorithm [16]. With sampling step T_s, the equivalent learning rate η_e is calculated in the following form: $\eta_e = \eta T_s$ (similarly $\mu_e = \mu T_s$). So, the value of learning rate η should be adapted to the sampling step. When the numerical step is very small (for example, $T_s = 20 \,\mu s$), transient estimation errors obtained during the estimation process are also very small, less than 1.5% (for nominal motor parameters used in the voltage and current models of the estimator). For such small numerical steps it is possible to apply a relatively high learning coefficient η in the BP algorithm (Eq. (10.74)), which ensures a good reconstruction of the motor speed (see Fig. 10.37a).

Generally, a lower sampling step T_s gives better speed reconstruction for lower value of equivalent learning rate η_e. However, in the DSP realization a limitation of the sampling step (due to available computation power) should be expected. The measured and estimated speed transients as well as estimation errors for nominal motor parameters and for doubled stator resistance are presented in Fig. 10.38.

However, for wrong motor parameters used in flux models, speed estimation errors occurred, similar to the case of simulation (see Fig. 10.36). Selected examples of such experimental results are presented in Figs. 10.38c and 10.38d.

10.6.2.2 Neural Modeling Method Based on Stator Current Error.
After rearranging the IM mathematical model described by Eq. (5.1)–(5.6), one obtains the equation for stator current:

$$\sigma x_s \frac{d\mathbf{i}_s}{dt} = \frac{x_M}{x_r}\left(\frac{T_N}{T_r} - j\omega_m\right)\mathbf{\psi}_r - \left(\frac{x_M^2 T_N}{x_r T_r} + r_s\right)\mathbf{i}_s + \mathbf{u}_s. \tag{10.76}$$

The rotor flux vector can be easily calculated based on Eq. (10.72). The stator current value obtained from Eq. (10.76) can be compared with its measured value and thus motor speed can be estimated with the help of the BP algorithm, similar to the scheme of Fig. 10.34a. The block diagram of the proposed modified speed estimation [18, 19] is presented in Fig. 10.39.

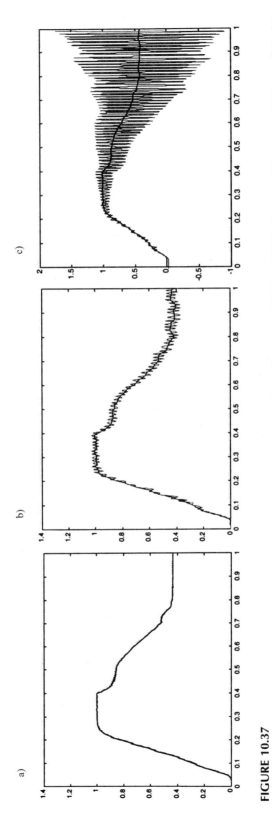

FIGURE 10.37
Rotor speed and its estimate for different sampling steps and learning rates: (a) $T_s = 20$ μs; $\eta_e = 0.04$ ($\eta = 2000$); (b) $T_s = 100$ μs, $\eta_e = 0.08$ ($\eta = 800$); (c) $T_s = 100$ μs, $\eta_e = 0.1$ ($\eta = 1000$).

10.6 ANN-BASED INDUCTION MOTOR SPEED ESTIMATION 391

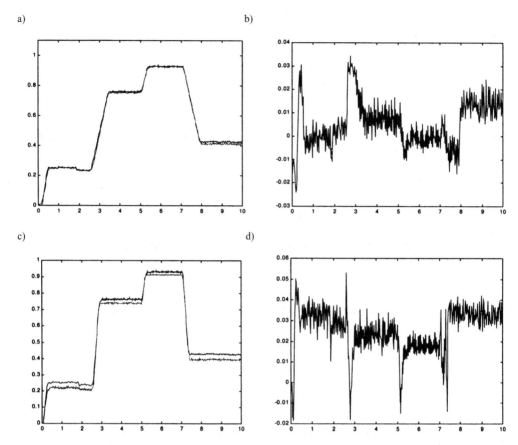

FIGURE 10.38
Experimental transients of the actual and estimated motor speed (a, c) as well as the estimation error (b, d) for nominal motor parameters (a, b) and for incorrect stator resistance $r_s = 2r_{sN}$ (c, d).

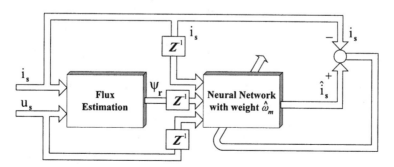

FIGURE 10.39
Block diagram of modified neural modeling based speed estimation.

Therefore, the discrete model of stator current is as follows:

$$\mathbf{i}_s(k) = \frac{x_M}{x_r \sigma x_s} \left(\frac{T_s}{T_r} - j\omega_m \frac{T_s}{T_N} \right) \mathbf{\psi}_r(k-1)$$
$$+ \left[1 - \frac{1}{\sigma x_s} \left(\frac{x_M^2 T_s}{x_r T_r} + r_s \frac{T_s}{T_N} \right) \right] \mathbf{i}_s(k-1) + \frac{T_s}{T_N \sigma x_s} \mathbf{u}_s(k-1). \quad (10.77)$$

This equation can be rewritten in ANN form, with one weight w_2 proportional to mechanical speed ω_m in the following way:

$$\mathbf{i}_s(k) = w_1 \mathbf{\psi}_r(k-1) + w_2 j \mathbf{\psi}_r(k-1) + w_3 \mathbf{i}_s(k-1) + w_4 \mathbf{u}_s(k-1). \quad (10.78)$$

So, the linear ANN can be designed, similarly as in the case of Eq. (10.73a). It is presented in Fig. 10.40.

In this case not flux, but current estimation error is used for ANN tuning:

$$\varepsilon(k) = \mathbf{i}_s(k) - \hat{\mathbf{i}}_s(k). \quad (10.79)$$

Therefore, the estimated speed can be calculated as follows:

$$w_2 = w_2 - \eta(-e_\alpha \psi_{ri\beta} + e_\beta \psi_{ri\alpha}) + \mu \Delta w_2$$
$$\hat{\omega}_m = -\frac{T_N}{T_s} \frac{x_r}{x_M} \sigma x_s w_2. \quad (10.80)$$

The presented method was tested experimentally and by simulation. The influence of learning rate on ANN estimator behavior in steady-state operation is shown in Fig. 10.41. It can be observed that the higher learning rate causes higher ripples in estimated speed. However, the dynamic behavior of the estimator with the smaller learning rate could not be correct and the estimated speed could not follow the real mechanical speed of the induction motor. This situation is presented in Fig. 10.42 (right), where the dynamic results of the ANN speed estimator are shown. It is possible to conclude that the learning rate should be a compromise between steady-state ripple and dynamic tracking speed performance. In the system, presented in this section, the value of the learning rate is assumed to be $\eta = 0.1$.

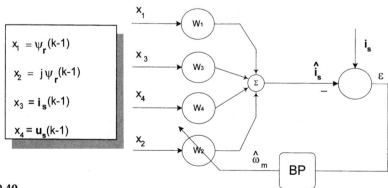

FIGURE 10.40
Structure of linear ANN used in speed estimation method from Eq. (10.77).

10.6 ANN-BASED INDUCTION MOTOR SPEED ESTIMATION

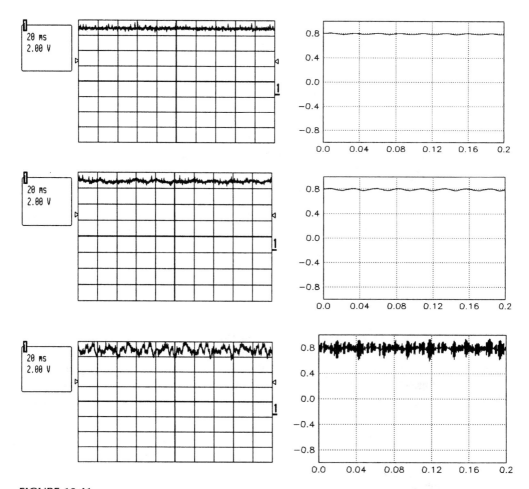

FIGURE 10.41
Influence of learning rate on the ANN estimator behavior in the steady state: $\eta = 0.02$, $\eta = 0.2$, $\eta = 2.0$ from upper to lower figure, respectively (*Left*: experimental results; *Right*: simulation results).

10.6.3 Off-Line Trained Speed Estimation Based on Neural Identification

10.6.3.1 Introduction. The ANN-based off-line identification procedure can be solved in two ways (Fig. 10.43):

1. With a feedforward multilayer ANN, which uses the measured inputs and outputs of the drive in few previous steps as inputs of the ANN—similar to the problem of neural modeling (see Section 10.6.2)
2. With a recurrent network (with feedback loop) which uses the estimated outputs of the network as its actual inputs

In the task of IM speed or flux estimation this concept can be used directly and ANN can be trained based on measured motor current and voltages (Fig. 10.43). The main problem, however, is the choice of a proper structure of ANN.

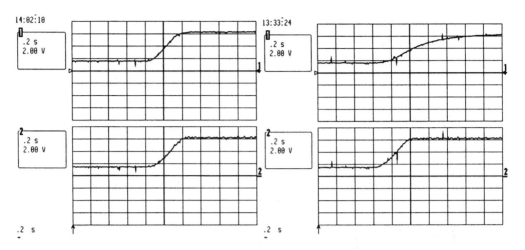

FIGURE 10.42
Influence of learning rate on the ANN estimator behavior in dynamic state of speed step response. Lower: speed from encoder; upper: speed from estimator. *Left*: results for $\eta = 0.01$; *Right*: results for $\eta = 0.001$.

A new architecture has been proposed that is somewhere between a feedforward multilayer perceptron-type architecture and a full recurrent network architecture [63]. For lack of a more generic name, this class of architectures is called a *locally recurrent–globally feedforward* (LRGF) architecture (Fig. 10.44). Based on the general concept presented in Fig. 10.44, ANN with different internal structures can be designed: *multilayer feedforward network, cascade network, local output feedback network, interlayer feedback network, global feedback network* [7]. Connecting each structure described above to another, a so-called compound network can be composed (Fig. 10.45).

10.6.3.2 Speed Identification Based on Stator Line Currents.
Theoretically, it is possible to obtain 256 different compound multilayer network structures. In [7] more than 100 ANN of different internal structures with nonlinear (sigmoidal) activation functions of neurons used in the hidden layers and linear output neuron were tested. These ANN were trained based on input/output training data obtained by simulations of the induction motor transients during line-start and load-torque T_L step changes, using MATLAB-SIMULINK and Neural Network

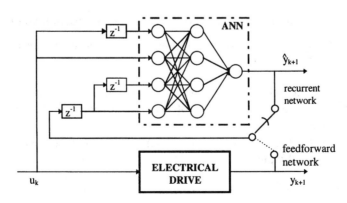

FIGURE 10.43
Identification based on ANN.

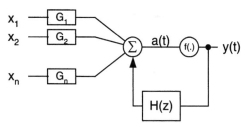

FIGURE 10.44
Generalized LRGF network architecture.

Toolbox. The backpropagation algorithm with Levenberg–Marquardt's modification was used for the training procedures. ANN were trained for constant and varying rotor resistance r_r values (0.8, 1.0, 1.2)r_{rN} and various load torque values: (0; 0.4; 0.8; 1.2)T_L. After training the networks were tested for regimes different from those used in training data. Typical speed estimation waveforms for two ANN architectures are shown in Fig. 10.46.

10.6.3.3 Speed Identification Based on Stator Line Currents and Additional Input Signals.
Significant improvement can be achieved when additional, preprocessed input signal is introduced to the network. As additional signal can be used: stator current magnitude, stator voltage magnitude, or frequency.

The stator current magnitude used as the additional signal has improved speed estimation not only with the help of an Elman network (Fig. 10.47, left side) but even for the feedforward network with a small number of neurons in the hidden layers (Fig. 10.47, right side).

Using two additional input signals, magnitudes of stator currents and voltages, the speed reconstruction can be significantly improved. As can be seen in experimental transients of Fig. 10.48, with a single hidden layer feedforward ANN architecture 4-5-1 the average error is in the range of 1–2%.

Generally, the following conclusions can be formulated:

- The best speed reconstruction is obtained for the ANN with additional input information including stator current magnitude and/or stator voltage magnitude
- The best results with additional input were obtained for two-hidden-layer feedforward and Elman networks
- A larger number of delayed input current values improves the speed estimation quality

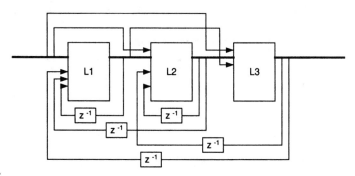

FIGURE 10.45
Compound multilayer network with all possible connections (L1, L2, 1st and 2nd hidden layers; L3, output layer).

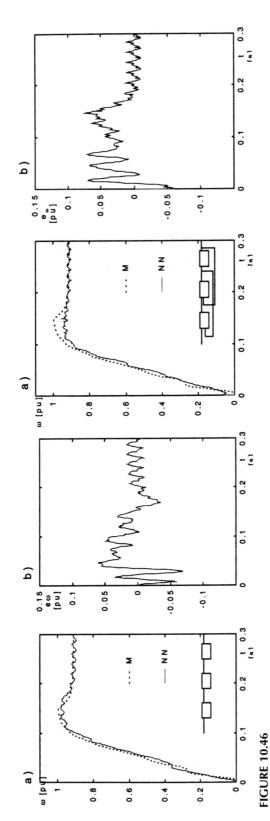

FIGURE 10.46

Transients of rotor speed (solid line), its neural estimate (dotted line) (a) and estimation error ε_ω (b) for 2-10-5-1 feedforward ANN with 1 pair of current delayed inputs (left side) and for 2-5-3-1 NN with interlayer connections and 2 pairs of current delayed inputs (right side).

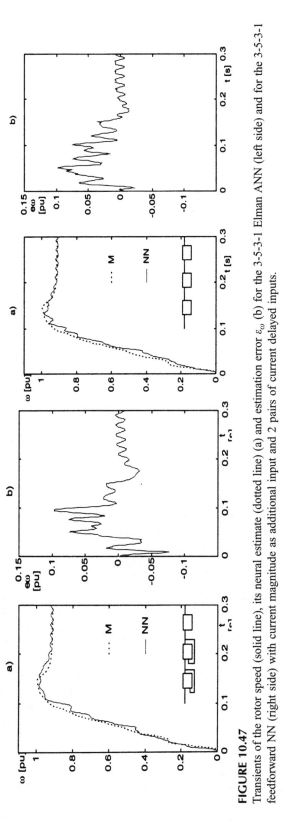

FIGURE 10.47

Transients of the rotor speed (solid line), its neural estimate (dotted line) (a) and estimation error ε_ω (b) for the 3-5-3-1 Elman ANN (left side) and for the 3-5-3-1 feedforward NN (right side) with current magnitude as additional input and 2 pairs of current delayed inputs.

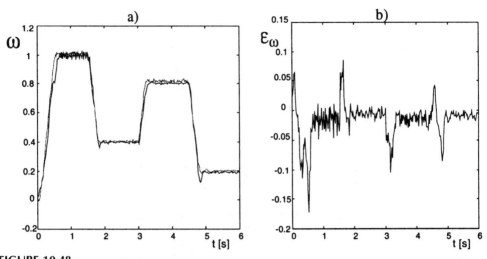

FIGURE 10.48
Experimental transients of the motor speed, its neural estimate (a) and the estimation error (b) during frequency changes for 4-5-1 feedforward ANN architecture.

10.7 ANN-BASED INDUCTION MOTOR FLUX ESTIMATION

Various methods of rotor flux estimation are used alike: simulators, flux observers, and Kalman filters [17]. Neural networks offer a good alternative solution for the magnitude and position of the rotor and/or stator flux estimation.

10.7.1 Flux Estimation Based on Stator Currents

For flux estimation the same approach demonstrated in the previous section in Fig. 10.43 can be used to design an optimal structure of ANN in different operating conditions of the induction motor. For the rotor flux vector estimation two different ANN are used: one for flux magnitude ψ_r, and the second for the flux position $\sin \gamma_\psi$. The feedforward backpropagation ANN networks of various internal structures were trained for different load torque transients: from no load during IM line-start operation to $1.2T_{LN}$ with 0.1 or 0.4 step, for constant motor parameters. Typical results of the 3-25-1 ANN with sigmoidal activation functions in the hidden layer and with linear activation function in the output layer for various sequences of load torque changes are presented in Fig. 10.49. Two components of the stator current vector in stator-oriented (α–β) coordinates and rotor speed were used as ANN input signals.

Similar results can be obtained for the network with two hidden layers 3-8-5-1. However, the ANN is not able to reconstruct correctly the transient process during line-start operation of the motor. Also, the greater number of the hidden layers does not improve the quality of the reconstruction process [6].

Better results can be obtained for the network with historical inputs (delayed current values). The ability of the two-hidden-layer 2-15-7-1 ANN trained with 500 input and output data vectors each and with four historical inputs in each current component vector is shown in Fig. 10.50. The load torque changes in the range $(0-1.2)T_{LN}$ were used in the training procedure.

10.7 ANN-BASED INDUCTION MOTOR FLUX ESTIMATION

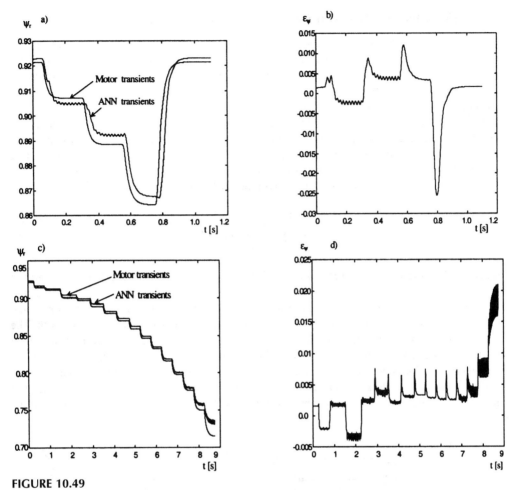

FIGURE 10.49
Transients of the rotor flux magnitude ψ_r, its neural estimate (a, c) and estimation error (b, d) during load change tests for the feedforward 3-25-1 ANN.

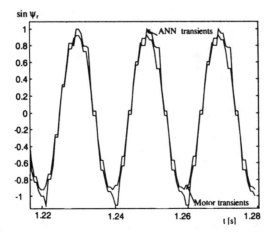

FIGURE 10.50
Rotor flux phase $\sin \gamma_\psi$ and its neural estimate for 2-15-7-1 feedforward ANN.

10.7.2 Flux Estimation Based on Stator Currents and Voltages

10.7.2.1 ANN Architecture. A cascade neural architecture for stator and rotor flux vectors estimation is presented in Fig. 10.51. The system consists of two ANNs. The first one estimates stator flux from the stator voltage and current data; the second one applies the stator current and already-estimated stator flux to the rotor flux estimation. Mathematically the first problem can be classified as a nonlinear dynamic system approximation and the second one as a function approximation. Both ANNs use single-hidden-layer architecture. Training data are generated from a simulated mathematical model operated in various working conditions. In contrast to methods based on backpropagation (Sections 10.6 and 10.7.1), here a dynamic ANN architecture is used where the number of neurons in the hidden layer may change dynamically during the training process. According to Fig. 10.51 two blocks of neural approximates are used:

$$\mathbf{\Psi}_s(k+1) = g^1[\mathbf{u}_s(k), \mathbf{i}_s(k)] \tag{10.81}$$

$$\mathbf{\Psi}_r(k) = g^2[\mathbf{\Psi}_s(k), \mathbf{i}_s(k)]. \tag{10.82}$$

10.7.2.2 Incremental Learning. In both ANNs one-hidden-layer architectures are used but the number of hidden neurons is not fixed in advance. Instead of backpropagation or Levenberg–Marquardt, so-called incremental learning for function approximation originated by Jones and Barron is used [3]. In each iteration only one neuron is optimized and added to the network. The iterative process is terminated when final conditions such as error level or number of hidden neurons are met. The incremental approximation scheme is shown in Fig. 10.52.

In every iteration, one hidden neuron is optimized and added to the network, but all other hidden neurons parameters remain unchanged. Then all output weights are recalculated. The stroked lines indicate connection parameters being recalculated when a neuron is added to the network.

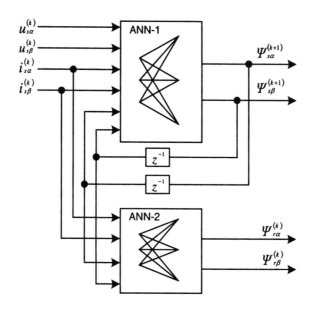

FIGURE 10.51
ANN cascade architecture for stator and rotor flux estimation where $u_{s\alpha}^{(k)}$, $u_{s\beta}^{(k)}$, $i_{s\alpha}^{(k)}$, $i_{s\beta}^{(k)}$, $\Psi_{s\alpha}^{(k)}$, $\Psi_{s\beta}^{(k)}$, $\Psi_{r\alpha}^{(k)}$, $\Psi_{r\beta}^{(k)}$, denote real and imaginary components of the stator voltage and current, respectively, and the stator and rotor real and imaginary components of the flux, respectively.

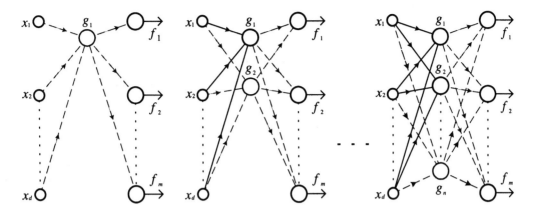

FIGURE 10.52
Illustration of incremental approximation concept shows in sequence situation when first, second, and nth neurons are added.

Hidden neuron functions are described by $g_i \Re^d \to \Re$, $i = 1, \ldots, n$. Usually the hidden layer is constructed from one type of neuron. Different functions are implemented via alteration of hidden neuron parameters. Thus one can assume that the hidden layer consists of the following neurons $g\Re^d \times \Re^p \to \Re$ where $\Re^d$ corresponds to the input space and $\Re^p$ is the hidden neuron parameter space, and $g_i(x) = g(x, a_i)$. The whole network computes functions of the form

$$f_n^1(x) = \sum_{i=1}^{n} w_i^1 g(x, a_i) \tag{10.83}$$

$$f_n^2(x) = \sum_{i=1}^{n} w_i^2 g(x, a_i) \tag{10.84}$$

$$\ldots$$

$$f_n^m(x) = \sum_{i=1}^{n} w_i^m g(x, a_i) \tag{10.85}$$

where w_i^k, $i = 1, \ldots, n$, are output weights of the kth function, $k = 1, \ldots, m$.

The described network can approximate any continuous function with any accuracy, provided that one may use as many hidden neurons as are needed and that the g function fulfills quite weak universal approximation property conditions [3]. When designing an approximation scheme one has to choose incrementally parameters of hidden neurons a_i, $i = 1, \ldots, n$ and all output weights w_i^k, $i = 1, \ldots, n$, $k = 1, \ldots, m$.

The function to be approximated is represented via examples, i.e., an input/output set $S = \{(x_i, y_i) \in \Re^d \times \Re^m\}_{i=1}^{N}$.

Let us define two matrices: the input examples matrix $X^{N,d} = [x_1, \ldots, x_N]^T$ and the output examples matrix $Y^{N,d} = [y_1, \ldots, y_N]^T$ and the neuron vector $G_i = [g_i(x_1), \ldots, g_i(x_N)]^T \in \Re^N$ and approximating functions vectors values F_n:

$$F_n = [F^1, \ldots, F^m] = \begin{bmatrix} f_n^1(x_1) & \cdots & f_n^m(x_1) \\ \cdots & \cdots & \cdots \\ f_n^1(x_N) & \cdots & f_n^m(x_N) \end{bmatrix} \in \Re^{N,m}. \tag{10.86}$$

Let us also define $H_n = [G_1, \ldots, G_n] \in \Re^{N,n}$ and $W_n = [w^1(n), \ldots, w^m(n)] \in \Re^{n,m}$. Then

$$F_n = H_n W_n. \tag{10.87}$$

Now to force $F_n \cong Y$, the best approximation is achieved when $W_n = H_n^+ Y$, where H_n^+ denotes the pseudoinverse. Thus approximating functions values on X, i.e., F_n could be written

$$F_n = P_n Y \tag{10.88}$$

where P_n is a projection matrix and $P_n = H_n H_n^+$. P_n can be calculated incrementally as

$$P_n = P_{n-1} + \tilde{z}_n \tilde{z}_n^T \tag{10.89}$$

where $z_n = (I - P_{n-1})G_n$, $\tilde{z}_n = z_n / \|z_n\|$, and maximum learning rate in every iteration is achieved if the following criterion of parameters selection in hidden units is adopted:

$$\sup_{G_n} ((F_n^j)^T \tilde{z}_n)^2 \tag{10.90}$$

where is F_n^j is the jth column of the F_n matrix. The only problem now is to compromise in every iteration on the choice of hidden neuron parameters when various F_n^j columns generate different requirements formulated by (10.89). We have been using λ_i weighting factors when more than one function is simultaneously approximated. This comes to the following criterion of maximization:

$$\sup_{G_n} \tilde{z}_n^T \left(\sum_{i=1}^m \lambda_i F_n^i (F_n^i)^T \right)^2 \tilde{z}_n \tag{10.91}$$

where

$$\lambda_i \in (0, 1), \quad i = 1, \ldots, m, \quad \sum_{i=1}^m \lambda_i = 1.$$

Note that $\sum_{i=1}^m \lambda_i F_n^i (F_n^i)^T$ is a fixed matrix and could be calculated once during the learning process.

So, if one denotes $A = \sum_{i=1}^m \lambda_i F_n^i (F_n^i)^T$, the incremental learning criterion could be written

$$\sup_{G_n} \tilde{z}_n^T A \tilde{z}_n. \tag{10.92}$$

Typical learning curves with incremental algorithms for ANN-1 and ANN-2 are presented in Fig. 10.53. In the example of Fig. 10.53, when the learning process was terminated, it appeared that 33 hidden neurons in the first network and 12 in the second one are needed.

In order to examine the generalization property of the neural system, the testing signal $\mathbf{u}_s$, different from the teaching, was applied. Typical plots of stator and rotor flux vectors are shown in Fig. 10.54. Very fast learning and avoidance of the local minimum problem (frequently found in the backpropagation algorithm) are the main advantages of the presented incremental approximation method.

10.8 NEURO-FUZZY TORQUE CONTROL (NF-TC) OF INDUCTION MOTOR

In this section, a controller based on the *adaptive neuro-fuzzy inference system* (ANFIS) for voltage source PWM inverter-fed induction motors is presented. This controller combines fuzzy logic and ANN for decoupled flux and torque control. The speed error $\omega_c - \omega_m$ is delivered to the PI speed controller, which generates the torque command m_c. In the applications where the mechanical speed sensor should be eliminated, the estimated value ω_{me} instead of measured ω_m is used. The stator flux amplitude ψ_{Sc} and the electromagnetic torque m_c are command signals,

10.8 NEURO-FUZZY TORQUE CONTROL (NF-TC) OF INDUCTION MOTOR

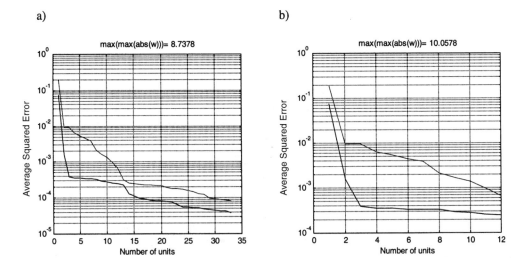

FIGURE 10.53
Average squared error versus number of hidden neurons for incremental learning (a) for ANN-1, (b) for ANN-2.

which are compared with the estimated ψ_S and m_e values respectively, giving instantaneous flux error ε_ψ and torque error ε_m, as shown in Fig. 10.55.

The error signals ε_ψ and ε_m are delivered to the neuro-fuzzy (NF) controller, which also uses information on the position (γ_S) of the actual stator flux vector. The NF controller determines the stator voltage command vector in polar coordinates $\mathbf{v}_c = [V_c, \varphi_{Vc}]$ for the voltage modulator, which finally generates the pulses S_a, S_b, S_c to control transistor switches of the PWM inverter. This scheme is similar to the direct torque control (DTC) structure of Fig. 9.4 (Chapter 9). However, instead of hysteresis the neuro-fuzzy controller is applied. Also, the switching table is replaced by the space vector modulator.

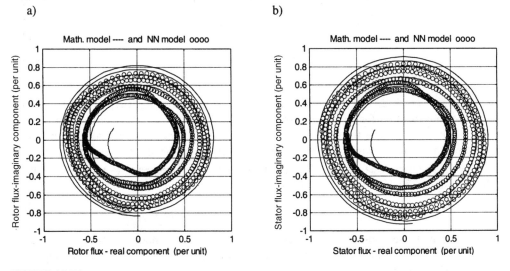

FIGURE 10.54
Stator flux vector (a) and rotor flux vector (b) for mathematical and ANN models.

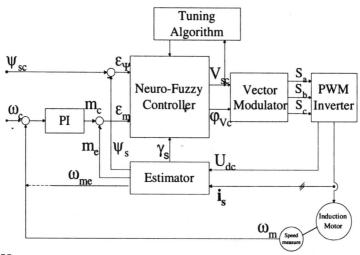

FIGURE 10.55
Neuro-fuzzy torque control of transistor PWM inverter-fed induction motor.

Example 10.3: Design of neuro-fuzzy controller for PWM inverter-fed induction motor

NF controller scheme. The internal structure of the NF controller of Fig. 10.55 is shown in Fig. 10.56. The system based on the ANFIS structure consists of five layers (see Section 10.4.2):

Layer 1: Sampled flux ε_ψ and torque ε_m errors multiplied by w_ψ and w_m weights as

$$\varepsilon'_\psi = w_\psi \varepsilon_\psi \qquad (10.93)$$
$$\varepsilon'_m = w_m \varepsilon_m \qquad (10.94)$$

are delivered to the membership functions in both inputs. To simplify the DSP calculations, the functions are triangular as shown in Fig. 10.57. The first part outputs are calculated based on

$$O^1_{mi} = \mu_{A_{mi}}(w_m \varepsilon_m) \qquad (10.95)$$
$$O^1_{\psi j} = \mu_{A_{\psi j}}(w_\psi \varepsilon_\psi) \qquad (10.96)$$

where O^1_{mi}, $O^1_{\psi j}$ are the first-layer output signals, $i = 1, 2, 3$ are the node numbers for the torque error, $j = 1, 2, 3$ are the node numbers for the flux error, $\mu_{A_{mi}}(w_m \varepsilon_m)$ is the triangular membership function for the torque error, $\mu_{A_{\psi j}}(w_\psi \varepsilon_\psi)$ is the triangular membership function for the flux error, w_ψ is the stator flux error input weight, and w_m is the torque error input weight. The number of membership function is I, J for torque and flux error, respectively.

Layer 2: The second layer calculates the minimum that corresponds to the classical fuzzy logic system. The calculation can be written as

$$w^2_k = \min\lfloor \mu_{A_{mi}}(w_m e_m), \mu_{A_{\psi j}}(w_\psi e_\psi) \rfloor, \qquad (10.97)$$

where w^2_k is the second-layer output signals, and $k = IJ$ is the node number for the present layer. Not every node is connected together. The connections are between outputs of membership functions with different input.

10.8 NEURO-FUZZY TORQUE CONTROL (NF-TC) OF INDUCTION MOTOR

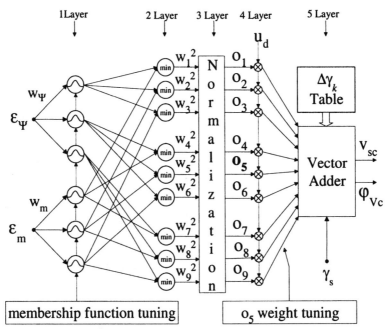

FIGURE 10.56
Neuro-fuzzy controller from Fig. 10.16.

Layer 3: In the third layer the output values are normalized in such a way that the following equation is fulfilled:

$$o_k^3 = \frac{w_k^2}{\sum_K w_k^2} \tag{10.98}$$

where o_k^3 are the third-layer output signals.

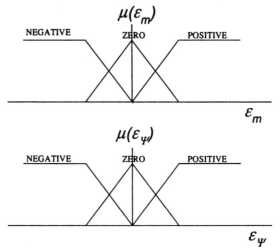

FIGURE 10.57
Triangular membership function sets.

Table 10.3 An Example of Reference Voltage Increment Angle Selection

ε_ψ	P	P	P	Z	Z	Z	N	N	N
ε_m	P	Z	N	P	Z	N	P	Z	N
$\Delta\gamma_k$	$+\dfrac{\pi}{4}$	0	$-\dfrac{\pi}{4}$	$+\dfrac{\pi}{2}$	$+\dfrac{\pi}{2}$	$-\dfrac{\pi}{2}$	$+\dfrac{3\pi}{4}$	$+\pi$	$-\dfrac{3\pi}{4}$

Layer 4: The weight is calculated in this layer as

$$v_{sck} = o_k^3 \cdot u_d \tag{10.99}$$

where v_{sck} is the amplitude of the kth component of the reference voltage vector. When the weight o_k^3 is active ($o_k^3 > 0$), then the regulator chooses the increment angle $\Delta\gamma_i$ value from Table 10.3. The increment $\Delta\gamma_k$ is not needed when $o_k^3 = 0$, because the multiplication $o_k^3 u_d$ is equal to zero.

Layer 5: The reference voltage vector $\mathbf{v}_{sc}$ is a vector sum of the reference voltage vector components:

$$\mathbf{v}_{sc} = \sum_{k}^{K} \mathbf{v}_{sck}, \tag{10.100}$$

where $\mathbf{v}_{sck} = v_{sck} e^{j\varphi_k}$ is the reference voltage component vector, and φ_k is the reference voltage component vector phase. The angle of the reference voltage vector is calculated as

$$\varphi_{V_c} = \gamma_s + \Delta\gamma_k \tag{10.101}$$

where γ_s is the actual angle of the stator flux vector, and $\Delta\gamma_k$ is the increment angle (from Table 10.3).

There are four nonzero output signals from the first layer (two for each input) during steady-state operation. The result is with four generated voltage vector components ($\mathbf{v}_{sck}$) in every sampling time. Vectors $\mathbf{v}_{sck}$ are added to each other and the result, the $\mathbf{v}_{sc}$ voltage vector, is

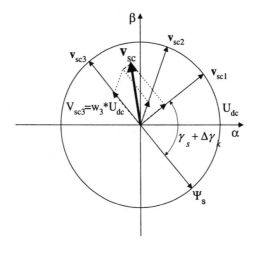

FIGURE 10.58
Reference voltage calculation method (only three out of four $\mathbf{v}_{sck}$ nonzero vectors are shown).

10.8 NEURO-FUZZY TORQUE CONTROL (NF-TC) OF INDUCTION MOTOR

delivered to the space vector modulator. An example of the reference voltage $\mathbf{v}_{sc}$ calculation is presented in Fig. 10.58 (for simplicity, instead of four, there are only three $\mathbf{v}_{sck}$ nonzero vectors used for illustration). The space vector modulator calculates switching states S_a, S_b, and S_c according to the well-known algorithm (Chapter 4, Section 4.2).

Construction of increment selection table. From the induction motor voltage equation

$$\mathbf{u}_s = r_s \mathbf{i}_s + T_N \frac{d\mathbf{\psi}_s}{dt} + j\omega_s \mathbf{\psi}_s \tag{10.102}$$

and the torque equation

$$m_e = \operatorname{Im}(\mathbf{\psi}_s^* \mathbf{i}_s), \tag{10.103}$$

analyzed in stator flux oriented system (Fig. 10.59) one obtains

$$\psi_s = \frac{1}{T_N} \int_0^t (u_s \cos \varphi) dt + \psi_{s0} \tag{10.104}$$

$$m_e = \psi_s \frac{u_s \sin \varphi - \omega_s \psi_s}{r_s} \tag{10.105}$$

with the assumption $u_{sx} \gg r_s i_{sx}$ and the increment angle $\Delta \gamma_k = \varphi$.

It can be seen from Eq. (10.105) that the output torque is dependent on the speed. Moreover, it can be seen that the x and y components of the stator voltage can control the stator flux and torque, respectively. Unfortunately, the torque is also dependent on the stator flux amplitude, which means that it is not decoupled from the stator flux. However, the NF-TC method controls the stator flux precisely, which further ensures decoupling. During the POSITIVE$_m$ or NEGATIVE$_m$ torque error and ZERO$_\psi$ flux error only torque control is needed. For the increment angle $\Delta \gamma_k = \pi/2$ and $\Delta \gamma_k = -\pi/2$ (see Table 10.3), Eqs. (10.104) and (10.105) will then transform to

$$\psi_s = \psi_{s0} \tag{10.106}$$

$$m_e = \psi_s \frac{\pm u_s - \omega_s \psi_s}{r_s}, \tag{10.107}$$

which results in torque control without changes of the stator flux amplitude.

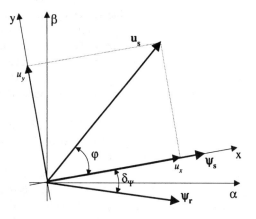

FIGURE 10.59
Fluxes and voltage angles in stator flux oriented system.

During the ZERO$_m$ torque error and POSITIVE$_\psi$ or NEGATIVE$_\psi$ flux, the angle $\Delta\gamma_k$ is set to 0 or 2π, and the flux is controlled according to the equation

$$\psi_s = \frac{1}{T_N}\int_0^t u_s dt + \psi_{s0}. \tag{10.108}$$

However, it has also an influence on the motor torque:

$$m_e = \frac{\omega_s \psi_s^2}{r_s}. \tag{10.109}$$

The increment angles for both nonzero errors of the flux and torque (pairs: POSITIVE$_\psi$ and POSITIVE$_m$, POSITIVE$_\psi$ and NEGATIVE$_m$, NEGATIVE$_\psi$ and POSITIVE$_m$, NEGATIVE$_\psi$ and NEGATIVE$_m$) can be chosen as a compromise between increments for separate flux and torque errors (last two points). For example, for the POSITIVE$_\psi$ flux error and ZERO$_m$ torque error and for POSITIVE$_m$ torque error and ZERO$_\psi$ flux error, there has been chosen respectively $\Delta\gamma_k = 0$ and $\Delta\gamma_k = /\pi/2$. This means that for the POSITIVE$_\psi$ flux error and POSITIVE$_m$ torque error, the middle value between 0 and $\pi/2$ increments is chosen and is equal to $\pi/4$. For ZERO$_m$ torque error and ZERO$_\psi$ flux error it is necessary to keep the reference speed without flux changes. The increment angle is then $\Delta\gamma_k = \pi/2$ and the torque is expressed by Eq. (10.109).

Self-tuning procedure. The controller can be tuned automatically by least square estimation algorithm (for output membership function) and back propagation algorithm (for output and input membership function) as in [59–61] for inverted pendulum stabilization. The NF-TC can be tuned in the same way. Here another simple and effective *off-line* tuning method is proposed. The NF-TC system contains three membership functions for each input. The tuning of the membership function width corresponds to scaling of the flux and torque errors. The scaling factors are w_ψ and w_m weights. The NF-TC is a nonlinear high-order system. Therefore, it is very helpful to use simulation for controller design. Figure 10.60 presents computed flux and torque error as function of the input weights. The surfaces have also been verified experimentally. It can be seen that there is a clearly defined minimum without any other local minimum points.

This is because, for small weights, the controller chooses a high value of the reference voltage amplitude, which results in high flux and torque errors (ripples). On the other hand, if the weights are too big, the steady-state errors increase. This tendency allows use of a simple gradient method to find the optimal working point (optimal w_ψ and w_m values), which guarantees minimal flux and torque errors. However, the torque and flux are not fully decoupled. It can be seen from Eqs. (10.104) and (10.105) that, for nonzero synchronous angular speed, the changes of the flux magnitudes ψ_s influence the motor torque, while the torque change does not influence the flux magnitude. That is why the flux error minimum should be found first, before searching the torque error minimum. The *off-line* tuning process of the system is presented in Fig. 10.61. Note that the tuning surfaces in Fig. 10.60 have a general validity. The numerical results may differ little according to motor parameters and supply voltage. Also, the final flux and torque errors depend on the chosen inverter switching frequency.

The NF-TC scheme guarantees very fast flux and torque responses. This is due to the lack of integration. Unfortunately, this property causes torque error in steady-state operation. One of the solutions is adding an integration block. However, as a consequence, the torque response will be slow. In the NF-TC, instead of integration, the weight o_5 can be used to reach zero torque error at steady state. This weight decides the amplitude of the reference voltage vector. Therefore,

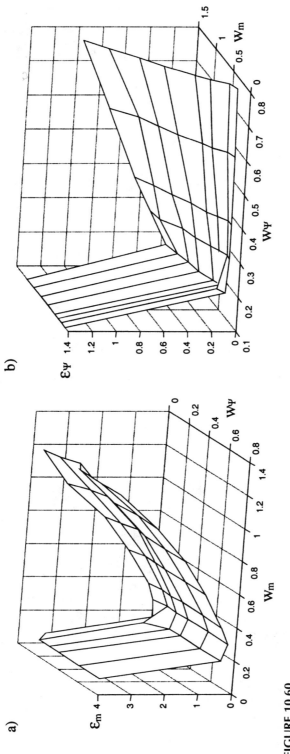

FIGURE 10.60
The torque (a) and flux (b) error tuning surfaces.

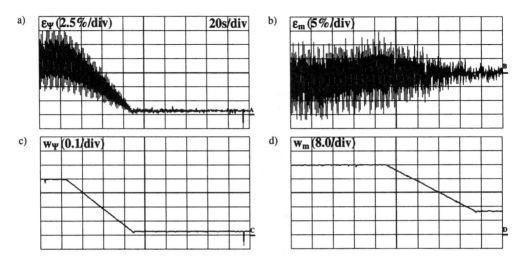

FIGURE 10.61
Experimental tuning of input weights w_ψ and w_m. (a) and (b) flux and torque error behavior, respectively; (c) and (d) flux and torque input weights during system tuning.

instead of calculating by the controller, the value of the output weight o_5 is calculated from the equation

$$o_5 = k_\omega \omega_m + k_m m_c + k_\psi (1 - \psi_{sc}) \qquad (10.110)$$

where k_ω, k_ψ, k_m are experimentally chosen factors to compensate for the steady-state torque error.

The steady-state operations of the tuned system are presented in Figs. 10.62a–c. The sampling time, 500 μs, gives flux and torque errors in the range of 1% and 3.5%, respectively. The stator

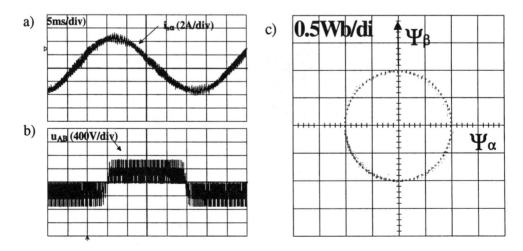

FIGURE 10.62
Experimental steady-state operation for the tuned NF-TC system, (a) stator current, (b) line-to-line voltage PWM, (c) stator flux vector.

current is not distorted by the sector changes, as in the conventional DTC method [38, 39, 64] and the stator flux trajectory is circular.

The motor magnetization process is presented in Fig. 10.63a. It is visible that the magnetization process takes about 10 sampling times (about 5 ms). The reference stator voltage chosen by the controller is parallel to the stator flux vector. It results in short torque distortion that is visible in the oscillogram.

The torque transients to the step changes are presented in Fig. 10.63b. It can be seen that for a constant stator flux amplitude, the flux and torque are fully decoupled and the flux amplitude is not distorted during torque steps. The stator current response is also presented in the figure. The torque response time is about 3 ms (Fig. 10.63c), which gives a similar dynamic as in the conventional DTC method [38, 39, 45, 64].

The four-quadrant speed transient for fast speed ramp reversal is presented in Fig. 10.63d. When a speed sensor is used the system is stable in the whole speed range (including zero speed

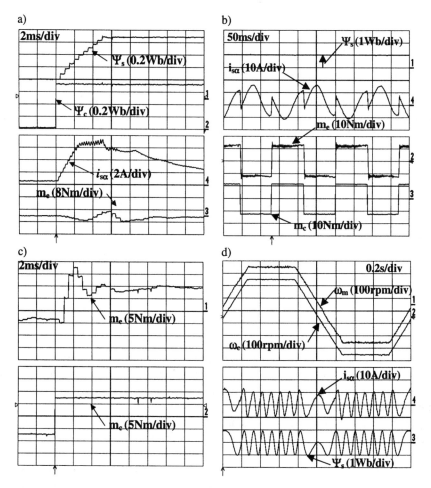

FIGURE 10.63
Experimental oscillograms for (a) motor magnetization at zero speed, (b) the torque transients to the step changes, (c) the torque transients to the small-signal step charge; (d) four-quadrant operation, Ψ_s, stator flux amplitude; Ψ_c, reference stator flux amplitude; $i_{s\alpha}$, stator current; m_e, output torque; m_c, reference torque; ω_{mc}, reference rotor speed; ω_m, rotor speed.

at full load). However, instabilities can only occur for sensorless operation in the zero-speed region. The lower speed control range depends on the flux and the speed estimators' quality. In the laboratory setup speed-sensorless stable operation has been observed over 1–2% of nominal speed.

Compared to conventional DTC the presented NF-TC scheme has the following features and advantages:

- Constant switching frequency and uni-polar motor voltage thanks to applied space vector modulator
- Lower sampling time
- Reliable start and low-speed operations
- Simple tuning procedure
- Torque and current harmonics mainly dependent on sampling time
- Fast torque and flux response

10.9 SENSORLESS FIELD-ORIENTED CONTROL OF INDUCTION MOTOR WITH ANN SPEED ESTIMATION

The basics of field-oriented control (FOC) have been presented in Chapter 5. The main feature of this approach is the coordinate transformation, which allows one to recalculate the decoupled field oriented coordinates i_{sx}, i_{sy} of the stator current vector into the fixed stator frame $i_{s\alpha}$, $i_{s\beta}$:

$$i_{s\alpha} = i_{sx} \cos \gamma_s - i_{sy} \sin \gamma_s \quad (10.111)$$

$$i_{s\beta} = i_{sx} \sin \gamma_s + i_{sy} \cos \gamma_s. \quad (10.111a)$$

In indirect FOC implementation (Fig. 10.64) the rotor flux position angle γ_s is calculated using the reference values of flux and torque currents (the slip frequency model) and the value of mechanical speed ω_m:

$$\gamma_s = \int (\omega_m + \omega_r) dt \quad (10.112)$$

with

$$\omega_r = \frac{1}{T_r} \frac{i_{syc}}{i_{sxc}}. \quad (10.113)$$

Instead of a mechanical speed sensor, the ANN speed estimator of Fig. 10.39 has been applied. The steady-state operation of the tuned FOC system is presented in Fig. 10.65. The sampling time in the experimental system has been set to 170 µs. The result has been obtained for half of the nominal speed (25 Hz).

The speed transients to the step changes are presented in Fig. 10.66. Figure 10.66a presents the startup of the motor in the low-speed region (small signal behavior), whereas Fig. 10.66b illustrates half-rated-speed reversal. It can be seen that, with constant flux current component i_{sx}, the flux amplitude is kept constant during torque transients. So, the flux and torque are dynamically decoupled. The torque response time is about 5 ms and is mainly determined by current control loop design [18, 31]. The dynamic behavior of the drive for a step change of the load torque is shown in Fig. 10.67. Also here the flux magnitude is kept constant, and speed and torque transients are well damped. The speed controller is designed according to the symmetry criterion [18].

10.9 SENSORLESS FIELD-ORIENTED CONTROL OF INDUCTION MOTOR 413

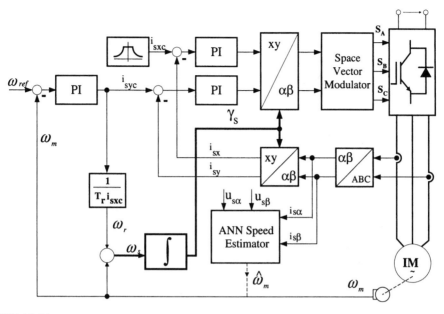

FIGURE 10.64
Indirect FOC of current controlled PWM inverter-fed induction motor (dashed line shows ANN based speed sensorless operation).

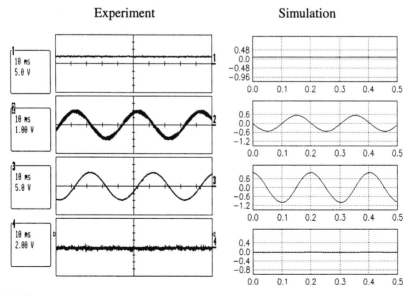

FIGURE 10.65
Steady-state operation of the induction motor controlled via FOC with ANN based speed estimator from Fig. 10.39 (ω_m). 1, estimated speed; 2, stator phase current; 3, rotor flux; 4, electromagnetic torque.

414 CHAPTER 10 / NEURAL NETWORKS AND FUZZY LOGIC CONTROL

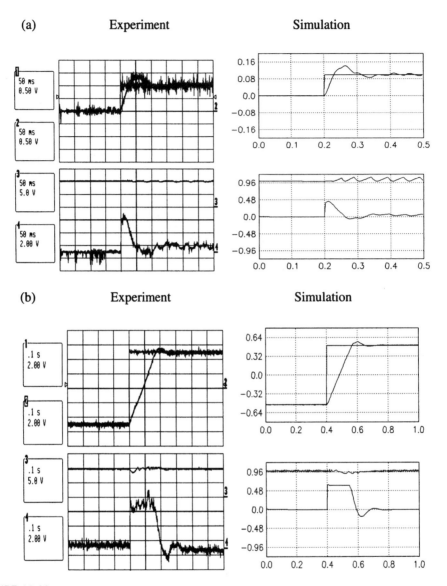

FIGURE 10.66
Speed step response for the induction motor controlled via FOC with ANN speed estimator from Fig. 10.39. (a) Startup $\omega_m = 0.0 \rightarrow 0.1$; (b) speed reversal $\omega_m = -0.5 \rightarrow 0.5$. 1, reference speed; 2, mechanical speed; 3, amplitude of rotor flux; 4, electromagnetic torque.

The main features of the presented FOC scheme can be summarized as follows:

- One of the simplest *on-line* operated ANN estimator schemes is based on combination of the voltage flux model and current estimator (Fig. 10.39).
- The speed sensorless FOC scheme operated with a new ANN speed estimator, instead of a mechanical speed sensor, shows very good performance in steady state and in dynamical states.

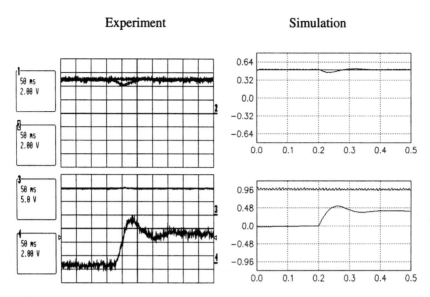

FIGURE 10.67
Response to load step change for the induction motor controlled via FOC with ANN speed estimator from Fig. 10.39 ($m_L = 0.0 \rightarrow 0.4$). 1, reference speed; 2, mechanical speed; 3, amplitude of rotor flux; 4, electromagnetic torque.

- The drive keeps all well-known advantages of FOC schemes: it allows the independent control of rotor flux and motor torque, and it has very fast torque response.
- The drive provides reliable start and stop as well as operation in the low-speed region (1–2% of nominal speed).

10.10 SUMMARY

A brief review of neural networks and fuzzy and neuro-fuzzy systems is given in this chapter. Several groups of applications which illustrate estimations and control in main areas of power electronics and drives are added to supplement the theoretical principles. There is no doubt that neural networks and fuzzy and neuro-fuzzy systems will offer a new interesting perspective for future research. At present, however, they represent only some alternative solutions to existing estimations and control techniques, and their specific applications areas in power electronics cannot be clearly defined. It is believed that thanks to continous developments in digital signal processing technology, artificial intelligence techniques will have a strong impact on power electronics control, estimation, and monitoring in coming decades.

REFERENCES

[1] T. Munakata, *Fundamentals of the New Artificial Intelligence. Beyond Traditional Paradigms.* Springer-Verlag, New York, 1998.
[2] B. K. Bose, Expert system, fuzzy logic and neural network applications in power electronics and motion control. *Proc. IEEE* **82**, 1303–1323 (1994).
[3] L. Grzesiak and B. Beliczyński, Simple neural cascade architecture for estimating of stator and rotor flux vectors. *Proc. European Power Electronics Conf.*, Lausanne, 1999.

[4] M. P. Kazmierkowski and T. Kowalska-Orłowska, ANN based estimation and control in converter-fed induction motor drives. In *Soft Computing in Industrial Electronics* (Osaka S. J. and Sztandera L, eds.), Physica Verlag, Heidelberg, 2002.

[5] T. Orlowska-Kowalska, Application of extended observer for flux and rotor time–constant estimation in induction motor drives. *IEE Proc., Part D* **136**, 324–330 (1989).

[6] T. Orlowska-Kowalska and C. T. Kowalski, Neural network based flux estimator for the induction motor. *Proc. PEMC '96*, Budapest, pp. 187–191, 1996.

[7] T. Orlowska-Kowalska and P. Migas, Analysis of the neural network structures for induction motor state variables estimation. *Conf. Proc. SPEEDAM '98*, Sorrento, P3.55–59, 1998.

[8] K. Simoes and B. K. Bose, Neural network based estimation of feedback signals for a vector controlled induction motor drive. *IEEE Trans. Industr. Appl.* **31**, 620–639 (1995).

[9] M. G. Simoes and B. K. Bose, Neural network based estimation of feedback signals for a vector controlled induction motor drive. *Proc. IEEE IAS Annual Meeting*, Denver, CO, pp. 471–479, 1994.

[10] A. K. P. Toh and E. P. Nowicki, A flux estimator for field oriented control of an induction motor using an artificial neural network. *Proc. IEEE-IAS '94*, pp. 585–592, 1994.

[11] A. P. Toh, E. P. Nowicki and F. Ashrafzadeh, A flux estimator for field oriented control of an induction motor using an artificial neural network. *Proc. IEEE IAS Annual Meeting*, Denver, CO, pp. 585–592, 1994.

[12] M. T. Wishart and R. G. Harley, Identification and control of induction machines using artificial neural networks. *Proc. IEEE IAS Annual Meeting*, Toronto, Canada, pp. 703–709, 1993.

[13] L. Ben-Brahim and R. Kurosawa R, Identification of induction motor speed using neural networks. *Proceedings of the Power Converter Conference PCC*, Yokohama, Japan, pp. 689–694, 1993.

[14] L. Ben-Brahim, S. Tadakuma, and A. Akdag, Speed control of induction motor without rational transducers, *IEEE Trans. Indust. Appl.* **35**, 844–850 (1999).

[15] T. C. Huang and M. A. El-Sharkawi, High performance speed and position tracking of induction motors using multi-layer fuzzy control. *IEEE Trans. Energy Conversion*, 353–358 (1996).

[16] T. Orlowska-Kowalska and P. Migas, Neural speed estimator for the induction motor drive. *Conf. Proc. EPEA/PEMC '98*, Praha, Vol. 8, 7.89–7.94, 1998.

[17] K. Rajashekara, A. Kawamura and K. Matsue, *Sensorless Control of AC Motor Drives. Speed and Position Sensorless Operation*. IEEE Press, Piscataway, NJ, 1996.

[18] D. L. Sobczuk, Application of artificial neural networks for control of PWM inverter-fed induction motor drives. Ph.D. thesis, Institute of Control and Industrial Electronics, Warsaw University of Technology, Warsaw, Poland, 1999.

[19] D. L. Sobczuk and P. Z. Grabowski, DSP implementation of neural network speed estimator for inverter fed induction motor. *Proc. IEEE Int. Conf. Industrial Electronics Control and Instrumentation IECON '98*, Aachen, Germany, pp. 981–985, 1998.

[20] L. A. Cabrera and M. E. Elbuluk, Tuning the stator resistance of induction motors using artificial neural network. *Proc. IEEE-PESC '95*, pp. 421–427, 1995.

[21] S. A. Mir and M. E. Elbuluk, Precision torque control in inverter fed induction machines using fuzzy logic. *Proc. IEEE-PESC*, pp. 396–401, 1995.

[22] B. Burton, F. Karman, R. G. Harley, T. G. Habetler, M. A. Brooke, and R. Poddar, Identification and control of induction motor stator currents using fast on-line random training of a neural network. *IEEE Trans. Indust. Appl.* **33**, 697–704 (1997).

[23] P. Vas, *Artificial-Intelligence-Based Electrical Machines and Drives*. Oxford University Press, New York, 1999.

[24] J. O. P. Pinto, B. K. Bose, L. E. Borges, and M. P. Kazmierkowski, A neural-network-based space-vector PWM controller for voltage inverer-fed induction motor drive. *IEEE Trans. Indust. Appl.* **36**, 1628–1636 (2000).

[25] Y. Y. Tzou, Fuzzy-tuning current-vector control of a 3-phase PWM inverter. *Proc. IEEE-PESC*, pp. 326–331, 1995.

[26] M. R. Buhl and R. D. Lorenz, Design and implementation of neural networks for digital current regulation of inverter drives. *IEEE-IAS Ann. Mtg., Conf. Rec.*, pp. 415–421, 1991.

[27] M. A. Dzieniakowski and M. P. Kazmierkowski, Self-tuned fuzzy PI current controller for PWM-VSI. *Proc. EPE '95*, Sevilla, pp. 1.308–1.313, 1995.

[28] F. Harashima, Y. Demizu, S. Kondo, and H. Hashimoto, Application of neural networks to power converter control. *IEEE IAS '89 Conf. Rec.*, San Diego, pp. 1087–1091, 1989.

[29] Y. Ito, T. Furuhashi, S. Okuma, and Y. Uchikawa, A digital current controller for a PWM inverter using a neural network and its stability. *IEEE-PESC Conf. Rec.*, San Antonio, pp. 219–224, 1990.

[30] M. P. Kazmierkowski, D. L. Sobczuk, and M. A. Dzieniakowski, Neural network current control of VS-PWM inverters. *Proc. EPE '95, Sevilla*, pp. 1.415–1.420, 1995.
[31] M. P. Kazmierkowski and L. Malesani, Current control techniques for three-phase voltage-source PWM converters: A survey. *IEEE Trans. Indust. Electron.* **45**, 691–703 (1998).
[32] S. S. Min, K. C. Lee, J. W. Song, and K. B. Cho, A fuzzy current controller for fieldoriented controlled induction machine by fuzzy rule. *IEEE PESC '92 Conf. Rec.*, Toledo, pp. 265–270, 1992.
[33] G. Cybenko, Approximations by superpositions of a sigmoidal function. *Mathematics of Control, Signals and Systems*, **2**, 303–314 (1989).
[34] S. Saetieo and D. A. Torrey, Fuzzy logic control of a space vector PWM current regulator for three phase power converters. *Proc. IEEE-APEC*, 1997.
[35] D. R. Seidl, D. A. Kaiser, and R. D. Lorenz, One-step optimal space vector PWM current regulation using a neural network. *IEEE-IAS Ann. Mtg., Conf. Rec.*, pp. 867–874, 1994.
[36] T. C. Huang and M. A. El-Sharkawi, Induction motor efficiency maximizer using multi-layer fuzzy control. *Int. Conf. Intelligent System Application to Power Systems*, Orlando, FL, Jan. 28–Feb. 2, 1996.
[37] Y. S. Kung, C. M. Liaw, and M. S. Ouyang, Adaptive speed control for induction motor drives using neural networks. *IEEE Trans. Indust. Electron.* **42**, 25–32 (1995).
[38] P. Z. Grabowski, Direct flux and torque neuro-fuzzy control of inverter-fed induction motor drives. Ph.D. thesis, Institute of Control and Industrial Electronics, Warsaw University of Technology, Warsaw, Poland.
[39] P. Z. Grabowski, M. P. Kazmierkowski, B. K. Bose, and F. Blaabjerg, A simple direct-torque neuro-fuzzy control of PWM-inverter-fed induction motor drive. *IEEE Trans. Indust. Electron.* **47**, 863–870 (2000).
[40] M. Cichowlas, D. Sobczuk, M. P. Kazmierkowski, and M. Malinowski, Novel artificial neural network (ANN) based current controller PWM rectifiers. *Proc. EPE-PEMC '00, Kosice, Slovak Republic*, Vol. 1, pp. 41–46, 2000.
[41] M. Y. Chow, *Methodologies of Neural Network and Fuzzy Logic Technologies for Motor Incipient Fault Detection*. World Scientific Co., Singapore.
[42] V. Valouch, Fuzzy space vector modulation based control of modified instantaneous power in active filter. *Proc. IEEE–ISIE*, Pretoria, South Africa, pp. 230–233, 1998.
[43] L. A. Belfore and A.-R. A. Arkadan, Modeling faulted switched reluctance motors using evolutionary neural networks, *IEEE Trans. Indust. Electron.* **42**, 226–233 (1997).
[44] P. V. Goode and M.-Y. Chow, Using a neural/fuzzy system to extract heuristic knowledge of incipient faults in induction motors: Part II—Application. *IEEE Trans. Indust. Electron.* **42**, 139–146 (1995).
[45] P. V. Goode and M.-Y. Chow, Using a neural/fuzzy system to extract heuristic knowledge of incipient faults in induction motors: Part I—Methodology. *IEEE Trans. Indust. Electron.* **42**, 131–138 (1995).
[46] R. R. Schoen, B. K. Lin, T. G. Habetler, J. H. Schlag, and S. Farag, An unsupervised on-line system for induction motor fault detection using stator current monitoring. *Proc. IEEE IAS Annual Meeting*, Denver, CO, ISBN 0-7803-1993-1 pp. 103–109, 1994.
[47] D. E. Rumelhart and J. L. McClelland J. L. (eds.), *Parallel Distributed Processing. Explorations in the Microstructure of Cognition*, Vol. 1: Foundations (Chapters 2, 8 and 11). MIT Press, 1986.
[48] T. Fukuda and T. Shibata, Theory and application of neural networks for industrial control. *IEEE Trans. Indust. Electron.* **39**, 472–489 (1992).
[49] T. Fukuda, T. Shibata, and M. Tokita, Neuromorphic control: adaption and learning. *IEEE Trans. Indust. Electron.* **39**, 497–503 (1992).
[50] B. Kosko, *Neural Networks and Fuzzy Systems: A Dynamical Approach to Machine Intelligence*. Prentice Hall, Englewood Cliffs, NJ, 1992.
[51] K. S. Narandera and K. Parthasarathy, Identification and control of dynamical systems using neural networks. *IEEE Trans. Neural Networks* **1**, 4–27 (1990).
[52] L. A. Zadeh, Fuzzy sets. *Information Control* **8**, 338–353 (1965).
[53] L. A. Zadeh, The concept of linguistic variable and its application to aproximate reasoning. *Information Sci.* **8**, 199–249 (1975).
[54] H. J. Zimmermann, *Fuzzy Set Theory and Its Application*. Kluwer Academic Publishers, Boston, 1991.
[55] C. C. Lee, Fuzzy logic in control systems: Fuzzy logic controller—Part I. *IEEE Trans. Systems, Man, and Cybernet.* **20**, 404–418 (1991).
[56] C. C. Lee, Fuzzy logic in control systems: Fuzzy logic controller—Part II. *IEEE Trans. Systems, Man, and Cybernet.* **20**, 419–435 (1990).
[57] E. H. Mamdani, Application of fuzzy algorithms for control of a simple dynamic plant. *Proc. IEEE* **121**, 1585–1588 (1974).

[58] S. Tzafestas and N. P. Papanikolopoulos, Incremental fuzzy expert PID control. *IEEE Trans. Indust. Electron.* **37**, 365–371 (1990).

[59] J.-S. R. Jang, Self-learning fuzzy controllers based on temporal back propagation. *IEEE Trans. Neural Networks*, **3**, 714–723 (1992).

[60] J.-S. R. Jang, ANFIS: Adaptive-network-based fuzzy inference system. *IEEE Trans. System, Man, Cybernet.* **23**, 665–684 (1993).

[61] J.-S. R. Jang and C.-T. Sun, Neuro-fuzzy modeling and control. *Proc. IEEE* **83**, 378–406 (1995).

[62] J. O. P. Pinto, B. K. Bose, L. E. Borges, and M. P. Kazmierkowski, A neural network based space vector PWM controller for voltage-fed inveter induction motor drive. *IEEE IAS-Annual Meeting '99*, pp. 2614–2622, 1999.

[63] A. C. Tsoi and A. Back, Locally recurrent globally feedforward networks: A critical review of architectures. *IEEE Trans. Neural Networks* **5**, 229–239 (1994).

[64] I. Takahashi and T. Noguchi, A new quick-response and high efficiency control strategy of an induction machine. *IEEE Trans. Indust. Appl.* **22**, 820–827, (1986).

[65] J. Tanomaru and S. Omatu, Towards effective neuromorphic controllers. *Proc. IEEE-IECON '91*, Kobe, Japan, pp. 1395–1400, 1991.

CHAPTER 11

Control of Three-Phase PWM Rectifiers

MARIUSZ MALINOWSKI and MARIAN P. KAZMIERKOWSKI
Warsaw University of Technology, Warsaw, Poland

LIST OF SYMBOLS

Symbols (General)

$x(t), x$ instantaneous value
X^*, x^* reference
$\underline{x}$ complex vector
$\underline{x}^*$ conjugate complex vector
$|X|$ magnitude (length) of function
$\Delta X, \Delta x$ deviation

Symbols (Special)

α phase angle of reference vector
φ phase angle of current
ω angular frequency
ε control phase angle
$\cos \varphi$ fundamental power factor
f frequency
$i(t), i$ instantaneous current
j imaginary unit
k_P, k_i proportional control part, integrating control part
$p(t), p$ instantaneous active power
$q(t), q$ instantaneous reactive power
t instantaneous time
$v(t), v$ instantaneous voltage

$\underline{\Psi}_L$	virtual line flux vector
$\Psi_{L\alpha}$	virtual line flux vector components in stationary α, β coordinates
$\Psi_{L\beta}$	virtual line flux vector components in the stationary α, β coordinates
Ψ_{Ld}	virtual line flux vector components in the synchronous d, q coordinates
Ψ_{Lq}	virtual line flux vector components in the synchronous d, q coordinates
$\underline{u}_L$	line voltage vector
$u_{L\alpha}$	line voltage vector components in the stationary α, β coordinates
$u_{L\beta}$	line voltage vector components in the stationary α, β coordinates
u_{Ld}	line voltage vector components in the synchronous d, q coordinates
u_{Lq}	line voltage vector components in the synchronous d, q coordinates
$\underline{i}_L$	line current vector
$i_{L\alpha}$	line current vector components in the stationary α, β coordinates
$i_{L\beta}$	line current vector components in the stationary α, β coordinates
i_{Ld}	line current vector components in the synchronous d, q coordinates
i_{Lq}	line current vector components in the synchronous d, q coordinates
$\underline{u}_S, \underline{u}_{conv}$	converter voltage vector
$u_{S\alpha}$	converter voltage vector components in the stationary α, β coordinates
$u_{S\beta}$	converter voltage vector components in the stationary α, β coordinates
u_{Sd}	converter voltage vector components in the synchronous d, q coordinates
u_{Sq}	converter voltage vector components in the synchronous d, q coordinates
S_a, S_b, S_c	switching state of the converter
C	capacitance
I	root mean square value of current
L	inductance
R	resistance
S	apparent power
P	active power
Q	reactive power

Subscripts

a, b, c	phases of three-phase system
d, q	direct and quadrature component
$\alpha, \beta, 0$	alpha, beta components and zero sequence component
L-L	line to line
$Load$	load
$conv$	converter
ref	reference
m	amplitude
rms	root mean square value

Abbreviations

ASD	adjustable speed drives
DPC	direct power control
DSP	digital signal processor
EMI	electromagnetic interference
IGBT	insulated gate bipolar transistor
PFC	power factor correction

PI   proportional integral (controller)
PLL   phase locked loop
PWM   pulse-width modulation
SVM   space vector modulation
THD   total harmonic distortion
UPF   unity power factor
VF   virtual flux
VF-DPC   virtual flux based direct power control
VFOC   virtual flux oriented control
VOC   voltage oriented control
VSI   voltage source inverter

11.1 OVERVIEW

11.1.1 Introduction

As has been observed in recent decades, an increasing portion of generated electric energy is converted through rectifiers, before it is used at the final load. In power electronic systems, especially, diode and thyristor rectifiers are commonly applied in the front end of dc-link power converters as an interface with the ac line power (grid) (Fig. 11.1). The rectifiers are nonlinear in nature and, consequently, generate harmonic currents in the ac line power. The high harmonic content of the line current and the resulting low power factor of the load cause a number of problems in the power distribution system:

- Voltage distortion and electromagnetic interface (EMI) affecting other users of the power system
- Increasing voltampere ratings of the power system equipment (generators, transformers, transmission lines, etc.)

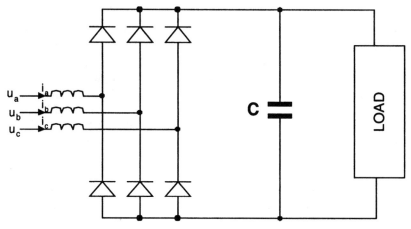

FIGURE 11.1
Diode rectifier.

Therefore, governments and international organizations have introduced new standards (in the United States, IEEE 519, and in Europe, IEC 61000-3) which limit the harmonic content of the current drawn from the power line by rectifiers. As a consequence many new switch-mode rectifier topologies that comply with the new standards have been developed.

In the area of variable speed ac drives, it is believed that three-phase PWM boost ac/dc converter will replace the diode rectifier. The resulting topology consist of two identical bridge PWM converters. The line-side converter operates as a rectifier in forward energy flow, and as an inverter in reverse energy flow. In further discussion assuming the forward energy flow as the basic mode of operation the line-side converter will be called a PWM rectifier. The ac side voltage of the PWM rectifier can be controlled in magnitude and phase so as to obtain sinusoidal line current at unity power factor (UPF). Although such a PWM rectifier/inverter (ac/dc/ac) system is expensive, and the control is complex, the topology is ideal for four-quadrant operation. Additionally, the PWM rectifier provides dc bus voltage stabilization and can also act as an active line conditioner (ALC) that compensates harmonics and reactive power at the point of common coupling of the distribution network. However, reducing the cost of the PWM rectifier is vital for its competitiveness with other front-end rectifiers. The cost of power switching devices (e.g., IGBT) and digital signal processors (DSPs) is generally decreasing and further reduction can be obtained by reducing the number of sensors. Sensorless control exhibits advantages such as improved reliability and lower installation costs.

11.1.2 Rectifier Topologies

A voltage source PWM inverter with diode front-end rectifier is one of the most common power configurations used in modern variable speed ac drives (Fig. 11.1). An uncontrolled diode rectifier has the advantage of being simple, robust, and low cost. However, it allows only unidirectional power flow. Therefore, energy returned from the motor must be dissipated on a power resistor controlled by a chopper connected across the dc link. A further restriction is that the maximum motor output voltage is always less than the supply voltage.

Equations (11.1) and (11.2) can be used to determine the order and magnitude of the harmonic currents drawn by a six-pulse diode rectifier:

$$h = 6k \pm 1, \qquad k = 1, 2, 3, \ldots \qquad (11.1)$$

$$\frac{I_h}{I_1} = 1/h. \qquad (11.2)$$

Harmonic orders as multiples of the fundamental frequency—5th, 7th, 11th, 13th, etc., with a 50 Hz fundamental—correspond to 250, 350, 550, and 650 Hz, respectively. The magnitude of the harmonics in per unit of the fundamental is the reciprocal of the harmonic order: 20% for the 5th, 14.3% for the 7th, etc. Equations (11.1) and (11.2) are calculated from the Fourier series for ideal square-wave current (critical assumption for infinite inductance on the input of the converter). Equation (11.1) is a fairly good description of the harmonic orders generally encountered. The magnitude of actual harmonic currents often differs from the relationship described in Eq. (11.2). The shape of the ac current depends on the input inductance of the converter (Fig. 11.2). The riple current is equal to $1/L$ times the integral of the dc ripple voltage. With infinite inductance the ripple current is zero and the flat-top wave of Fig. 11.2d results. The full description of harmonic calculation in a six-pulse converter can be found in [1].

Besides the six-pulse bridge rectifier a few other rectifier topologies are known [2–3]. Some of them are presented in Fig. 11.3. The topology of Fig. 11.3a presents a simple solution of boost-type converter with the possibility to increase dc output voltage. This is an important feature for ASD's converters giving maximum motor output voltage. The main drawback of this solution is stress on the components and low-frequency distortion of the input current. The next

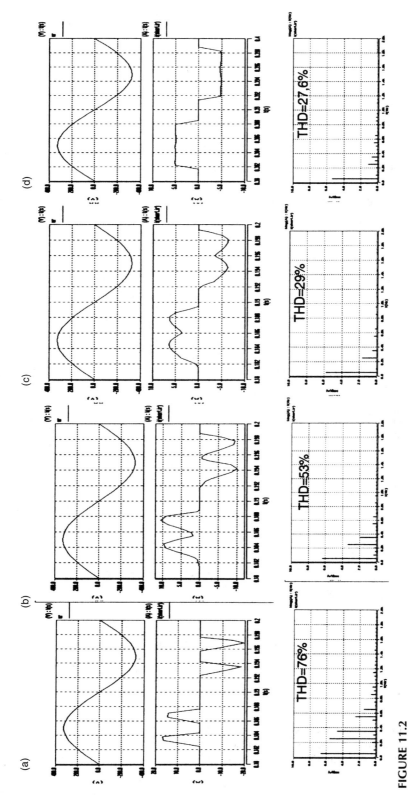

FIGURE 11.2
Simulation results of diode rectifier at different input inductance (from 0 to infinity).

topologies (b) and (c) use PWM rectifier modules with a very low current rating (20–25% level of rms current comparable with (e) topology). Hence they have a low cost potential and provide only the possibility of regenerative braking mode (b) or active filtering (c). Figure 11.3d presents a three-level converter called a *Vienna* rectifier [4]. The main advantage is low switch voltage, but nontypical switches are required. Figure 11.3e presents the most popular topology used in ASD, UPS, and more recently as a PWM rectifier. This universal topology has the advantage of using a low-cost three-phase module with a bidirectional energy flow capability. Among its disadvantages are a high per-unit current rating, poor immunity to shoot-through faults, and high switching losses. The features of all topologies are compared in Table 11.1.

The last topology is most promising and therefore was chosen by most global companies (Siemens, ABB, and others) [5, 6]. In a dc distributed power system (Fig. 11.4) or ac/dc/ac converter, the ac power is first transformed into dc thanks to a three-phase PWM rectifier. It provides UPF and low current harmonic content. The converters connected to the dc bus provide

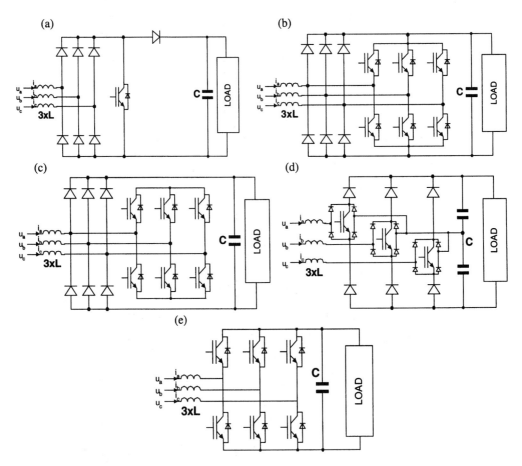

FIGURE 11.3
Basic topologies of switch-mode three-phase rectifiers. (a) Simple boost-type converter; (b) diode rectifier with PWM regenerative braking rectifier; (c) diode rectifier with PWM active filtering rectifier; (d) Vienna rectifier (3-level converter); (e) PWM reversible rectifier (2-level converter).

Table 11.1 Features of Three-Phase Rectifiers

Feature topology	Regulation of dc output voltage	Low harmonic distortion of line current	Near sinusoidal current waveforms	Power factor correction	Bidirectional power flow	Remarks
Diode rectifier	−	−	−	−	−	
Rec(a)	+	−	−	+	−	
Rec(b)	−	−	−	−	+	
Rec(c)	−	+	+	+	−	UPF
Rec(d)	+	+	+	+	−	UPF
Rec(e)	+	+	+	+	+	UPF

further desired conversion for the loads, such as adjustable speed drives for induction motors (IM) and permanent magnet synchronous motors (PMSM), dc/dc converters, multidrive operation, etc.

11.1.3 Control Strategies

Control of active PWM rectifiers can be considered as a dual problem with vector control of an induction motor (Fig. 11.5) [7]. The speed control loop of the vector drive corresponds to the dc-link voltage control, and the reference angle between the stator current and the rotor flux is replaced by the reference angle of the line voltage. Various control strategies have been proposed in recent works on this type of PWM converter. Although these control strategies can achieve the same main goals, such as the high power factor and near-sinusoidal current wave forms, their principles differ. Particularly, the voltage oriented control (VOC), which guarantees high

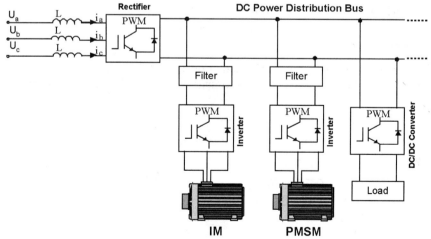

FIGURE 11.4
Dc distributed power system.

426 CHAPTER 11 / CONTROL OF THREE-PHASE PWM RECTIFIERS

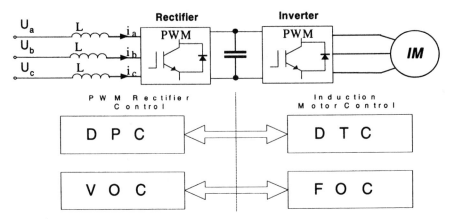

FIGURE 11.5
Relationship between control of PWM line rectifier and PWM inverter-fed IM.

dynamics and static performance via an internal current control loop, has become very popular and has constantly been developed and improved [8, 38]. Consequently, the final configuration and performance of the VOC system largely depends on the quality of the applied current control strategy [13]. Another control strategy called direct power control (DPC) is based on the instantaneous active and reactive power control loops [14, 15]. In DPC there are no internal current control loops and no PWM modulator block, because the converter switching states are selected by a switching table based on the instantaneous errors between the commanded and estimated values of active and reactive power. Therefore, the key point of the DPC implementation is a correct and fast estimation of the active and reactive line power.

The control techniques for a PWM rectifier can be generally classified as voltage-based and virtual flux-based, as shown in Fig. 11.6. The virtual flux-based method corresponds to a direct analogy of IM control.

11.2 OPERATION OF PWM RECTIFIER

11.2.1 Introduction

Figure 11.7b shows a single-phase representation of the rectifier circuit (Fig. 11.7a). L and R represents the line inductor. $\underline{u}_L$ is the line voltage and $\underline{u}_S$ is the bridge converter voltage

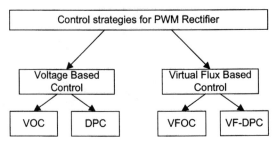

FIGURE 11.6
Classification of control methods for PWM rectifier.

11.2 OPERATION OF PWM RECTIFIER

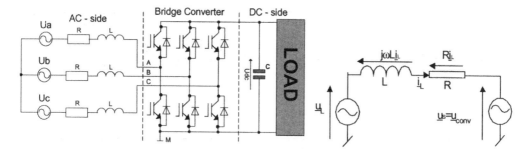

FIGURE 11.7
Simplified representation of three-phase PWM rectifier for bidirectional power flow.

controllable from the dc side. Depending on the modulation index, the magnitude of u_S can take up to the maximum value allowed by the modulator and the dc voltage level.

Figure 11.8 presents general phasor diagram and both rectification and regenerating phasor diagrams when unity power factor is required. The figure shows that the voltage vector u_S is higher during regeneration (up to 3%) than in the rectifier mode. It means that these two modes are not symmetrical [16].

The main circuit of a PWM rectifier (Fig. 11.7a) consists of three legs with IGBT transistors or, in the case of high power, GTO tyrystors. The ac side voltage can be represented with eight possible switching states (Fig. 11.9) (six active and two zero) described by

$$u_{k+1} = \begin{cases} (2/3)u_{dc}e^{jk\pi/3} & \text{for} \quad k = 0, \ldots, 5 \\ 0 \end{cases} \qquad (11.3)$$

Inductors connected between the input of the rectifier and the line are an integral part of the circuit. It brings the current source character of the input circuit and provides the boost feature of

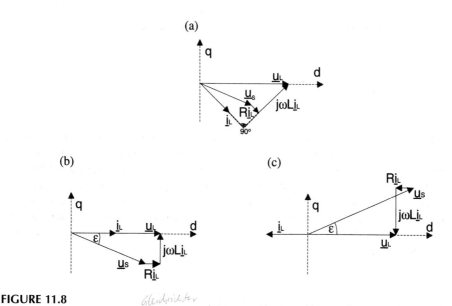

FIGURE 11.8
Phasor diagram for the PWM rectifier. (a) General phasor diagram; (b) rectification at unity power factor; (c) inversion at unity power factor.

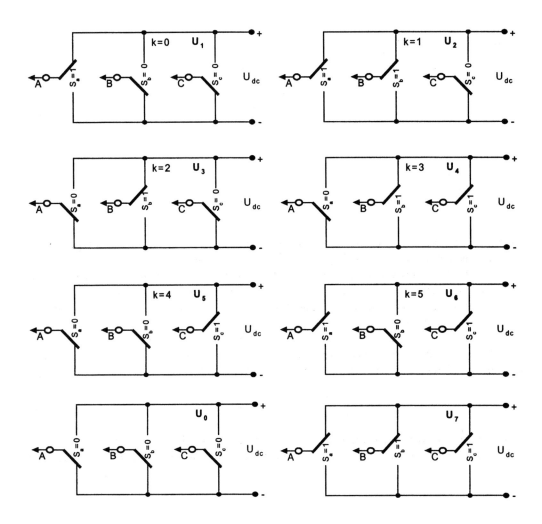

FIGURE 11.9
Switching states of a PWM bridge converter.

the converter. The line current i_L is controlled by the voltage drop across the inductance L interconnecting the two voltage sources (line and converter). It means that the inductance voltage u_I equals the difference between the line voltage u_L and the converter voltage u_S. When we control the phase angle ε and the amplitude of converter voltage u_S, we control indirectly the phase and amplitude of the line current. In this way, the average value and sign of the dc current is subject to control that is proportional to the active power conducted through the converter. The reactive power can be controlled independently with a shift of the fundamental harmonic current I_L with respect to voltage U_L.

11.2.2 Mathematical Description of PWM Rectifier

The basic relationship between vectors of the PWM rectifier is presented in Fig. 11.10.

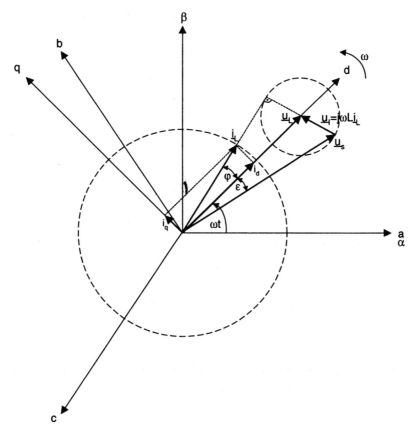

FIGURE 11.10
Relationship between vectors in a PWM rectifier.

11.2.2.1 Description of Line Voltages and Currents. Three-phase line voltage and the fundamental line current are

$$u_a = E_m \cos \omega t \tag{11.4a}$$

$$u_b = E_m \cos\left(\omega t + \frac{2\pi}{3}\right) \tag{11.4b}$$

$$u_c = E_m \cos\left(\omega t - \frac{2\pi}{3}\right) \tag{11.4c}$$

$$i_a = I_m \cos(\omega t + \varphi). \tag{11.5a}$$

$$i_b = I_m \cos\left(\omega t + \frac{2\pi}{3} + \varphi\right) \tag{11.5b}$$

$$i_c = I_m \cos\left(\omega t - \frac{2\pi}{3} + \varphi\right) \tag{11.5c}$$

where $E_m(I_m)$ and ω are amplitudes of the phase voltage (current) and angular frequency, respectively, with the assumption

$$i_a + i_b + i_c \equiv 0. \tag{11.6}$$

We can transform equations (11.4) to the α–β system thanks to Eqs. (11.82), and the input voltages in the α–β stationary frame are expressed by

$$u_{L\alpha} = \sqrt{\frac{3}{2}} E_m \cos(\omega t) \tag{11.7a}$$

$$u_{L\beta} = \sqrt{\frac{3}{2}} \sin(\omega t). \tag{11.7b}$$

The input voltages in the synchronous d–q coordinates (Fig. 11.10) are expressed by

$$\begin{bmatrix} u_{Ld} \\ u_{Lq} \end{bmatrix} = \begin{bmatrix} \sqrt{\frac{3}{2}} E_m \\ 0 \end{bmatrix} = \begin{bmatrix} \sqrt{u_{L\alpha}^2 + u_{L\beta}^2} \\ 0 \end{bmatrix} \tag{11.8}$$

11.2.2.2 Description of Input Voltage in PWM Rectifier.
Line-to-line voltages of a PWM rectifier can be described with the help of Fig. 11.9 as

$$u_{Sab} = (S_a - S_b) \cdot u_{dc} \tag{11.9a}$$

$$u_{Sbc} = (S_b - S_c) \cdot u_{dc} \tag{11.9b}$$

$$u_{Sca} = (S_c - S_a) \cdot u_{dc} \tag{11.9c}$$

and the phase voltages are equal:

$$u_{Sa} = f_a \cdot u_{dc} \tag{11.10a}$$

$$u_{Sb} = f_b \cdot u_{dc} \tag{11.10b}$$

$$u_{Sc} = f_c \cdot u_{dc} \tag{11.10c}$$

where

$$f_a = \frac{2S_a - (S_b + S_c)}{3} \tag{11.11a}$$

$$f_b = \frac{2S_b - (S_a + S_c)}{3} \tag{11.11b}$$

$$f_c = \frac{2S_c - (S_a + S_b)}{3}. \tag{11.11c}$$

The f_a, f_b, f_c are assumed to be 0, $\pm 1/3$, and $\pm 2/3$.

11.2.2.3 Description of PWM Rectifier

Model of Three-Phase PWM Rectifier. The voltage equations for a balanced three-phase system without a neutral connection can be written as (Fig. 11.8b)

$$\underline{u_L} = \underline{u_i} + \underline{u_{conv}} \tag{11.12}$$

$$\underline{u_L} = R\underline{i_L} + \frac{d\underline{i_L}}{dt}L + \underline{u_{conv}} \tag{11.13}$$

$$\begin{bmatrix} u_a \\ u_b \\ u_c \end{bmatrix} = R \begin{bmatrix} i_a \\ i_b \\ i_c \end{bmatrix} + L\frac{d}{dt}\begin{bmatrix} i_a \\ i_b \\ i_c \end{bmatrix} + \begin{bmatrix} u_{Sa} \\ u_{Sb} \\ u_{Sc} \end{bmatrix} \tag{11.14}$$

and additionally for currents,

$$C\frac{du_{dc}}{dt} = S_a i_a + S_b i_b + S_c i_c - i_{dc}. \tag{11.15}$$

The combination of equations (11.10), (11.11), (11.14), and (11.15) can be represented as a three-phase block diagram (Fig. 11.11) [17].

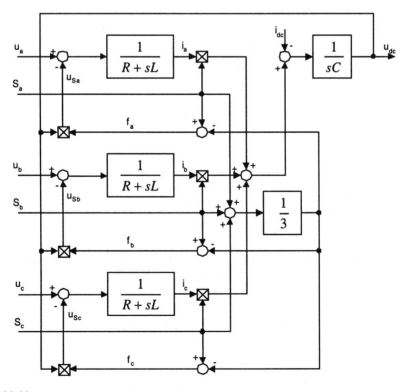

FIGURE 11.11
Block diagram of a voltage source PWM rectifier in natural three-phase coordinates.

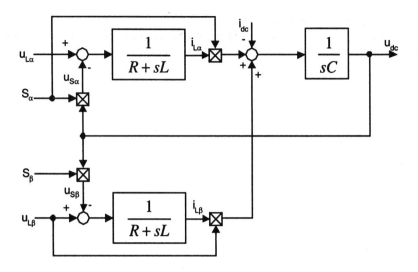

FIGURE 11.12
Block diagram of voltage source PWM rectifier in stationary α–β coordinates.

Model of PWM Rectifier in Stationary Coordinates (α–β). The voltage equations in the stationary α–β coordinates are obtained by applying (11.82) to (11.14) and (11.15) and are written as

$$\begin{bmatrix} u_{L\alpha} \\ u_{L\beta} \end{bmatrix} = R \begin{bmatrix} i_{L\alpha} \\ i_{L\beta} \end{bmatrix} + L \frac{d}{dt} \begin{bmatrix} i_{L\alpha} \\ i_{L\beta} \end{bmatrix} + \begin{bmatrix} u_{S_\alpha} \\ u_{S_\beta} \end{bmatrix} \quad (11.16)$$

and

$$C \frac{du_{dc}}{dt} = (i_{L\alpha} S_\alpha + i_{L\beta} S_\beta) - i_{dc}. \quad (11.17)$$

A block diagram of the α–β model is presented in Fig. 11.12.

Model of PWM Rectifier in Synchronous Rotating Coordinates (d–q). The equations in the synchronous d–q coordinates (Fig. 11.13) are

$$u_{Ld} = R i_{Ld} + L \frac{di_{Ld}}{dt} - \omega L i_{Lq} + u_{Sd} \quad (11.18a)$$

$$u_{Lq} = R i_{Lq} + L \frac{di_{Lq}}{dt} + \omega L i_{Ld} + u_{Sq} \quad (11.18b)$$

$$C \frac{du_{dc}}{dt} = (i_{Ld} S_d + i_{Lq} S_q) - i_{dc}. \quad (11.19)$$

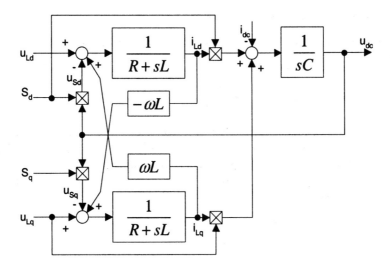

FIGURE 11.13
Block diagram of voltage source PWM rectifier in synchronous d–q coordinates.

R can be practically neglected because the voltage drop on the resistance is much lower than the voltage drop on the inductance, which gives simplified equations (11.13), (11.14), (11.16), (11.18).

$$\underline{u_L} = \frac{d\underline{i_L}}{dt}L + \underline{u_{conv}} \qquad (11.20)$$

$$\begin{bmatrix} u_a \\ u_b \\ u_c \end{bmatrix} = L\frac{d}{dt}\begin{bmatrix} i_a \\ i_b \\ i_c \end{bmatrix} + \begin{bmatrix} u_{Sa} \\ u_{Sb} \\ u_{Sc} \end{bmatrix} \qquad (11.21)$$

$$\begin{bmatrix} u_{L\alpha} \\ u_{L\beta} \end{bmatrix} = L\frac{d}{dt}\begin{bmatrix} i_{L\alpha} \\ i_{L\beta} \end{bmatrix} + \begin{bmatrix} u_{S_\alpha} \\ u_{S_\beta} \end{bmatrix} \qquad (11.22)$$

$$u_{Ld} = L\frac{di_{Ld}}{dt} - \omega L i_{Lq} + u_{Sd} \qquad (11.23a)$$

$$u_{Lq} = L\frac{di_{Lq}}{dt} + \omega L i_{Ld} + u_{Sq}. \qquad (11.23b)$$

The active and reactive power supplied from the source is given by

$$p = \operatorname{Re}\{\underline{u} \cdot \underline{i}^*\} = u_\alpha i_\alpha + u_\beta i_\beta = u_a i_a + u_b i_b + u_c i_c \qquad (11.24)$$

$$q = \operatorname{Im}\{\underline{u} \cdot \underline{i}^*\} = u_\beta i_\alpha - u_\alpha i_\beta = \frac{1}{\sqrt{3}}(u_{bc}i_a + u_{ca}i_b + u_{ab}i_c). \qquad (11.25)$$

11.2.3 Steady-State Properties: Limitation

For proper operation of a PWM rectifier a minimum dc-link voltage is required. Generally it can be determined by the peak of line-to-line supply voltage:

$$V_{dc\,min} > V_{LN(rms)} * \sqrt{3} * \sqrt{2} = 2,45 * V_{LN(rms)}. \tag{11.26}$$

This is a true definition but does not apply in all situations. Other publications [18, 19] define minimum voltage but do not take into account line current (power) and line inductors. The determination of this voltage more complicated and is presented in [20].

Equations (11.23) can be transformed to vector form in synchronous d–q coordinates defining the derivative of the current as

$$L\frac{di_{Ldq}}{dt} = \underline{u}_{Ldq} - j\omega L i_{Ldq} - \underline{u}_{Sdq}. \tag{11.27}$$

Equation (11.27) defines the direction and rate of current vector movement. Six active vectors (U_{1-6}) of input voltage in PWM rectifier rotate clockwise in synchronous d–q coordinates. For vectors U_0, U_1, U_2, U_3, U_4, U_5, U_6, and U_7 the current derivatives are denoted respectively as U_{p0}, U_{p1}, U_{p2}, U_{p3}, U_{p4}, U_{p5}, U_{p6}, and U_{p7} (Fig. 11.14).

The full current control is possible when the current is kept in the specified error area (Fig. 11.15). Figures 11.14 and 11.15 demonstrate that any vectors can force the current vector inside the error area when the angle created by vectors U_{p1} and U_{p2} is $\xi \leq \pi$. It results from the trigonometrical condition that vectors U_{p1}, U_{p2}, U_1, and U_2 form an equilateral triangle for

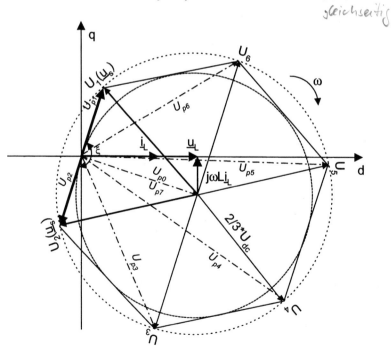

FIGURE 11.14
Instantaneous position of vectors.

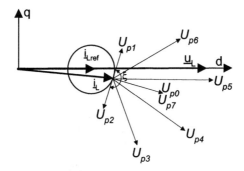

FIGURE 11.15
Limit condition for correct operation of PWM rectifier.

$\xi = \pi$ where $\underline{u}_{Ldq} - j\omega L\underline{i}_{Ldq}$ is an altitude. Therefore, from the simple trigonometrical relationship, it is possible to define the boundary condition as:

$$|\underline{u}_{Ldq} - j\omega L\underline{i}_{Ldq}| = \frac{\sqrt{3}}{2}\underline{u}_{sdq} \tag{11.28}$$

after transformation, assuming that $\underline{u}_{Sdq} = 2/3U_{dc}$, $\underline{u}_{Ldq} = E_m$, and $\underline{i}_{Ldq} = i_{Ld}$ (for UPF) we get the condition for minimal dc-link voltage:

$$u_{dc} > \sqrt{3[E_m^2 + (\omega L i_{Ld})^2]} \quad \text{and} \quad \xi > \pi. \tag{11.29}$$

This equation shows the relation between supply voltage (usually constant), output dc voltage, current (load), and inductance. It also means that the sum of the vector $\underline{u}_{Ldq} - j\omega L\underline{i}_{Ldq}$ should not exceed the linear region of modulation, i.e., the circle inscribed in the hexagon (see Chapter 4, Section 4.2).

The inductor has to be designed carefully because low inductance will give a high current ripple and will make the design more dependent on the line impedance. A high value of inductance will give a low current ripple, but simultaneously reduce the operation range of the rectifier. The voltage drop across the inductance controls the current. This voltage drop is controlled by the voltage of the rectifier but its maximal value is limited by the dc-link voltage. Consequently, a high current (high power) through the inductance requires either a high dc-link voltage or a low inductance (low impedance). Therefore, after transformation of Eq. (11.29) the maximal inductance can be determined as

$$L < \frac{\sqrt{\frac{u_{dc}^2}{3} - E_m^2}}{\omega_1 i_{Ld}}. \tag{11.30}$$

11.3 DIRECT POWER CONTROL

11.3.1 Block Scheme of DPC

The main idea of DPC proposed in [15] and next developed by [14, 37] is similar to the well-known direct torque control (DTC) for induction motors. Instead of torque and stator flux the instantaneous active (p) and reactive (q) powers are controlled (Fig. 11.16).

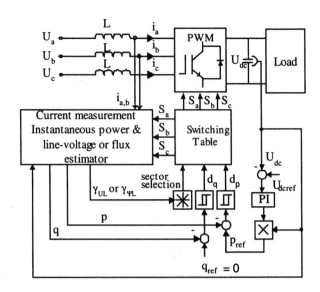

FIGURE 11.16
Block scheme of PWM rectifier.

The commands of reactive power q_{ref} (set to zero for unity power factor) and active power p_{ref} (delivered from the outer PI-DC voltage controller) are compared with the estimated q and p values (see Sections 11.3.2 and 11.3.3), in reactive and active power hysteresis controllers, respectively. The digitized output signal of the reactive power controller is defined as

$$d_q = 1 \quad \text{for} \quad q < q_{ref} - H_q \tag{11.31a}$$

$$d_q = 0 \quad \text{for} \quad q > q_{ref} + H_q, \tag{11.31b}$$

and similarly that of the active power controller as

$$d_p = 1 \quad \text{for} \quad p < p_{ref} - H_p \tag{11.32a}$$

$$d_p = 0 \quad \text{for} \quad p > p_{ref} + H_p, \tag{11.32b}$$

where H_q and H_p are the hysteresis bands.

The digitized variables d_p, d_q and the voltage vector position $\gamma_{UL} = \text{arctg}(u_{L\alpha}/u_{L\beta})$ or flux vector position $\gamma_{\psi L} = \text{arctg}(\psi_{L\alpha}/\psi_{L\beta})$ form a digital word, which by accessing the address of the lookup table selects the appropriate voltage vector according to the switching table (described in Section 11.3.4).

The region of the voltage or flux vector position is divided into 12 sectors, as shown in Fig. 11.17, and the sectors can be numerically expressed as

$$(n-2)\frac{\pi}{6} \leq \gamma_n < (n-1)\frac{\pi}{6}, \quad n = 1, 2, \ldots, 12 \tag{11.33}$$
$$\text{or } (n-5)\frac{\pi}{6} \leq \gamma_n < (n-4)\frac{\pi}{6}$$

Note that the sampling frequency has to be a few times higher than the average switching frequency. This very simple solution allows precise control of instantaneous active and reactive power and errors are limited only by the hysteresis band. No transformation into rotating coordinates is needed and the equations are easily implemented. This method deals with instantaneous variables; therefore, estimated values contain not only a fundamental but also harmonic components. This feature also improves the total power factor and efficiency [14].

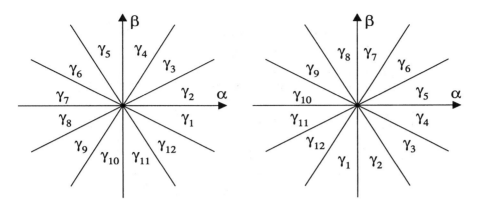

FIGURE 11.17
Sector selection for (left) DPC and (right) VF-DPC.

Further improvements can be achieved by using a sector detection with a PLL (phase-locked loop) generator instead of a zero crossing voltage detector. This guarantees sector detection that is very stable and free of disturbances, even under operation with distorted and unbalanced line voltages.

11.3.2 Instantaneous Power Estimation Based on Voltage

The main idea of voltage-based power estimation for DPC was proposed in [14, 15]. The instantaneous active and reactive power are defined by the product of the three phase voltages and currents by equations (11.24) and (11.25). The instantaneous values of active (p) and reactive power (q) in ac voltage sensorless system are estimated by Eqs. (11.34) and (11.35). The active power p is the scalar product of the current and the voltage, whereas the reactive power q is calculated as their vector product. The first part of both equations represents power in the inductance and the second part is the power of the rectifier.

$$p = L\left(\frac{di_a}{dt}i_a + \frac{di_b}{dt}i_b + \frac{di_c}{dt}i_c\right) + U_{dc}(S_a i_a + S_b i_b + S_c i_c) \quad (11.34)$$

$$q = \frac{1}{\sqrt{3}}\left\{3L\left(\frac{di_a}{dt}i_c - \frac{di_c}{dt}i_a\right) - U_{dc}[S_a(i_b - i_c) + S_b(i_c - i_a) + S_c(i_a - i_b)]\right\}. \quad (11.35)$$

As can be seen in (11.34) and (11.35), the equations have to be changed according to the switching state of the converter, and both equations require a knowledge of the line inductance L. Supply voltage usually is constant; therefore the instantaneous active and reactive power are proportional to i_{Ld} and i_{Lq}.

The ac-line voltage sector is necessary to read the switching table; therefore knowledge of the line voltage is essential. However, once the estimated values of active and reactive power are calculated and the ac-line currents are known, the line voltage can easily be calculated from instantaneous power theory as

$$\begin{bmatrix} u_{L\alpha} \\ u_{L\beta} \end{bmatrix} = \frac{1}{i_{L\alpha}^2 + i_{L\beta}^2}\begin{bmatrix} i_{L\alpha} & -i_{L\beta} \\ i_{L\beta} & i_{L\alpha} \end{bmatrix}\begin{bmatrix} p \\ q \end{bmatrix}. \quad (11.36)$$

The instantaneous power and ac voltage estimators are shown in Fig. 11.18. In spite of its simplicity, this power estimation method has several disadvantages:

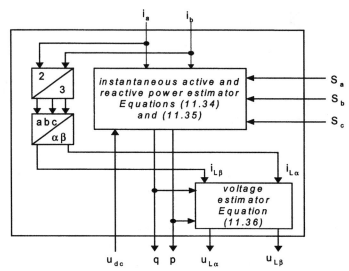

FIGURE 11.18
Instantaneous power estimation based on line voltage

- High values of the line inductance and sampling frequency are needed (an important point for the estimator, because a smooth shape of current is needed).
- Power estimation depends on the switching state. Therefore, calculation of the power and voltage should be avoided at the moment of switching, because of high estimation errors.

11.3.3 Instantaneous Power Estimation Based on Virtual Flux

The virtual flux (VF)-based approach has been proposed to improve the VOC [21, 22]. Here it will be applied for instantaneous power estimation, where voltages imposed by the line power in combination with the ac side inductors are assumed to be quantities related to a virtual ac motor as shown in Section 11.6.2.

With the definitions

$$\underline{\Psi}_L = \int \underline{u}_L dt \qquad (11.37)$$

where

$$\underline{u}_L = \begin{bmatrix} u_{L\alpha} \\ u_{L\beta} \end{bmatrix} = \sqrt{2/3} \begin{bmatrix} 1 & 1/2 \\ 0 & \sqrt{3}/2 \end{bmatrix} \begin{bmatrix} u_{ab} \\ u_{bc} \end{bmatrix} \qquad (11.38)$$

$$\underline{\Psi}_L = \begin{bmatrix} \Psi_{L\alpha} \\ \Psi_{L\beta} \end{bmatrix} = \begin{bmatrix} \int u_{L\alpha} dt \\ \int u_{L\beta} dt \end{bmatrix} \qquad (11.39)$$

$$\underline{i}_L = \begin{bmatrix} i_{L\alpha} \\ i_{L\beta} \end{bmatrix} = \sqrt{2/3} \begin{bmatrix} 3/2 & 0 \\ \sqrt{3}/2 & \sqrt{3} \end{bmatrix} \begin{bmatrix} i_a \\ i_b \end{bmatrix} \qquad (11.40)$$

$$\underline{u}_s = \underline{u}_{conv} = \begin{bmatrix} u_{s\alpha} \\ u_{s\beta} \end{bmatrix} = \sqrt{2/3} \begin{bmatrix} 1 & -1/2 & -1/2 \\ 0 & \sqrt{3}/2 & -\sqrt{3}/2 \end{bmatrix} \begin{bmatrix} u_{AM} \\ u_{BM} \\ u_{CM} \end{bmatrix}, \qquad (11.41)$$

the voltage equation can be written as

$$\underline{u}_L = R\underline{i}_L + \frac{d}{dt}(L\underline{i}_L + \underline{\Psi}_s). \tag{11.42a}$$

In practice, R can be neglected, giving

$$\underline{u}_L = L\frac{d\underline{i}_L}{dt} + \frac{d}{dt}\underline{\Psi}_s = L\frac{d\underline{i}_L}{dt} + \underline{u}_s. \tag{11.42b}$$

Using complex notation, the instantaneous power can be calculated as

$$p = \text{Re}(\underline{u}_L \cdot \underline{i}_L^*) \tag{11.43a}$$
$$q = \text{Im}(\underline{u}_L \cdot \underline{i}_L^*) \tag{11.43b}$$

where * denotes the conjugate line current vector. The line voltage can be expressed by the virtual flux as

$$\underline{u}_L = \frac{d}{dt}\underline{\Psi}_L = \frac{d}{dt}(\Psi_L e^{j\omega t}) = \frac{d\Psi_L}{dt}e^{j\omega t} + j\omega\Psi_L e^{j\omega t} = \frac{d\Psi_L}{dt}e^{j\omega t} + j\omega\underline{\Psi}_L \tag{11.44}$$

where $\underline{\Psi}_L$ denotes the space vector and Ψ_L its amplitude. For the virtual flux oriented d–q coordinates (Fig. 11.33), $\underline{\Psi}_L = \Psi_{Ld}$, and the instantaneous active power can be calculated from (11.43a) and (11.44) as

$$p = \frac{d\Psi_{Ld}}{dt}i_{Ld} + \omega\Psi_{Ld}i_{Lq}. \tag{11.45}$$

For sinusoidal and balanced line voltages, Eq. (11.45) is reduced to

$$\frac{d\Psi_{Ld}}{dt} = 0 \tag{11.46}$$
$$p = \omega\Psi_{Ld}i_{Lq}, \tag{11.47}$$

which means that only the current components orthogonal to the flux $\underline{\Psi}_L$ vector, produce the instantaneous active power. Similarly, the instantaneous reactive power can be calculated as

$$q = -\frac{d\Psi_{Ld}}{dt}i_{Lq} + \omega\Psi_{Ld}i_{Ld} \tag{11.48}$$

and with (11.46) it is reduced to

$$q = \omega\Psi_{Ld}i_{Ld} \tag{11.49}$$

However, to avoid coordinate transformation into d–q coordinates, the power estimator for the DPC system should use stator-oriented quantities, in α–β coordinates (see Fig. 11.33). Using (11.43) and (11.44),

$$\underline{u}_L = \frac{d\Psi_L}{dt}\bigg|_\alpha + j\frac{d\Psi_L}{dt}\bigg|_\beta + j\omega(\Psi_{L\alpha} + j\Psi_{L\beta}) \tag{11.50}$$

$$\underline{u}_L\underline{i}_L^* = \left\{\frac{d\Psi_L}{dt}\bigg|_\alpha + j\frac{d\Psi_L}{dt}\bigg|_\beta + j\omega(\Psi_{L\alpha} + j\Psi_{L\beta})\right\}(i_{L\alpha} - ji_{L\beta}). \tag{11.51}$$

That gives

$$p = \left\{\frac{d\Psi_L}{dt}\bigg|_\alpha i_{L\alpha} + \frac{d\Psi_L}{dt}\bigg|_\beta i_{L\beta} + \omega(\Psi_{L\alpha}i_{L\beta} - \Psi_{L\beta}i_{L\alpha})\right\} \tag{11.52a}$$

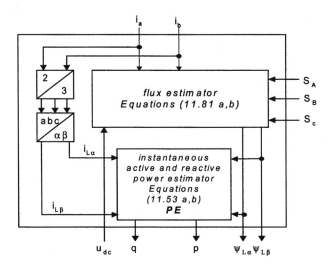

FIGURE 11.19.
Instantaneous power estimator based on virtual flux.

and

$$q = \left\{ -\frac{d\Psi_L}{dt}\bigg|_\alpha i_{L\beta} + \frac{d\Psi_L}{dt}\bigg|_\beta i_{L\alpha} + \omega(\Psi_{L\alpha} i_{L\alpha} + \Psi_{L\beta} i_{L\beta}) \right\}. \tag{11.52b}$$

For sinusoidal and balanced line voltage the derivatives of the flux amplitudes are zero. The instantaneous active and reactive powers can be computed as [23]

$$p = \omega \cdot (\Psi_{L\alpha} i_{L\beta} - \Psi_{L\beta} i_{L\alpha}) \tag{11.53a}$$

$$q = \omega \cdot (\Psi_{L\alpha} i_{L\alpha} + \Psi_{L\beta} i_{L\beta}). \tag{11.53b}$$

The measured line currents i_a, i_b and the estimated virtual flux components $\Psi_{L\alpha}$, $\Psi_{L\beta}$ are delivered to the instantaneous power estimator block (PE) as depicted in Fig. 11.19.

11.3.4 Switching Table

The instantaneous active and reactive power depends on the position of the converter voltage vector, which has an indirect influence for voltage on the inductance and the phase and amplitude of the line current. Therefore a different pattern of switching table can be applied to direct control (DTC, DPC) with influence for control conditions such as instantaneous power and current ripple, switching frequency, and dynamic performance. There are also some papers that proposed a method to compose different switching tables for DTC, but we cannot find many references for DPC. More switching table techniques exist for drives for the sake of a wide range of output frequency and dynamic demands [24–27]. The selection of vector is made so that the error between q and q_{ref} should be within the limits [Eqs. (11.31) and (11.32)]. It depends not only on the error of the amplitude but also on the direction of q, as shown in Fig. 11.20.

Some behavior of direct control is not satisfactory. For instance when the instantaneous reactive power vector is close to one of the sector boundaries, two of four possible active vectors are wrong. These wrong vectors can only change the instantaneous active power without correction of the reactive power error, which is easily visible on the current. A few methods to

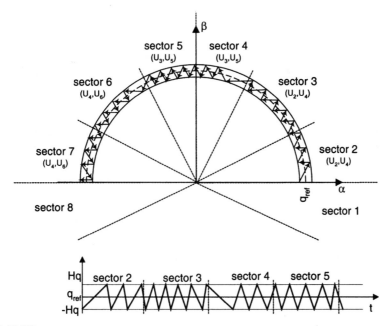

FIGURE 11.20
Selection of voltage vectors for **q**.

improve DPC behavior in the sector borders have been proposed; one of them is to add more sectors or hysteresis levels. Therefore switching tables are generally constructed for the sake of difference in

- Number of sectors
- Dynamic performance
- Two- and three-level hysteresis controllers

11.3.4.1 Number of Sectors. Most often the vector plane is divided into 6 (11.54) or 12 (11.55) sections:

$$(2n - 3)\frac{\pi}{6} \leq \gamma_n < (2n - 1)\frac{\pi}{6} \qquad n = 1, 2, \ldots, 6 \qquad (11.54)$$

$$(n - 2)\frac{\pi}{6} \leq \gamma_n < (n - 1)\frac{\pi}{6} \qquad n = 1, 2, \ldots, 12 \qquad (11.55)$$

11.3.4.2 Hysteresis Controllers. The amplitudes of the instantaneous active and reactive hysteresis band have a relevant effect on the converter performance. In particular, the harmonic current distortion, the average converter switching frequency, the power pulsation, and the losses are strongly affected by the amplitudes of the bands. The controllers proposed by [14] in classical DPC are two-level comparators for instantaneous active and reactive power (Fig 11.21a). Three-level comparators can provide further improvements. The most common combinations of hysteresis controllers for active and reactive power are presented in Fig. 11.21.

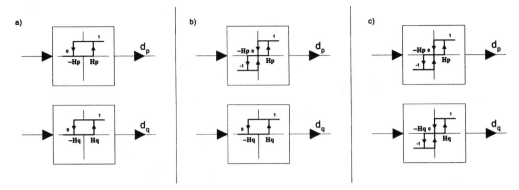

FIGURE 11.21
Hysteresis controllers. (a) Two level; (b) mixed two- and three-level; (c) three-level.

The two-level hysteresis controllers for instantaneous reactive power can be written as

$$\text{if} \quad \Delta q > H_q \quad \text{then} \quad d_q = 1$$
$$\text{if} \quad H_q \leq \Delta q \leq H_q \quad \text{and} \quad d\Delta q/dt > 0 \quad \text{then} \quad d_q = 0$$
$$\text{if} \quad -H_q \leq \Delta q \leq H_q \quad \text{and} \quad d\Delta q/dt < 0 \quad \text{then} \quad d_q = 1$$
$$\text{if} \quad \Delta q < -H_q \quad \text{then} \quad d_q = 0.$$

The three-level hysteresis controllers for instantaneous active power can be written as a sum of two-level hysteresis, as

$$\text{if} \quad \Delta p > H_p \quad \text{then} \quad d_p = 1$$
$$\text{if} \quad 0 \leq \Delta p \leq H_p \quad \text{and} \quad d\Delta p/dt > 0 \quad \text{then} \quad d_q = 0$$
$$\text{if} \quad 0 \leq \Delta p \leq H_p \quad \text{and} \quad d\Delta p/dt < 0 \quad \text{then} \quad d_p = 1$$
$$\text{if} \quad -H_q \leq \Delta p \leq 0 \quad \text{and} \quad d\Delta p/dt > 0 \quad \text{then} \quad d_p = -1$$
$$\text{if} \quad -H_q \leq \Delta p \leq 0 \quad \text{and} \quad d\Delta p/dt < 0 \quad \text{then} \quad d_p = 0$$
$$\text{if} \quad \Delta p < -H_p \quad \text{then} \quad d_p = -1$$

11.3.4.3 Features of Switching Table. General features of switching table and hysteresis controllers:

- The switching frequency depends on the hysteresis wide of the active and reactive power comparators.
- By using three-level comparators, the zero vectors are naturally and systematically selected. Thus the number of switchings is considerably smaller than in the system with two-level hysteresis comparators.
- Zero vectors decrease switching frequency but provide short-circuits for the line-to-line voltage.
- Zero vectors $U_0(000)$ and $U_7(111)$ should be appropriately chosen.
- For DPC only the neighbor vectors should be selected that decrease dynamics but provide low current and power ripples (low THD).

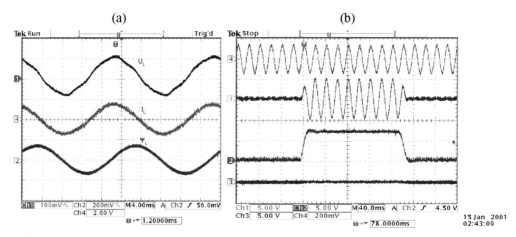

FIGURE 11.22
Experimental waveforms with distorted line voltage for VF-DPC. (a) Steady state. From top: line voltage, line currents (5 A/div) and virtual flux. (b) Transient of the step change of the load (startup of converter). From top: line voltages, line currents (5 A/div), instantaneous active (2 kW/div), and reactive power.

- A switching table with PLL (phase-locked loop) sector detection guarantees a very stable operation that is free of disturbances, even under distorted and unbalanced line voltages.
- Twelve sectors provide more accurate voltage vector selection.

Experimental results for virtual flux based DPC are shown in Fig. 11.22.

11.3.5 Summary

The presented DPC system constitutes a viable alternative to the VOC (see Section 11.4) of PWM rectifiers, but conventional solutions shown by [14] possess several disadvantages:

- Because the estimated values are changed every time according to the switching state of the converter, it is important to have high sampling frequency. This requires a very fast microprocessor and A/D converters. Good performance is obtained at 80 kHz sampling frequency. It means that results precisely depend on sampling time.
- Because the switching frequency is not constant, a high value of inductance is needed (about 10%). (This is an important point for the line voltage estimation because a smooth shape of current is needed.)
- The wide range of the variable switching frequency can result in trouble when designing the necessary input filter.
- Calculation of power and voltage should be avoided at the moment of switching because it gives high errors of the estimated values.

Based on duality with a PWM inverter-fed induction motor, a new method of instantaneous active and reactive power calculation has been proposed. This method uses the estimated virtual flux (VF) vector instead of the line voltage vector in the control. Consequently voltage-sensorless line power estimation is much less noisy thanks to the natural low-pass behavior of the integrator used in the calculation algorithm. Also, differentiation of the line current is avoided in this scheme. So, the presented VF-DPC of PWM rectifiers has the following features and advantages:

- No line voltage sensors required.
- Simple and noise robust power estimation algorithm, easy to implement in a DSP.
- Lower sampling frequency (as conventional DPC [14]).
- Sinusoidal line currents (low THD).
- No separate PWM voltage modulation block.
- No current regulation loops.
- Coordinate transformation and PI controllers not required.
- High dynamic, decoupled active and reactive power control.
- Power and voltage estimation gives the possibility to obtain instantaneous variables with all harmonic components, which has an influence for improvement of total power factor and efficiency.

The typical disadvantages are:

- Variable switching frequency.
- Solution requires a fast microprocessor and A/D converters.

As shown in this section, thanks to duality phenomena, an experience with the high-performance decoupled PWM inverter-fed induction motor control can be used to improve properties of the PWM rectifier control.

11.4 VOLTAGE AND VIRTUAL FLUX ORIENTED CONTROL

11.4.1 Introduction

Similarly to FOC of an induction motor [7], voltage oriented control (VOC) and virtual flux oriented control (VFOC) for line-side PWM rectifier is based on coordinate transformation between stationary $\alpha-\beta$ and synchronous rotating $d-q$ reference systems. Both strategies guarantee fast transient response and high static performance via an internal current control loops. Consequently, the final configuration and performance of the system largely depends on the quality of the applied current control strategy [13]. The easiest solution is hysteresis current control, which provides a fast dynamic response, good accuracy, no dc offset, and high robustness. However, the major problem of hysteresis control is that its average switching frequency varies with the dc load current, which makes the switching pattern uneven and random, thus resulting in additional stress on switching devices and difficulties of LC input filter design.

Therefore, several strategies are reported in the literature to improve performance of current control [13, 28]. Among the presented regulators, the widely used scheme for high-performance current control is the $d-q$ synchronous controller, where the currents being regulated are dc quantities, which eliminates steady-state error.

11.4 VOLTAGE AND VIRTUAL FLUX ORIENTED CONTROL

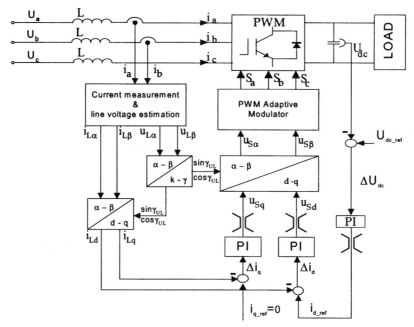

FIGURE 11.23
Block scheme of ac voltage sensorless VOC.

11.4.2 Block Scheme of VOC

The conventional control system uses closed-loop current control in a rotating reference frame; the voltage oriented control (VOC) scheme is shown in Fig. 11.23.

A characteristic feature for this current controller is the processing of signals in two coordinate systems. The first is the stationary α–β and the second is the synchronously rotating d–q coordinate system. Three-phase measured values are converted to equivalent two-phase system α–β and then are transformed to the rotating coordinate system in a block α–β/d–q:

$$\begin{bmatrix} k_d \\ k_q \end{bmatrix} = \begin{bmatrix} \cos\gamma_{UL} & \sin\gamma_{UL} \\ -\sin\gamma_{UL} & \cos\gamma_{UL} \end{bmatrix} \begin{bmatrix} k_\alpha \\ k_\beta \end{bmatrix} \quad (11.56a)$$

Thanks to this type of transformation the control values are dc signals. An inverse transformation d–q/α–β is achieved on the output of the control system, and it gives the rectifier reference signals in stationary coordinates:

$$\begin{bmatrix} k_\alpha \\ k_\alpha \end{bmatrix} = \begin{bmatrix} \cos\gamma_{UL} & -\sin\gamma_{UL} \\ \sin\gamma_{UL} & \cos\gamma_{UL} \end{bmatrix} \begin{bmatrix} k_d \\ k_q \end{bmatrix} \quad (11.56b)$$

For both coordinate transformations the angle of the voltage vector γ_{UL} is defined as

$$\sin\gamma_{UL} = u_{L\beta}/\sqrt{(u_{L\alpha})^2 + (u_{L\beta})^2} \quad (11.57a)$$

$$\cos\gamma_{UL} = u_{L\alpha}/\sqrt{(u_{L\alpha})^2 + (u_{L\beta})^2}. \quad (11.57b)$$

In voltage-oriented d–q coordinates, the ac line current vector $\underline{i}_L$ is split into two rectangular components $\underline{i}_L = [i_{Ld}, i_{Lq}]$ (Fig. 11.24). The component i_{Lq} determines reactive power, whereas i_{Ld} decides active power flow. Thus the reactive and the active power can be controlled

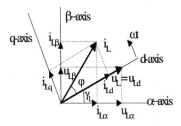

FIGURE 11.24
Vector diagram of VOC. Coordinate transformation of current, line, and rectifier voltage from stationary α–β coordinates to rotating d–q coordinates.

independently. The UPF condition is met when the line current vector, i_L, is aligned with the line voltage vector, u_L (Fig. 11.8b). By placing the d-axis of the rotating coordinates on the line voltage vector a simplified dynamic model can be obtained.

The voltage equations in the d–q synchronous reference frame in accordance with Eqs. (11.18) are as follows:

$$u_{Ld} = R \cdot i_{Ld} + L \frac{di_{Ld}}{dt} + u_{Sd} - \omega_s \cdot L \cdot i_{Lq} \tag{11.58}$$

$$0 = u_{Lq} = R \cdot i_{Lq} + L \frac{di_{Lq}}{dt} + u_{Sq} + \omega_s \cdot L \cdot i_{Ld}. \tag{11.59}$$

Regarding Fig. 11.23, the q-axis current is set to zero in all conditions for unity power factor control while the reference current i_{Ld} is set by the dc-link voltage controller and controls the active power flow between the supply and the dc-link. For $R \approx 0$ Eqs. (11.58) and (11.59) can be reduced to

$$E_m = L \frac{di_{Ld}}{dt} + u_{Sd} - \omega_s \cdot L \cdot i_{Lq} \tag{11.60}$$

$$0 = L \frac{di_{Lq}}{dt} + u_{Sq} + \omega_s \cdot L \cdot i_{Ld}. \tag{11.61}$$

Assuming that the q-axis current is well regulated to zero, the following equations hold true:

$$E_m = L \frac{di_{Ld}}{dt} + u_{Sd} \tag{11.62}$$

$$0 = u_{Sq} + \omega_s \cdot L \cdot i_{Ld}. \tag{11.63}$$

As a current controller, the PI-type can be used. However, the PI current controller has no satisfactory tracing performance especially for the coupled system described by Eqs. (11.60) and (11.61). Therefore for a high-performance application with accuracy current tracking at a dynamic state the decoupled controller diagram for the PWM rectifier should be applied, as shown in Fig. 11.25 [29]:

$$u_{Sd} = \omega L i_{Lq} + E_m + \Delta u_d \tag{11.64}$$

$$u_{Sq} = -\omega L i_{Ld} + \Delta u_q \tag{11.65}$$

where Δ are the output signals of the current controllers:

$$\Delta u_d = k_p(i_d^* - i_d) + k_i \int (i_d^* - i_d) dt \tag{11.66}$$

$$\Delta u_q = k_p(i_q^* - i_q) + k_i \int (i_q^* - i_q) dt. \tag{11.67}$$

11.4 VOLTAGE AND VIRTUAL FLUX ORIENTED CONTROL

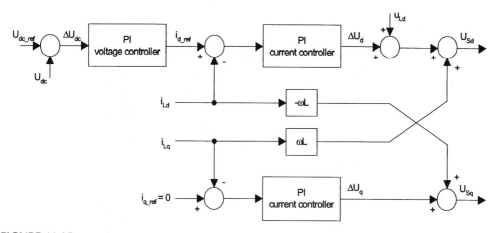

FIGURE 11.25
Decoupled current control of PWM rectifier.

Remark: For detailed design procedure of synchronous current controller, see Example 4.2 in Section 4.3.3.3.

The output signals from PI controllers after $dq/\alpha\beta$ transformation (Eq. (11.56b)) are used for switching signal generation by a space vector modulator (SVM) (see Section 4.2). The waveforms for VOC are shown in Fig. 11.26.

11.4.3 Block Scheme of VFOC

The concept of virtual flux (VF) can also be applied to improve the VOC scheme, because disturbances superimposed onto the line voltage influence directly the coordinate transformation in control system (11.57). Sometimes this is only solved by phase-locked loops (PLLs), but the quality of the controlled system depends on how effectively the PLL's have been designed [30]. Therefore, it is easier to replace the angle of the line voltage vector γ_{UL} by the angle of the VF vector $\gamma_{\Psi L}$, because $\gamma_{\Psi L}$ is less sensitive than γ_{UL} to disturbances in the line voltage, thanks to the natural low-pass behavior of the integrators in (11.81) (because nth harmonics are reduced by a factor $1/k$. For this reason, it is not necessary to implement PLLs to achieve robustness in the flux-oriented scheme, since $\underline{\Psi}_L$ rotates much more smoothly than $\underline{u}_L$. The angular displacement of line flux vector $\underline{\Psi}_L$ in α–β coordinates is defined as

$$\sin \gamma_{\Psi L} = \Psi_{L\beta}/\sqrt{(\Psi_{L\alpha})^2 + (\Psi_{L\beta})^2} \tag{11.68a}$$

$$\cos \gamma_{\Psi L} = \Psi_{L\alpha}/\sqrt{(\Psi_{L\alpha})^2 + (\Psi_{L\beta})^2}. \tag{11.68b}$$

The virtual flux oriented control (VFOC) scheme is shown in Fig. 11.27. The vector of virtual flux lags the voltage vector by 90° (Fig. 11.28). Therefore, for the UPF condition, the d-component of the current vector, i_L, should be zero.

In the virtual flux oriented coordinates Eqs. (11.58) and (11.59) are transformed into

$$u_{Lq} = \frac{di_{Lq}}{dt} + u_{Sq} + \omega \cdot L \cdot i_{Ld} \tag{11.69}$$

$$0 = L\frac{di_{Ld}}{dt} + u_{Sd} - \omega \cdot L \cdot i_{Lq} \tag{11.70}$$

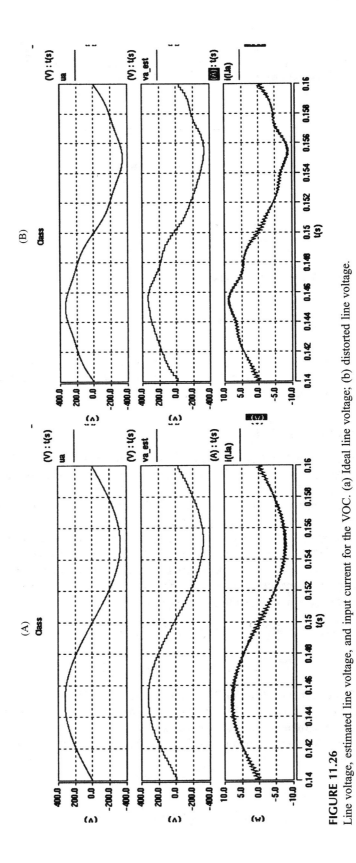

FIGURE 11.26
Line voltage, estimated line voltage, and input current for the VOC. (a) Ideal line voltage; (b) distorted line voltage.

11.4 VOLTAGE AND VIRTUAL FLUX ORIENTED CONTROL

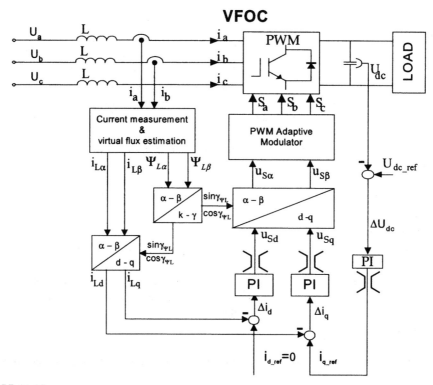

FIGURE 11.27
Block scheme of VFOC.

and for $i_{Ld} = 0$ Eqs. (11.69) and (11.70) can be described as

$$u_{Lq} = L\frac{di_{Lq}}{dt} + u_{Sq} \tag{11.71}$$

$$0 = u_{Sd} - \omega \cdot L \cdot i_{Lq}. \tag{11.72}$$

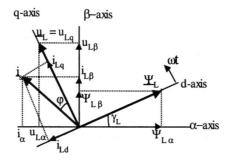

FIGURE 11.28
Vector diagram of VFOC. Coordinate transformation of line and rectifier voltage and current from fixed α–β coordinates to rotating d–q coordinates.

11.4.4 Summary

It is shown by simulations and experimental results that line voltage estimators perform very well even under unbalanced and predistorted conditions. Furthermore, the current follows the voltage fairly well with VOC control strategies that provide a high value of the total power factor. However, sometimes sinusoidal currents are desired even under unbalanced and predistorted conditions because sinusoidal current do not produce nonsinusoidal voltage drops across the line impedances. For the conventional VOC scheme some compensating algorithms exists [31] or the concept of virtual flux (VF) can be applied to improve the VOC scheme.

VOC with line voltage estimation and VFOC with a virtual flux estimator, compared to DPC, exhibits some advantages:

- Low sampling frequency (cheaper A/D converters and microcontrollers) for good performance, e.g., 5 kHz
- Fixed switching frequency (easier design of the input filter)
- Possible implementation of modern PWM techniques (see Section 4.2)

Moreover, the VFOC provide improved rectifier control under nonideal line voltage condition, because ac voltage sensorless operation is much less noisy thanks to the natural low-pass behavior of the integrator used in the flux estimator.

There are also some disadvantages for both control strategies:

- Coupling occurs between active and reactive components and some decoupling solution is required
- Coordinate transformation and PI controllers are required

11.5 SENSORLESS OPERATION

11.5.1 Introduction

Normally, the PWM rectifier needs three kinds of sensors:

- Dc-voltage sensor (1 sensor)
- Ac-line current sensors (2 or 3 sensors)
- Ac-line voltage sensors (2 or 3 sensors)

The sensorless methods provide technical and economical advantages to the system, such as simplification, isolation between the power circuit and controller, reliability, and cost effectiveness. The possibility to reduce the number of the expensive sensors have been studied especially in the field of motor drive application [32], but the rectifier application differs from the inverter operation for the following reasons:

- Zero vector will short the line power
- The line operates at constant frequency 50 Hz and synchronization is necessary

The most used solutions for reducing of sensors include:

- Ac voltage and current sensorless

- Ac current sensorless
- Ac voltage sensorless

11.5.2 Ac Voltage and Current Sensorless

Reduction of current sensors especially for ac drives is well known [32]. The two-phase currents may be estimated based on information on dc link current and reference voltage vector in every PWM period. No full protection is the main practical problem in the system. Particularly for the PWM rectifier, the zero vectors (U_0, U_7) present no current in the dc-link and three line phases are short circuited simultaneously. A new improved method presented in [33, 34] is to sample the dc-link current few a times in one switching period. The basic principle of current reconstruction is shown in Fig. 11.29 together with a voltage vector's patterns determining the direction of current flow. One active voltage vector is needed to reconstruct one phase current and another voltage vector is used to reconstruct a second phase current using values measured from the dc current sensor. The relationship between the applied active vectors and the phase currents measured from the dc link sensor is shown in Table 11.2, which is based on eight voltage vectors composed of six active vectors and two zero vectors.

The main problem of ac current estimation is based on minimum pulse-time for dc-link current sampling. It appears when either of two active vectors is not present, or is applied only for a short time. In such a case, it is impossible to reconstruct the phase current. This occurs in the case of reference voltage vectors passing one of the six possible active vectors or a low modulation index (Fig. 11.30). The minimum short time to obtain a correct estimation depends on the rapidness of the system, delays, cable length, and *dead-time* [34]. The way to solve the problem is to adjust the PWM pulses or to allow that no current information is present in some time period. Therefore improved compensation consists of calculating the errors, which are introduced by the PWM pulse adjustment and then compensating for this error in the next switching period.

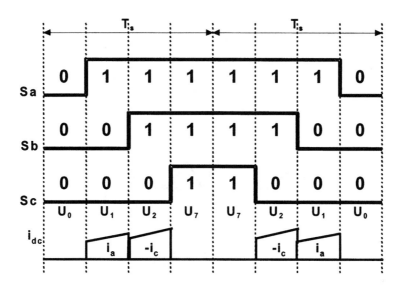

FIGURE 11.29
PWM signals and dc link in current sector I.

Table 11.2 Relationship between Voltage Vector of Converter, Dc-Link Current and Line Currents

Voltage vector	Dc link current i_{dc}
$U_1(100)$	$+i_a$
$U_2(110)$	$-i_c$
$U_3(010)$	$+i_b$
$U_4(011)$	$-i_a$
$U_5(001)$	$+i_c$
$U_6(101)$	$-i_b$
$U_0(000)$	0
$U_7(111)$	0

The ac voltage and current sensorless methods in spite of cost reduction possess several disadvantages: higher current ripple; problems with the DPWM and overmodulation mode; sampling is presented few times per switching state, which is not technically convenient; unbalance and startup condition are not reported.

11.5.3 Ac Current Sensorless

This very simple solution is based on measuring the voltage on the inductor (u_I) in two lines. Supply voltage can be estimated with the assumption that the voltage on the inductance is equal to the line voltage when the zero vector occurs in the converter (Fig. 11.31).

On the basis of the voltage on the inductor described in Eq. (11.73),

$$u_{IR} = L\frac{di_{LR}}{dt}, \tag{11.73}$$

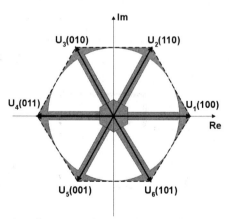

FIGURE 11.30
Voltage vector area requiring the adjustment of PWM signals, when a reference voltage passes one of possible six active vectors and in case of low modulation index and overmodulation.

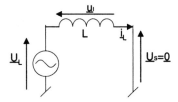

FIGURE 11.31
PWM rectifier circuit when the zero voltage is applied.

the line current can be calculated as

$$i_{LR} = \frac{1}{L}\int u_{IR}dt. \qquad (11.74)$$

Thanks to Eq. (11.74) the observed current will not be affected by derivation noise, but it directly reduces the dynamic of the control. This leads to problems with overcurrent protection.

11.5.4 Ac Voltage Sensorless

Previous solutions present some overvoltage and overcurrent protection troubles. Therefore the dc-voltage and the ac-line current sensors are an important part of the overvoltage and overcurrent protection, while it is possible to replace the ac-line voltage sensors with a line voltage estimator or virtual flux estimator, which is described in the next section.

11.6 VOLTAGE AND VIRTUAL FLUX ESTIMATION

11.6.1 Line Voltage Estimation

An important requirement for a voltage estimator is to estimate the voltage correctly under unbalanced conditions and preexisting harmonic voltage distortion. Not only should the fundamental component be estimated correctly, but also the harmonic components and the voltage imbalance. It gives a higher total power factor [14]. It is possible to calculate the voltage across the inductance by current differentiation. The line voltage can then be estimated by adding the rectifier voltage reference to the calculated voltage drop across the inductor. However, this approach has the disadvantage that the current is differentiated and noise in the current signal is gained through the differentiation. To prevent this a voltage estimator based on the power estimator of [14] can be applied. In [14] the current is sampled and the power is estimated several times in every switching state.

In conventional space vector modulation (SVM) for three-phase voltage source converters, the currents are sampled during the zero-vector states because no switching noise is present and a filter in the current feedback for the current control loops can be avoided. Using Eqs. (11.75) and (11.76) the estimated power in this special case (zero states) can be expressed as

$$p = L\left(\frac{di_a}{dt}i_a + \frac{di_b}{dt}i_b + \frac{di_c}{dt}i_c\right) = 0 \quad (11.75)$$

$$q = \frac{3L}{\sqrt{3}}\left(\frac{di_a}{dt}i_c - \frac{di_c}{dt}i_a\right). \quad (11.76)$$

It should be noted that in this special case it is only possible to estimate the reactive power in the inductor. Since p and q are dc values, it is possible to prevent the noise of the differentiated current by use of a simple (digital) low-pass filter. This ensures robust and noise-insensitive performance of the voltage estimator.

Based on instantaneous power theory, the estimated voltages across the inductance are

$$\begin{bmatrix} u_{I\alpha} \\ u_{I\beta} \end{bmatrix} = \frac{1}{i_{L\alpha}^2 + i_{L\beta}^2} \begin{bmatrix} i_{L\alpha} & -i_{L\beta} \\ i_{L\beta} & i_{L\alpha} \end{bmatrix} \begin{bmatrix} 0 \\ q \end{bmatrix} \quad (11.77)$$

where $u_{I\alpha}$, $u_{I\beta}$ are the estimated values of the three-phase voltages across the inductance L, in the fixed α–β coordinates.

The estimated line voltage u_{est} can now be found by adding the voltage reference of the PWM rectifier to the estimated inductor voltage [35]:

$$\vec{u}_{est} = \vec{u}_{conv} + \vec{u}_I. \quad (11.78)$$

11.6.2 Virtual Flux Estimation

The line voltage in combination with the ac side inductors are assumed to be quantities related to a virtual ac motor as shown in Fig. 11.32. Thus, R and L represent the stator resistance and the stator leakage inductance of the virtual motor, and phase-to-phase line voltages U_{ab}, U_{bc}, U_{ca} would be induced by a virtual air gap flux. In other words, the integration of the phase-to-phase voltages leads to a virtual line flux vector $\underline{\Psi}_L$, in stationary α–β coordinates (Fig. 11.33).

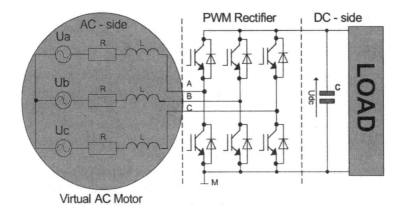

FIGURE 11.32
Representation of a three-phase PWM rectifier system for bidirectional power flow.

11.6 VOLTAGE AND VIRTUAL FLUX ESTIMATION

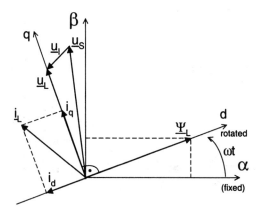

FIGURE 11.33
Reference coordinates and vectors. $\underline{\Psi}_L$, virtual line flux vector; $\underline{u}_S$, converter voltage vector; $\underline{u}_L$, line voltage vector; $\underline{u}_I$, inductance voltage vector; $\underline{i}_L$, line current vector.

A virtual flux equation can be presented as [36] (Fig. 11.34), similarly to Eq. (11.78):

$$\underline{\psi}_{est} = \underline{\psi}_{conv} + \underline{\psi}_L. \tag{11.79}$$

Based on the measured dc-link voltage U_{dc} and the converter switch states S_a, S_b, S_c the converter voltages are estimated as follows:

$$u_{S\alpha} = \sqrt{\frac{2}{3}} U_{dc}\left(S_a - \frac{1}{2}(S_b + S_c)\right) \tag{11.80a}$$

$$u_{S\beta} = \frac{1}{\sqrt{2}} U_{dc}(S_b - S_c) \tag{11.80b}$$

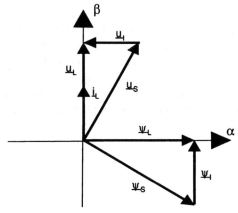

FIGURE 11.34
Relation between voltage and flux vectors.

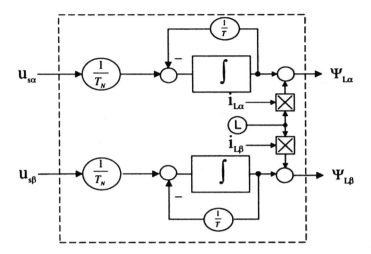

FIGURE 11.35
Block scheme of virtual flux estimator.

Then, the virtual flux $\underline{\Psi}_L$ components are calculated from (11.80):

$$\Psi_{L\alpha} = \int \left(u_{S\alpha} + L\frac{di_{L\alpha}}{dt} \right) dt \qquad (11.81a)$$

$$\Psi_{L\beta} = \int \left(u_{S\beta} + L\frac{di_{L\beta}}{dt} \right) dt. \qquad (11.81b)$$

The virtual flux component calculation is shown in Fig. 11.35.

11.7 CONCLUSION

Dc link-based power conversion is in a stage of transition from diode to controlled rectifier in the front end. An elegant solution is to use a PWM rectifier/inverter system. The most popular control strategies for PWM inverter-fed induction motors are FOC and DTC, which correspond to VOC and DPC for PWM line rectifiers, respectively. Thanks to duality phenomena, experience with decoupled induction motor control can be used to improve performance of the PWM rectifier control. The main features of four basic control strategies for PWM rectifiers are summarized in Table 11.3.

11.8 APPENDIX

For the typical three-phase system without neutral wire, zero sequence component i_0 of the current system does not exist ($i_a + i_b + i_c = 0$). This finally gives a simple realization of signal

Table 11.3 Performances of Control Techniques for PWM Rectifiers

Technique	Advantages	Disadvantages
VOC	• Fixed switching frequency (easier design of the input filter) • Advanced PWM strategies can be used • Cheaper A/D converters	• Coordinate transformation and decoupling between active and reactive components is required • Complex algorithm • Input power factor lower than that for DPC
DPC	• No separate PWM voltage modulation block • No current regulation loops • No coordinate transformation • Good dynamics • Simple algorithm • Decoupled active and reactive power control • Instantaneous variables with all harmonic components estimated (improvement of the power factor and efficiency)	• High values of the inductance and sampling frequency are needed (important point for the estimator, because smooth shape of the current waveform is needed) • Power and voltage estimation should be avoided at the moment of switching (it yields high errors) • Variable switching frequency • Fast microprocessor and A/D converters required
VFOC	• Fixed switching frequency • Advanced PWM strategies can be used • Cheaper A/D converters	• Coordinate transformation and decoupling between active and reactive components is required • Complex algorithm • Input power factor lower than that for VF-DPC
VF-DPC	• Simple and noise-resistant power estimation algorithm, easy to implement in a DSP • Lower sampling frequency than that for DPC • Low THD of line currents at a distorted and unbalanced supply voltage (sinusoidal line currents) • No separate PWM voltage modulation block • No current regulation loops • No coordinate transformation • Good dynamics • Simple algorithm • Decoupled active and reactive power control	• Variable switching frequency • Fast microprocessor and A/D converters required

processing, thanks to only two signals in the α–β coordinate, which is the main advantage of $abc/\alpha\beta$ transformation. With this assumption,

$$\begin{bmatrix} u_\alpha \\ u_\beta \end{bmatrix} = \sqrt{\frac{2}{3}} \begin{bmatrix} 1 & -1/2 & -1/2 \\ 0 & \sqrt{3}/2 & -\sqrt{3}/2 \end{bmatrix} \begin{bmatrix} u_a \\ u_b \\ u_c \end{bmatrix} \quad (11.82)$$

and

$$\begin{bmatrix} i_\alpha \\ i_\beta \end{bmatrix} = \sqrt{\frac{2}{3}} \begin{bmatrix} 1 & -1/2 & -1/2 \\ 0 & \sqrt{3}/2 & -\sqrt{3}/2 \end{bmatrix} \begin{bmatrix} i_a \\ i_b \\ i_c \end{bmatrix}. \quad (11.83)$$

REFERENCES

[1] D. E. Rice, A detailed analysis of six-pulse converter harmonic currents. *IEEE Trans. Indust. Appl.* **30**, 294–304 (1994).
[2] J. C. Salmon, Operating a three-phase diode rectifier with a low-input current distortion using a series-connected dual boost converter. *IEEE Trans. Power Electron.* **11**, 592–603 (1996).
[3] J. C. Salmon, Reliable 3-phase PWM boost rectifiers employing a stacked dual boost converter subtopology. *IEEE Trans. Indust. Appl.* **32**, 542–551 (1996).
[4] J. W. Kolar, F. Stogerer, J. Minibock, and H. Ertl, A new concept for reconstruction of the input phase currents of a three-phase/switch/level PWM (VIENNA) rectifier based on neutral point current measurement. *Proc. IEEE-PESC Conf.*, pp. 139–146, 2000.
[5] ABB, *ACS600 catalog*, 2000.
[6] Siemens, *Simovert Masterdrives Vector Control Catalog*, 2000.
[7] M. P. Kaźmierkowski and H. Tunia, *Automatic Control of Converter-Fed Drives*. Elsevier, 1994.
[8] M. P. Kazmierkowski, M. A. Dzieniakowski, and W. Sulkowski, The three phase current controlled transistor dc link PWM converter for bi-directional power flow. *Proc. PEMC Conf.*, Budapest, pp. 465–469, 1990.
[9] H. Kohlmeier, O. Niermeyer, and D. Schroder, High dynamic four quadrant ac-motor drive with improved power-factor and on-line optimized pulse pattern with PROMC. *Proc. EPE Conf.*, Brussels, pp. 3.173–3.178, 1985.
[10] O. Niermeyer and D. Schroder, Ac-motor drive with regenerative braking and reduced supply line distortion. *Proc. EPE Conf.*, Aachen, pp. 1021–1026, 1989.
[11] B. T. Ooi, J. C. Salmon, J. W. Dixon, and A. B. Kulkarni, A 3-phase controlled current PWM converter with leading power factor. *Proc. IEEE-IAS Conf.*, pp. 1008–1014, 1985.
[12] B. T. Ooi, J. W. Dixon, A. B. Kulkarni, and M. Nishimoto, An integrated ac drive system using a controlled current PWM rectifier/inverter link. *Proc. IEEE-PESC Conf.*, pp. 494–501, 1986.
[13] M. P. Kazmierkowski and L. Malesani, Current control techniques for three-phase voltage-source PWM converters: a survey. *IEEE Trans. Indust. Electron.* **45**, 691–703 (1998).
[14] T. Noguchi, H. Tomiki, S. Kondo, and I. Takahashi, Direct power control of PWM converter without power-source voltage sensors. *IEEE Trans. Indust. Appl.* **34**, 473–479 (1998).
[15] T. Ohnishi, Three-phase PWM converter/inverter by means of instantaneous active and reactive power control. *Proc. IEEE-IECON Conf.*, pp. 819-824, 1991.
[16] D. Zhou and D. Rouaud, Regulation and design issues of A PWM three-phase rectifier. *Proc. IEEE-IECON Conf.*, pp. 485–489, 1999.
[17] V. Blasko and V. Kaura, A new mathematical model and control of a three-phase ac–dc voltage source converter. *IEEE Trans. Power Electron.* **12**, 116–122 (1997).
[18] S. Bhowmik, R. Spee, G. C. Alexander, and J. H. R. Enslin, New simplified control algorithm for synchronous rectifiers. *Proc. IEEE-IECON Conf.*, pp. 494–499, 1995.
[19] S. Bhowmik, A. van Zyl, R. Spee, and J. H. R. Enslin, Sensorless current control for active rectifiers. *Proc. IEEE-IAS Conf.*, pp. 898–905, 1996.
[20] A. Sikorski, An ac/dc converter with current vector modulator. *Electr. Power Qual. Utilisation* **6**, 29–40 (2000).

[21] J. L. Duarte, A. Van Zwam, C. Wijnands, and A. Vandenput, Reference frames fit for controlling PWM rectifiers. *IEEE Trans. Indust. Electron.* **46**, 628–630 (1999).

[22] P. J. M. Smidt and J. L. Duarte, A unity power factor converter without current measurement. *Proc. EPE Conf.*, Sevilla, pp. 3.275–3.280, 1995.

[23] M. Malinowski, M. P. Kaźmierkowski, S. Hansen, F. Blaabjerg, and G. D. Marques, Virtual flux based direct power control of three-phase PWM rectifier. *IEEE Trans. Indust. Appl.* **37**, 1019–1027 (2001).

[24] T. G. Habetler and D. M. Divan, Control strategies for direct torque control. *IEEE Trans. Indust. Appl.* **28**, 1045–1053 (1992).

[25] M. P. Kaźmierkowski and W. Sulkowski, A novel control scheme for transistor PWM inverter-fed induction motor drive. *IEEE Trans. Indust. Electron.* **38**, 41–47 (1991).

[26] C. Lascu, I. Boldea, and F. Blaabjerg, A modified direct torque control for induction motor sensorless drive. *IEEE Trans. Indust. Appl.* **36**, 122–130 (2000).

[27] I. Takahashi and T. Noguchi, A new quick response and high efficiency control strategy of induction motor. *Proc. IEEE-IAS Conf.*, pp. 496–502, 1985.

[28] M. Cichowlas, D. L. Sobczuk, M. P. Kaźmierkowski, and M. Malinowski, Novel artificial neural network based current controller for PWM rectifiers. *Proc. EPE-PEMC Conf.*, Kosice, pp. 1.41–1.46, 2000.

[29] B. H. Kwon, J. H. Youm, and J. W. Lim, A line-voltage-sensorless synchronous rectifier. *IEEE Trans. Power Electron.* **14**, 966–972 (1999).

[30] P. Barrass and M. Cade, PWM rectifier using indirect voltage sensing. *IEE Proc.—Electr. Power Appl.* **146**, 539–544 (1999).

[31] S. J. Huang and J. C. Wu, A control algorithm for three-phase three-wired active power filters under nonideal mains voltage. *IEEE Trans. Power Electron.* **14**, 753–760 (1999).

[32] F. Blaabjerg, J. K. Pedersen, U. Jaeger, and P. Thoegersen, Single current sensor technique in the dc link of three-phase PWM-VS inverters: a review and a novel solution. *IEEE Trans. Indust. Appl.* **33**, 1241–1253 (1997).

[33] W. C. Lee, T. J. Kweon, D. S. Hyun, and T. K. Lee A novel control of three-phase PWM rectifier using single current sensor. *Proc. IEEE-PESC Conf.*, 1999.

[34] B. Andersen, T. Holmgaard, J. G. Nielsen, and F. Blaabjerg, Active three-phase rectifier with only one current sensor in the dc-link. *Proc. PEDS Conf.*, pp. 69–74, 1999.

[35] S. Hansen, M. Malinowski, F. Blaabjerg, and M. P. Kazmierkowski, Control strategies for PWM rectifiers without line voltage sensors. *Proc. IEEE-APEC Conf.*, Vol. 2, pp. 832–839, 2000.

[36] M. Weinhold, A new control scheme for optimal operation of a three-phase voltage dc link PWM converter. *Proc. PCIM Conf.*, pp. 371–3833, 1991.

[37] V. Manninen, Application of direct torque control modulation technology to a line converter. *Proc. EPE Conf.*, Sevilla, pp. 1.292–1.296, 1995.

[38] R. Barlik and M. Nowak, Three-phase PWM rectifier with power factor correction. *Proc. EPN'2000*, Zielona Góra, pp. 57–80, 2000 (in Polish).

CHAPTER 12

Power Quality and Adjustable Speed Drives

STEFFAN HANSEN and PETER NIELSEN
Danfoss Drives A/S, Grasten, Denmark

During recent years we have seen increased focus on power quality. In many places production and distribution of electric power have been split into separate business units. Distribution companies have to supply electric power to the end users, and of course the quality of that service is of great importance when negotiating the price.

Discussing power quality is a complex issue. Power quality is measured in terms of such things as line interruptions, sags, brown-outs, flicker, transients, phase unbalance, and distortion. For all devices in the grid there is a general issue of immunity and emission regarding all these power quality parameters.

The increased use of power electronic equipment is of great importance when dealing with power quality. Most of the nonlinear currents in the utility grid today are caused by the input stage of power electronic converters. Power electronics create problems, but at the same time they also solve many of the power quality related problems. An adjustable speed drive (ASD), for example, feeds harmonic currents into the grid (compared to a directly line operated motor), affecting the power quality in a negative manner by distorting the supply voltage. But at the same time the soft-starting capability of the ASD prevents large surge currents when starting up the motor, and thus reduces voltage sags in the grid. This example illustrates the complexity of the power-quality discussion very well: it is not always black or white and power quality should be viewed in a broad perspective on a system level.

Power quality issues should not be dealt with on an equipment level, but a system-wide solution should be sought. Economically this is often a more attractive solution than suboptimizations many places in the system. Anyhow, the most feasible technical solutions are not always seized because of practical and economical reasons. It is not always possible to decide for total system solutions.

A wide range of commercial products for improvement of power quality is available on the market today. Passive as well as active solutions such as filters, line voltage restorers, and uninterruptible power supplies are becoming more and more common as the focus on power quality is increased.

National and international standards such as IEEE 519 and EN 61000 are setting limits to power quality related parameters as harmonic currents and voltages. In general, limits to harmonic current emission point toward suboptimization of the performance of a single piece of equipment, whereas limits to voltage distortion point towards a system-level optimization. Voltage distortion is the product of harmonic current and harmonic impedance of the supply system and thus cannot be predicted from knowledge of the equipment performance alone. The voltage distortion caused by a given nonlinear load is highly dependent on the grid in which the load is installed.

About 56% of electrical power is used for electric motors, of which about 10% are controlled by ASDs which are becoming commodities in more and more applications today [1]. This chapter focuses on the power quality aspects of three-phase ASDs connected in low-voltage networks. Emissions and immunity toward the different power quality parameters are discussed and practical examples are given. Some of the most common clean-power interfaces for the input stage are reviewed and system solutions are pointed out as well.

12.1 EFFECT OF POOR UTILITY POWER QUALITY

Power quality is characterized by a large amount of parameters such as different distortion factors, measures of unbalance, availability of supply, and many, many more. Some parameters describe power system behavior whereas others are related to the individual pieces of equipment connected to the grid.

The power factor λ is one of the key performance parameters for equipment. The power factor of a piece of equipment is defined as the ratio of the active power P to the apparent power S:

$$\lambda = \frac{P}{S}. \tag{12.1}$$

The power factor can also be expressed as the active fundamental current over the rms current. As the rms current determines the rating of cables, fuses, and switchgear, the power factor is a good measure of how hard the equipment loads the system. By definition, the power factor is in the range from zero to one. The power factor definition may also be applied to an entire system or subsystem.

The apparent power S can be larger than the active power for two reasons:

(a) A phase shift φ between current and voltage causing a reactive power flow, Q

(b) A contribution to the rms current by harmonic components in the current

If we assume voltage and current to be purely sinusoidal, the power-factor definition reduces to $\lambda = \cos(\varphi)$, also known as displacement power factor. But in distorted systems it should always be kept in mind that the harmonic currents contribute to the rms current and thus the power factor.

The effect of the harmonic currents is frequently taken into account by calculating the total harmonic distortion (THD). The definition of THD is based upon the Fourier expansion of nonsinusoidal waveforms. For a distorted current waveform, the total harmonic distortion is defined as

$$\text{THD} = \sqrt{\sum_{h=2}^{h=h_{max}} \left(\frac{I_h}{I_1}\right)^2} \cdot 100\% \quad (12.2)$$

where I_h is the amplitude of the harmonic current of hth order, I_1 is the fundamental component of the waveform (50 or 60 Hz component), and h_{max} is the maximum number of harmonics to be included (typically 40 or 50). The THD is used as performance index for distorted voltages as well. It is common to multiply the THD by 100% to obtain a percentage of distortion.

From the definition that the rms value of a distorted current equals

$$I_{rms} = \sqrt{\sum_{h=1}^{h=h_{max}} I_h^2} \quad (12.3)$$

we can derive the following relationship between current THD and rms current:

$$I_{rms} = I_1 \sqrt{1 + \text{THD}^2}. \quad (12.4)$$

Using this relationship we can generalize the power-factor definition to take harmonic currents into account as well, only knowing the THD of the current:

$$\lambda = \frac{P}{S} = \frac{I_1 \cos(\varphi)}{I_{rms}} = \frac{\cos(\varphi)}{\sqrt{1 + \text{THD}^2}}. \quad (12.5)$$

The diode rectifier has the feature that the fundamental current is almost in phase with the supply voltage. This means that the displacement power factor is almost 1, or $\cos(\varphi) = 1$. For an ASD this means that the reactive power into the motor is not seen from the line side. The reactive power is circulated through the inverter switches and not fed to the motor from the distribution transformer. In this way the ASD actually lowers the system load by cancelling the reactive power from the motor. The trade-off is that the current is no longer sinusoidal, so the ASD decreases the fundamental component but increases the harmonic components.

The resulting power factor may easily be higher with the ASD than for a directly line-operated motor as shown in Fig. 12.1.

But, as mentioned before, power quality is more than just power factor and THD calculations. The list of phenomena related to power quality is long and complex. A few selected topics are briefly discussed in the following sections.

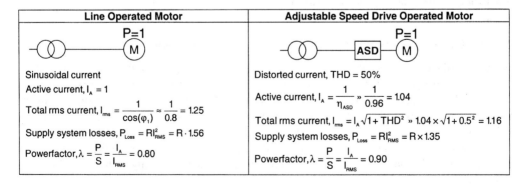

FIGURE 12.1
Power-factor of line operating motor and ASD controlled motor.

On a system level, the calculation of voltage distortion caused by the harmonic current is very important. All equipment connected to the system has to operate under the distorted voltage. Therefore the voltage distortion is the most important parameter from a system point of view. Many standards such as IEEE 519 set limits to the total harmonic distortion of the voltage.

The harmonic voltage drop depends upon the impedance at that harmonic frequency. The voltage distortion depends upon the harmonic spectrum and not upon the THD of the current alone. One ampere fifth harmonic causes, for example, a reactive voltage drop five times larger than 1 ampere fundamental current. Thus, the voltage distortion can be recalculated as

$$V_{\text{THD}} = \sqrt{\sum_{h=2}^{\infty}\left(\frac{hX_1 I_h}{V_1}\right)^2} \times 100\% = \frac{X_1}{V_1}\sqrt{\sum_{h=2}^{\infty}(hI_h)^2} \times 100\% = \frac{1}{I_{\text{sc}}}\sqrt{\sum_{h=2}^{\infty}(I_h h)^2} \times 100\%. \quad (12.6)$$

For a given nonlinear load it is convenient to define a "harmonic constant," Hc, as [2]

$$\text{Hc} = \sqrt{\sum_{h=2}^{\infty}\left(h\frac{I_h}{I_1}\right)^2} \times 100\%. \quad (12.7)$$

The harmonic constant depends, of course, upon the rectifier topology. Typical values for the harmonic constant are given later in this chapter. Using the definition of the harmonic constant, the voltage THD is calculated by the following expression:

$$V_{\text{THD}} = \text{Hc}\frac{P_{\text{ASD}}}{S_{\text{SC}}}[\%]. \quad (12.8)$$

The lower the harmonic constant, the lower the current distortion and thus the voltage distortion.

The ratio of short circuit power and rated equipment input power is commonly known as the short circuit ratio [3]:

$$R_{\text{sce}} = \frac{S_{\text{SC}}}{P_{\text{ASD}}}. \quad (12.9)$$

Thus we can define the voltage distortion by knowing only the short-circuit ratio and the harmonic constant:

$$V_{\text{THD}} = \frac{H_C}{R_{\text{sce}}}. \quad (12.10)$$

Knowing the rectifier topology and the short-circuit ratio it is possible to make a very simple calculation of the resulting voltage distortion in the network.

12.1.1 Effect of Utility Voltage Unbalance on ASD Equipment

Voltage unbalance often occurs in supply systems. The main cause is single-phase loads that are not evenly distributed across all three phases. When a three-phase rectifier is connected to an unbalanced grid, some undesired effects occur. First of all, the rectifier starts to draw/generate third harmonic currents from the supply. From the dc link the unbalance results in a ripple voltage (of twice the supply frequency) due to difference in crest voltages of the three phases. This influences the conduction intervals of the diodes and the input current waveform is further distorted. Typically, the current has two uneven "humps" when the supply voltage is unbalanced (Fig. 12.2).

The diode rectifier is quite insensitive to voltage unbalance and also has a smoothing effect of the voltage unbalance. The phase having the lowest voltage also has to deliver the lowest current,

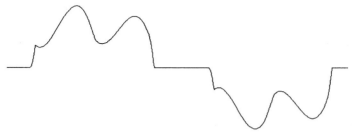

FIGURE 12.2
Current drawn by a rectifier connected to an unbalanced grid.

as the conduction interval of this phase is smaller than for the other phases. Thus, the load is reduced on the phase which has the lowest voltage.

12.1.2 Effect of Voltage Sags on ASDs

Voltage sags are a commonly known phenomenon in supply systems (Fig. 12.3). A voltage sag (or dip) is a disturbance where the rms value of the line voltage is reduced for a period ranging from one half-cycle of the voltage to 500 ms. Shorter occurrences are regarded as transient disturbances. Occurrences during longer than 500 ms are defined as an undervoltage condition.

A typical cause of voltage sags is the direct line start of large induction motors that normally draw 5 to 7 times their rated current during startup. Short circuits in other branches of the supply system are also a common origin of voltage sags. Also, loose or defective wiring can cause voltage sags due to increased system impedance.

For ASDs or other equipment with rectifier front ends, the sag will sooner or later result in a loss of dc-link voltage. Of course the size of the dc-link capacitor and the load of the equipment will determine when the capacitor "runs out of energy" and the dc voltage decreases to a level determined by the sagging line voltage. A critical situation occurs when the dc-link voltage reaches a level that makes the ASD trip on an undervoltage condition. Adjusting this limit to a low level will increase the immunity of the ASD toward line sags. Other precautions such as reducing the load could also be taken. For example, in an HVAC installation, reduced speed is often preferred to a trip.

12.1.3 Phase Loss and Line Interruptions

An important feature of ASDs is how they react during short line interruptions or loss of a single input phase. These situations affect the ASD application all the way to the mechanical system

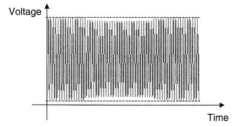

FIGURE 12.3
Typical voltage sag recorded in a low-voltage network.

and are therefore of extreme importance to the end user of the equipment. In the case of complete loss of supply voltage for a shorter period it is often an advantage to have motors running on ASDs compared to line operation. If the voltage is back before the rotor flux of the still rotating machinery has decreased to zero, there is a risk that the line voltage is in counterphase to the emf of the machine when reconnected. In this case the machine will draw currents far beyond nominal values from the line and might blow fuses or damage other equipment.

Using an intelligent ASD in such situations, it is possible to make a flying start of the machine as it is possible to generate an output voltage of the ASD at any angle. Once the position of the counter-emf is known, the motor can be ramped smoothly back to its operating point.

In the case of loss of a single phase for a shorter or longer period, the ASD is still able to produce balanced and sinusoidal voltages to the motor. Depending on the capacitance of the dc link the output voltage might be limited in this mode of operation. Another possibility is to limit the output power of the ASD during the phase loss to avoid overloading the two healthy phases (or the phase legs of the diode bridge in these two phases).

12.1.4 Effect of Capacitor Switching Transients on ASDs

In induction motors and other large linear loads current is usually lagging voltage. It is common practice to improve the displacement power factor ($\cos(\varphi)$) of electrical supply systems by insertion of capacitor banks in the network. The capacitor banks are switched in and out according to the load situation of the system.

The capacitor banks have some undesired side effects that should be taken carefully into account when adding nonlinear loads to a system with power-factor correcting capacitor banks. The combination of the (inductive) short-circuit impedance at the point of installation and the capacitors creates a parallel resonance circuit. The harmonic resonance of this circuit can be calculated as [5]

$$H_r = \sqrt{\frac{S_{cap}}{S_{sc}}} \qquad (12.11)$$

where S_{cap} is the total three-phase kVA rating of the capacitor bank, S_{sc} is the three-phase short-circuit power (kVA) at the point of installation, and H_r is the resonance expressed by its corresponding harmonic order. If the harmonic resonance is close to one of the characteristic harmonics of the nonlinear load, large currents may flow in the resonance circuit and lead to malfunction or breakdown of the equipment.

Another aspect of the capacitor banks is the overvoltages generated when switching the capacitor banks in and out. The presence of a diode-bridge rectifier close to the capacitor bank will have a positive effect on these switching overvoltages as the dc-link capacitor will absorb some of the energy from the overvoltage and in this way reduce the overvoltage on the line and protect other equipment in the network. An example of this is shown in Fig. 12.4.

Again we see an example where it is not easy to determine whether the diode rectifier is good or bad for the power quality. Resonance problems may occur, but on the other hand the diode bridge limits the transient overvoltages caused by switching the capacitors.

The following sections put focus on different three-phase rectifier topologies and evaluate their harmonic performance. Focus is upon current THD and harmonic constant. Different system-level solutions minimizing voltage distortion are also presented.

12.2 HARMONIC CURRENT GENERATION OF STANDARD ASD

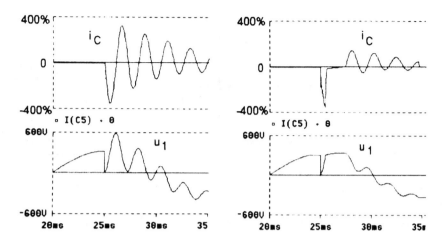

FIGURE 12.4
The effect of a diode-bridge rectifier on capacitor switching transients. Left: without diode bridge. Right: with diode bridge [6].

12.2 HARMONIC CURRENT GENERATION OF STANDARD ASDs

The diode rectifier is a highly nonlinear load and it is therefore considered to be the main source of harmonic currents in today's power system. The circuit diagrams of a typical single-phase and three-phase diode rectifier are shown in Fig. 12.5, and the typical ac-line currents of these rectifier configurations are shown in Fig. 12.6. It is clearly seen that the current is not sinusoidal.

It should be mentioned that there is also other equipment that causes harmonic currents, but its influence on the overall power system is limited compared to that of the diode rectifier. However, this equipment may have significant influence on the local power system.

12.2.1 Harmonic Currents Generated by the Diode Rectifier

Even though the diode rectifier is a simple topology, predicting the harmonic currents generated by the diode rectifier is quite difficult. The reason for this is that the line currents are highly dependent on the line impedance and the impedance in the dc-link in the form of the dc-link inductance and dc-link capacitor. In this section the harmonic currents of the single- and three-

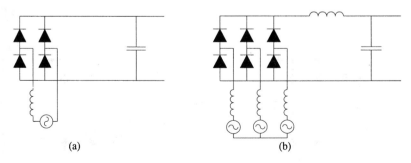

FIGURE 12.5
(a) Single-phase diode rectifier with a capacitor in the dc-link for smoothing the dc voltage. (b) Three-phase diode rectifier with dc-link capacitor and a dc-link inductance for smoothing the input current.

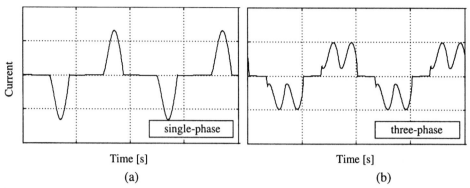

FIGURE 12.6
Typical line current of (a) single-phase diode rectifier, (b) three-phase diode rectifier.

phase diode rectifier are shown as a function of the line and dc-link impedance. These numbers might very well be used to calculate the resulting harmonic voltage distortion as described in Section 12.1.

It is assumed that the diode rectifier is of the voltage stiff type. This is a fair assumption since most of the diode rectifiers used in todays power electronic converters are of this type. This is also the case for the single-phase diode rectifier.

Some three-phase diode rectifiers have an additional inductance in the dc-link to suppress harmonic currents and to ensure continuous current supply into the dc-link capacitor. This has also the advantage of increased lifetime of the capacitors used in the dc-link. However, some manufacturers choose to omit the dc-link inductance and offer additional ac-reactance. Single-phase diode rectifiers tend to be low-cost and low-power. Therefore, a dc-link inductance normally is not used. So, the harmonic current distortion of three diode rectifier topologies is discussed:

- The basic three-phase diode rectifier without any dc-link inductance (Fig. 12.7a)
- The three-phase diode rectifier with a 3% dc-link inductance (Fig. 12.7b)
- The basic single-phase diode rectifier without any dc-link inductance (Fig. 12.8)

Note that $L_{dc} = 3\%$ is at the lower end of the average value used in the industry (usually 3–5%).

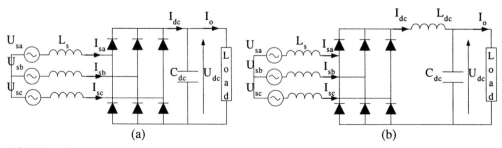

FIGURE 12.7
The three-phase diode rectifier (a) without any additional inductance and (b) with a 3% dc-link inductance.

12.2 HARMONIC CURRENT GENERATION OF STANDARD ASD

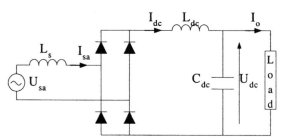

FIGURE 12.8
The basic single-phase diode rectifier.

12.2.1.1 The Single-Phase Diode Rectifier. Figure 12.9 shows the Hc and THD$_i$ values of the single-phase diode rectifier for varying short circuit ratio. The lowest THD$_i$ value of the single-phase diode rectifier is 61%. The lowest Hc value equals 225%.

12.2.1.2 The Three-Phase Diode Rectifier. Figure 12.10 shows the Hc and THD$_i$ values of the three-phase diode rectifier with and without a 3% dc-link inductance and varying short-circuit ratio. The lowest THD$_i$ value of both rectifiers is close to 25%. The lowest Hc value equals 150%.

In Fig. 12.10a the THD$_i$ and Hc are increasing more rapidly from $R_{sce} = 20$–80 than above 80. This is due to the different conduction modes of the basic diode rectifier. Below a short-circuit ratio of 80 the current of the basic three-phase diode rectifier is discontinuous (DCM) as shown in Fig. 12.11. Above this ratio the current becomes continuous (CCM). In between these conduction modes (in a very limited short-circuit ratio range) there is another discontinuous mode, where the commutation is started between the diodes, but the current is still discontinuous (DCCM II). For the three-phase diode rectifier with sufficient dc-link inductance the current is independent of the short-circuit ratio.

Figures 12.9 and 12.10 clearly show that the Hc and THD$_i$ values are highly dependent on the ac impedance. Therefore, it is important to know basic system parameters such as the short-circuit power at the connection point and the nominal load of the diode rectifier before it is possible to estimate the harmonic current distortion of the diode rectifier.

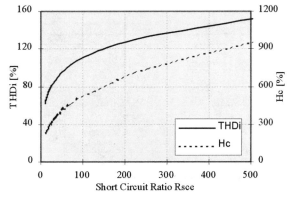

FIGURE 12.9
Hc and THD$_i$ values for the single-phase diode rectifier with varying short-circuit ratio.

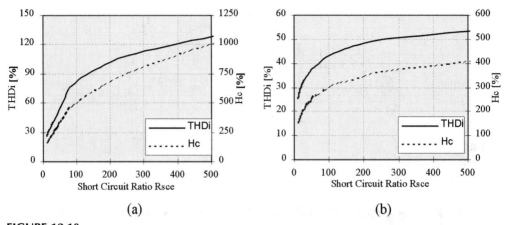

FIGURE 12.10
Hc and THD_i values for the three-phase diode rectifier as a function of the short-circuit ratio without (a) and with (b) 3% dc-link inductance.

12.3 CLEAN POWER UTILITY INTERFACE RECTIFIERS FOR ASDs

A large number of rectifier topologies are known to reduce the line-side harmonic currents of ASDs compared to the diode rectifier. The most popular topologies for ASDs are described in the following.

12.3.1 Twelve-Pulse Rectifier

Multipulse (especially 12-pulse) rectifiers are frequently used today to reduce the harmonic line currents of ASDs. The characteristic harmonics of the diode rectifier are in general expressed as

$$h = (np \pm 1), \qquad n = 1, 2, 3, \ldots, \tag{12.12}$$

where p is the pulse number, which is defined as the number of nonsimultaneous commutations per period.

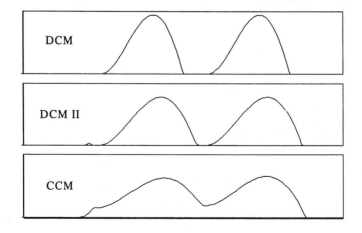

FIGURE 12.11
The three modes of the basic three-phase diode rectifier: discontinuous conduction mode (DCM), discontinuous conduction mode II (DCM II), continuous conduction mode (CCM).

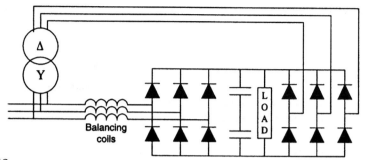

FIGURE 12.12
Twelve-pulse configuration with one transformer.

The input current of a 6-pulse diode rectifier therefore has 5th, 7th, 11th, 13th, etc., harmonic components, whereas the harmonic components of a 12-pulse diode rectifier are 11th, 13th, 23rd, 25th, etc. Normally, the 5th and 7th harmonic currents of a 6-pulse rectifier are dominant. Because the 12-pulse topology ideally eliminates the 5th and 7th harmonic components the harmonic current distortion is significantly reduced compared to a 6-pulse rectifier. The input current harmonics of a 18-pulse rectifier ideally has 17th, 19th, 35th, 37th, etc., components. These rectifiers are normally considered to be clean power converters, because the THD_i is less than 5% under normal operating conditions.

To obtain multipulse performance a minimum of two 6-pulse converters (12-pulse) must be supplied from voltages with different phase shift. A possible 12-pulse configuration with one phase-shifting transformer is shown in Fig. 12.12 and the resulting line-currents are shown in Fig. 12.13.

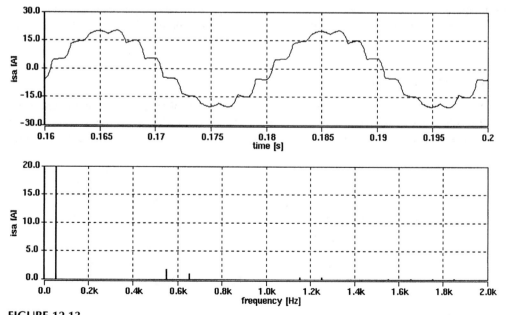

FIGURE 12.13
Line-current and Fourier series of the simulated parallel 12-pulse rectifier with $R_{sce} = 20$. $THD_i = 10.5\%$ and $Hc = 138\%$.

The required phase shift can be calculated as a function of number of six- pulse converters used:

$$\text{phase shift}° = \frac{60}{\text{number of converters}}. \quad (12.13)$$

According to this equation a 12-pulse configuration requires a phase shift of 30°, an 18-pulse configuration one of 20° and a 24-pulse configuration requires a phase shift of 15°.

The main problem of the 12-pulse rectifier shown is that for optimal performance the two converters must share the current equally. In the case of Fig. 12.12, with only one transformer, there should be added extra reactance to the 6-pulse rectifier without a transformer. This extra reactance (balancing coils) should equal the leakage reactance of the transformer. However, adding only balancing coils to ensure equal current sharing is not recommended. To ensure some degree of equal current sharing even under nonideal conditions a center-tapped current transformer, the so-called interphase transformer (IPT), is put into the dc-link.

The interphase transformer is a center-tapped current transformer where the primary and secondary side have equal numbers of windings. Ideally, for the flux to be zero, both currents have to be equal.

By introducing the IPT some of the current-sharing problems can be solved, but because the interphase transformer is an ac device, unequal current sharing caused by unequal dc-voltage output of the two rectifiers cannot be solved. Basically the IPT is only ensuring equal ripple current of both converters. Therefore it is still a requirement that the voltage drop of both rectifiers are equal. This includes the voltage drop of the diodes as well as the voltage drop of the transformer/balancing coils.

As with the 6-pulse rectifier, the 12-pulse rectifier is highly dependent on the line impedance and input voltage quality. And it is known that a predistorted grid results in unequal current sharing of the two parallel rectifier bridges. At predistorted voltage some 5th and 7th harmonic current distortion must therefore be expected. Experience shows that a $THD_i = 10\text{--}20\%$ and $Hc = 125\text{--}175\%$ can be expected as realistic harmonic performance. There are solutions to overcome the increased harmonic distortion at voltage unbalance and predistortion, such as by using a three-winding transformer with galvanic isolation and putting the two diode rectifiers in series.

12.3.2 Active Rectifier

Because of the capabilities to regenerate power, near sinusoidal input current and controllable dc-link voltage the active rectifier, as shown in Fig. 12.14, is popular in high-performance ASDs where frequent acceleration and deacceleration are needed. Due to the harmonic limiting

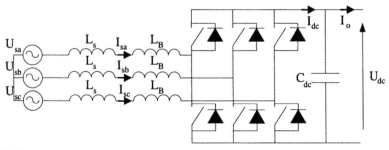

FIGURE 12.14
The active rectifier.

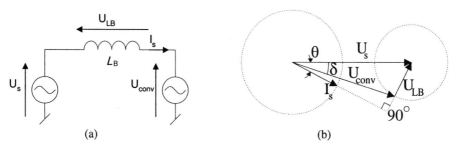

FIGURE 12.15
(a) Simple diagram of two voltage sources connected via an inductance L_B. (b) General vector diagram of the active rectifier.

standards and the increased focus on harmonic currents and voltages in general, the active rectifier may replace the diode rectifier in other applications as well.

The power topology of the active rectifier is identical to that of the PWM–voltage source converter (VSC). It is obvious that the control strategy has a significant impact on the harmonic currents generated.

The basic control of the active PWM rectifier is explained in Fig. 12.15, where the line current I_s is controlled by the voltage drop across an inductance L_B interconnecting the two voltage supplies (source and converter). The inductance voltage (U_{LB}) equals the difference between the line voltage (U_s) and the converter voltage (U_{conv}). It can also be seen in Figs. 12.15 and 12.16 that it is possible to control both the active and the reactive power flow.

The conventional voltage oriented control strategy (VOC) [7, 8] in the rotating d–q axis reference frame is one of the most popular control strategies for active rectifiers, as shown in Fig. 12.17. The advantage of the rotating d–q axis frame is that the controlled quantities such as

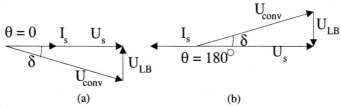

FIGURE 12.16
Vector diagram for unity power factor for both (a) rectification mode and (b) inversion mode.

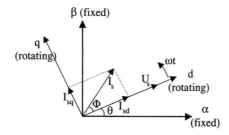

FIGURE 12.17
Coordinate transformation of line and rectifier voltage and current from fixed α–β coordinates to rotating d–q coordinates.

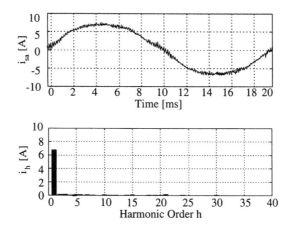

FIGURE 12.18
Recorded line current of an active rectifier laboratory setup, together with the Fourier spectrum of the line current. The current $THD_i = 4.1\%$.

voltages and currents become dc values. This simplifies the expressions for control purposes and simple linear controllers can be used.

By placing the d-axis of the rotating frame on the line voltage vector a simplified dynamic model can be obtained in the rotating frame. The q voltage equals zero by definition, and hence in order to have a unity displacement power factor the q current has to be controlled to zero as well.

12.3.3 Future Perspectives

As shown previously the harmonic current of the active rectifier can almost be controlled to zero so that sinusoidal line current can be obtained (Fig. 12.18). However, the active rectifier is switching on the grid and LCL filters should therefore be used instead of only a boost inductance as done in the experimental results presented earlier. Problems with switching noise without LCL filters have been reported [9]. So far only a few papers have reported work with the design of the LCL filter and the influence of the filter on the control of the active rectifier. Also, very few papers discuss working with EMI-related problems of the active rectifier, such as the increased common mode and differential mode noise compared to the diode rectifier. It is believed that these problems in the future will receive more focus.

Because of the increased power range of ASDs during recent years and the active rectifier's capability of power regeneration it is believed that the active rectifier becomes a state-of-the-art solution in applications where power regeneration is beneficial, such as hoisting or high-inertia applications.

12.4 SYSTEM-LEVEL HARMONIC REDUCTION TECHNIQUES

Instead of reducing the harmonic current emission of the ASDs on the equipment level, it is possible to reduce the harmonic distortion on a system level. Large industrial plants, for example, have shunt capacitor banks for displacement power factor correction, and since capacitors have low impedance to currents with higher frequencies it is obvious to use these capacitor banks for filtering harmonic currents. The power source impedance can often be represented by a simple inductive reactance and the simplest filter approach is just to connect the shunt capacitor bank directly to the grid. However, this can be very dangerous because of parallel resonance as mentioned in Section 12.1.4. Also, this may not be satisfactory because a large capacitor rating would be required to provide low impedance at the 5th harmonic. And using a capacitor bank with more VAr compensation than used in the plant would result in leading power factor and possible overvoltage. Because of the complex nature of the passive filter design, passive filters are normally only used in large industrial plants (in the MVA range) or on the utility side to attenuate harmonic distortion.

12.4.1 Active Filters

One way to overcome the disadvantages of the passive shunt filter is to use an active filter. The active filter is an emerging technology and several manufacturers are offering active filter for harmonic current reduction in ASD applications. This section reviews some of the most important features. Also some basic control strategies are reviewed because these have significant influence on the harmonic currents.

The active filter is controlled to draw a compensating current i_{af} from the utility, so that it cancels current harmonics on the ac side of the diode rectifier as shown in Fig. 12.19.

The most widespread topology is the shunt active filter consisting of six active switches, e.g., IGBTs. Again, the power topology of the active filter is basically the same as that of the active rectifier (i.e., VSC). Lately also active series filters and hybrid systems combining active and passive filters have been considered. These topologies are not covered here, but a review can be found in [10].

Obviously, the control strategy chosen for the active filter has a significant impact on the remaining system harmonic distortion level. There are several different control strategies. The strategy used will depend on the application of the active filter.

One of the first control strategies that made active filters interesting in real applications and out of the laboratory stage was the $p-q$ theory by [11]. The basic principles of this control strategy are still widely used.

Transforming the three-phase voltages U_{sa}, U_{sb}, and U_{sc} and the three-phase load currents i_{La}, i_{Lb}, and i_{Lc} into the stationary $\alpha-\beta$ reference frame, the instantaneous real power p_L and instantaneous imaginary power q_L of the load can be calculated as

$$\begin{bmatrix} p_L \\ q_L \end{bmatrix} = \begin{bmatrix} u_\alpha & u_\beta \\ -u_\beta & u_\alpha \end{bmatrix} \begin{bmatrix} i_\alpha \\ i_\beta \end{bmatrix}. \tag{12.14}$$

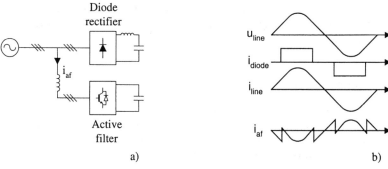

FIGURE 12.19
(a) Shunt active filter. (b) The theoretical current of an active filter for achieving a sinusoidal line current.

The instantaneous power p_L and q_L are normally divided into three components:

$$p_L = p_{L,dc} + p_{L,lf} + p_{L,hf}$$
$$q_L = q_{L,dc} + q_{L,lf} + q_{L,hf} \qquad (12.15)$$

where

$p_{L,dc}, q_{L,dc}$: dc components
$p_{L,lf}, q_{L,lf}$: low-frequency components
$p_{L,hf}, q_{L,hf}$: high-frequency components

The fundamental instantaneous real and imaginary powers are the dc components, whereas the negative sequence current is the low-frequency component (2 × fundamental frequency). The harmonics are to be found in the higher frequency components. The different components are extracted by the use of high-pass filters. The command currents $i_{ca,ref}$, $i_{cb,ref}$ and $i_{cc,ref}$ can now be found by the following:

$$\begin{bmatrix} i_{ca,ref} \\ i_{cb,ref} \\ i_{cc,ref} \end{bmatrix} = \sqrt{\frac{2}{3}} \begin{bmatrix} 1 & 0 \\ -\frac{1}{2} & \frac{\sqrt{3}}{2} \\ -\frac{1}{2} & \frac{\sqrt{3}}{2} \end{bmatrix} \begin{bmatrix} u_\alpha & u_\beta \\ -u_\beta & u_\alpha \end{bmatrix}^{-1} \begin{bmatrix} p_{ref} \\ q_{ref} \end{bmatrix}. \qquad (12.16)$$

The instantaneous real and imaginary power references p_{ref} and q_{ref} depend upon what the filter must compensate for. If only harmonics are compensated, p_{ref} and q_{ref} are equal to the high-frequency components of the instantaneous real power p_L and instantaneous imaginary power q_L.

The p–q theory control of an active filter is highly dependent on the fast response of the current controller. Normally this problem is overcome by the use of a high switching and sample frequency (20–40 kHz) which results in high bandwidth of the current controller. However, this also leads to high switching losses.

If the harmonic currents are considered stationary (which is a fair assumption for an ASD application) it is possible to overcome this problem by the use of selective harmonic control strategies. Two different approaches, the FFT approach and the transformation approach, are frequently used for selective harmonic control. The FFT approach [12] basically determines the harmonics of the previous period by Fourier series and injects the detected harmonics with an opposite phase angle. In the transformation approach [13] the harmonic currents are transformed into individual rotating reference frames where the individual harmonics become dc quantities. The dc signal errors are easily controlled to zero with linear controllers such as the PI controller. In [14] these two control strategies are compared along with the p–q control strategy. Both strategies are found to be superior to the p–q control with respect to performance at low switching frequencies (6 kHz).

12.4.2 Phase Multiplication

In applications with multiple converters, a solution combining several separated standard 6-pulse diode rectifiers, the so-called quasi 12-pulse topology, can be interesting. The quasi 12-pulse topology is well known. However, the topology has the reputation of being sensitive to uneven load distribution and variations. The reason for this may be that the quasi 12-pulse rectifier is not very well documented in the literature and very few papers have exploited the possibilities of this

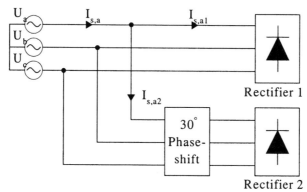

FIGURE 12.20
Quasi 12-pulse scheme with a 30° phase-shifting transformer.

very simple scheme [15]. In this section the harmonic performance of the quasi 12-pulse rectifier scheme is discussed by using the following example.

On a 1 MVA 10/0.4 kV supply transformer ($e_x = 5\%$ and $e_r = 1\%$) two 100 kW (input power) ASDs are connected with a standard 6-pulse diode rectifier. Rectifier 1 and rectifier 2 have a built-in dc-link inductance of 3%. Without any further measures, the total harmonic current distortion $THD_i = 38\%$ and the total harmonic voltage distortion $THD_v = 3.6\%$ measured at the secondary side of the supply transformer. To reduce the harmonic distortion, rectifier 2 is phase shifted by 30° as shown in Fig. 12.20.

Figure 12.21 shows the system current (i_{sa}) and the corresponding Fourier spectrum when both rectifiers are fully loaded; the $THD_i = 10.5\%$. Figure 12.22 shows the current THD_i as a function of the load of both converter groups. As shown in Fig. 12.22 the current THD_i is less than 20% in the main operating area, namely between 40% and 100% of both converter groups. A THD_i of close to 10% is achieved in the operating area between 60% and 100%.

Intuitively, one would assume that both the voltage and current THD are smallest when both transformers are equally loaded. But Fig. 12.23 shows that this is not necessarily true. The current THD_i is smallest when both converter groups are fully loaded, whereas the lowest voltage THD_v is achieved at low load. The maximum voltage distortion is achieved when either both converter groups are fully loaded or only one converter group is fully loaded and the other converter group is unloaded. It should be noted that the maximum voltage distortion is cut by almost a factor of 2 compared with the voltage distortion obtained without the phase shifting transformer ($THD_v = 3.6\%$).

As shown, the quasi 12-pulse has an excellent performance even under wide load variations. The reason for this is that the harmonic currents of the three-phase diode rectifier are almost constant in the CCM as shown in [15].

12.4.3 Mixing Single- and Three-Phase Nonlinear Loads

It has been shown by [16] that the 5th and 7th harmonic current of single-phase and three-phase diode rectifiers often are in counter phase. This knowledge can be used to reduce the system harmonic current distortion by mixing single- and three-phase diode rectifiers. In this section an example is shown to illustrate the impact of mixing single- and three-phase diode rectifiers. In general it can be difficult to predict the cancellation effect, especially when taking the impedance and load dependency of the harmonic currents into account.

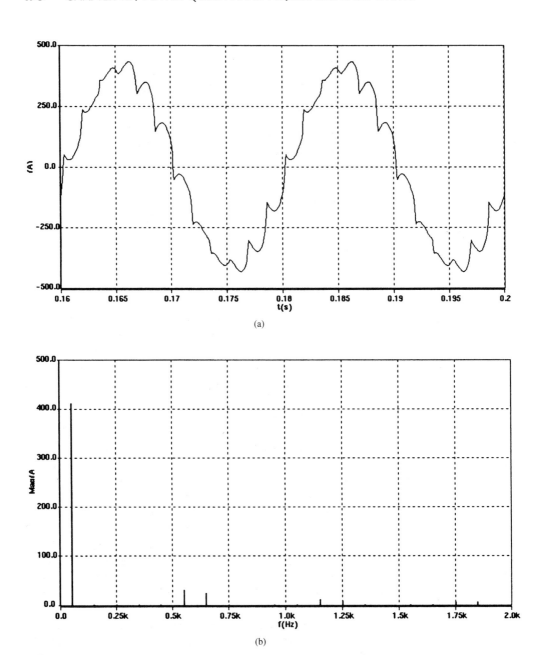

FIGURE 12.21
Quasi 12-pulse scheme. (a) Simulated system current, i_{sa}. (b) Fourier spectrum of i_{sa}. Both rectifier 1 and rectifier 2 are fully loaded.

12.4.3.1 Example of Harmonic Mitigation by Mixing Single- and Three-Phase Diode Rectifiers.
A plant with a 1 MVA distribution transformer is simulated. The MV line is assumed sinusoidal and balanced. The transformer is loaded with some single-phase diode rectifiers (the total load is 170 kW) and a 170 kW three-phase diode rectifier. The three-phase rectifier is located near the transformer with a 50 m, 90 mm^2 copper cable. The single-phase rectifier loads are evenly distributed on the three-phases with a 200 m, 50 mm^2 copper cable. It is assumed that

12.4 SYSTEM-LEVEL HARMONIC REDUCTION TECHNIQUES 479

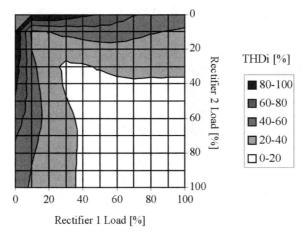

FIGURE 12.22
Contour plot of the current THD$_i$ as a function of the load of both rectifiers in a quasi 12-pulse scheme.

the single-phase rectifiers are plugged in the wall sockets and therefore a long cable is used for the single-phase rectifiers. Figure 12.24 shows the simulated system.

The reactance of the cable is 0.07 Ω/km and the capacitive effects are ignored. The impedance of the cables is shown by their per-unit values related to the transformer. The fundamental voltage drop across the long cable is about 7% at 170 kW single-phase rectifier load. The impedance of the cable is the dominant short-circuit impedance as seen from the single-phase rectifiers.

Figure 12.25a shows a simulation result of the currents drawn by the two rectifier groups. The currents add up in the secondary winding of the transformer, which is shown in Fig. 12.25b.

Intuitively, it is seen that the two waveforms are supporting each other well. The single-phase current has a "valley filling" effect on the three-phase current, and the resulting waveform looks more sinusoidal than either of the two individual currents.

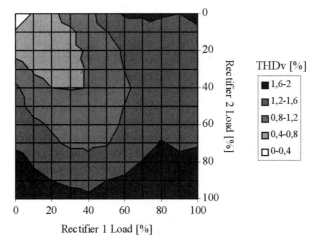

FIGURE 12.23
Contour plot of the voltage THD$_v$ as a function of the load of both rectifiers in a quasi 12-pulse scheme.

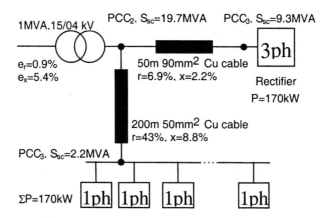

FIGURE 12.24
Simulated system with transformer, cable, and load. The impedance of the cables is shown by their per-unit values related to the transformer.

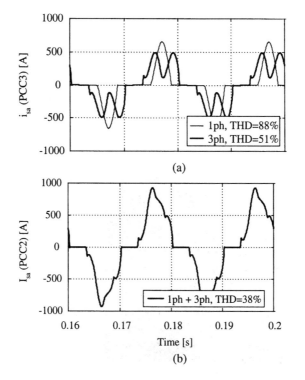

FIGURE 12.25
Simulated currents drawn in the system. (a) Rectifier currents drawn at the PCC_3. (b) Total current in the secondary windings of the transformer (PCC_2).

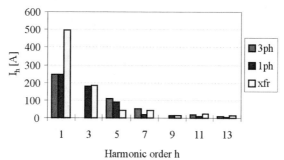

FIGURE 12.26
Harmonic spectrum of rectifier (1ph and 3ph) and transformer currents (xfr).

The total harmonic current distortion of the three-phase current is 51% and 88% for the single-phase. When the currents are added in the transformer the resulting distortion is only 38%. This reduction in the distortion is mainly due to 5th harmonic cancellation, as it can be seen more clearly from Fig. 12.26. Figure 12.26 shows the harmonic spectrum of the three currents of Fig. 12.25.

The fundamental components of the two rectifier loads are in phase and thus add up arithmetically in the transformer. The third harmonic is not present in the spectrum of the three-phase load and therefore the third harmonic component of the single-phase rectifier is seen directly in the transformer. The interesting part is to observe what happens with the 5th harmonic current. In this case a 110 A current from the three-phase rectifier is seen and 90 A from the single-phase rectifier. On the transformer only about 45 A is seen. This is only about 20% of the arithmetical sum of the two rectifier contributions. The 7th harmonic component in the transformer equals less than 60% of the arithmetical sum.

It is concluded in [16] that adding a three-phase rectifier to an existing single-phase load will not increase the current THD_i at the transformer but actually lower the THD_i and thereby lower the losses in the transformer.

12.5 CONCLUSIONS

The purpose of this chapter is to give an introduction to power quality, especially in relation to the use of adjustable speed drives. Some of the problems of running ASDs under nonideal situations are outlined and a range of methods to limit harmonic currents are presented. This chapter has shown numerous examples that power quality is a very complex matter to discuss. It is not always possible to see things as either black or white, especially when system-level performance is taken into account.

It is concluded that a standard ASD with a diode rectifier is quite robust to most of the disturbances reviewed, but of course the effect of the disturbances must be taken into account when designing the rectifier. In some cases the ASD even has a positive effect on the power quality, such as suppressing overvoltages from switching capacitors or cancelling reactive power from line-operated motors.

It is also shown how ASDs are generating harmonic currents into the grid. These harmonic currents are causing harmonic voltage drops across the supply impedance. These voltage drops are resulting in a distorted voltage in the grid that other connected equipment has to accept. A very simple method to calculate the voltage THD caused by an ASD has been presented.

A detailed description of the harmonic currents from the three-phase rectifier is given. The impact of alternative front-end solutions such as multi-pulse converters and active rectifiers is also described. But harmonic current is not only interesting on the equipment level; to get a view of the voltage distortion in a system, all loads have to be taken into account. Therefore different system-level solutions such as quasi 12-pulse and mixing single- and three-phase rectifier loads are also reviewed. These system-level solutions are often economically attractive as the optimization is made on a system level rather than suboptimizations on each piece of equipment.

This chapter has mainly focused on the harmonic currents generated by different rectifiers and the voltage distortion caused by these harmonic currents. Different solutions to mitigate the harmonic currents from rectifiers are shown, passive solutions as well as active. Also some system-level solutions where the harmonic performance is improved by a marginal (or no) effort and cost for the total installation have been briefly reviewed.

There are many cases where a piece of equipment is bad for one power quality-related parameter but good for another; for example, a diode bridge rectifier that cancels reactive power flow and introduces harmonic currents. In some cases it may be an advantage and in other cases a disadvantage. Different pieces of equipment generating harmonic currents at the same order, but in counter phase, are resulting in an improved system power factor compared to the power factor of each piece of equipment individually. This shows that great care should always be taken when evaluating different solutions. To reach the right conclusions the largest possible part of the system should be taken into account.

REFERENCES

[1] F. Abrahamsen, Energy optimal control of induction motor drives. Ph.D. dissertation, Aalborg University, Institute of Energy Technology, 2000, ISBN 87-89179-26-9.
[2] D. A. Paice, *Power Electronic Converter Harmonics*. IEEE Press, 1996, ISBN 0-7803-1137-X.
[3] IEC/EN 61000-3-2/A14, Electromagnetic compatibility, part 3: Limits, section 2: Limits for harmonic current emissions (equipment input current up to and including 16 A per phase), 1995/2000.
[4] http://grouper.ieee.org/groups/sag/ (Nov. 2001).
[5] IEEE Std 519-1992, *IEEE Recommended Practice and Requirement for Harmonic Control in Electrical Power Systems*. IEEE, 1993, ISBN 1-55937-239-7.
[6] M. Fender, Vergleichende Untersuchungen der Netzrückwirkungen von Umrichtern mit Zwischenkreis bei Beachtung Realer Industrieller Anschlußstrukturen. Ph.D. dissertation, Technische Universität Dresden, 1997 (German).
[7] F. Blaabjerg and J. K. Pedersen, An integrated high power factor three phase AC-DC-AC converter for AC-machines implemented in one microcontroller. *Proc. IEEE PESC Conf.*, 1993, pp. 285–292.
[8] S. Hansen, M. Malinowski, F. Blaabjerg, and M. Kazmierkowski, Sensorless control strategies for PWM rectifier. *Proc. IEEE APEC Conf.*, 2000, Vol. 2, pp. 832–839.
[9] W. A. Hill and S. C. Kapoor, Effect of two-level PWM source on plant power system harmonics. *Proc. IEEE IAS Conf.*, 1998, Vol. 2, pp. 1300–1306.
[10] H. Akagi, New trends in active filters for power conditioning. *IEEE Trans. Indust. Appl.* **32**, 1312–1322 (1996).
[11] H. Akagi, Y. Kanazawa, and A. Nabae, Instantaneous reactive power compensators comprising switching devices without energy storage components. *IEEE Trans. Indust. Appl.* **20**, 625–630 (1984).
[12] F. Abrahamsen and A. David, Adjustable speed drives with active filtering capability for harmonic current compensation. *Proc. IEEE PESC Conf.*, 1995, Vol. 2, pp. 1137–1143.
[13] S. Jeong and M. Woo, DSP-based active power filter with predictive current control. *IEEE Trans. Indust. Electron.* **44**, 329–336 (1997).
[14] J. Svensson and R. Ottersten, Shunt active filtering of vector current-controlled VSC at a moderate switching frequency. *IEEE Trans. Indust. Appl.* **35**, 1083–1090 (1999).
[15] S. Hansen, U. Borup, and F. Blaabjerg, Quasi 12-pulse rectifier for adjustable speed drives. *Proc. IEEE APEC Conf.*, 2001, Vol. 2, pp. 827–834.
[16] S. Hansen, P. Nielsen, and F. Blaabjerg, Harmonic cancellation by mixing nonlinear single-phase and three-phase loads. *IEEE Trans. Indust. Appl.* **36**, 152–159 (2000).

CHAPTER 13

Wind Turbine Systems

LARS HELLE and FREDE BLAABJERG
Institute of Energy Technology, Aalborg University, Aalborg, Denmark

13.1 OVERVIEW

For centuries the wind has been used to grind grain and although present applications powered by the wind have other purposes than grinding grain, almost any wind-powered machine—no matter what job it does—is still called a *windmill*. In the 1920s and 1930s, before electric wires were stretched to every community, small wind generators were used to power lights and appliances. At the instance of the growth in the worldwide infrastructure with widely distributed electrical power, the use of wind generators has been almost suspended for several decades. Among others, a consequence of the oil crisis of the 1970s is that the global energy policy of today is toward renewable energy resources and for that reason the windmill has begun its renaissance.

Since the mid-1980s the worldwide installed wind turbine power has increased dramatically and several international forecasts expect the growth to continue. Figure 13.1a shows the accumulated worldwide installed wind power from 1982 to 1999 [1]. Supporting these forecasts is a number of national and international energy programs that proclaim a high utilization of wind power. Among these, the European Commission has scheduled 12% penetration of renewable energy by the year 2010 [3] and the objective for the United States is 10,000 MW of installed capacity by the year 2010 [4]. These high political ambitions along with fast progress in generator concepts, semiconductor devices, and solid materials have founded a strong basis for the development of large and cost-competitive wind turbines. Figure 13.1b shows the annual average size in kilowatts for wind turbines installed in Denmark in the period from 1982 to 1999 and Fig. 13.1c shows the estimated costs of wind generated electricity in Denmark during the past 20 years [2]. The calculations of Fig. 13.1c are based on 20 years depreciation, 5% interest rates, and a siting in roughness class 1.

This chapter surveys the wind turbine technology as it formed in the past, as it appears in the beginning of the 21st century and as it might develop in the future.

Figure 13.2 shows an overview of the general power conversion from wind power to electrical power available to the consumer. The power represented by the wind is converted into rotational

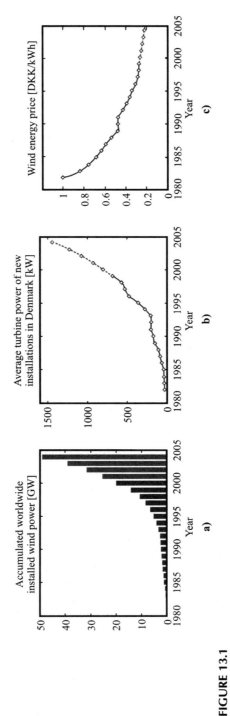

FIGURE 13.1
Trends in the field of wind turbines. (a) Accumulated worldwide installed wind power. (b) The average turbine power for wind turbines installed in Denmark from 1982 to 1999. (c) Estimated costs of wind generated electricity in Denmark. The dotted lines indicate a prognosis.

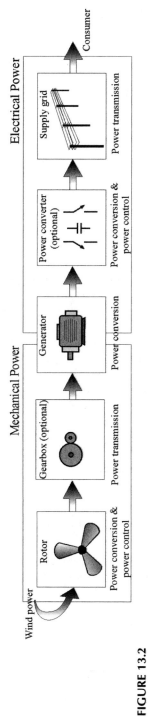

FIGURE 13.2
Power conversion from wind power to electrical power available to the consumer.

power by the rotor of the wind turbine. This power conversion is in some turbine configurations partly controllable. The rotational power is then transferred to the generator, either directly or through a gearbox to step up the rotor speed. The mechanical power is then converted into electrical power by means of a generator. From the generator, the electrical power is transferred to the supply grid either directly or through an electrical power conversion stage. From the supply grid, the power finally becomes available for the consumer. The main purpose of this chapter is to present the different concepts for converting the mechanical power at the shaft of the generator to electrical power available at the supply grid. However, to form a foundation to understand the problems associated with different electrical concepts a short description of the conversion of wind power to rotational power is provided.

13.1.1 Wind Energy

Considering the wind as a source of energy for the public supply grid, two main issues have to be faced:

- The unpredictability of the macro-scale airflow, i.e., average wind speed
- Problems caused by the micro-scale air flow, i.e., rapid wind speed changes such as wind gusts

The macro-scale airflow is imposed by local low-pressure and high-pressure zones and is characterized by slowly varying conditions. Thus, the macro-scale airflow may represent the average wind speed. However, because of the movements of these low- and high-pressure zones, the long-term prognosis availability of wind power is impossible to predict. Hence, as long as no efficient and cost-competitive energy storage is available, wind energy can only be used as a supplement to conventional energy production, typically based on coal, gas, and nuclear energy. Another problem in the utilization of the wind energy is the micro-scale airflow caused by, e.g., obstacles in the terrain. This micro-scale airflow creates fast fluctuations in the wind speed at a given site and hence fast transients in the available power. As will be shown later, these wind speed fluctuations may cause a number of problems, both in the turbine construction and in the supply grid. Figure 13.3 shows a typical measure of the wind speed, measured at the nacelle of an on-shore wind turbine.

The instantaneous power P_{wind} available in the wind flowing through an area A_v can be described by

$$P_{wind} = \frac{1}{2} \rho_{air} \cdot A_v \cdot v_w^3 \tag{13.1}$$

where ρ_{air} is the mass density of air and v_w is the velocity of the wind. Since a full utilization of this wind power requires the wind speed to be zero after passing the turbine, full utilization of the power described by (13.1) is not possible. Actually, the theoretical maximum power extraction ratio, the so-called Betz limit, is 59%. The derivation of this limit is beyond the scope of this chapter, but for further details see [5]. In practice the actual wind extraction ratio, described by the power performance coefficient C_p, will be below the Betz limit and it is influenced by several factors, among these, the blade design and the ratio between wind speed and rotor tip speed. The power transmitted to the hub of the wind turbine can be expressed as

$$P_{tur} = \frac{1}{2} C_p(\lambda) \cdot \rho_{air} \cdot A_v \cdot v_w^3 \tag{13.2}$$

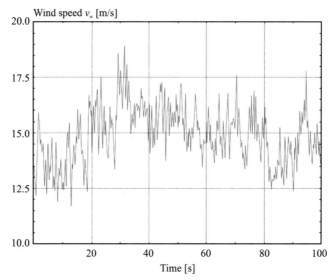

FIGURE 13.3
Typical wind speed at the hub of an on-shore wind turbine.

The power performance coefficient varies considerably for various designs, but in general it is a function of the blade tip speed ratio λ. The blade tip speed ratio is defined as

$$\lambda = \frac{v_{tip}}{v_w} = \frac{r_{rt}\omega_{rt}}{v_w} \tag{13.3}$$

where v_{tip} is the blade tip speed, r_{rt} is the radius of the propeller, and ω_{rt} is the angular velocity of the propeller. Figure 13.4 shows a typical relation between the power performance coefficient and the tip speed ratio.

A typical value for the maximum power performance coefficient in Fig. 13.4, denoted by $\hat{C}_p$, is 0.48–0.5.

13.1.2 Power Control

Wind turbines are designed to produce electrical energy as cheaply as possible and therefore they are generally designed to yield maximum output at wind speeds around 15 meters per second. In the case of stronger winds it is necessary to waste a part of the excess energy of the wind in order

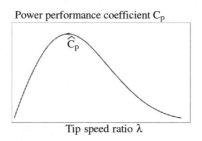

FIGURE 13.4
Power performance coefficient C_p versus tip speed ratio λ.

to avoid damaging the wind turbine. All wind turbines are therefore designed with some sort of power control. The control of the power extracted from the wind can be done in several ways, although stall and pitch control (or a combination) seem to be the prevalent methods in modern wind turbines (both constant speed solutions and variable speed solutions). However, for completeness, a short description of known power control methods is provided.

Ailerons: Some older wind turbines use ailerons (flaps) to control the power of the rotor. Ailerons are moveable flaps along the blade trailing edge—as known from airplanes.

Twistable tip: Instead of moveable flaps, some turbines use a method where the tip of the turbine blades can be turned, changing the aerodynamic performance of the blade and thereby increasing the friction in the rotational direction.

Yaw control: Another theoretical possibility is to yaw the rotor partly out of the wind to decrease power. This technique of yaw control is in practice used only for very small wind turbines (1 kW or less), as it subjects the rotor to cyclically varying stress that may damage the entire structure.

Pitch: Basically, the functionality of a pitch-controlled wind turbine consists in its ability to change the power performance coefficient C_p by turning the rotor blades around their longitudinal axis. On a pitch-controlled wind turbine the turbine's electronic controller checks the output power of the wind turbine and whenever the output power becomes too high, the rotor blades are pitched slightly out of the wind. Conversely, the blades are turned back into the wind whenever the wind drops again. During normal operation the blades will pitch a fraction of a degree at a time—and the rotor will be turning at the same time. Although pitch control seems to be a simple task, its practical realization requires some engineering efforts in order to make sure that the blades pitch the desired angle and that all blades are in angle synchronisation. Figure 13.5a shows the power performance coefficient versus tip speed ratio for various pitch angles. Based on this figure, the functionality of the block diagram in Fig. 13.5b, representing a pitch control system, becomes more obvious. The pitch mechanism can be operated either by the use of electrical actuators or hydraulic actuators where the latter seems to be the prevalent method. Figure 13.6a shows a typical average power profile for an 850 kW pitch controlled wind turbine.

Stall: A simpler power control mechanism is the passive stall regulation (or just stall regulation). In a stall-regulated wind turbine, the rotor blades are mounted onto the hub at a fixed angle—contrary to the pitch-controlled turbine. Hence, it is the aerodynamic performance of the blades that provides the power control. The blades are designed in such a way that at the moment the output power reaches the nominal power, turbulence at the back of the blades occurs, thus reducing the power extracted from the wind. To ensure a gradually occurring stall rather than an abrupt stall, the blades of a stall-regulated turbine are slightly twisted along their longitudinal axis, thereby providing stall to occur gradually. The basic advantage of stall control is that one avoids moving parts in the rotor itself and a complex control system. On the other hand, stall control represents a very complex aerodynamic design problem. Figure 13.6b shows a typical average power profile for a stall-regulated wind turbine. Compared with the pitch controlled wind turbine it appears that at low wind speeds, their performance is almost identical. In the power-limiting zone, i.e., wind speeds above nominal wind speed, the stall- regulated turbine shows a slight power overshoot with a decreasing output power as wind speed increases. The pitch-controlled counterpart has almost constant power extraction at high wind speeds.

Active stall: An increasing number of new and larger wind turbines are being developed with an active-stall power control mechanism. The active stall machines resemble pitch-controlled

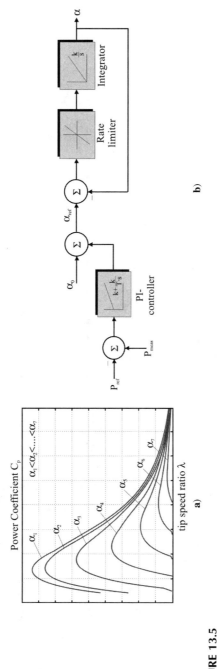

FIGURE 13.5

Pitch control of a wind turbine. (a) Power coefficient versus tip speed ratio for various pitch angles. (b) Block diagram of the pitch control system.

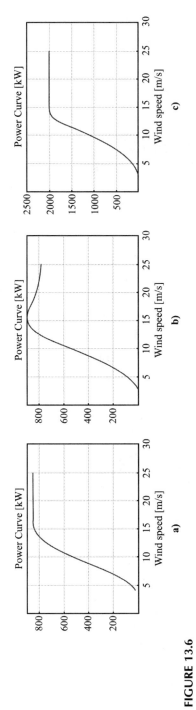

FIGURE 13.6
Output power curves versus wind speed for different control methods. (a) Pitch control. (b) Stall control. (c) Active stall control.

machines, since they are able to pitch their blades. At low wind speeds, the active stall controlled turbine will typically be programmed to pitch the blades like a conventional pitch controlled machine in order to get the maximum power extraction from the wind. At and above rated wind speed, i.e., rated power, the active stall regulated turbine behaves in an opposite manner as the pitch-controlled turbine. Instead of pitching the blades out of the wind, the attack angle of the blades is further increased, thus provoking a stall situation. Compared to passive stall control, the active stall-controlled wind turbine shows the same flat power performance as the pitch-controlled turbine. Further, because of the pitch-controlled blades, the rated power level can be tuned precisely, eliminating effects of differences in air density, blade-surface contamination, etc. In this way, the uncertainties in the rated power level (typical for passive stall control) can be avoided. Consequently, active stall control guarantees maximum power output for all environmental conditions without overloading the drive train of the turbine. The pitch mechanism is usually operated using either hydraulics or electric actuators.

At this point, it should be noted that none of the discussed power control principles can be made fast enough to track the fast power transients from wind gusts (cf. Fig. 13.3).

13.1.3 Model of the Mechanical Transmission System

When describing the behavior and performance of a wind turbine in relation to the grid, it is often convenient to have a model of the mechanical system, rotor blade to generator, because some of the undesired effects measured at the grid connection are caused by vibrations in the mechanical structure. Based on Fig. 13.2, two models have to be considered, one including a gearbox and one with direct drive, although both models end up with more or less the same expression. Figure 13.7 shows the mechanical models of the drive train of a wind turbine.

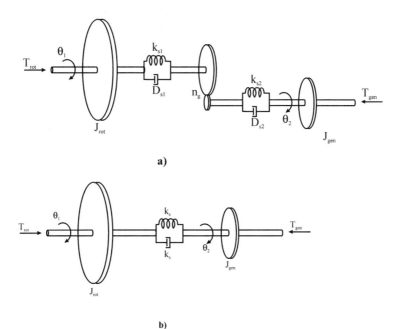

FIGURE 13.7
Mechanical models of the drive trains. (a) Drive train including a gearbox. (b) Gearless design.

Figure 13.7a illustrates the model when the turbine incorporates a gearbox and Fig. 13.7b illustrates the model of a gearless design. The transfer function for the system shown in Fig. 13.7a becomes

$$\theta_1 - n_g \cdot \theta_2 = \frac{T_{rot} - n_g \cdot T_{gen}}{s^2(J_{rot}) + s(D_{s1} + n_g^2 D_{s2}) + (k_s + n_g^2 \cdot k_s)} \tag{13.4}$$

where s is the Laplace operator, n_g is the gearbox ratio, T_{rot} is the torque arising from the wind, T_{gen} is the torque generated by the generator, D_s and k_s are the viscous damping coefficient and the torsional spring rate, respectively, and θ_1 and θ_2 are the angular positions of the rotor shaft and the generator shaft, respectively. Applying the same procedure for the gearless design in Fig. 13.7b the following equation is obtained:

$$\theta_1 - \theta_2 = \frac{T_{rot} - T_{gen}}{s^2(J_{rot}) + s \cdot D_s + k_s}. \tag{13.5}$$

13.1.4 General Structure of a Wind Turbine

In the history of wind turbines several concepts have been proposed, including the vertical axis Darrieus turbine, the Chalk multiblade turbine, and the horizontal two-blade turbine. However, at present, the horizontal three-blade turbine is the overall dominating topology, although principles described in the remaining part of the chapter can be applied to any turbine configuration. Figure 13.8 shows the general structure of the nacelle for a horizontal three-blade grid-connected wind turbine. (It should be noted that the structure may deviate for the different concepts presented in later sections; the presence of the gearbox and the size of the generator especially depend on the considered concept).

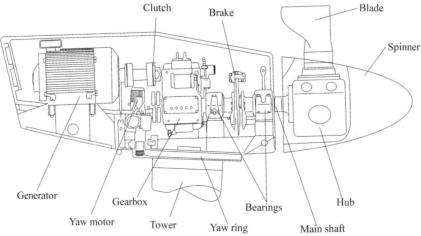

FIGURE 13.8
Typical structure of the nacelle for a horizontal three-blade grid-connected wind turbine. (The structure may deviate for the different concepts presented in later sections.)

13.2 CONSTANT SPEED WIND TURBINES

The majority of the presently installed wind turbines operate at constant (or near constant) speed. This implies that regardless of the wind speed, the angular speed of the rotor is fixed and determined by the frequency at the supply grid, the gear ratio, and the generator layout. In general, constant-speed solutions are characterized by simple and reliable construction on the electrical parts while the mechanical parts are subject to higher stresses and additional safety factors must be incorporated in the mechanical design. Further, constant-speed wind turbines have a certain impact on the supply grid, especially in the areas with weak supply grids and a high penetration of wind energy.

13.2.1 Topology Overview

For the constant-speed wind turbine, the induction generator (IG) and the wound rotor synchronous generator (SG) have been applied, where the majority have been based on the induction generator. Figure 13.9 illustrates the two topologies.

Figure 13.9a illustrates a constant-speed wind turbine based on the squirrel-cage induction generator. To compensate for the reactive power consumption of the induction generator, a capacitor bank (normally stepwise controllable) is inserted in parallel with the generator in order to obtain about unity power factor. Further, to reduce the mechanical stress and reduce the interaction between supply grid and turbine during connection and startup of the turbine, a soft starter is incorporated. Figure 13.9b shows a solution based on a wound synchronous generator where the converter is coupled to the rotor winding, thereby providing the magnetization of the generator.

13.2.2 Squirrel-Cage Induction Generator

So far the squirrel-cage induction generator has been the prevalent choice—actually to such an extent that the induction generator seems to be a de facto standard in constant-speed wind turbines. The reasons for this popularity are mainly due to its simplicity, high efficiency, and low maintenance requirements, which generally are restricted to bearing lubrication only. In a basic configuration, where the induction generator is coupled directly to the supply grid, the wind turbine will have a very high impact on the supply grid because of the necessity to obtain the excitation current from the supply grid. Also, because of the steep torque speed characteristic of an induction generator, the fluctuations in the wind power will to some extent be transferred

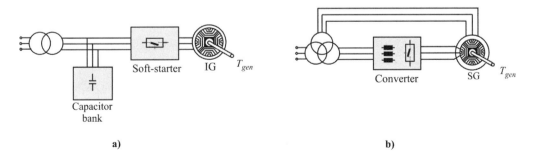

FIGURE 13.9

Constant-speed wind turbine schemes. (a) Induction generator (IG). (b) Wound rotor synchronous generator (SG).

directly to the supply grid. These transients become especially critical during connection of the wind turbine to the grid.

To overcome these two problems, wind turbines based on a squirrel-cage induction generator are often equipped with a soft-starter mechanism and an installation for reactive power compensation.

13.2.2.1 Soft-Starter Function. Connecting a wind turbine directly to the utility grid will cause the same transient inrush phenomena as known from connecting a conventional induction motor to the supply grid. To reduce the effects of a start-up situation, constant speed wind turbines are typically equipped with some kind of a soft-starter mechanism. Figure 13.10 shows the configuration of an active controlled soft starter and the associated current and voltage waveforms during startup.

After startup, the thyristors are typically shorted out by mechanical contactors connected in parallel with the back-to-back connected thyristor pairs, to eliminate the power losses in the thyristors due to the conduction voltage drop across the thyristors.

The inrush problem becomes even more critical when a whole wind turbine park is considered. In the case of a power failure and a subsequent power recovery, it is essential that the wind turbines in the park is disconnected before the power recovers. To prevent the reconnection from causing a very large grid disturbance or even a new power failure, the wind turbines often have to be reconnected in smaller groups in order to reduce the total inrush currents.

13.2.2.2 Reactive Power Consumption. When connecting an induction generator to the supply grid, the supply grid has to supply the reactive power to the generator, thereby decreasing the power factor of the wind turbine. Figure 13.11a shows a typical power factor versus load for an uncompensated induction generator. From a utility supply operator's point of view, this is an unacceptable loading condition, because the reactive power consumption causes losses in the supply grid. To reduce the reactive power consumption, capacitors are often connected in parallel with the individual turbine. Figure 13.11b shows a power factor versus load for a wind turbine with fixed compensation. Figure 13.11 is based on data for a 225-kW Siemens generator installed in a Vestas wind turbine [6].

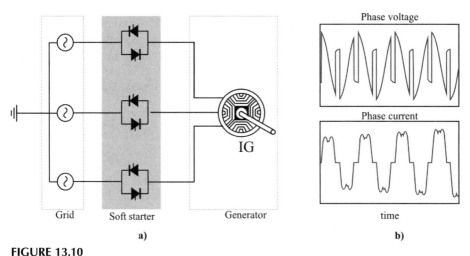

FIGURE 13.10

Soft-starter in a wind turbine. (a) Topology. (b) Voltage and current waveforms during startup.

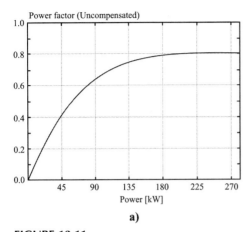 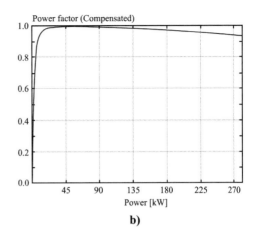

FIGURE 13.11
Power factor versus active power production. (a) Uncompensated wind turbine generator. (b) Compensated wind turbine generator.

In wind farms, the compensation can be done as a common compensation and the compensation can be made stepwise controllable in order to achieve an accurate compensation in different load conditions.

13.2.3 Wound-Rotor Synchronous Generator

A more rarely used type of generator in constant speed wind turbines is the wound-rotor synchronous generator (SG). The stator windings of the synchronous generator are connected directly to the supply grid and hence the rotational speed is strictly fixed by the frequency of the supply grid. The rotor of the synchronous generator carries the field system, provided with a winding carrying dc current. In the case of a field winding, the rotor currents must be supplied either by a brushless exciter with a rotating rectifier or by using slip rings and brushes. During transient and subtransient load steps the synchronous generator may have problems by keeping in synchronism with the frequency at the supply grid, and the only way to help the machine stay in synchronism is by controlling the magnetization of the machine. Further, because of the strictly fixed rotor speed, power transients transmitted to the supply grid are even more pronounced than for the induction generator. Besides the problems related to the strictly fixed speed, the synchronous generator also offers some advantageous features compared to the fixed speed induction generator. Designing the synchronous generator with a suitable number of poles, the SG can be used for direct drive applications without any gearbox. Further, since the magnetization power is provided by the excitation circuit of the rotor, the synchronous generator does not need any further reactive power compensation systems such as a capacitor battery at the grid side of the turbine.

13.2.4 Problems Related to Constant-Speed Operation

Among others, a reason for the prevalent use of fixed-speed wind turbines is the simple and reliable generator construction that for small wind turbines seems to be the most competitive concept in terms of cost per kilowatt-hour. In large wind turbines and particular in wind turbine farms, the problems with fixed-speed operation become more and more significant. As shown in the previous sections, to deal with these problems several precautions have to be taken, which might reduce the reliability and competitiveness of the fixed-speed system. Some of the drawbacks associated with the fixed-speed wind turbine are summarized here.

Energy capture: One problem concerning the design of a constant-speed wind turbine is the choice of a nominal wind speed at which the wind turbine produces its rated power. This problem arises from the fact that the energy capture of the wind is a nonlinear function depending on the ratio between wind speed and rotor tip speed as described in Eq. (13.2). The problem concerning the energy capture from constant-speed wind turbines is visualized in Fig. 13.12b, where the power transmitted to the hub shaft versus rotor speed is plotted for different wind speeds, $v_1 \ldots v_4$. From Fig. 13.12b it appears that at wind speeds above and below the rated wind speed, the energy capture does not reach the maximum value. In almost any literature treating variable speed wind turbines, this statement is one of the major arguments for the use of variable speed wind turbines. However, Fig. 13.12b gives information only about the power extracted from the wind and not about the net power delivered to the supply grid. Because of the more complex structure of a variable-speed wind turbine, the power losses from mechanical power to electrical power might be higher, thereby wasting some of the gained power. Presently, the available literature does not provide any unambiguous reports of whether or not the net energy yield is significantly higher for a variable-speed wind turbines.

Mechanical stress: Another problem concerning the fixed-speed wind turbine is the design of the mechanical system. Because of the almost fixed speed of the wind turbine, fluctuations in the wind power are converted to torque pulsations, which cause mechanical stress. To avoid breakdowns the drive train and gearbox of a fixed-speed wind turbine must be able to withstand the absolute peak loading conditions, and consequently additional safety factors need to be incorporated into the design [7].

Power quality: The power generated from a fixed-speed wind turbine is sensitive to fluctuations in the wind. Because of the steep speed–torque characteristics of an induction generator, any change in the wind speed is transmitted through the drive train on to the grid [7]. An improvement of the power quality is the pitch control that to a certain extent compensates slow variations in the wind by pitching the rotor blades and thereby changing the power performance coefficient C_p. The pitch control is not able to compensate for gusts and the fast periodic torque pulsations that occur at the frequency at which the blades pass the tower. The rapidly changing wind power may create an objectionable voltage flicker, which

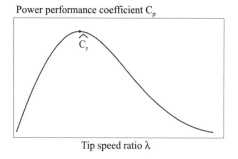
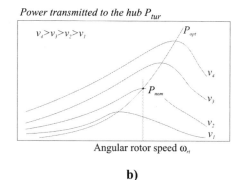

a) b)

FIGURE 13.12
(a) Power performance coefficient C_p versus tip speed ratio λ. (b) The power transmitted to the hub shaft at different wind speeds.

causes annoyances to the human eye in the form of disturbances in the light. Another power quality problem of the fixed-speed wind turbine (based on induction generators) is the reactive power consumption. Many of the electrical networks to which wind farms are connected are weak with high source impedances. The output power of a constant-speed wind turbine changes constantly with the wind condition, resulting in voltage fluctuations at the point of connection. Because of these voltage fluctuations the constant-speed wind turbine draws varying amounts of reactive power from the utility grid, which increases both the voltage fluctuations and the line losses. To improve the power quality of wind turbines, large reactive components, actively controlled as well as passive, are often used to compensate the reactive power consumption. To get an impression of the size of the compensation installation, [8] treats a static VAr compensator for a 24 MW wind turbine farm and it is found that the necessary installation amounts to 8.8 MVAr.

13.3 VARIABLE SPEED WIND TURBINES

As the size of wind turbines increases and the penetration of wind energy in certain areas increases, the inherent problems of the constant speed wind turbines becomes more and more pronounced, especially in areas with relatively weak supply grids. To overcome these problems, the trend in modern wind turbine technology is doubtless toward variable-speed concepts. However, the introduction of variable-speed wind turbines increases the number of applicable generator types and further introduces several degrees of freedom in the combination of generator type and power converter type. Hence, presently no variable-speed wind turbine concept seems to occupy the "de facto standard" position as the induction generator does in the constant-speed concept. This section surveys the most promising concepts and highlights their respective features. To summarize the common feature of all the variable-speed wind turbines, the power equation for a variable speed wind turbine is written

$$P_{tur} = J_{tur}\omega_r \frac{d\omega_r}{dt} + P_r + P_s \tag{13.6}$$

where J_{tur} is the inertia of the rotating parts, P_r is the rotor power, P_s is the stator power, and ω_r is the angular speed. From (13.6) it appears that the main feature of a variable-speed wind turbine is the ability to store and extract energy in the rotating parts by letting the rotor accelerate or decelerate, thereby providing a filter between the input (wind power) and the output (grid power).

13.3.1 Topology Overview

Before a more detailed description of the different variable-speed wind turbine solutions, an overview is provided by Fig. 13.13, illustrating the topologies, which at present are believed to be the most promising solutions. However, the literature on variable speed wind turbines covers several alternatives, and these will briefly be described in Section 13.3.6.

Figures 13.13a and 13.13b are both based on the simple and reliable squirrel-cage induction generator, while Figs. 13.13c and 13.13d are of the wound-rotor induction generator type. The last two solutions are based on the synchronous generator, where Fig. 13.13e is externally magnetized and Fig. 13.13f is magnetized by the use of permanent magnets.

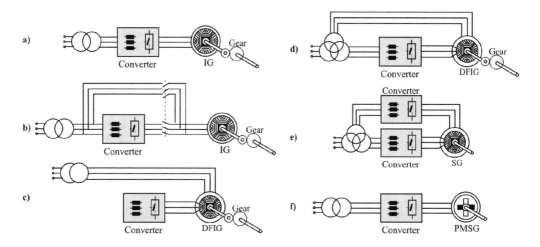

FIGURE 13.13
Topologies for use in variable-speed wind turbines. (a) Induction generator (IG) with full-scale converter. (b) Induction generator (IG) with bypass able converter (reduced VA rating for the converter). (c) Doubly fed induction generator with rotor resistance control. (d) Doubly fed induction generator (DFIG) with rotor-connected converter. (e) Externally magnetized synchronous generator (SG). (f) Synchronous generator with permanent magnets (PMSG).

13.3.2 Induction Generator

A simple method for obtaining a full variable-speed wind turbine system is to apply a bidirectional power converter (back-to-back) to a conventional squirrel-cage induction generator. Figure 13.14 illustrates a wind turbine based on this concept.

The variable-frequency, variable-voltage power generated by the machine is converted to fixed-frequency, fixed-voltage power by the use of a power converter and supplied to the utility grid. The power converter supplies the lagging excitation current to the machine while the reactive power supplied to the utility grid can be controlled independently. An advantage of the variable-speed wind turbine based on the induction generator is that control techniques such as field weakening at light loads can be applied in order to reduce the iron losses of the machine. On the other hand, the present solution requires a step-up gear and further, the power converter has to handle the rated power of the turbine. To reduce the VA ratings of the power converter, the system in Fig. 13.13b may be used, where the power converter can be bypassed when the turbine reaches a certain power level. When bypassing the power converter, the wind turbine becomes a

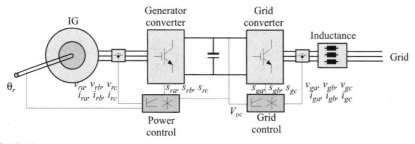

FIGURE 13.14
Induction generator, controlled by a back-to-back PWM-voltage source inverter.

fixed-speed turbine and hence some of the problems associated with fixed-speed operation are reintroduced. However, dependent on the configuration of the power converter, the power converter may be used as a reactive power compensator whenever the converter is bypassed, thereby avoiding the need for a large capacitor battery. In Fig. 13.14 the converter is presented as a back-to-back two-level converter, but the converter could theoretically be any three-phase power converter allowing bidirectional power flow. As an alternative to the two-level back-to-back converter, [9] describes a solution based on a current-source converter. The following examples of generator inverter control and rotor inverter control are, however, based on the system illustrated by Fig. 13.14.

13.3.2.1 Generator-Inverter Control.
Figure 13.15 shows an example of realizing the generator inverter control scheme, using the direct torque control (DTC) principle [10].

Input to the generator-inverter control scheme is the desired electromechanical torque at the generator shaft. From the torque reference, the flux reference ψ_{ref} is calculated—alternatively, the flux reference level may be set manually. Based on measurements of the generator voltages and currents, the actual flux and torque are estimated and compared with their respective reference values. The error is input to a hysteresis controller. Based on the output from these two hysteresis controllers the desired switch vector is chosen. For further details regarding direct torque control, see Chapter 9. In order to track the maximum power point, cf. Fig. 13.12, the torque reference command T_{ref} could be obtained by the use of a fuzzy-logic controller as described in [11] or by the use of a torque command generator as described in U.S. Patent No. 6,137,187.

13.3.2.2 Grid-Inverter Control.
Figure 13.16 shows an example of realizing the grid control scheme [11]. To control the active power through the grid inverter, the dc-link voltage is measured and compared to the actual dc-link voltage. The dc-link voltage error is fed to a PI-controller having a power reference P_{ref} as output. To provide a fast response of the grid inverter control, the measured power from the generator $P_{DC,meas}$, measured in the dc-link, is fed forward. The power error is input to another PI controller, giving the active current reference $i_{gq,ref}$ as output. The active current reference is compared to the actual active current i_{gq} all transformed to the synchronously rotating reference frame. The current error is then fed into a third PI controller having the voltage $v_{gq,ref}$ as output. Similarly, the reactive current reference $i_{gd,ref}$ is compared to the actual reactive current and the error is input to a PI controller having the voltage $v_{gd,ref}$ as

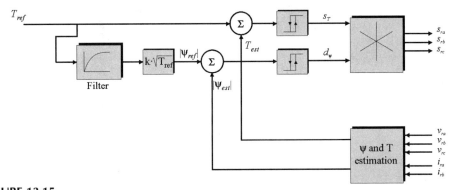

FIGURE 13.15
Example of generator-inverter control concept based on the direct torque control (DTC) principle.

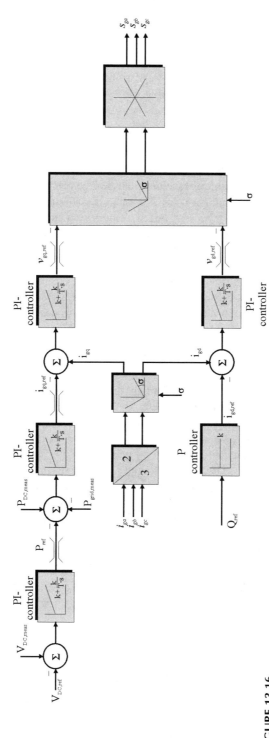

FIGURE 13.16
Example of a grid inverter control scheme.

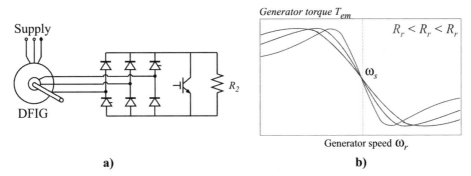

FIGURE 13.17
Rotor resistance control of an induction generator. (a) Topology for improving the speed range by adjusting the rotor resistance. (b) The torque–speed characteristic of the machine with rotor resistance control.

output. The two voltages $v_{gq,ref}$ and $v_{gd,ref}$ are transformed back to the stationary reference frame and the gate signals s_{ga}, s_{gb}, and s_{gc} are calculated.

13.3.3 Doubly Fed Induction Generator

Compared to the squirrel-cage induction generator, the main difference is that the doubly fed induction generator provides access to the rotor windings, thereby giving the possibility of impressing a rotor voltage. By this, power can be extracted or impressed to the rotor circuit and the generator can be magnetized from either the stator circuit or the rotor circuit. Basically, two different methods for speed control can be applied to the doubly fed induction generator: the rotor impedance control scheme, which is the most simple, and the more complex static slip power recovery scheme. Figure 13.13c illustrates the rotor impedance control while Fig. 13.13d illustrates the static slip-power recovery scheme.

13.3.3.1 Rotor Impedance Control. The use of adjustable resistors for starting and speed control of wound-rotor slip-ring induction machines is well treated in the literature [12, 13]. In [14] a partly controlled three-phase rectifier is connected to the rotor circuit and loaded either with fixed resistances or without them, while the stator is connected directly to the supply grid. By controlling the equivalent rotor resistance the speed range of the machine is improved. Figure 13.17a shows the circuit topology of the rotor resistance control.

Figure 13.17b illustrates the principles in the rotor resistance control. By proper control of the rotor resistance, torque pulsations caused by variations in the wind power can be reduced at the expense of speed pulsations and additional losses. Figure 13.18 shows a block diagram representation of a rotor resistance control scheme [15]. The resistance R_1 is the internal winding resistance, R_2 is the external controllable resistance, and d is the duty-cycle of the transistor, cf. Fig. 13.17a.

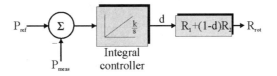

FIGURE 13.18
Block diagram representing a control scheme for a rotor resistance control.

In [16], rotor resistance control is extended to rotor impedance control, offering the opportunity to partly improve the power factor. The system suffers from lower efficiency at improved power factor. To avoid the brushes [16] proposes a controllable rotor circuit rotating on the generator shaft. This concept is among others used by Vestas Wind Systems in their Optislip concept. The advantages of this concept are a simple circuit topology and improved operating speed range compared to the squirrel-cage induction generator. To a certain extent this topology can reduce the mechanical stresses and power fluctuations caused by wind gusts. The stator of the machine is connected directly to the supply grid and the power semiconductors are rated only to handle the slip power. The disadvantages are that the operating speed range is limited to a few percent above synchronous speed (for generation mode), and only poor control of active and reactive power is obtained. The slip power is dissipated in the adjustable rotor resistance, and furthermore the presence of the step-up gear increases the cost and weight of the system, and causes a slight decrease in system efficiency.

To give an impression of the problems regarding the power dissipation, a 1 MW wind turbine is considered. In certain conditions, the turbine operates at a slip, s, 5% above synchronous speed and the power P_r to be dissipated in the rotor circuit becomes:

$$P_r = \frac{s \cdot P_m}{1-s} = 47.6 \text{ kW} \tag{13.7}$$

where P_m is the mechanical input power.

13.3.3.2 Converter Control.
A more elegant and progressing concept is shown in Fig. 13.19 where the doubly-fed induction generator is controlled by a bidirectional power converter—in the present illustration a back-to-back two-level converter. By this scheme, the generator can be operated as a generator at both sub- and supersynchronous speed and the speed range depends only on the converter ratings. Besides nice features such as variable-speed operation, active and reactive power control, and fractional power conversion through the converter, the system suffers from the inevitable need for slip rings, which may increase the maintenance of the system and decrease its reliability. Further, the system comprises a step-up gear.

The two inverters—the grid side inverter and the rotor side inverter—in Fig. 13.19 can be controlled independently, and by a proper control the power factor at the grid side can be

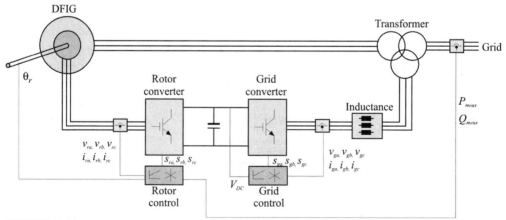

FIGURE 13.19
Doubly fed induction generator with a back-to-back PWM-voltage source converter connected to the rotor circuit.

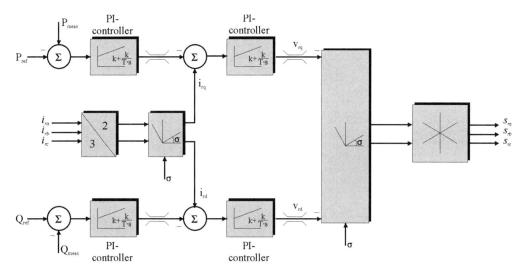

FIGURE 13.20
Control structure for active and reactive power control in a doubly fed induction generator.

controlled to unity or any desired value. By more sophisticated control schemes the system can be used for active compensation of grid-side harmonics.

13.3.3.3 Rotor-Converter Control Scheme. Several control structures may be applied for controlling the rotor-side inverter, and the control structure in Fig. 13.20 is only shown as an example.

Inputs to the control structure in Fig. 13.20 are the desired active and reactive power along with the measured rotor currents and measured active and reactive power. The error between measured active/reactive power and reference values are fed into a PI-controller giving the current references $i_{rd,ref}$ and $i_{rq,ref}$. These current references are compared to the actual currents and the errors are inputs to a set of PI-controllers giving the voltage references $v_{rq,ref}$ and $v_{rd,ref}$ as outputs. From these voltage references the control signals s_{ra}, s_{rb}, and s_{rc} for controlling the rotor inverter are calculated. The detection of the slip angle σ can either be done by the use of an optical encoder or by position-sensorless schemes. It is beyond the scope of this chapter to go into details about position-sensorless schemes for doubly fed induction machines but [18] covers this issue in detail.

13.3.3.4 Grid-Inverter Control Scheme. Besides the realization of a grid controller in Fig. 13.16, another structure is illustrated in Fig. 13.21.

Figure 13.21 illustrates a block diagram of a grid side controller. Inputs to this control diagram are the measured dc-link voltage V_{DC}, the three measured line-currents i_{ga}, i_{gb}, and i_{gc}, and the measured angle σ of the grid voltage. Further inputs are the desired dc-link voltage $V_{DC,ref}$ and the reactive power controlling current $i_{gd,ref}$. The measured dc-link voltage is compared to the actual dc-link voltage and the error is fed into a PI-controller. The output from this PI-controller is the active current reference $i_{gq,ref}$, which is compared with the actual active current—transformed into the synchronous rotating reference frame and fed into another PI-controller. Output from this controller is the active voltage reference $v_{gq,ref}$. The reactive power is controlled by the current reference $i_{gd,ref}$. Compared to the control structure in Fig. 13.16, the scheme in Fig. 13.22 does not incorporate any feed forward of the generator power, whereby a

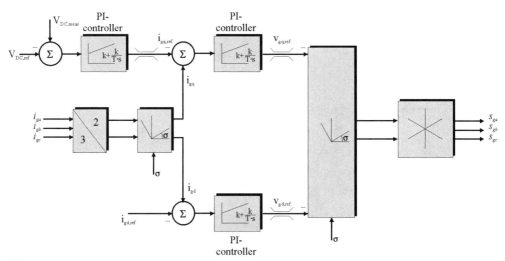

FIGURE 13.21
Example of a control structure for the grid-side controller in a doubly fed induction generator.

more simple structure is obtained. However, this is done at the cost of lower dynamic performance.

13.3.4 Wound Rotor Synchronous Generator

A common reported solution to avoid the step up gear between the propeller and the generator is the use of a multipole wound-rotor synchronous generator. Figure 13.22 illustrates a synchronous generator, controlled by a back-to-back voltage source converter. The grid inverter control can be realized identically to either Fig. 13.16 or 13.21.

The externally magnetized multipole synchronous generator is commercially used by the German wind turbine manufacturer Enercon and the Dutch wind turbine manufacturer Lagerwey in their large wind turbines. Compared to the solutions based on the squirrel-cage induction generator and the wound-rotor induction generator, the main advantage is the possibility to avoid the step-up gear, by which the reliability and the noise level of the system may be reduced. The price to be paid for a gearless design is a large and heavy generator construction and a converter that has to handle the full power of the system.

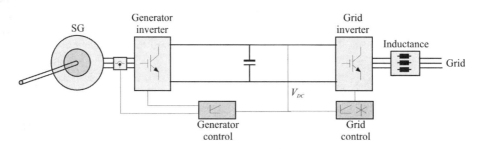

FIGURE 13.22
Synchronous generator with a back-to-back PWM-VSI.

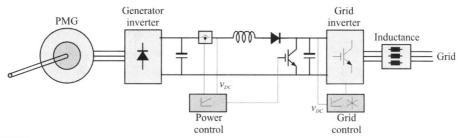

FIGURE 13.23
Permanent-magnet synchronous generator with diode rectifier and three-phase grid inverter.

13.3.5 Permanent-Magnet Synchronous Generator

Another configuration suitable for variable-speed wind turbines with gearless design is the multipole permanent-magnet synchronous generator (PMG). A system employing a PMG is illustrated in Fig. 13.23.

By the use of permanent magnets, the slip rings (or brushless exciter) are avoided, but at additional costs due to the prices of permanent magnets. Further, the weight and volume are still high.

13.3.6 Other Solutions

Besides the common solutions just treated, several alternatives exist, although these only are believed to be of minor importance in the overall wind turbine market.

13.3.6.1 Switched Reluctance Generators. In literature considering the switched reluctance generator for wind turbine applications, the most common argument is the high efficiency, the reduced costs, due to the simple construction of the generator [19] and the opportunity of eliminating the step-up gear. Further advantages are lower diameter than a direct driven synchronous generator, simpler converter, and higher power-to-weight ratio [20]. The disadvantages are a relatively high VA rating of the converter, and high converter losses due to the high amount of field energy which has to be supplied and removed for each stroke. The VA rating of the power converter might be reduced by implementing permanent magnets in the generator [21], but clearly this increases the costs and the complexity of the generator. Furthermore, the mechanical stresses of the generator are high due to the high torque ripple. The full power of the generating system has to be handled by the power converter. It is believed that the switched reluctance generator will be of interest mainly for use in small wind turbines for household applications and the like.

13.3.6.2 Cascaded Generators. Except for the doubly fed induction generator, a major drawback of the generator systems treated in the previous sections is that the power converter has to handle the rated power of the system. One way to avoid a full-rated power converter and at the same time achieve a brushless solution is by the use of a cascaded generator. Among others, the brushless doubly fed reluctance generator (BDFRG), the brushless doubly fed induction generator (BDFiG), and the cascaded doubly fed induction generator (CDFIG) all have these properties, and in addition they offer the opportunity of having full active and reactive power control. Figure 13.24 shows the topology of these three cascaded generators.

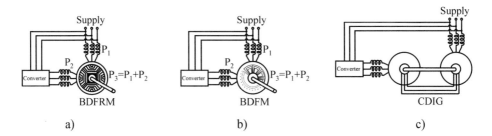

FIGURE 13.24
Cascaded generator topologies. (a) The brushless doubly fed induction generator. (b) The brushless doubly fed reluctance generator. (c) The cascaded doubly fed induction generator.

In [22] a BDFIG variable-speed wind generating system with a speed range between 1200 and 2000 rpm is developed. In this system, the converter is rated to handle 25% of the full power. It is stated that the generator efficiency is comparable to that of a conventional induction generator while the efficiency of the converter is higher than the efficiency of a full-scale converter, giving an overall higher efficiency. Further advantages are compact design (almost comparable to a conventional IG) [7] and better harmonic characteristics because most of the power is generated directly to the grid [23]. A wind power generating system based upon the BDFRG is described in [24]. The advantages are a higher generator efficiency compared to the BDFIG due to the absent of copper losses in the rotor circuit, enhanced reliability, and reduced costs, also achieved because of the absence of rotor windings and brushes. Finally, the controllability and flexibility of the generating system are accentuated. In [25] it is mentioned that the design of the rotor is quite complex and it is a compromise among complexity, efficiency, and torque per volume. The third of the reported cascaded generators in wind turbine generating systems is the CDIG [26]. The CDIG consists in principle of two doubly fed slip-ring induction generators where the two rotors are mechanically and electrically connected (no brushes are in use). In [26] two equal-sized slip-ring induction generators are used. By this arrangement the power converter has to handle 50% of the rated power while full active and reactive power control is achieved. A drawback of this method is that the axial length of the generator is higher than other generators [7].

13.4 SYSTEM SOLUTIONS: WIND FARMS

Previously, wind turbines were sited on an individual basis or in small concentrations making it most economical to operate each turbine as a single unit. Today and in the future, wind turbines will be sited in remote areas (including off-shore sites) and in large concentrations counting up to several hundred megawatts of installed power. This opens up new technical opportunities for designing and controlling the wind turbines, but at the same time increases the demand for reliability, availability, and grid impact.

13.4.1 HVDC Link Based on Group Connection

Figure 13.25 illustrates a simple park solution based on a high-voltage dc (HVDC) link, where several wind turbines are connected to the same power converter. At the public supply grid, the wind turbine park is connected through another power converter, thereby having the possibility

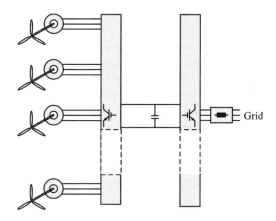

FIGURE 13.25
Wind turbine park solutions based on group connection of a wind turbine park to an HVDC link through a local ac network.

of a long, high-voltage dc-link cable. In this concept, all the turbines connected to the same converter will rotate with (or near) the same angular velocity, thereby giving up some of the features in the variable-speed concept. For instance, because of different wind speeds over the entire park site, the converter frequency has to match the average wind speed. Further, since wind gusts do not appear simultaneously at all wind turbines, it is not possible to store these fast transients in the rotating parts and hence the power delivered by each wind turbine will show a high content of power transients.

The turbines in Fig. 13.25 could be either wound-rotor synchronous machines or squirrel-cage induction machines.

13.4.2 HVDC Link Based on Individual Connection

A second solution based on the HVDC principle is illustrated in Fig. 13.26. In this concept, the park is still connected through a central power converter, but instead of having a local ac network

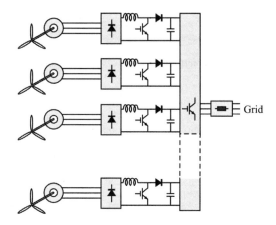

FIGURE 13.26
Wind turbine park solution based on individual connection of wind turbines to HVDC link.

within the farm, a common dc-link bus provides the power transmission. Each wind turbine contains a power converter connected to the common dc-link bus. This park solution provides all the features of the variable-speed concept since each turbine can be controlled independently. Concerning reliability and availability it appears that only one power converter is in a critical position (the converter at the public supply grid side), while faults in a converter in an individual turbine have only a fractional effect on the overall power production.

The generators in Fig. 13.26 are illustrated as synchronous generators but if each diode rectifier and boost converter is replaced by a voltage source inverter, the generators could be squirrel cage induction generators as well.

13.5 CONCLUSION

This chapter has provided a brief and comprehensive survey of topologies and control concepts used within the field of modern wind turbines. After an introduction, presenting the basics for converting the kinetic energy in the wind to rotational kinetic energy available at the generator, the constant-speed wind turbines, which so far have been the prevalent choice, were discussed. To understand the trend in modern wind turbine technology, which is toward variable-speed wind turbines, the problems associated with constant-speed operation were discussed and the way the variable-speed concept may improve the performance of the turbines were outlined.

By introducing the variable-speed opportunity in wind turbines, several different competitive converter and generator concepts are available, and at present no variable-speed topology occupies a de facto standard position in the market. Section 13.3 was dedicated to discuss some of the most promising topologies regarding converter concepts, generator concepts, and control schemes.

Another upcoming trend in wind turbines seems to be park solutions, where turbine control and utility grid interaction is centralized. Two different concepts have been presented, but it is believed that several alternatives will enter the market in the future.

REFERENCES

[1] BTM Consults Aps, *International Wind Energy Development—World Market Update 2000*. ISBN 87-987788-0-3, 2001. BTM Consults Aps, Ringkøbing, Denmark.
[2] H. Bindner, New wind turbine designs—Challenges and trends. *Proc. Nordic Wind Power Conf.*, pp. 117–123, March 2000.
[3] L. H. Hansen, P. H. Madsen, F. Blaabjerg, H. C. Christensen, U. Lindhard, and K. Eskildsen, Generators and power electronics technology for wind turbines. *Proc. IECON 2001*, pp. 2000–2005, 2001.
[4] International Energy Agency. IEA, Wind Energy Annual Report 1998.
[5] J. F. Walker and N. Jenkins, *Wind Energy Technology*. John Wiley & Sons, 1997. ISBN: 0-471-96044-6.
[6] H. Bindner and A. Hansen, Dobbelt styrbar 3-bladet vindmølle: Sammenligning mellem pitchreguleret vindmølle og pitchreguleret vindmølle med variabelt omløbstal. *Risø-R- 1072(DA)*, December 1998.
[7] G. T. van der Toorn, R. C. Healey, and C. I. McClay, A feasibility of using a BDFM for variable speed wind turbine applications. *Int. Conf. Electrical Machines*, Vol. 3, pp. 1711–1716, September 1998.
[8] K. H. Søbrink, R. Stöber, F. Schettler, K. Bergmann, N. Jenkins, J. Ekanayake, Z. Saad-Saoud, M. L. Lisboa, G. Strbac, J. K. Pedersen, and K. O. H. Pedersen, *Power Quality Improvements of Wind Farms*. Søndergaard bogtryk og offset, 1998. ISBN: 87-90707-05-2.
[9] M. Salo, P. Puttonen, and H. Tuusa, A vector controlled current-source PWM-converter for a wind power application. *Int. Power Electron. Conf.*, Vol. 3, pp. 1603–1608, April 2000.
[10] I. Takahashi and T. Noguchi, A new quick-response and high efficiency control strategy of an induction motor. *IEEE Trans. Industr. Appl.* **IA-22**, (1986).

[11] M. G. Simões, B. K. Bose, R. J. Spiegel, Design and performance evaluation of a fuzzy-logic-based variable-speed wind generation system. *IEEE Trans. Industr. Appl.* **33**, 956–965 (1997).
[12] A. E. Fitzgerald, C. Kingsley, and S. Umans, *Electric Machinery*. McGraw-Hill, New York, 1983. ISBN: 0-7021134-5.
[13] M. G. Say, *Alternating Current Machines*. Pitman, 1976. ISBN: 0- 273-36197-X.
[14] I. Volčkov and M. Petrović, Start and speed control of a slip ring induction motor by valves operating in the on-off mode in the rotor circuit. *Int. Conf. Electrical Machines*, Vol. 2, pp. 851–859, September 1980.
[15] R. Tirumala and N. Mohan, Dynamic simulation and comparison of slip ring induction generators used for wind energy generation. *Int. Power Electron. Conf.*, Vol. 3, pp. 1597–1602, April 2000.
[16] A. K. Wallace and J. A. Oliver, Variable-speed generation controlled by passive elements. *Int. Conf. Electrical Machines*, Vol. 3, pp. 1554–1559, September 1998.
[17] M. Yamamoto and O. Motoyoshi, Active and reactive power control for doubly-fed wound rotor induction generator. *IEEE Trans. Power Electron.* **6**, 624–629 (1991).
[18] E. Bogalecka, Power control of a doubly fed induction generator without speed or position sensor. *Eur. Conf. Power Electron. Appl.*, Vol. 8, pp. 224–228, September 1993.
[19] K. Liu, M. Stiebler, and S. Güngör, Design of a switched reluctance generator for direct-driven wind energy system. *Universities Power Eng. Conf.*, Vol. 1, pp. 411–414, September 1998.
[20] R. Cardenas, W. F. Ray, and G. M. Asher, Transputer-based control of a switched reluctance generator for wind energy applications. *Eur. Conf. Power Electron. Appl.*, Vol. 3, pp. 69–74, September 1995.
[21] I. Haouara, A. Tounzi, and F. Piriou, Study of a variable reluctance generator for wind power conversion. *Eur. Conf. Power Electron. Appl.*, Vol. 2, pp. 631–636, September 1997.
[22] C. S. Brune, R. Spée, and A. K. Wallace, Experimental evaluation of a variable-speed doubly-fed wind-power generation system. *IEEE Trans. Industr. Appl.* **30**, 648–655 (1994).
[23] R. Spée, S. Bhowmik, and J. H. R. Enslin, Adaptive control strategies for variable-speed doubly-fed wind generation systems. *IEEE Indust. Appl. Soc. Annual Meeting*, Vol. 1, pp. 545–552, October 1994.
[24] L. Xu and Y. Tang, A novel wind-power generating system using field orientation controlled doubly-exited brushless reluctance machine. *IEEE Industr. Appl. Soc. Annual Meeting*, Vol. 1, pp. 408–413, October 1992.
[25] R. E. Betz, Synchronous reluctance and brushless doubly fed reluctance machines. Course material in the Danfoss visiting professor program, Aalborg University, November 1998.
[26] B. Hopfensperger, D. J. Atkinson, and R. A. Lakin, Application of vector control to the cascaded induction machine for wind power generation schemes. *Eur. Conf. Power Electron. Appl.*, Vol. 2, pp. 701–706, September 1997.

Index

A

ac choppers, 18
Acoustic noise, 109
Activation functions, 353–354
Active filters, 475–476
Active clamping, 48–49
Active rectifiers, 472–474
ac to ac converters, 15–24
ac to dc converters, 4–15
ac voltage and current sensorless, 451–453
ac voltage controllers, 15
Adaptive neuro-fuzzy inference system (ANFIS), 369–373, 402–412
Adaptive space vector modulation (ASVM), 107–109
Adjacent state current modulator (ASCM), 57–58
Adjustable speed drives (ASDs)
  effect of capacitor switching transients on, 466
  effect of short line interruptions of input phase, 465–466
  effect of utility voltage unbalance on, 464–465
  effect of voltage sags on, 465
  harmonic current generation of standard, 467–469
  system-level harmonic reduction techniques, 474–481
  topologies, 470–474
Analog hysteresis, 150–151
Analytical model solution, 219–220
Artificial intelligence (AI)
  applications, 352
  defined, 351
  groups of, 351
Artificial neural networks (ANNs)
  activation functions, 353–354
  backpropagation, 362–363
  feedforward, 356–363
  field-oriented control, 412–415
  generalized delta learning rule, 360–362
  induction motor flux estimation, 398–402
  induction motor speed estimation, 386–398
  learning methods, 356–357
  linear filter, 355
  multilayer neural networks (MNN), 355–356, 363–364
  neural identification, 393–398
  neural modeling, 386–393
  neurocontrol, 363–368
  neuron layers, 355
  neurons, 352–354
  off-line trained neural comparator, 380–382
  pulse width modulation control, closed-loop, 375–386
  pulse width modulation control, open- loop, 373–374
  resonant dc link (RDCL) converters, 379–380
  topologies, 354–356
  Widrow–Hoff (standard delta) learning rule, 357–360
ASDs. *See* Adjustable speed drives
Auxiliary resonant commutated pole (ARCP) inverters, 39

B

Backpropagation, 362–363
Base units, 202
Betz limited, 486
Bidirectional switches
  commutation techniques, 65–69
  topologies, 64–65
Blade tip ratio, 487
Blanking time, 29
Block diagrams, 163–167
Boost-type converters, 14, 28
Brushless doubly fed inductance generator (BDFIG), 505–506
Brushless doubly fed reluctance generator (BDFRG), 505–506
Buck-type converters, 14, 25

C

Capacitor switching transients, affects on adjustable speed drives, 466–467
Carrier-based pulse width modulation, 92–93, 98
Cascaded doubly fed induction generator (CDFIG), 505–506
Cascaded generators, 505–506
Choppers, 3, 24
Circulating current-conducting dual converter, 9, 12
Circulating current-free dual converter, 9
Clamp circuit, 79–80
Clamping methods, 48–49
Closed-loop pulse width modulation. *See* Pulse width modulation, closed-loop
Coenergy
  inductances and, 258–261
  torque, 254–258
Common-collector (CC) unidirectional switches, 65
Common-emitter (CE) unidirectional switches, 65
Commutation techniques, bidirectional switch, 65–69
Conduction losses, 104
Conservation of energy equation, for a machine, 254, 257
Constant current in d-axis control (CCDAC), 283–284
Constant mutual flux linkages (CMFL) control strategies, 226, 227, 236–237, 240, 241
Constant power loss (CPL). *See* Permanent magnet synchronous machine (PMSM), constant power loss (CPL) control system and
Constant-slip frequency, 172
Constant switching frequency predictive controllers, 142–145
Constant torque region, 171
Continuous pulse width modulation, 92
Converter control, 502–503
Converter loss, 210
  model, 214–215
Correlation method, 185
Cross-magnetization, 271
Current controlled pulse width modulation (CC-PWM) converters, 114–117
Current-controlled R-FOC, 180–189
Current error compensation, 116
Current feedback loop, 114
Current reference generator, 297–298
Current sign detection methods, 69
Current sign four-step commutation, 67–68
Current-source inverters (CSIs), 28, 40–42
Current-source PWM rectifier, 14
Cyclic loads, constant power loss and, 241–247
Cycloconverters, 15, 21–22

D

Damping factor, 123–124
dc to ac converters, 28–42
dc to dc converters, 24–28
Dead time, 29, 451
  current commutation, 65–66
Decoupling
  control, 132, 148
  of multiscalar model, 197–198
Degrees of freedom, 119
Delta current modulator (DCM), 57
Delta modulation (DM), 150, 379
Delta modulation current controllers (DM-CCs), 150
Diode embedded unidirectional switches, 65
Diode rectifiers, harmonic currents generated by, 467–469
Direct method, 373
Direct power control (DPC), 426
  block scheme, 435–437
  instantaneous power estimation based on virtual flux, 438–440
  instantaneous power estimation based on voltage, 437–438
  switching table, 440–443
Direct torque control (DTC), 69–70, 162
  matrix converters and, 76
Direct torque and flux control (DTFC), development of, 302
Direct torque and flux control, induction motors
  conclusions, 324
  experimental results, 320–324
  flux, torque, and speed estimator, 314–316
  flux/torque coordination in, 306–308
  principles, 302–306
  space vector modulation system, 313–314
  stator resistance estimation, 324–326
  torque and flux controllers, 316–320
Direct torque and flux control, permanent magnet synchronous motors, 326
  flux and torque observer, 327–328
  initial rotor position detection, 330–331
  rotor position and speed observer, 329–330
Direct torque and flux control, reluctance synchronous motors
  flux estimator, 332–333
  experimental setup, 335–336
  space vector modulation system sensorless drive, 331–332, 333–335
  speed estimator, 333
  test results and conclusions, 336–343
  torque estimator, 333
Direct torque and flux control, synchronous motors
  current source inverter drive, 345–348
  flux/torque coordination in, 310–313
  large motor drives, 343–348
  principles, 308–310
  voltage source converter drives, 343–344
Discontinuous pulse width modulation (DPWM), 92–93, 96, 107
Discrete hysteresis, 150–151

INDEX   513

Discrete pulse modulation (DPM), 45–46, 47
    types of, 56–58
Discrete time model, 143–144
Distorted power factor, 462
Distortion factor, 106–107
Doubly fed induction generators, 501–504
    brushless, 505–506
    cascaded, 505–506
Doubly fed reluctance generator, brushless, 505–506
DPC. *See* Direct power control
dq model, 264–268
DTC. *See* Direct torque control
DTFC. *See* Direct torque and flux control
Dual converters, 8–9
Duty ratio, 3

E

Eddy current losses, 271, 272
Electromagnetic noise, 109
Energy capture, 496
Energy optimal control
    control methods, comparison of, 222–223
    flux level selection, 212–218
    model-based control, 219–221
    motor drive loss minimization, 209–212, 217–218
    search control, 221
    simple state control, 218
Energy shaping design, 198
Equivalent circuits, 167–170
Euler–Lagrange equations, 162, 198, 199
Extinction angle, 16

F

Feedback linearization control (FLC), 161, 190
    block scheme and basic principles, 191–194
    versus field oriented control, 194–195
Feedforward artificial neural networks, 356–363
Field coordinates, 161
Field oriented control (FOC), 161, 176
    artificial neural network speed estimation, 412–415
    current-controlled, 180–189
    rotor-flux-oriented, 177–180
    stator-flux-oriented, 178–180
    voltage controlled, 189–190
Field weakening
    classical, 289–290
    maximum power, 291–293
Firing angle, 6
First-quadrant choppers, 25–26
Flux current, 176
Flux level selection, 212–218
Flux linkage, 168–170
Flux vector estimation, 187–189
FOC. *See* Field oriented control
Four-quadrant choppers, 27

Four-step commutation, 67–68
Full-wave rectifiers, 5
Fuzzy logic control, 368–369, 382–386
Fuzzy membership function, 369

G

Generalized delta learning rule, 360–362
General learning architecture, 365
Generator-inverter control, 499
Grid-inverter control, 499, 501, 503–504
Grid loss, 210

H

Hard computing, 352
Hard-switching converters, 37
Harmonic constant, 464
Harmonic current generation of standard adjustable
        speed drives, 467–469
Harmonic distortion, total, 462–464
Harmonic motor losses, 213–214
Harmonic reduction techniques, system-level,
        474–481
High-voltage dc links, 506–508
Hysteresis, iron losses, 271–272
Hysteresis current controllers, 146–148, 441–442
    analog and discrete, 150–151

I

IGBTs (insulated gate bipolar transistors), 5
    reverse blocking (RBIGBT), 64–65
Indirect method, 373–374
Indirect modulation, 72
Indirect space vector modulation, 72–75
Inductances
    coenergy and, 258–261
    leakage, 271
    SYNCREL, 261–264
Induction generators
    doubly fed, 501–504
    variable speed wind turbines and, 498–501
Induction motor flux estimation, artificial neural
        network
    based on stator currents, 398–399
    based on stator currents and voltages, 400–402
Induction motors (IM)
    See also Direct torque and flux control (DTFC),
        induction motors
    advantages of, 161
    basic theory of, 162–172
    control methods, types of, 172–174
    data, 203–205
    energy optimal control of, 209–224

514 INDEX

Induction motors (IM) (*Continued*)
    feedback linearization control, 190–195
    field-oriented control, 176–190
    flux/torque coordination in, 306–308
    model, 132–136
    multiscalar control, 195–198
    passivity-based control, 198–201
    per unit systems, 202–203
    scalar control, 172, 174–176
    vector control, 161, 172, 174
Induction motor speed estimation, artificial neural network
    based on neural identification, 393–398
    based on neural modeling, 386–393
Input filter, 78–79
Input-output decoupling, 161
Input voltages, compensating for unbalanced and distorted, 76–78
Internal model controllers (IMCs), 118–119, 136
Inverters
    conduction loss, 214–215
    current-source, 28, 40–42
    switching loss, 215
    total loss, 215
    voltage-source, 28–40
Iron losses, 271–272

K

Kirchloff's voltage law, 254

L

L_d lookup table, 294
Leakage inductance, 271
Linear controllers
    basic structures, 117–120
    design rules, 120–121
Linear iron circuit, 258
Linearization control, feedback, 190–195
Linearization of multiscalar model, 197–198
Line-to-line output voltages, 32
Line-to-neutral output voltages, 32
Line voltage estimation, 453–454
Loss minimization, motor drive, 209–212, 217–218

M

Magnitude control ratio, 3
Matrix converters, 15–16, 22–24
    bidirectional switch commutation techniques, 65–69
    bidirectional switch topologies, 64–65
    compensating for unbalanced and distorted input voltages, 76–78
    direct torque control and, 76

    implementation aspects, 78–82
    limited ride-through capability, 82–84
    modulation techniques, 69–76
    motors, next generation, 84–86
    overview of, 61–64
    power module with bidirectional switches, 82
Maximum efficiency (ME) control system, 226, 227, 234–235, 240, 241, 278–279
Maximum power factor control (MPFC), 279–281
Maximum torque per ampere control (MTPAC), 278–279
Maximum torque per unit current (MTPC) control strategy, 226, 227, 236, 240, 241
Mechanical stress, 496
Model-based control, 219–221
Modulation techniques
    for matrix converters, 69–75
    for resonant dc link (RDCL) converters, 56–58
MOS controlled thyristors (MCT), 64
Motor drive loss minimization, 209–212, 217–218
Motor loss, 210
    model, 213–214
Multilayer neural networks (MNN), 355–356, 363–364
Multilevel inverters, 36
Multiscalar control, 195–198
Mutual inductances, 263

N

Neural identification, 393–398
Neural modeling, 386–393
Neurocontrol (neuromorphic control), 363
    adaptive, 367–368
    direct adaptive control, 365–366
    feedback error training, 366
    indirect adaptive control, 366–367
    inverse and direct inverse control, 364–365
Neuro-fuzzy inference system, adaptive, 369–373, 402–412
Nonlinear on-off controllers, 145
    delta modulation, 150
    hysteresis, 146–148, 150–151
    on-line optimization, 149
Non-short circuit method, 54–56
Notch commutated three-phase PWM converters, 50
Numerical model solution, 220–221

O

Off-line learning, 357
On-line adaptation, 185
On-line optimization, 149
Open-loop constant volts/Hz control, 174–176
Open-loop pulse width modulation. *See* Pulse width modulation, open-loop

Open-loop transfer function, 130
Operating quadrants, 8
Overlap current commutation, 66
Overmodulation, 98–104

P

Parallel resonant dc link converters, 45–46
   active clamp, 48–49
   passive clamp, 48
   voltage peak control, 49
Parallel resonant dc link converters, pulse width modulation, 49
   modified ACRDCL for, 51–53
   notch commutated three-phase, 50
   zero switching loss, 50–51
Passive clamping, 48
Passivity-based control (PBC), 162, 198
   model of, 199–200
   observerless, 200–201
Permanent magnet synchronous machine (PMSM)
   See also Direct torque and flux control (DTFC), permanent magnet synchronous motors
   background, 226
   electrical equations, 228
   future work, 247–248
   literature review, 226–227
   model with losses, 228–229
   power loss estimator, 242–247
   prototype, 248
   total power loss equation, 229
   variable-speed wind turbines and, 505
Permanent magnet synchronous machine, constant power loss (CPL) control system and, 225
   base speed, 231, 234
   comparisons, 229–231, 237–241
   cyclic loads, 241–247
   flux-weakening operating region, 231
   implementation scheme, 232–233
   parameter dependency, 231–232
   rationale for, 227–228
Permanent magnet synchronous machine, torque control strategies
   air gap flux linkages versus speed, 240–241
   base emf versus speed, 240
   base speed, 241
   comparison of strategies based on constant power loss, 237–241
   constant mutual flux linkages, 226, 227, 236–237
   current versus speed, 240
   implementation complexities, 241
   maximum efficiency, 226, 227, 234–235
   maximum torque per unit current, 226, 227, 236
   power factor versus speed, 241
   power versus speed, 240
   speed range before flux weakening, 241
   torque per current versus speed, 240
   torque versus speed envelope, 240
   unity power factor, 226, 227, 236
   zero d-axis current, 226–227, 235
Per unit systems, 202–203
Phase control, 2–3
   converters, 16
   rectifiers, 4, 6, 13
Phase-controlled three-phase ac voltage controllers
   delta, 21
   fully controlled, 17–21
   half controlled, 21
   pulse width modulation, 21
Phase multiplication, 476–477
Phasor diagrams, 167–170
PI current controllers
   constant switching frequency predictive, 142–145
   conventional, 117–118
   ramp comparison, 121–125
   state feedback, 141–142
   stationary resonant, 125–141
   stationary vector, 125
   synchronous vector, 125
PMSM. *See* Permanent magnet synchronous machine
Pole commutated resonant dc link converters, 47–48
Power control, wind turbine systems and, 487–491
Power control methods, 488, 491
Power diodes, 1
Power electronic converters
   ac to ac converters, 15–24
   ac to dc converters, 4–15
   dc to ac converters, 28–42
   dc to dc converters, 24–28
   principles of electric power conditioning, 1–4
Power factor, 279–281, 462
Power interruptions, affects on adjustable speed drives, 465–466
Power MOSFETs, 5
Power quality
   constant-speed wind turbines and, 496–497
   effects of poor utility, 462–466
   harmonic current generation of standard adjustable speed drives, 467–469
   products for improving, 461–462
   role of, 461
   system-level harmonic reduction techniques, 474–481
   topologies for adjustable speed drives to improve, 470–474
Power-up circuit, 80–82
Probability density function (pdf), 112
Pulse density modulation (PDM), 379
Pulse width modulation (PWM), 2, 3
   ac chopper, 21
   ac voltage controller, 21
   dc chopper, 25
   mode, 30
   parallel resonant converters, 49–53
   phase-controlled ac voltage controllers, 21

Pulse width modulation (PWM) (*Continued*)
   pole commutated converters, 47
   rectifiers, 4–5, 13–15
Pulse width modulation, closed-loop
   artificial neural networks (ANNs), 375–386
   basic requirements and definitions, 114–117
   linear controllers, 117–121
   nonlinear on-off controllers, 145–151
   PI current controllers, 121–145
Pulse width modulation, open-loop
   adaptive space vector modulation, 107–109
   artificial neural networks, 373–374
   basic requirements and definitions, 90–91
   carrier-based, 92–93, 98
   distortion factor, 106–107
   overmodulation, 98–104
   performance criteria, 104–107
   random, 109–113
   range of linear operation, 104
   space vector modulation, 93–98, 373–374
   switching losses, 104–106
Pulse width modulation, three-phase voltage source converters and
   closed-loop, 114–151
   generalized/hybrid, 89
   open-loop, 90–114
   overview, 89–90
Pulse width modulation rectifiers, three-phase
   comparisons of control techniques, 457
   control strategies, 425–426
   description of rectifier, 431–433
   direct power control, 435–444
   line voltages and currents, 429–430
   mathematical description, 428–433
   output voltage, 430
   overview, 421–422
   representation of, 426–428
   sensorless operation, 450–453
   steady-state properties, 434–435
   topologies, 422–425
   virtual flux oriented control (VFOC), 444, 447–450
   voltage and virtual flux estimation, 453–456
   voltage oriented control (VOC), 425–426, 444–447, 450
PWM. *See* Pulse width modulation

R

Ramp comparison current controllers, 121–125
Random carrier frequency (RCF) modulation, 112
Random number generator (RNG), 112
Random pulse position (RPP) modulation, 111
Random pulse width modulation, 109–113
Recall mode, 357
Rectifiers
   active, 472–474
   circulating current-conducting dual converter, 9, 12
   circulating current-free dual converter, 9
   full-wave, 5
   mixing, 477–481
   operating quadrants, 8
   phase-controlled, 4, 6, 13
   pulse width modulated, 4–5, 13–15
   single-phase, 5, 469
   six-pulse, 5, 6
   three-phase, 5, 469
   twelve-pulse, 470–472, 477
   two-pulse, 5
   uncontrolled, 4, 6, 13
Reluctance machines. *See also* Direct torque and flux control (DTFC), reluctance synchronous motors; Synchronous reluctance machines (SYNCREL)
Resonance, 37–39
   maintaining, 53–56
Resonant controllers, stationary, 126–141
Resonant dc link (RDCL) inverters/ converters, 37–39
   active clamped, 48–49, 51–53
   artificial neural networks, 379–380
   maintaining resonance, 53–56
   modulation strategies, 56–58
   parallel, 45–46, 48–53
   pole commutated, 47–48
   series, 46–47
Reverse blocking IGBT (RBIGBT), 64–65
Ride-through capability, matrix converters and, 82–84
Ripple factor, 6
Rise time, 137
Rotating vectors, modulation with, 71
Rotor and stator slot effects, 273
Rotor-converter control, 503
Rotor flux error, neural modeling based on, 386–389
Rotor-flux-oriented control (R-FOC), 177–180
   current-controlled, 180–189
   direct, 187
   flux vector estimation, 187–189
   indirect, 180–184
   parameter adaptation, 184–187
Rotor flux vector estimator, 189
Rotor impedance control, 501–502
Rotor voltage equation
   in rotor-flux coordinates, 177
   in stator-flux coordinates, 178

S

Saturation, 270–271, 293
Scalar control, 172, 174–176
Scalar modulation, 70–71
Search control, 221
Self-inductances, 263
Self-organization, 357
Series resonant dc link converters, 46–47
Short circuit method, 53–54
Shot-through, 29, 65

INDEX   517

Sigma delta modulator, 58
Simple state control, 218
Single-phase inverters
　full-bridge, 29
　half-bridge, 28–29
Single-phase rectifiers, 5, 469
Sinusoidal pulse width modulation, 92
Six-pulse rectifiers, 5, 6
Snubber circuits, 45
Soft computing, 352
Soft-switching converters, 37–39, 45
Space vector equations, 162–163
Space vector model, SYNCREL, 268–269
Space vector modulation (SVM), 69–70, 89
　adaptive, 107–109
　basics of, 93–96
　indirect, 72–75
　open-loop PWM and, 93–98
　symmetrical zero states, 96
　three-phase, 96
　two-phase, 96, 98
　variants of, 98
Space vectors, voltage, 32–35, 37
Square-wave mode, 29, 32
Squirrel-cage induction generator, 493–495
Standard delta learning rule, 357–360
State feedback, 296–297
State feedback controllers, 119–120, 141–142
State space method, 119
Static ac switch, 17
Stationary frame values, 265–266, 268
Stationary resonant controllers, 126–141
Stationary vector controllers, 126
Stator current error, neural modeling based on, 389–393
Stator-flux-oriented control (S-FOC), 178–180
　voltage controlled, 189–190
Stator flux vector estimators, 187–188
Stator line currents
　flux estimation based on, 398–402
　speed identification based on, 394–398
Stator slot effects, 273
Stator voltage equation
　in rotor-flux coordinates, 177–178
　in stator-flux coordinates, 179
Stator voltage vector, 174
Steady-state characteristics, 171–172
Steady-state properties, 434–435
Step-down converters, 14, 25
Step-up converters, 14, 28
Suboscillation method, 89
Supervised learning, 356
SVM. *See* Space vector modulation
Switched reluctance generators, 505
Switched reluctance machine (SRM), 251
Switching frequency, 3
　predictive algorithm, 149
Switching frequency controllers

　constant average, 147–148
　variable, 146–147
Switching functions/variables, 2
Switching intervals, 32
Switching losses, 104–106
Switching table, 440–443
Switch mode operation, 89
Symmetrical zero states, 96
Synchronous motors
　See also Direct torque and flux control (DTFC), synchronous motors
　direct torque and flux control (DTFC) and, 308–310
　flux/torque coordination in, 310–313
　wound-rotor generator, 495, 504
Synchronous reluctance machines (SYNCREL)
　cageless, 252
　coenergy and inductances, 258–261
　coenergy and torque, 254–258
　current reference generator, 297–298
　doubly excited machines, 260–261
　dq model, 264–268
　drive system, major components, 294–298
　inductances, 261–264
　iron losses, 271–272
　L_d lookup table, 294
　leakage inductance, 271
　line start, 252
　pros and cons of, 251–252
　rotor and stator slot effects, 273
　rotor designs, 252–253
　saturation, 270–271, 293
　space vector model, 268–269
　state feedback, 296–297
　torque estimator, 295
　velocity observer, 295–296
Synchronous reluctance machines, control properties, 273
　break frequencies interrelationships, 285
　comparisons, 284–289
　constant current in d-axis control (CCDAC), 283–284
　field weakening, 289–293
　maximum efficiency (ME), 278–279
　maximum power factor control (MPFC), 279–281
　maximum rate of change of torque control (MRCTC), 281–283
　maximum torque per ampere control (MTPAC), 278–279
　normalized forms, 274–278
　torque interrelationships, 284–285, 286–289
Synchronous vector controllers, 126

T

Three-level neutral-clamped inverters, 36–37
Three-phase inverters
　current-source, 40

518 INDEX

Three phase inverters (*Continued*)
  full-bridge, 30, 32–34, 35
  incomplete-bridge, 30–32, 35
Three-phase rectifiers, 5, 469
Three-phase space vector modulation, 96
Three-phase voltage source converters. *See* Pulse width modulation, three-phase voltage source converters and
Thyristor converters, 46, 47
  MOS controlled, 64
Torque, coenergy and, 254–258
Torque control strategies. *See* Permanent magnet synchronous machine, torque control strategies
Torque estimator, 295
Torque-producing current, 176
Total harmonic distortion (THD), 462–464
Trajectory tracking control, 149
Transient reactance, 170
Transmission loss, 210
Triacs, 16
Triangular carrier signal, 122–123
Twelve-pulse rectifiers, 470–472, 477
Twin air gap technique, 262–264
Two degrees of freedom (TDF) controllers, 119
Two-level inverters, 36
Two-phase space vector modulation, 96, 99
Two-pulse rectifiers, 5
Two-quadrant choppers, 27
Two-step commutation, 68–69

U

Uncontrolled rectifiers, 4, 6, 13
Unity power factor (UPF) control strategy, 226, 227, 236, 240, 241
Unsupervised learning, 357

V

Variation theory, 162
Vector control, 161, 172, 174, 301
Velocity observer, 295–296
Venturini modulation, 69, 70–71
Virtual flux
  estimation, 454–456
  instantaneous power estimation based on, 438–440
Virtual flux oriented control (VFOC), 444, 447–450
Voltage
  control, 23
  estimation, 453–454
  instantaneous power estimation based on, 437–438
  modulation, 116
  sags and affects on adjustable speed drives, 465
  unbalances and affects on adjustable speed drives, 464–465
Voltage controlled S-FOC, 189–190
Voltage oriented control (VOC), 425–426, 444–447

Voltage peak control, 49
Voltage-source inverters (VSIs)
  hardt-switching converters, 37
  multilevel, 36
  PWM mode, 30
  single-phase, full-bridge, 29
  single-phase, half-bridge, 28–29
  soft-switching converters, 37–39
  square-wave mode, 29, 32
  three-level neutral-clamped, 36–37
  three-phase, current-source, 40
  three-phase, full-bridge, 30, 32–34, 35
  three-phase, incomplete-bridge, 30–32, 35
  two-level, 36
  voltage sources, 39–40
Voltage-source pulse width modulation rectifiers, 14
Voltage space vectors, 32–35, 37

W

Widrow–Hoff (standard delta) learning rule, 357–360
Wind farms, 506–508
Winding function technique, 262
Windmills, 483
Wind turbine systems
  model of mechanical transmission system, 491–492
  power control, 487–491
  renewed interest in, 483
  structure of, 492
  wind energy, 486–487
Wind turbine systems, constant speed
  problems with, 495–497
  reactive power consumption, 494–495
  soft-starter function, 494
  squirrel-cage induction generator, 493–495
  topology, 493
  wound-rotor synchronous generator, 495
Wind turbine systems, variable speed
  cascaded generators, 505–506
  doubly fed induction generator, 501–504
  induction generator, 498–501
  permanent-magnet synchronous generator, 505
  switched reluctance generators, 505
  topology, 497–498
  wound-rotor synchronous generator, 504
Workless forces, 198
Wound-rotor synchronous generator, 495, 504

Z

Zero current switching (ZCS), 46
Zero *d*-axis current (ZDAC) control strategy, 226–227, 235, 240, 241
Zero sequence signal (ZSS), carrier-based PWM with, 92–93
Zero switching loss PWM converters, 50–51
Zero voltage switching (ZVS), 45, 47